Handbook of Nanomaterials

Handbook of Nanomaterials

Edited by **Rich Falcon**

NY RESEARCH PRESS

New York

Published by NY Research Press,
23 West, 55th Street, Suite 816,
New York, NY 10019, USA
www.nyresearchpress.com

Handbook of Nanomaterials
Edited by Rich Falcon

International Standard Book Number: 978-1-63238-258-0 (Hardback)

Printed in China.

Contents

Preface

Every book is a source of knowledge and this one is no exception. The idea that led to the conceptualization of this book was the fact that the world is advancing rapidly; which makes it crucial to document the progress in every field. I am aware that a lot of data is already available, yet, there is a lot more to learn. Hence, I accepted the responsibility of editing this book and contributing my knowledge to the community.

This book aims to serve as a sourcebook for all advancements in nanomaterials. This book deals with researches regarding metal oxide nano structures and composites, and polymer materials. It even explains characterization, processing and fabrication of nanomaterials. These topics have utilized inventive methods of preparation, development, and continuous changes in multidimensional ways. It is a significant book for scientific organizations, governmental research-centers, academic libraries, and the overall research and expansion of nanomaterials in general. It has been created for worldwide audience with diverse backgrounds and education.

While editing this book, I had multiple visions for it. Then I finally narrowed down to make every chapter a sole standing text explaining a particular topic, so that they can be used independently. However, the umbrella subject sinews them into a common theme. This makes the book a unique platform of knowledge.

I would like to give the major credit of this book to the experts from every corner of the world, who took the time to share their expertise with us. Also, I owe the completion of this book to the never-ending support of my family, who supported me throughout the project.

Editor

Part 1

Un-Doped and Doped Nanostructure Materials

1

Iron Oxide Nanoparticles

Mohammed M. Rahman[1], Sher Bahadar Khan[1,2],
Aslam Jamal[3], Mohd Faisal[3] and Abdullah M. Aisiri[1,2]
[1]*The Center of Excellence for Advanced Materials Research, King Abdulaziz University, Jeddah*
[2]*Chemistry Department, Faculty of Science, King Abdulaziz University, Jeddah*
[3]*Centre for Advanced Materials and Nano-Engineering (CAMNE),*
Department of Chemistry, Faculty of Sciences and Arts, Najran University, Najran
Kingdom of Saudi Arabia

1. Introduction

A **semiconductor** is a material with electrical conductivity owing to the electron flow (as opposed to ionic conductivity) intermediate in magnitude between that of a conductor and an insulator. This means conductivity roughly in the range of 10^3 to 10^{-8} siemens per centimeter. It is well known that semiconductor has governed a significant role in progressing research in nanoscience and nanotechnology, leading to novel classes of semiconductor nanomaterials which are capable of nano-particles and wide range of synthesis and applications. Semiconductor materials are also the foundation of modern electronics, including radio, computers, telephones, and many other devices. Such devices include transistors, solar cells, many kinds of diodes including the light-emitting diode, the silicon controlled rectifier, and digital and analog integrated circuits. Similarly, semiconductor solar photovoltaic panels directly convert light energy into electrical energy. In a metallic conductor, current is carried by the flow of electrons. In semiconductors, current is often schematized as being carried either by the flow of electrons or by the flow of positively charged "holes" in the electron structure of the material. Actually, however, in both cases only electron movements are involved. Common semiconducting materials are crystalline solids, but amorphous and liquid semiconductors are known. These include hydrogenated amorphous silicon and mixtures of arsenic, selenium and tellurium in a variety of proportions. Such compounds share with better known semiconductors intermediate conductivity and a rapid variation of conductivity with temperature, as well as occasional negative resistance. Such disordered materials lack the rigid crystalline structure of conventional semiconductors such as silicon and are generally used in thin film structures, which are less demanding for as concerns the electronic quality of the material and thus are relatively insensitive to impurities and radiation damage. Organic semiconductors, that is, organic materials with properties resembling conventional semiconductors, are also known. Silicon is used to create most semiconductors commercially. Dozens of other materials are used, including germanium, gallium arsenide, and silicon carbide. A pure semiconductor is often called an "intrinsic" semiconductor. The electronic properties and the conductivity of a semiconductor can be changed in a controlled manner by adding very small quantities of other elements, called "dopants", to the intrinsic material. In crystalline silicon typically this is achieved by adding impurities of boron or phosphorus to the melt and then allowing the melt to solidify into the crystal.

An iron oxide nanoparticle, in nanotechnology, a particle is defined as a small object that behaves as a whole unit in terms of its transport and properties. Particles are further classified according to size: in terms of diameter, fine particles cover a range between 100 and 2500 nanometers. On the other hand, ultrafine particles are sized between 1 and 100 nanometers. Similar to ultrafine particles, nanoparticles are sized between 1 and 100 nanometers. Nanoparticles may or may not exhibit size-related properties that differ significantly from those observed in fine particles or bulk materials (Buzea et al., 2007). Although the size of most molecules would fit into the above outline, individual molecules are usually not referred to as nanoparticles. Nanoclusters have at least one dimension between 1 and 10 nanometers and a narrow size distribution. Nanopowders (Fahlman, 2007) are agglomerates of ultrafine particles, nanoparticles, or nanoclusters. Nanometer-sized single crystals, or single-domain ultrafine particles, are often referred to as nanocrystals. Nanoparticle research is currently an area of intense scientific interest due to a wide variety of potential applications in biomedical, optical and electronic fields.

Iron oxides are chemical compounds composed of iron and oxygen. Altogether, there are sixteen known iron oxides and oxyhydroxides (Cornell & Schwertmann, 2003). The uses of these various oxides and hydroxides are tremendously diverse ranging from pigments in ceramic glaze, to use in thermite.

Oxides:

- iron(II) oxide, wüstite (FeO)
- iron(II,III) oxide, magnetite (Fe_3O_4)
- iron(III) oxide (Fe_2O_3)
    - alpha phase, hematite (α-Fe_2O_3)
    - beta phase, (β-Fe_2O_3)
    - gamma phase, maghemite (γ-Fe_2O_3)
    - epsilon phase, (ε-Fe_2O_3)

Hydroxides:

- iron(II) hydroxide ($Fe(OH)_2$)
- iron(III) hydroxide ($Fe(OH)_3$), (bernalite)

Oxide/hydroxide:

- goethite (α-FeOOH),
- akaganéite (β-FeOOH),
- lepidocrocite (γ-FeOOH),
- feroxyhyte (δ-FeOOH),
- ferrihydrite ($Fe_5HO_8 \cdot 4H_2O$ approx.), or $5Fe_2O_3.9H_2O$, better recast as $FeOOH.0.4H_2O$
- high-pressure FeOOH
- schwertmannite (ideally $Fe_8O_8(OH)_6(SO) \cdot nH_2O$ or $Fe^{3+}{}_{16}O_{16}(OH,SO_4)_{12\text{-}13} \cdot 10\text{-}12H_2O$)

Beta phase iron oxide: Cubic face centered, metastable, at temperatures above 500 °C converts to alpha phase. It can be prepared by reduction of hematite by carbon, pyrolysis of iron(III) chloride solution, or thermal decomposition of iron(III) sulfate.

Iron(III) oxide or **ferric oxide** is the inorganic compound with the formula Fe_2O_3. It is of one of the three main oxides of iron, the other two being iron(II) oxide (FeO), which is rare, and iron(II,III) oxide (Fe_3O_4), which also occurs naturally as the mineral magnetite. As the mineral known as hematite, Fe_2O_3 is the main source of the iron for the steel industry. Fe_2O_3 is paramagnetic, reddish brown, and readily attacked by acids. Rust is often called iron(III) oxide, and to some extent, this label is useful, because rust shares several properties and has a similar composition. To a chemist, rust is considered an ill-defined material, described as hydrated ferric oxide.

2. Literature survey

Semiconducting nano-materials have attracted much attention because of their unique properties and potential applications in all areas of science (Kumar & Singhal, 2007). The simplest synthetic route to nanomaterials is probably self aggregation, in which ordered aggregates are formed in a spontaneous process (Whitesides & Boncheva, 2002). However, it is still a big challenge to develop simple and reliable synthetic way for low-dimensional metal oxide nanostructures with designed chemical components and controlled morphologies which strongly affect the properties of nano-materials (Dale & Huber, 2009). In recent years, nanosized and nanostructures of super-paramagnetic iron oxides have been investigated extensively because of their wide applications in nano-fields such as ferro-fluids (Raj et al., 1995), magnetocaloric refrigeration (McMichael et al., 1992), biotechnology, and in vivo bio-medical field (Rahman et al., 2011). These materials offer several potential biomaterial (Rahman et al., 2011) as well as biomedical (Rahman et al., 2009) applications in magnetically controlled drug delivery, magnetic resonance imaging as contrast agent, tissue repair, and detoxification of biological fluids (Garcia et al., 2004; Music et al., 1997). It has been extensively studied in diverse fields including catalysis (Brown et al., 1998) environment protection (Chen et al., 2005), magnetic storage media (Zeng et al., 2002), clinical diagnosis, and treatment (Jordan et al., 2003). Nanomaterials may also be utilized in different technological requests, viz. refrigeration systems, medical imaging, drug targeting, other biological applications, and catalysis (Kesavan et al., 1999). Reducing the crystal dimension and enhancing the surface area of the sensing materials are an optional approach to improve the responses, since the reduction/oxidation reactions are mainly activated by the active surfaces area.

Removal of organic pollutants (especially AO dye) in water has been a significant issue in wastewater treatment because of the non-biodegradable nature of these pollutants. The degradation methods, such as, adsorption using activated carbon and coagulation using coagulants, actively convert the aqueous dyes to the solid state leaving the contaminant intact. During the past decades, many researchers have put focus on searching for a direct and effective method to solve this problem (Rahman et al., 2006; Saquib et al., 2008). Recently the ecological contamination and inadvertent leakage of detrimental gases and liquids, semiconducting metal oxide chemical sensors have attracted the interest of ecologists, technologists, environmentalist, and others (Pare et al., 2008). One of the industrial gases in liquid form of interest is ammonia as an ammonium hydroxide because of its toxic and polluting nature. Because of its wide variety of applications including the production of nitrogenous fertilizers and other nitrogenous chemicals, as well as an industrial refrigerant, its global production is in excess of 100 million tons per annum. If highly concentrated it might lead to severe burns on our skin, eyes, throat, or lungs causing permanent blindness and lung disease. In addition, ammonia, hydroxide aerosols have a sun blocking function, corrosive natures, and the fume produced results in temperature reduction. Hence early detection and monitoring of presence of ammonium hydroxide in a wide range of industrial applications is desired. Solution phase ammonia sensors based on optical, electrical, and chemical detection have focused significant attention due to the possibility to operate at room temperature, to measure low level concentrations with fast response time and applicable to detect trace level ammonia solution in environments. In order to detect NH_4OH, there have been lots of efforts in developing a variety of chemical sensors such as electrochemical sensor and chemi-resistive sensors (Ballun et al., 2003), and optical sensors (Christie et al., 2003).

Many sensing research works have been performed with the metal oxide nanostructures for the detection of various chemicals such as hydrazine, acetone, ethanol etc. The chemical sensing by metal oxide thin films utilize mainly the properties of thin film formed by the physi-sorption and chemisorptions techniques. The chemical detection is based on the current changes of the fabricated thin films caused by the chemical components of the reacting system in aqueous medium (Ansari et al., 2008). Here, the main efforts are focused on detecting the minimum ammonium hydroxide concentration necessary for electrochemical detection. Low-dimension of NPs in metal oxide nanostructures allows very sensitive transduction of the liquid/surface interactions into a change in the electrochemical properties. The possibility to form a variety of structural morphologies metal oxide NPs offers many opportunities of tuning the chemical sensing properties. Taking into account their applications, low-dimensional NPs with iron oxide is prepared by hydrothermal technique to fabricate a simple and efficient chemical sensors consisting on a side-polished gold surface; and to test the chemical sensing performance on ammonium hydroxide at room temperature. To best of our knowledge, this is the first report for detection of ammonium hydroxide (in liquid phase) using simple and reliable I-V technique in short response time. Additionally we investigated the photo-catalytic degradation of organic dye named AO (water soluble) with as-grown iron oxide NPs in aqueous system under 250W mercury lamp.

3. Experimental sections

3.1 Materials and methods

Ferric chloride, urea, ammonia solution (25%), acridine orange, butyl carbitol acetate, ethyl acetate, monosodium phosphate, disodium phosphate, and all other chemicals used were of analytical grade and purchased from Sigma-Aldrich Company. The λ_{max} (404 nm) of as-grown Iron particles was measured using UV/visible spectroscopy Lamda-950, Perkin Elmer, Germany. FT-IR spectra were recorded on a Spectrum-100 FT-IR spectrophotometer in the mid-IR range in KBr media purchased from Perkin Elmer, Germany. Raman station 400 was used to measure the Raman shift of as-grown iron NPs using radiation source (Ar^+ laser line, λ; 513.4 nm), which was purchased from Perkin Elmer, Germany. Morphology of iron oxide NPs were recorded on FE-SEM instrument (FESEM; JSM-7600F, Japan). Elemental analysis was investigated using EDS from JEOL, Japan. The morphologies, sizes, and structures of NPs were executed by HR-TEM (TEM; JEM-2100F, Japan). HR-TEM sample were prepared as follows; the synthesized as-grown iron NPs were dispersed into ethanol under ultrasonic vibration for 5 minutes. Then the HR-TEM film is dipped in the solution and dry for investigation. The powder X-ray diffraction (XRD) patterns were taken on a X-ray diffractometer (XRD; X'Pert Explorer, PANalytical diffractometer) equipped with Cu-$K\alpha_1$ radiation (λ =1.5406 nm) using a generator voltage of 45 kV and a generator current of 40 mA were applied for the determination. I-V technique is employed by using Electrometer (Kethley, 6517A, Electrometer, USA).

3.2 Hydrothermal methods

Hydrothermal synthesis includes the various techniques of crystallizing substances from high-temperature aqueous solutions at high vapor pressures; also termed "hydrothermal method". The term "hydrothermal" is of geologic origin. Geochemists and mineralogists have studied hydrothermal phase equilibria since the beginning of the twentieth century.

George W. Morey at the Carnegie Institution and later, Percy W. Bridgman at Harvard University did much of the work to lay the foundations necessary to containment of reactive media in the temperature and pressure range where most of the hydrothermal work is conducted. Hydrothermal synthesis can be defined as a method of synthesis of single crystals that depends on the solubility of minerals in hot water under high pressure. The crystal growth is performed in an apparatus consisting of a steel pressure vessel called autoclave, in which a nutrient is supplied along with water. A gradient of temperature is maintained at the opposite ends of the growth chamber so that the hotter end dissolves the nutrient and the cooler end causes seeds to take additional growth. Possible advantages of the hydrothermal method over other types of crystal growth include the ability to create crystalline phases which are not stable at the melting point. Also, materials which have a high vapor pressure near their melting points can also be grown by the hydrothermal method. The method is also particularly suitable for the growth of large good-quality crystals while maintaining good control over their composition. Disadvantages of the method include the need of expensive autoclaves, and the impossibility of observing the crystal as it grows (O'Donoghue, 1983). A large number of compounds belonging to practically all classes have been synthesized under hydrothermal conditions: elements, simple and complex oxides, tungstates, molybdates, carbonates, silicates, germanates etc. Hydrothermal synthesis is commonly used to grow synthetic quartz, gems and other single crystals with commercial value. Some of the crystals that have been efficiently grown are emeralds, rubies, quartz, alexandrite and others. The method has proved to be extremely efficient both in the search for new compounds with specific physical properties and in the systematic physicochemical investigation of intricate multicomponent systems at elevated temperatures and pressures.

3.3 Synthesis of Iron oxide NPs

Low dimensional iron oxide NPs have been synthesized by adding uni-molar concentration of ferric chloride and urea into hydrothermal cell (Teflon line autoclave). $FeCl_3$ and urea were slowly dissolved into the de-ionized water separately to make 0.5 M concentration at room temperature. Then the solution was mixed gently and stirred until mix properly. The solution pH was slowly adjusted using ammonia solution drop wise to approximately 9.66. Then the mixture was put in hydrothermal cell (Teflon line autoclave) to put in the oven for 6 hours heated up and maintained at 150°C. Then the solution was washed with acetone and kept for drying at room temperature. The as-grown iron oxide products were characterized in detail in terms of their structural and optical properties.

The development of Fe_2O_3 nanoparticles can be well explained stranded on the chemical reactions concerned and crystal growth behaviors of iron oxide. For the synthesis of Fe_2O_3 NPs, ferric chloride ($FeCl_3$) and NH_4OH (in presence of urea) were mixed under continuous stirring at 150°C. During the reaction method, the NH_4OH performs in major rules, like control the pH value of the solution as well as resource to supply hydroxyl ions to the solution. The $FeCl_3$ reacts with NH_4OH and forms FeOOH, which, upon heating, further produce into Fe^{2+} and OH^- ions, which consequently assists in the development of Fe_2O_3 ions according to the chemical reactions **(i)-(iii)**.

$$NH_4OH_{(aq)} \rightarrow NH_4^+{}_{(aq)} + OH^-{}_{(aq)} \quad (i)$$

$$FeCl_{3(s)} \rightarrow Fe^{2+}{}_{(aq)} + 3Cl^-{}_{(aq)} \quad (ii)$$

$$Fe^{2+}_{(aq)} + 2OH^{-}_{(aq)} \rightarrow FeOOH_{(aq)} + H^{+} \quad \text{(iii)}$$

The FeOOH finally, however dissociates to the formation of Fe_2O_3 nuclei according to the reactions **(iv)-(v).**

$$FeCl_{3(aq)} + 2NH_4OH_{(aq)} \rightarrow FeOOH_{(aq)} + 2NH_4^{+}{}_{(aq)} + H^{+}{}_{(aq)} + 3Cl^{-}{}_{(aq)} \quad \text{(iv)}$$

$$FeOOH_{(aq)} \rightarrow Fe_2O_{3(s)} + H_2O_{(l)} \quad \text{(v)}$$

The initially formed Fe_2O_3 nuclei perform as building blocks for the development of final products. With reaction time under the appropriate heating conditions in hydrothermal method, the Fe_2O_3 nuclei concentration enlarges which escorts the construction of desired nanoparticle products. As the Fe_2O_3 NPs are prepared by the well accretion of low-dimensional particles, therefore it is supposed that the fundamental entity for the configuration of Fe_2O_3 structure is nanoparticles.

3.4 Photocatalyst

In chemistry, photocatalysis is the acceleration of a photoreaction in the presence of a catalyst. In catalyzed photolysis, light is absorbed by an adsorbed substrate. In photogenerated catalysis, the photocatalytic activity (PCA) depends on the ability of the catalyst to create electron–hole pairs, which generate free radicals (hydroxyl radicals: OH) able to undergo secondary reactions. Its comprehension has been made possible ever since the discovery of water electrolysis by means of the titanium dioxide. Commercial application of the process is called advanced oxidation process (AOP). There are several methods of achieving AOP's that can but do not necessarily involve TiO_2 or even the use of UV light. Generally the defining factor is the production and use of the hydroxyl radical.

3.5 Photocatalytic degradation of AO with NPs

Photo-catalytic degradation of AO using Iron oxide NPs was executed by optical absorption spectroscopy. The catalytic reaction was carried out in a 250.0 ml beaker, which contain 150.0 ml of AO dye solution (0.03 mM) and 150.0 mg of catalyst. Prior to irradiation, the solution was stirred and bubbled with oxygen for at least 15 min in the dark to allow equilibrium of the system so that loss of compound owing to the adsorption can be taken into account. The suspension was continuously purged with oxygen bubbling throughout the experiment. Irradiation was carried out using 250 W Mercury lamps. Samples (5.0 ml) were composed before and at usual intervals during the irradiation and AO solution were isolated from the photo-catalyst by centrifugation before analysis. The degradation was investigated by measuring the absorbance using UV-visible spectrophotometer (Lambda 950). The absorbance of AO (0.03 mM) was followed at 491 nm wavelength. All solutions were prepared and organized with de-ionized water in room conditions.

3.6 Fabrication of gold electrode using NPs

Gold electrode (surface area 0.0216 cm^2) is fabricated with as-grown iron NPs where butyl carbitol acetate (BCA) and ethyl acetate (EA) as a fabricating agent. Then it is kept in the oven at 50°C for 2 hours until the film is completely uniform, stable, and dried. Phosphate buffer solution (PBS, 0.1M, pH 7.0) is prepared by mixing 0.2M Na_2HPO_4 and 0.2M NaH_2PO_4 solution in 100.0ml de-ionize water. The as-grown iron oxide NPs were employed for the detection of ammonia in liquid phase. The thin film of NPs were made with conducting binders and embedded on the gold electrode, which is presented in the Scheme 1. The Pd and gold electrodes are used as counter and working electrodes, which are presented in Scheme 1A and Scheme 1B respectively. The fabricated electrode was kept in the oven at low temperature (60.0°C) for 2 hours to dry and uniform the film completely (Scheme 1B). The ammonium hydroxide was used as a target chemical in the liquid phase for the total measurement. The electrical response of target compound has been measured using I-V method, which is shown in Scheme 1C.

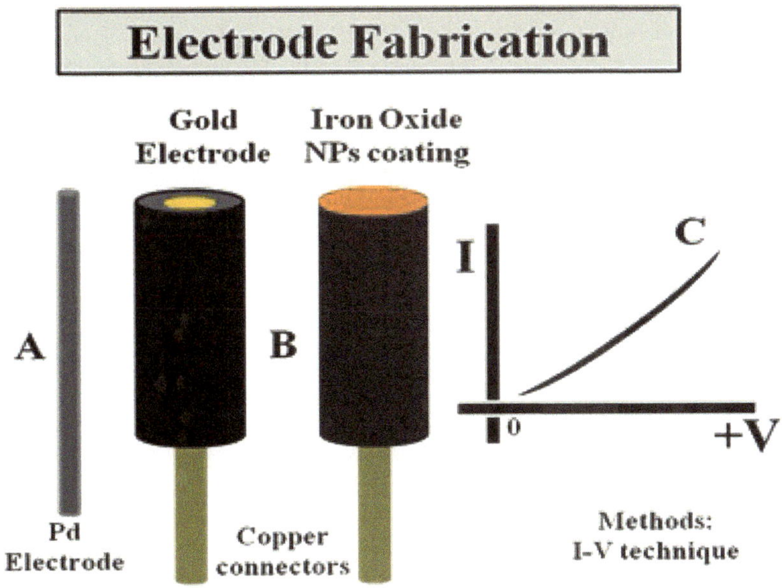

Scheme 1. Fabrication of chemical sensors using as-grown β-Fe_2O_3 NPs and its detection methodology

3.7 Lower detection limit (LOD)

In analytical chemistry, the **detection limit**, lower limit of detection, or **LOD** (limit of detection), is the lowest quantity of a substance that can be distinguished from the absence of that substance (a blank value) within a stated confidence limit (generally 1%) (MacDougall & Crummett, 1980). The detection limit is estimated from the mean of the blank, the standard deviation of the blank and some confidence factor. Another consideration that affects the detection limit is the accuracy of the model used to predict concentration from the raw analytical signal. There are a number of different "detection limits" that are commonly used. These include the instrument detection limit (IDL), the method detection limit (MDL), the practical quantification limit (PQL), and the limit of quantification (LOQ). Even when the same terminology is used, there can be differences in the LOD according to nuances of what definition is used and what type of noise contributes to the measurement and calibration (Long & Winefordner, 1983).

3.8 Linear dynamic range (LDR)

The detector response is said to be linear if the difference in response for two concentrations of a given compound is proportional to the difference in concentration of the two samples. Such response appears as a straight line in the calibration curve. The linear dynamic range of a detector is the maximum linear response divided by the detector noise. Most detectors eventually become non-linear as sample size is increased and this upper point is usually well defined. The chromatographer should know where this occurs to avoid errors in quantification.

3.9 Detection of ammonium hydroxide by I-V technique

A cell is constructed consisting of NPs coated gold electrode as a working electrode and Pd wire is used a counter electrode. Ammonium hydroxide solution is diluted at different concentrations in DI water and used as a target chemical. Amount of 0.1M phosphate buffer solution was kept constant as 20.0 ml throughout the investigation. Solution is prepared with various concentrations of ammonium hydroxide as 77.0 μM to 7.7 M (25% ammonia solution). The sensitivity is calculated from the ratio of voltage versus current of the calibration plot. Electrometer is used as a voltage sources for I-V measurement in simple two electrode system.

4. Properties of nanoparticles

4.1 Optical properties of β-Fe_2O_3 NPs

The optical property of the β-Fe_2O_3 nanoparticles is one of the important characteristics for the evaluation of its optical and photocatalytic activity. UV/visible absorption are a method in which the outer electrons of atoms or molecules absorb radiant energy and undergo transitions to high energy levels. In this procedure, the spectrum obtained owing to optical absorption can be analyzed to acquire the energy band gap of the semiconductor nanomaterials. The optical absorption measurement was carried out at room conditions. Therefore, the nanomaterial may be helpful for the development of non-linear optical sensors in this wavelength region, as the lack of absorption peaks is the major prerequisite for the nanomaterials to confirm non-linear properties. The absorption spectrum of as-grown iron oxide NPs solution is presented in **Fig. 1**(A). It displays an onset of absorption maxima at 404.0 nm in visible range between 200 to 800 nm wavelengths. The lambda maxima of as grown NPs are quite different to those observed earlier due to α-Fe_2O_3 and γ -Fe_2O_3 (Cherepy et al., 1998) but are very similar to those reported for β-Fe_2O_3 (Cornell & Schwertmann, 2003). It shows a broad absorption band around 404.0 nm indicating the formation of low dimensional β-Fe_2O_3 NPs having reddish colors. Band gap energy is calculated on the basis of the maximum absorption band (404nm) of β-Fe_2O_3 NPs and obtained to be 3.06931 eV, according to following equation (vi).

$$E_{bg} = \frac{1240}{\lambda} (eV) \qquad \text{(vi)}$$

Where E_{bg} is the band-gap energy and λ_{max} is the wavelength (404.0 nm) of the nanoparticles.

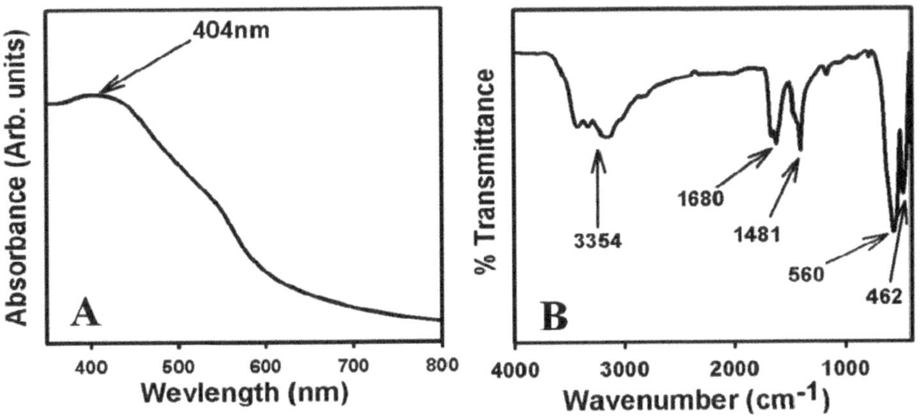

Fig. 1. (A) UV-vis absorption and (B) FT-IR spectra of as-grown Fe_2O_3 NPs.

Infrared spectroscopy (IR spectroscopy) is the spectroscopy that deals with the infrared region of the electromagnetic spectrum that is light with a longer wavelength and lower frequency than visible light. It covers a range of techniques, mostly based on absorption spectroscopy. As with all spectroscopic techniques, it can be used to identify and study chemicals. A common laboratory instrument that uses this technique is a Fourier transform infrared (FT-IR) spectrometer. The infrared portion of the electromagnetic spectrum is usually divided into three regions; the near-, mid- and far- infrared, named for their relation to the visible spectrum. The higher energy near-IR, approximately 14000–4000 cm^{-1} (0.8–2.5 μm wavelength) can excite overtone or harmonic vibrations. The mid-infrared, approximately 4000–400 cm^{-1} (2.5–25 μm) may be used to study the fundamental vibrations and associated rotational-vibrational structure [**Fig. 2**]. The far-infrared, approximately 400–10 cm^{-1} (25–1000 μm), lying adjacent to the microwave region, has low energy and may be used for rotational spectroscopy. The names and classifications of these subregions are conventions, and are only loosely based on the relative molecular or electromagnetic properties. **Fourier transform infrared (FT-IR)** spectroscopy is a measurement technique that allows one to record infrared spectra. Infrared light is guided through an interferometer and then through the sample (or vice versa). A moving mirror inside the apparatus alters the distribution of infrared light that passes through the interferometer. The signal directly recorded, called an "interferogram", represents light output as a function of mirror position. A data-processing technique called Fourier transform turns this raw data into the desired result (the sample's spectrum): Light output as a function of infrared wavelength (or equivalently, wavenumber). There is an alternate method for taking spectra (the "dispersive" or "scanning monochromator" method), where one wavelength at a time passes through the sample. The dispersive method is more common in UV-Vis spectroscopy, but is less practical in the infrared than the FTIR method. One reason that FTIR is favored is called "Fellgett's advantage" or the "multiplex advantage": The information at all frequencies is collected simultaneously, improving both speed and signal-to-noise ratio. Another is called "Jacquinot's Throughput Advantage": A dispersive measurement requires detecting much lower light levels than an FTIR measurement. There are other advantages, as well as some disadvantages, but virtually all modern infrared spectrometers are FTIR instruments.

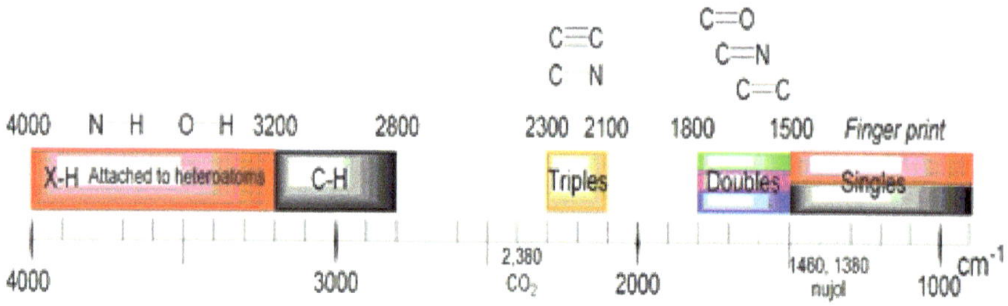

Fig. 2. FT-IR wave range for fundamental vibrations and associated rotational-vibrational frequencies

Fig. 1(B) represents the FT-IR spectrum of the as- grown NPs. It displays several bands at 462, 560, 1481, 1680, and 3354 cm^{-1}. The observed vibration bands may be assigned to Fe–O–Fe stretching vibration (462 and 560 cm^{-1}), O=C=O stretching vibration (1481 cm^{-1}), O-H stretching (3354 cm^{-1}), and O-H bending vibration (1680 cm^{-1}). The absorption bands at 1481, 1680, and 3354 cm^{-1} normally comes from carbon dioxide and water which generally nanomaterials absorbed from the environment due to their mesophorous structure. The observed vibration bands at low frequencies regions suggest the formation of β-Fe_2O_3.

4.2 Structural properties of of β-Fe_2O_3 NPs

Raman spectroscopy is a spectroscopic method utilized to reveal vibrational, rotational, and other low-frequency phases in a Raman active compounds. It depends on inelastic scattering, or Raman scattering, of monochromatic light, generally from a laser in the visible, near infrared, or near ultraviolet range [**Fig. 3**]. The laser light communicates with molecular vibrations, phonons or other excitations in the modes, showing in the energy of the laser photons being shifted up or down. The shift in energy represents information regarding the phonon modes in the system, where Infrared spectroscopy yields similar, but complementary information. Raman spectroscopy is generally established and utilized in material chemistry, since the information is specific to the chemical bonds and symmetry of metal-oxygen stretching or vibrational modes.

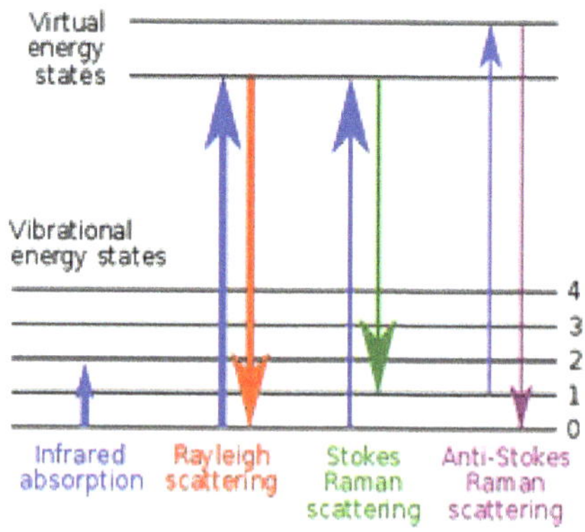

Fig. 3. Energy level diagram showing the states involved in Raman signal. The line thickness is roughly proportional to the signal strength from the different transitions.

Fig. 4A shows the Raman spectrum where main features of the wave-number are observed at about 221, 290, 410, 500, and 608 cm^{-1} for Fe-O stretching vibration. These large bands may be assigned to a magnetite phase of iron oxide NPs. At 608 cm^{-1}, higher wave-number shift is observed due to the various dimensional effects of the NPs.

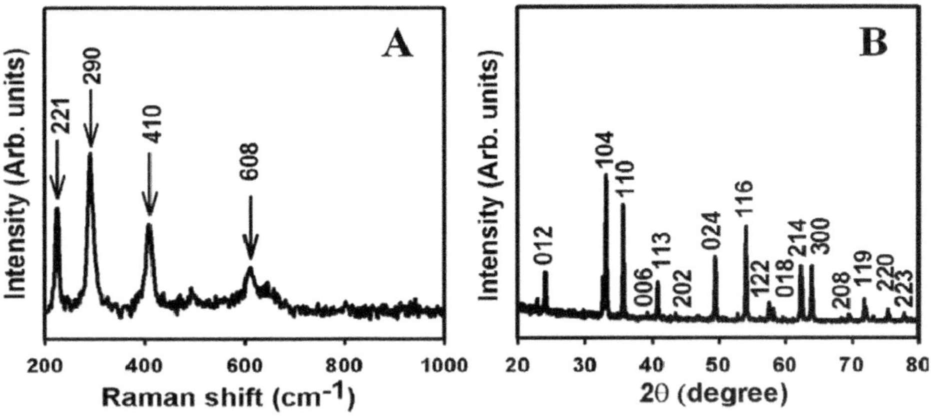

Fig. 4. (A) Raman spectra and (B) XRD patterns of as-grown iron oxide NPs.

X-ray scattering techniques are a family of non-destructive analytical techniques which reveal information about the crystallographic structure, chemical composition, and physical properties of materials and thin films. These techniques are based on observing the scattered intensity of an X-ray beam hitting a sample as a function of incident and scattered angle, polarization, and wavelength or energy. X-ray diffraction yields the atomic structure of materials and is based on the elastic scattering of X-rays from the electron clouds of the individual atoms in the system. The most comprehensive description of scattering from crystals is given by the dynamical theory of diffraction (Azároff et al., 1974). X-ray diffraction can be used to determine which iron oxide compounds are present in NPs by calculating or comparing with the standard value of lattice parameters, crystal structures and crystallinity. **Fig. 4**B shows typical crystallinity of the synthesized as-grown iron oxide NPs and their aggregates. All the reflection peaks in this pattern were found to match with the β-Fe_2O_3 phase having rhombohedral geometry [JCPDF # 089-2810]. The lattice parameters are a = 5.04, c = 13.75, point group: R-3c, and Radiation: CuKα1 (λ= 1.5406). The sample showed the major characteristic peaks for as grown crystalline metallic iron at 2θ values of 24.2(012), 32.2(104), 35.7(110), 39.4(006), 40.9(113), 43.5(202), 49.5(024), 54.1(116), 57.3(122), 57.6(018), 62.5(214), 64.0(300), 69.6(208), 72.0(119), 75.4(220), and 78.0(223) degrees. These indicate that the as-grown iron oxide NPs is well crystalline β-Fe_2O_3.

As scanning electron microscopes (SEM) have evolved the electron beam cross section has become smaller and smaller increasing magnification several fold. A field-emission cathode in the electron gun of a scanning electron microscope provides narrower probing beams at low as well as high electron energy, resulting in both improved spatial resolution and minimized sample charging and damage. Advantages of SEM are (a) semiconductor device cross section analyses for gate widths, gate oxides, film thicknesses, and construction details, (b) advanced coating thickness and structure uniformity determination, and (c) small contamination feature geometry and elemental composition measurement. High resolution

FE-SEM images of as-grown low dimensional iron oxide NPs are presented in Figure 5. Figure 5A to 5E shows the low to high magnified images of NPs grown by hydrothermal process. It is clear from the FE-SEM images that the synthesized products are NPs, which grown in a very high-density and possessed almost uniform shape presented in Figure 5A to 5D. Figure 5E exhibits the high-resolution FE-SEM image of the synthesized NPs which reflected that most of the NPs possessing spherical shapes. The diameter of iron NPs is calculated in the range of 50-90 nm where the average diameter of iron oxide NPs is close to 60 ± 10 nm.

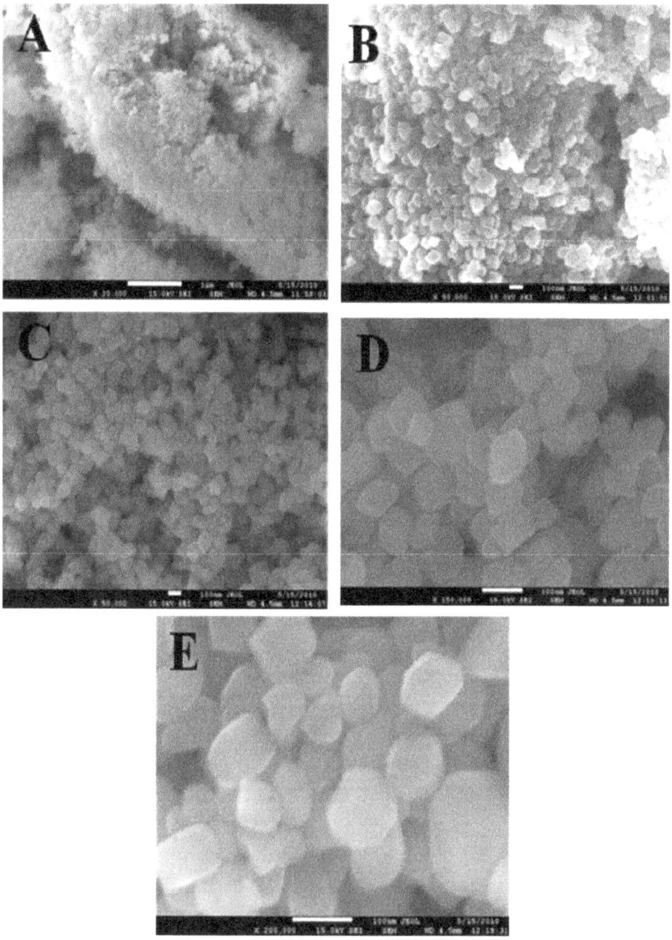

Fig. 5. (A)-(E) Low to High resolution FE-SEM images of as-grown iron oxide NPs

Energy-dispersive X-ray spectroscopy (EDS or EDX) is an analytical technique used for the elemental analysis or chemical characterization of a sample. It is one of the variants of X-ray fluorescence spectroscopy which relies on the investigation of a sample through interactions between electromagnetic radiation and matter, analyzing X-rays emitted by the matter in response to being hit with charged particles. Its characterization capabilities are due in large part to the fundamental principle that each element has a unique atomic structure allowing X-rays that are characteristic of an element's atomic structure to be identified uniquely from one another. To stimulate the emission of characteristic X-rays from a specimen, a high-energy beam of charged particles such as electrons or protons or a beam of X-rays, is focused into the sample being studied. At rest, an atom within the sample contains ground state (or unexcited) electrons in discrete energy levels or electron shells bound to the nucleus. The incident beam may excite an electron in an inner shell, ejecting it from the shell while creating an electron hole where the electron was. An electron from an outer, higher-energy shell then fills the hole, and the difference in energy between the higher-energy shell and the lower energy shell may be released in the form of an X-ray. The number and energy of the X-rays emitted from a specimen can be measured by an energy-dispersive spectrometer. As the energy of the X-rays is characteristic of the difference in energy between the two shells, and of the atomic structure of the element from which they were emitted, this allows the elemental composition of the specimen to be measured. The electron dispersive spectroscopy (EDS) analysis of these particles indicates the presence of Fe and O composition in the pure as-grown iron oxide NPs (Fe_2O_3). It is clearly displayed that as-grown synthesize materials contained only iron and oxygen elements, which presented in **Fig. 6**. The composition of iron and oxygen is 54.11% and 45.88% respectively. No other peak related with any impurity has been detected in the EDS, which confirms that the as-grown NPs are composed only with iron and oxygen.

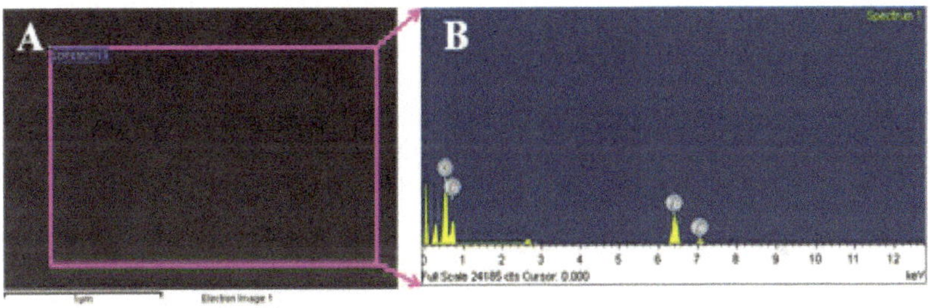

C

Element	Weight%	Atomic%
O K	45.89	74.75
Fe K	54.11	25.25
Totals	100.00	

Fig. 6. EDS of as-grown iron oxide NPs. (A) selected area, (B) EDS outcomes, and (C) elemental compositions (weight% and atomic%).

Transmission electron microscopy (TEM) is a microscopy technique whereby a beam of electrons is transmitted through an ultra thin specimen, interacting with the specimen as it passes through. An image is formed from the interaction of the electrons transmitted through the specimen; the image is magnified and focused onto an imaging device, such as a fluorescent screen, on a layer of photographic film, or to be detected by a sensor such as a CCD camera. TEMs are capable of imaging at a significantly higher resolution than light microscopes, owing to the small de Broglie wavelength of electrons. This enables the instrument's user to examine fine detail-even as small as a single column of atoms, which is tens of thousands times smaller than the smallest resolvable object in a light microscope. TEM forms a major analysis method in a range of scientific fields, in both physical and biological sciences. TEMs find application in cancer research, virology, materials science as well as pollution and semiconductor research. At smaller magnifications TEM image contrast is due to absorption of electrons in the material, due to the thickness and composition of the material. At higher magnifications complex wave interactions modulate the intensity of the image, requiring expert analysis of observed images. Alternate modes of use allow for the TEM to observe modulations in chemical identity, crystal orientation, electronic structure and sample induced electron phase shift as well as the regular absorption based imaging. The first TEM was built by **Max Knoll** and **Ernst Ruska** in 1931, with this group developing the first TEM with resolving power greater than that of light in 1933 and the first commercial TEM in 1939.

High-resolution transmission electron microscopy (**HRTEM**) is an imaging mode of the transmission electron microscope (TEM) that allows the imaging of the crystallographic structure of a sample at an atomic scale (Spence, 1988). Because of its high resolution, it is an invaluable tool to study nanoscale properties of crystalline material such as semiconductors and metals. At present, the highest resolution realized is 0.8 angstroms (0.08 nm) with microscopes such as the OAM at NCEM. Ongoing research and development such as efforts in the framework of TEAM will soon push the resolution of HRTEM to 0.5 Å. At these small scales, individual atoms and crystalline defects can be imaged. Since all crystal structures are 3-dimensional, it may be necessary to combine several views of the crystal, taken from different angles, into a 3D map. This technique is called electron crystallography. One of the difficulties with HRTEM is that image formation relies on phase-contrast. In phase-contrast imaging, contrast is not necessarily intuitively interpretable as the image is influenced by strong aberrations of the imaging lenses in the microscope. One major aberration is caused by focus and astigmatism, which often can be estimated from the Fourier transform of the HR-TEM image.

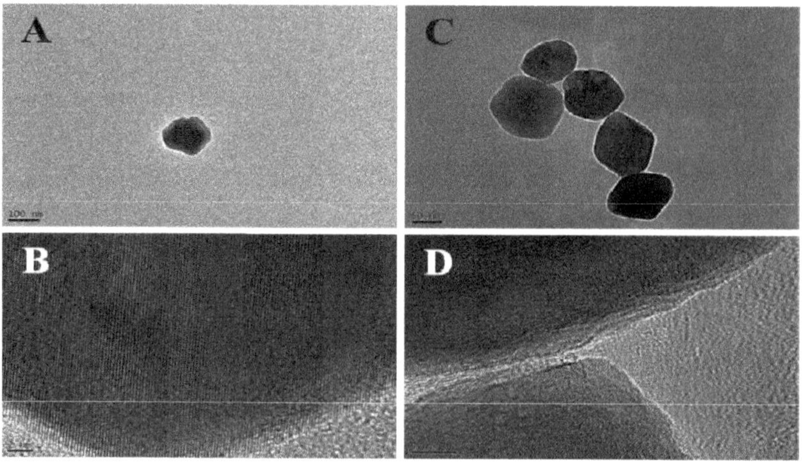

Fig. 7. (A and C)TEM and (B and D) HR-TEM images of iron oxide NPs

The morphologies of particles were observed by TEM, which are presented in the Fig. 7A and Fig. 7C. TEM photographs showed spherical NPs of iron oxide which appeared as deep dark spots. The diameter of as-grown iron oxide is in the range of 50 ~ 90 nm. A typical HR-TEM image of the iron NPs is shown in Figure 7B and 7D. It appears to be spherical in shape and fairly uniform in size with a mean diameter of 60 ± 10 nm. Most of the NPs are spherical in shape and the dimensions are well consistent with the FE-SEM observations. It is clearly revealed and calculated from the HR-TEM image that the distance between two lattice fringes is about 0.31 nm.

5. Potential applications of developed β-Fe_2O_3NPs

5.1 Photo-degradation of acridine orange (AO)

A **dye** can generally be described as a colored substance that has an affinity to the substrate to which it is being applied. The dye is generally applied in an aqueous solution, and may require a mordant to improve the fastness of the dye on the fiber. Both dyes and pigments appear to be colored because they absorb some wavelengths of light preferentially. In contrast with a dye, a pigment generally is insoluble, and has no affinity for the substrate. Some dyes can be precipitated with an inert salt to produce a lake pigment, and based on the salt used they could be aluminum lake, calcium lake or barium lake pigments. The first human-made (synthetic) organic dye, mauveine, was discovered by William Henry Perkin in 1856. Many thousands of synthetic dyes have since been prepared (Zollinger, 2003; Hunger, 2003). Synthetic dyes quickly replaced the traditional natural dyes. They cost less, they offered a vast range of new colors, and they imparted better properties to the dyed materials (Garfield, 2000). Dyes are now classified according to how they are used in the dyeing process.

Acridine orange is a nucleic acid selective fluorescent cationic dye useful for cell cycle determination. It is cell-permeable, and interacts with DNA and RNA by intercalation or electrostatic attractions respectively. When bound to DNA, it is very similar spectrally to fluorescein, with an excitation maximum at 502 nm and an emission maximum at 525 nm (green). When it associates with RNA, the excitation maximum shifts to 460 nm (blue) and the emission maximum shifts to 650 nm (red). Acridine orange will also enter acidic compartments such as lysosomes and become protonated and sequestered. In these low pH conditions, the dye will emit orange light when excited by blue light. Thus, acridine orange can be used to identify engulfed apoptotic cells, because it will fluoresce upon engulfment. The dye is often used in epifluorescence microscopy. Acridine orange is prepared from coal tar and creosote oil. Acridine orange can be used in conjunction with ethidium bromide to differentiate between viable, apoptotic and necrotic cells. Additionally, Acridine orange may be used on blood samples to fluoresce bacterial DNA, aiding in clinical diagnosis of bacterial infection once serum and debris have been filtered. Acridine orange can be used in the assessment of sperm chromatin quality.

Photodegradation is degradation of a photodegradable molecule caused by the absorption of photons, particularly those wavelengths found in sunlight, such as infrared radiation, visible light, and ultraviolet light. However, other forms of electromagnetic radiation can cause photodegradation. Photodegradation includes photodissociation, the breakup of molecules into smaller pieces by photons. It also includes the change of a molecule's shape to make it irreversibly altered, such as the denaturing of proteins, and the addition of other atoms or molecules. A common photodegradation reaction is oxidation. This type of photodegradation is used by some drinking water and wastewater facilities to destroy pollutants. Photodegradation in the environment is part of the process by which ambergris evolves from its fatty precursor.

The molecular structure and absorbance spectra of AO are presented in the Figure 8A and 8B respectively. The photo-degradation of the AO dye occurs predominantly on the β-Fe_2O_3 NPs surface. The extent of degradation of the AO dye was measured by monitoring its concentration with and without NP after reaching the degradation. Figure 8C shows the gradual decreasing in absorption spectra for the degradation of AO dye as a function of photo-irradiation time. It is observed that the irradiation of an aqueous suspension of AO dye in the presence of iron oxide NP leads to decrease in absorption spectra at 491 nm.

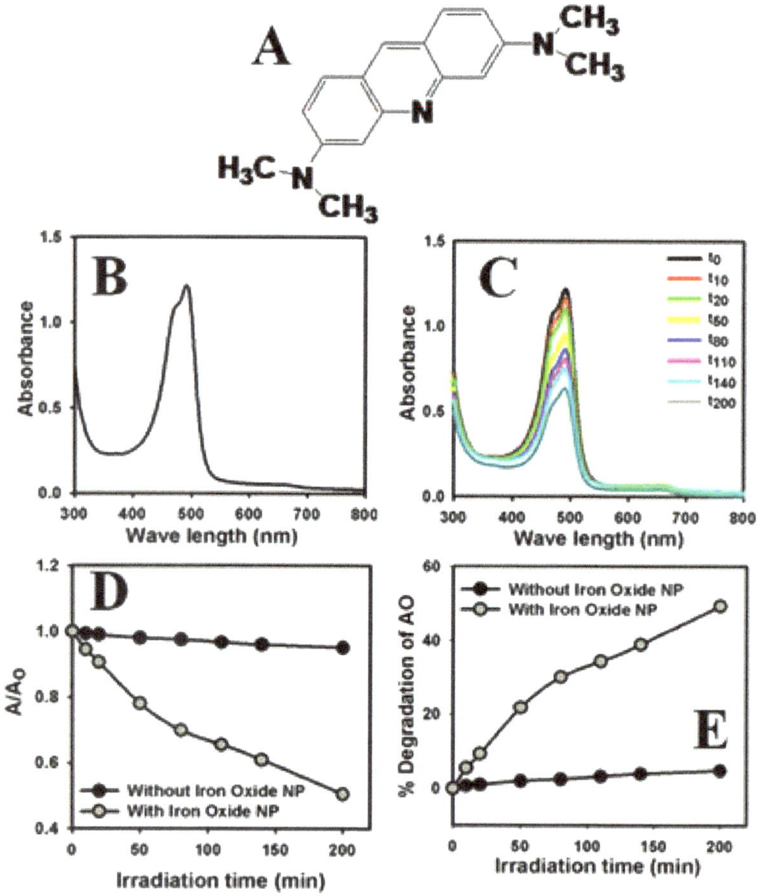

Fig. 8. Photocatalytic degradation of AO using iron oxide NPs. (A) Molecular structure of AO; (B) Spectrum of aqueous solution of AO; (C) Spectrum of AO at different time interval; (D) Change in absorbance; (E) % degradation in different time intervals of AO in presence and absence of as-grown iron oxide NPs.

The absorbance spectra at 491 nm is significantly decreased with increasing in exposure time and gradually decrease until 200 min which concludes that the AO dye has de-colorization property with iron oxide NPs close to 50%. The decrease in absorption intensity vs irradiation time for the AO in the presence and absence of iron oxide NPs is shown in the Figure 8D. Figure 8E shows a plot for the percent degradation versus irradiation time (min) for the oxygen saturated aqueous suspension of AO in the presence and absence of NPs. Figure 8E shows that around 50% degradation of the AO dye takes place after 200 min of irradiation in the presence of catalytic iron oxide NPs. The schematic representation of AO degradation is presented in Scheme 2.

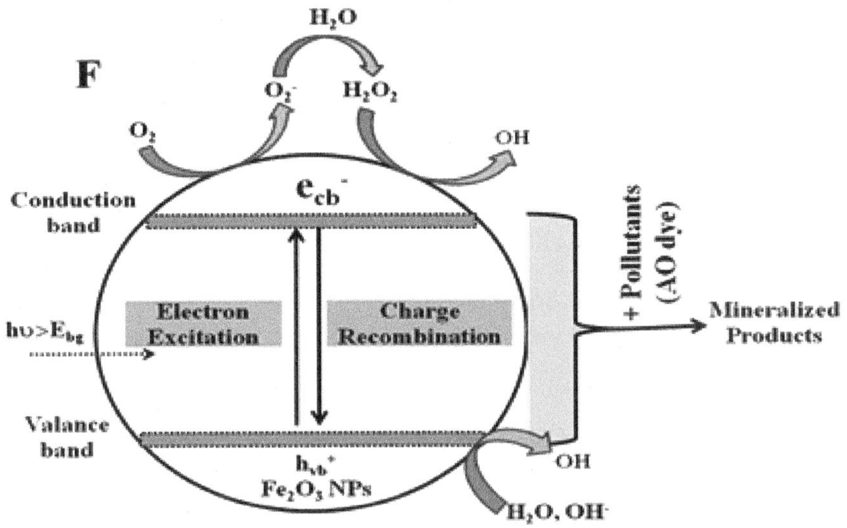

Scheme 2. Schematic representation of AO degradation mechanism

The experimental degradation rate constant is obtained from the initial slope acquired by linear regression from a plot of the natural logarithm (ln) of absorbance of the AO as a function of exposure time, i.e., first-order degradation kinetics. This rate constant is used to evaluate the degradation rates for the decomposition of AO using the formula appended below (vii):

$$-d[A]/dt = kc^n \qquad \text{(vii)}$$

Where, k is the rate constant, c the concentration of the pollutant, and n the order of reaction. The rate of degradation is 9.9×10^{-5} mole.L^{-1}.Min^{-1}. The degradation of AO clearly demonstrated that prepared iron oxide (MO_x) NPs possess significant photocatalytic activity and these NPs could be beneficial photocatalysts for the removal of colored dyes.

Mechanism of heterogeneous photocatalysis has been discussed extensively in literature. When a semiconductor such as MO_x absorbs a photon of energy equal to or greater than its band gap, there will be the formation of electron/hole pair (**Scheme 2**). If charge separation is maintained, the electron and hole may migrate to the metal oxide surface, which can eventually bring about redox reactions of the organic substrates dissolved in water in the presence of oxygen. During the photo-catalytic oxidation processes hydroxyl radicals (·OH) and superoxide radical anions ($O_2^{\cdot -}$) are supposed to be the main oxidizing species. These oxidative retorts would consequence in the mineralization of the dye.

Alternatively, direct absorption of light by the dye named AO, can lead to charge injection from the excited state of the AO to the conduction band of the semiconductor as summarized in the following equations (viii-xii):

$$MO_x + h\upsilon \rightarrow e^-_{cb} + h^+_{vb} \quad \text{(viii)}$$

$$O_2 + e^-_{cb} \rightarrow {}^{\cdot}O^-_2 \quad \text{(ix)}$$

$$H_2O + h^+_{vb} \rightarrow {}^{\cdot}OH + H^+ \quad \text{(x)}$$

$$AO_{ads} + h\upsilon \rightarrow AO_{ads}^* \quad \text{(xi)}$$

$$AO_{ads}^* + MO_x \rightarrow AO_{ads}^{+} + {}^{\cdot}MO_2\,(e^-) \quad \text{(xii)}$$

The degradation process was initiated by the photolysis of MO_x-oxide/hydroxyl species, and accelerated by mercury beam irradiation, due to enhance photolysis of NPs species, which enhances the regeneration of metal precursor with concomitant production of oxide free radicals. Hydroxyl radicals ($OH^{\cdot}$) and superoxide radical anions ($O_2^{\cdot -}$) contribute to the oxidation process by attacking the dye molecules and would results in the bleaching of the AO dye. The results above clearly indicate that the prepared NPs shows a good photo-catalytic activity, thus it can be used as a photo-catalyst for the treatment of wastewater.

5.2 Chemical sensors

A **sensor** is a device that measures a physical quantity and converts it into a signal which can be read by an observer or by an instrument. For example, a mercury-in-glass thermometer converts the measured temperature into expansion and contraction of a liquid which can be read on a calibrated glass tube. A thermocouple converts temperature to an output voltage which can be read by a voltmeter. For accuracy, most sensors are calibrated against known standards. Nanotechnology can enable sensors to detect very small amounts of chemical vapors or liquids. Various types of detecting elements, such as semiconductor materials, carbon nanotubes, iron oxide nanoparticles, zinc oxide nanowires or palladium nanoparticles can be used in nanotechnology-based chemical sensors. These detecting elements change their electrical characteristics, such as resistance or capacitance, when they absorb a gas/liquid molecule on it. Because of the small size of nanoparticles/nanotubes/nanowires, a few gas or liquid molecules are sufficient to change the electrical properties of the sensing elements. This allows the detection of a very low concentration of chemical vapors or liquids. This goal is to have small, inexpensive sensors that can sniff out chemicals just as dogs are used in airports to smell the vapors given off by explosives or drugs. The capability of producing small, inexpensive sensors that can quickly identify chemicals provides a king of nano-bloodhound that doesn't need sleep or exercise which can be useful in a number of ways. An obvious application is to mount these sensors throughout an airport, or any facility with security concerns, to check for vapors given off by explosive devices.

The potential application of β-Fe_2O_3 nanoparticles as chemical sensors has been explored for detecting and quantifying hazardous chemicals, which are not environmental friendly as well as safe. Development of this nanostructure material as chemi-sensors is in the primitive stage. The β-Fe_2O_3 NPs chemi-sensors have advantages such as stability in air, non-toxicity, chemical stability, electrochemical activity, ease to fabricate, and bio-safe characteristics. As in the case of chemical sensors, the principle of operation is that the current response in I-V technique of β-Fe_2O_3 NPs drastically changes when aqueous ammonia are adsorbed. The β-Fe_2O_3 nano-materials were used for fabrication of chemical sensor, where aqueous ammonia was considered as target analyte. The emaciated-film of nanomaterials sensor were fabricated with conducting agents and embedded on the gold electrode surface, which is presented in the Scheme 3(A). The fabricated electrode was accumulated in the oven at low temperature (60.0 °C) for 2 hours to dry, stable, and uniforms the film totally. The electrical responses of target analyte were executed using I-V method, which is presented in Scheme 3(B). Hypothetical, experimental, and mechanism of I-V responses of chemical sensor having β-Fe_2O_3 NPs thin film as a function of current versus potential for aqueous ammonia, which is shown in Scheme 3(B), Scheme 3(C), and Scheme 3(D) respectively. The time delaying of electrometer was kept for 1.0 sec. A significant increase in the current value with applied potential is clearly demonstrated.

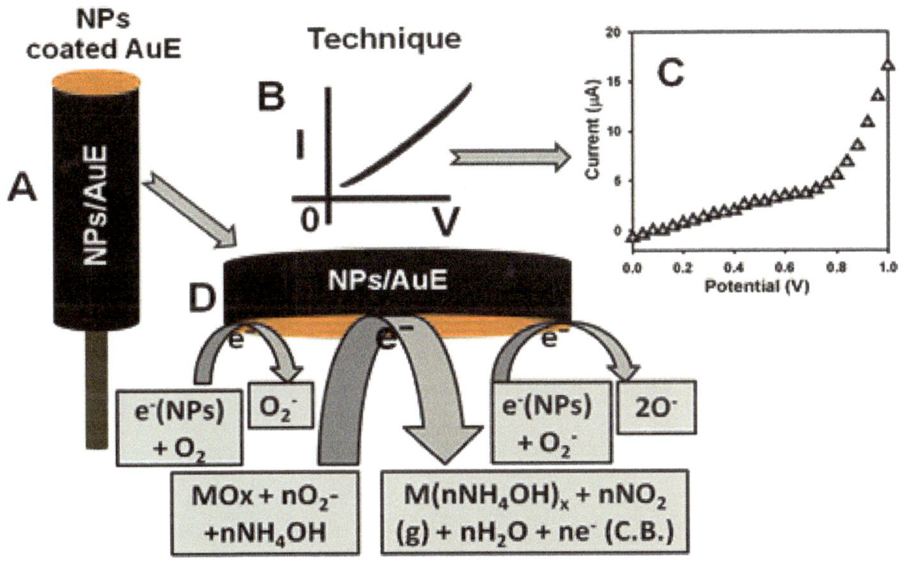

Scheme 3. Schematic view of (A) fabrication of AuE with β-Fe_2O_3 NPs and coating agents, (B) detection I-V method (theoretical), (C) experimental result of I-V plot, (D) proposed mechanisms of aqueous ammonia detection in presence of semiconductor β-Fe_2O_3 nanomaterials.

I-V responses of chemical sensor having β-Fe_2O_3 thin film as a function of current versus potential for the liquid ammonia, which is exposed in Figure 9. A significant increase in the current value with applied potential is clearly demonstrated, where the time delaying for electrometer is kept 1.0 sec. Fig. 9A represents the current changing before (dark-dotted) and after (gray-dotted) coating was made with NPs on gold electrode. The gray-dotted and dark-dotted curves indicated the response of the film before and after injecting 100.0 μl chemicals in bulk solution respectively, which is presented in Fig. 9B. Significant increase of surface film current is measured after every injection of the target component. 77.0 μM concentration of aqueous ammonia is initially taken into the electrochemical cell and added the higher concentration (10 times) in each injection from the stock analyte solution. During I-V measurement, it is injected low to high analyte concentrations (77.0 μM to 7.7 M) gradually into the 20.0 ml phosphate buffer solution in every step. Each I-V response to varying concentration of aqueous ammonia from 7.7 M to 77.0 μM on thin iron oxide NPs coating is presented in the Fig. 9C. It shows the current response changed of the iron oxide NPs as a function of ammonium hydroxide concentration in room conditions. It is observed that at lower to higher concentration of target compound, the current increase gradually. A wide range of analyte concentration was preferred to study the possible analytical parameters, which is calculated in 77.0 μM to 0.77 M. The calibration curve is plotted from the variation of analyte concentrations, which is revealed in the Fig. 9D. The sensitivity is calculated from the calibration curve, which is close to 0.5305 ± 0.02 $\mu Acm^{-2}mM^{-1}$. The linear dynamic range of this sensor exhibits from 77.0 μM to 0.77 M and the detection limit was around 21.8 ± 0.1 μM [3×noise(N)/slope(S)].

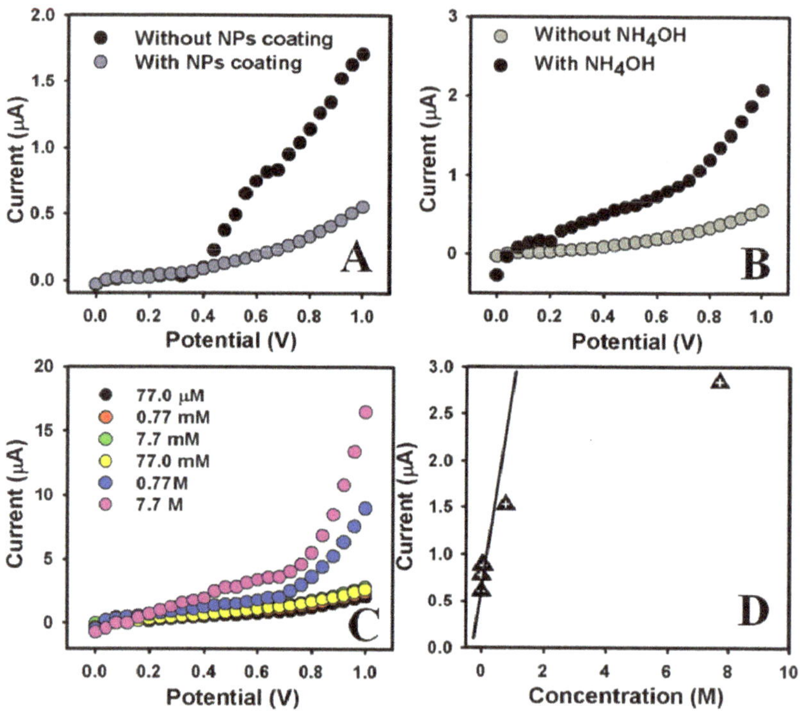

Fig. 9. I-V curves of (A) with and without coating of NPs; (B) with and without ammonium hydroxide (77.0 μM) using NPs coated gold electrode; (C) Concentration variation of ammonium hydroxide; and (D) calibration curve of ammonium hydroxide sensors

The response time was around 10.0 sec for the NPs coated-electrode to reach saturated steady state current. The high sensitivity of film can be attributed to the good absorption (porous surfaces fabricated with coating) and adsorption ability, high catalytic activity, and good biocompatibility of the iron oxide NPs. The estimated sensitivity of the fabricated sensor is relatively higher than previously reported ammonium hydroxide sensors based on other composite or materials modified electrodes (Raj et al., 2010). The aqueous ammonia sensing method of β-Fe_2O_3 NPs sensor is based on the semiconductor oxides, due to oxidation or reduction of the metal-oxide itself, in accordance with the dissolved oxygen (O_2) in bulk-solution or surface-air of the surrounding atmosphere (Scheme 4A), according to the equations (xiii-xiv).

$$\text{electron } (e^-, \beta\text{-}Fe_2O_3 \text{ NPs}) + O_2 \rightarrow O_2^- \quad \text{(xiii)}$$

$$\text{electron } (e^-, \beta\text{-}Fe_2O_3 \text{ NPs}) + O_2^- \rightarrow 2O^- \quad \text{(xiv)}$$

These reactions take place in bulk-solution or air/liquid interface or surrounding air due to the low carrier concentration which increased the resistances. The aqueous ammonia sensitivity headed to β-Fe_2O_3 NPs (i.e. MO_x) could be attributed to the high oxygen deficiency and defect density leads to increase oxygen adsorption. Larger the amount of oxygen adsorbed on the MO_x NPs surfaces, larger would be the oxidizing capability and faster would be the oxidation of analytes.

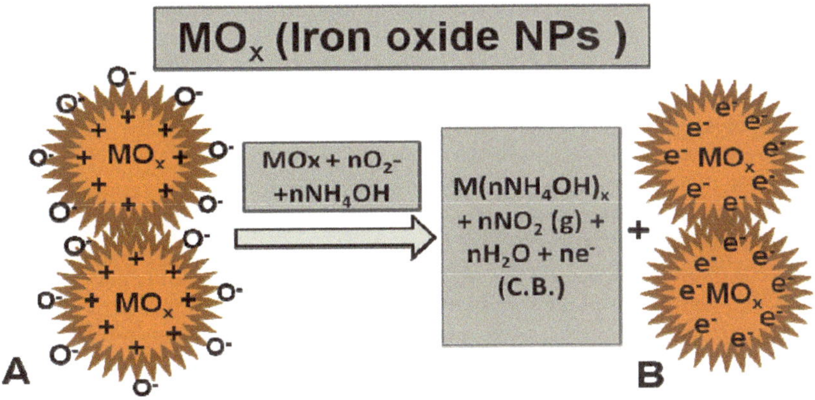

Scheme 4. Mechanism of β-Fe_2O_3 NPs chemical sensors at room conditions.

The reactivity of aqueous ammonia would be very large as compared and contrasted to other chemical with the semiconductor surfaces under the similar condition (Patil et al., 2007). When NH_4OH reacts with the adsorbed oxygen on the surface of the film, it acquires oxidized to nitrogen oxide gas and metal ammonium hydroxide, liberating free electrons in the conduction band (Scheme 4B), which could be articulated through the following reactions (xv-xvi):

$$MO_x + nNH_4OH + nO_2- \rightarrow M(nNH_4OH)x + nNO_2 \text{ (g)} + nH_2O + ne^- \text{ (C.B.)} \quad \text{(xv)}$$

$$4NH_4OH + 7O_2- \rightarrow 4NO_2 + 10H_2O + 15e- \quad \text{(xvi)}$$

These reactions corresponded to oxidation of the reducing carriers. These processes increased the carrier concentration and hence decreased the resistance on exposure to reducing liquids/analytes. At the room condition, the disclosure of metal oxide surface to reducing liquid/analytes results in a surface mediated combustion process. The exclusion of iono-sorbed oxygen amplifies the electron concentration and thus the surface conductance of the coating-film (Mujumdar, 2009). The reducing analyte (aq. NH_3) donates electrons to β-Fe_2O_3 NPs surface. Therefore, resistance is decreased, or conductance is increased. This is the reason why the analyte response (current response) increases with increasing potential. On the other way, due to large surface area, the NPs of provide a supportive nano-environment for the chemical detection with fast and good capacity. The high sensitivity of NPs provides high electron communication features which progress the direct electron movement between the active sites of NPs and GE. The modified thin film had a good stability and reproducibility. As for the nanomaterials, β-Fe_2O_3 NPs provide a path to a new generation of chemical sensors, but a premeditate effort has to be expended for nanostructures to be taken critically for large scale applications, and to achieving high appliance density with accessibility to individual sensors. Reliable methods for fabricating, assembling and integrating building blocks onto sensors need to be explored.

6. Conclusion

Finally, the present work provides a hydrothermal method to synthesize low-dimensional iron oxide NPs, which was characterized using several conventional techniques like XRD, FE-SEM, EDS, HR-TEM, UV/visible, FT-IR, and Raman spectroscopy. The detailed morphological characterizations by FE-SEM and TEM revealed that the synthesized NPs possess almost spherical shape with typical diameters of ~ 60 ± 10 nm. The optical properties of as-grown β-Fe_2O_3 NPs were investigated by UV-visible absorption which shows the presence of characteristic β-Fe_2O_3 peak in the spectrum. It was also investigated the photo-catalytic degradation with Acridine orange under UV radiation sources and found the 50% degradation held with as-grown β-Fe_2O_3 NPs. The as-grown NPs were employed for the detection of aqueous ammonia in solution phase as chemical sensors. The performance of the proposed aqueous ammonia sensor using low-dimensional β-Fe_2O_3 NPs film was excellent in terms of sensitivity, lower detection limit, and large linear dynamic ranges in short response time. This new approach is also introduced a new route for efficient chemical sensor development to control the environmental toxicity as well as carcinogenicity, and can also play an important role in environmental and health care fields.

7. Acknowledgment

We are greatly acknowledged to Najran University for their financial supports and research facilities. Centre for Advanced Materials and Nano-Engineering (CAMNE), Najran University, Najran is also greatly acknowledged for their chemicals and nanomaterials analyses.

8. References

Ansari, SG., Ansari, ZA., Wahab, R., Kim, YS., Khang, G., Shin, HS. (2008). Glucose sensor based on nano-baskets of tin oxide templated in porous alumina by plasma enhanced CVD. *Biosens. Bioelectron*. Vol. 23, pp. 1838.

Azároff, L. V.; R. Kaplow, N. Kato, R. J. Weiss, A. J. C. Wilson, R. A. Young (1974). X-ray diffraction. McGraw-Hill.

Ballun, G., Hajdu, F., Harsanyi, G. (2003). Highly sensitive ammonia sensor. IEEE 26th Int. Spring Sem. *Elect. Tech.* pp. 471.

Buzea, C., Pacheco, I., Robbie, K. (2007). "Nanomaterials and Nanoparticles: Sources and Toxicity". *Biointerphases* Vol. 2, pp. MR17.

Brown, ASC., Hargreaves, JSJ., Rijniersce, B. (1998) A study of the structural and catalytic effects of sulfation on iron oxide catalysts prepared from goethite and ferrihydrite precursors for methane oxidation. *Catal. Lett.* Vol. 53, pp. 7.

Chen, J., Xu, LN., Li, WY., Gou, XI. (2005) α-Fe_2O_3 Nanotubes in Gas Sensor and Lithium-Ion Battery Applications. *Adv. Mater.* Vol. 17, pp. 582.

Cherepy, NJ., Liston, DB., Lovejoy, JA., Deng, H., Zhang, ZJ. (1998). Ultrafast Studies of Photoexcited Electron Dynamics in γ- and α-Fe_2O_3 Semiconductor Nanoparticles. *J. Phys. Chem. B* Vol. 102, pp. 770.

Compendium of Chemical Terminology, 2nd ed. (1997) the "Gold Book". Online corrected version: (2006).

Cornell, RM., Schwertmann, U. (2003) *The Iron Oxides-Structure, Properties, Reactions, Occurrences and Uses.* Darmstadt: Wiley-VCH GmbH & Co. KGaA.

Cornell, RM., Schwertmann, U. (2003). *The iron oxides: structure, properties, reactions, occurrences and uses.* Wiley VCH.

Christie, S., Scorsone, E., Persaud, K., Kvasnik, F. (2003). Remote detection of gaseous ammonia using the near infrared transmission properties of polyaniline. *Sens. Actuators B* Vol. 90, pp. 163.

Dale, L., Huber. (2009) Synthesis, Properties, and Applications of iron NPs. *Small.* Vol. 1, pp. 482.

Fahlman, BD. (2007). *Materials Chemistry.* Springer. pp. 282.

Kumar, A., Singhal, A. (2007) Synthesis of colloidal β-Fe_2O_3 nanostructures—influence of addition of Co^{2+} on their morphology and magnetic behavior. *Nanotechn.* Vol. 18, pp. 475703.

Garcıa, KE., Morales, AL., Barrero, CE., Arroyave, CE., Greneche, JM. (2004) Magnetic and crystal structure in akaganeite nanoparticle. *Physica B.* Vol. 354, pp. 187.

Hunger, K., ed. (2003). Industrial Dyes. Chemistry, Properties, Applications. Weinheim: Wiley-VCH.

Garfield, S. (2000). *Mauve: How One Man Invented a Color That Changed the World.* Faber and Faber. ISBN 0-393-02005-3.

Jordan, A., Scholz, R., Maier-Hauff, K., Johannsen, M., Wust, P., Nadobny, J., Schirra, H., Schmidt, H., Galatsis, K., Cukrov, L., Wlodarski, W., McCormick, P., Kalantar-zadeh, K., Comini, E., Sberveglieri, G. (2003) p- and n-type Fe-doped SnO_2 gas sensors fabricated by the mechanochemical processing technique. *Sens. Actuators B.* Vol. 93, pp. 562.

Kesavan, V., Sivanand, Chandrasekaran, PSS., Koltypin, Y., Gedanken, A. (1999) Catalytic aerobic oxidation of cycloalkanes with nanostructured amorphous metals and alloys. *Angew. Chem. Int. Ed.* Vol. 38, pp. 3521.

Long, GL., Winefordner, JD. (1983). Linearization of electron capture detector response to strongly responding compounds. *Anal. Chem.* Vol. 55, pp. 713.

MacDougall, D., Crummett, WB. (1980). Guidelines for Data Acquisition and Data Quality Evaluation in Environmental Chemistry. *Anal. Chem.* Vol. 52, pp. 2242.

McMichael, RD., Shull, RD., Swartzendruber, LJ., Bennett, LH. (1992) Magnetocaloric effect in superparamagnets. *J. Magn. Magn. Mater.* Vol. 111, pp. 29.

Mujumdar, S. (2009). Synthesis and characterisation of SnO_2 films obtained by a wet chemical process. *Mat. Sci. Poland.* Vol. 27, pp. 123.

Music, S., Saric, A., Popovic, S. (1997) Effect of urotropin on the formation of β-FeOOH. *J. Mol. Struct.* Vol. 410-411, pp. 153.

O'Donoghue, M. (1983). A guide to Man-made Gemstones. *Great Britain: Van Nostrand Reinhold Company.* pp. 40.

Patil, DR., Patil, LA., Amalnerkar, PP. (2007). Ethanol gas sensing properties of Al_2O_3-doped ZnO thick film resistors. *Bull. Mat. Sci.* Vol. 30, pp. 553.

Pare, B., Jonnalagadd, SB., Tomar, H., Singh, P., Bhagwat, BW. (2008). ZnO assisted photocatalytic degradation of acridine orange in aqueous solution using visible irradiation. *Desalination* Vol. 232, pp. 80.

Raj, VB., Nimal, AT., Parmar, Y., Sharma, MU., Sreenivas, K., Gupta, V. (2010). Cross-sensitivity and selectivity studies on ZnO surface acoustic wave ammonia sensor. *Sens. Actuators B* Vol. 147, pp. 517.

Raj, K., Moskowitz, B., Casciari, R. (1995) Advances in ferrofluid technology. *J. Magn. Magn. Mater.* Vol. 149, pp. 174.

Rahman, MM., Jamal, A., Khan, SB., Faisal. M. (2011) Fabrication of Highly Sensitive Ethanol Chemical Sensor Based on Sm-Doped Co_3O_4 Nano-Kernel by Solution Method. J. Phys. Chem. C. Vol. 115 pp. 9503.

Rahman, MM., Jamal, A., Khan, SB., Faisal. M. (2011) Cu-doped ZnO Based Nanostructured Materials for Sensitive Chemical Sensor Applications. ACS App. Mater. Inter. Vol. 3 pp. 1346

Rahman, MM., Umar, A., Sawada, K. (2009) Development of Amperometric Glucose Biosensor Based on Glucose Oxidase Enzyme Immobilized with Multi-Walled Carbon Nanotubes at Low Potential Sens. Actuator: B Vol. 137 pp. 327.

Rahman, MM., Jeon, IC., Hasnat, MA., Samed, AJF. (2006) Xenon Beam Radiation Assisted Degradation of an Organic dye in Aqueous Medium. *Environ. Sci. Technol.* Vol. 1 pp. 210.

Saquib, M., Tariq, MA., Faisal, M., Muneer, M. (2008). Photocatalytic degradation of two selected dye derivatives in aqueous suspensions of titanium dioxide. Desalination. *Desalination* Vol. 219, pp.301.

Spence, JCH. (1988) [1980]. *Experimental high-resolution electron microscopy.* New York: Oxford U. Press. ISBN 0195054059.

Whitesides, GM., Boncheva, M. (2002) Beyond molecules: Self-assembly of mesoscopic and macroscopic components. *Proc. Natl. Acad. Sci. USA.* Vol. 99, pp. 4769.

Zeng, H., Li, J., Liu, JP., Wang, ZP., Sun, SH. (2002). Exchanged-coupled nanocomposite magnets via nanoparticle self-assembly. *Nature* Vol. 420, pp. 395.

Zollinger, H. 3rd ed. (2003). *Color Chemistry. Synthesis, Properties and Applications of Organic Dyes and Pigments,* Weinheim: Wiley-VCH.

2

Nucleation and Growth of TiO_2 Nanoparticles

H. Mehranpour[1], Masoud Askari[1] and M. Sasani Ghamsari[2]
[1]*Department of Material Science and Engineering, Sharif University of Technology, Tehran,*
[2]*Solid State Lasers Research Group, Laser & Optics Research School, NSTRI, Tehran, Iran*

1. Introduction

Crystallization is one of the most important and interesting phenomena in human life. For instance, many purifications and productions of materials are based on crystallization. In pharmacology, more than 90% of all pharmaceutical products are used in the crystalline form of solid material. Hence, the controlling of the crystallization process has a critical role on the properties of products and allows manufacturers to prepare materials with desired and reproducible properties. The recent interest in nanocrystals and other types of nanomaterials is a further illustration of the crystallization importance in the science and technology (Mehranpour et al., 2010). It is well known that at the nanometer scale, the optical, the electronic, and the catalytic properties of nanomaterials are highly sensitive to their size and shape (Finney & Finke, 2008). Consequently, the crystallization process (nucleation & growth) plays an important role in determining the crystal structure, shape, size and size distribution of the nanomaterials. Therefore, a theoretical approach to understand the mechanism of nanocrystal formation provides a greater control over the size, shape, and composition of nanocrystals and results an ability to tune the abovementioned properties simply by varying the crystallization conditions. One of the main goals of this theoretical approach of the nanocluster science is the ability to prepare nanocrystals that have very narrow size distributions (so-called "near-monodisperse"). That is the main reason to know why the nucleation and the growth mechanism of nanocrystals in solution have been the subject of increasing study. On the other hand, the agreement between the theoretical models and experimental results is not very good. This is a general shortcoming of nucleation theories. As a result, the nucleation theories in general have a little predictive power. It seems that a main goal of nucleation study is to improve the accuracy of theory (Finney & Finke, 2008). Titanium dioxide (TiO_2) is one of the semiconductor materials with unique photocatalytic properties that make it an interesting candidate for different applications such as white pigment, cosmetics, catalysis support and a photocatalyst. During last two decades, this act has been highlighted that the physical and the chemical properties of nano-TiO_2 depend on its size, morphology and crystalline polymorph strongly. That is well known that these specifications of TiO_2 nanopowders are controlled by crystallization and phase transformation. Since the employed synthesis method has critical role on the crystallization and phase transformation of obtained material, many attempts have been done to use a procedure which has enough potential to prepare TiO_2 nanopowders with designed properties. Researchers made a lot of efforts to innovate the best techniques that can be applied for this purpose. Despite the several attempts that have been served to prepare titanium oxide nanoparticles with very high uniformity in size and morphology, a deep understanding of nucleation and growth mechanism needs to make a better control on the physical and the chemical properties of synthesized TiO_2 nanomaterial. LaMer theory described the kinetic formation of particles which are controlled by diffusion of elements (particles, ions,..). In 1999 Tadao Sugimoto and his model described the formation of nanoparticles that their nucleation and growth comfort with LaMer diagram and controlled by diffusion. In this chapter, at first, the kinetic of nucleation and the growth of particles will be discussed generally and some of scientific reports will be analyzed. Then, the kinetic of nucleation and the growth mechanism of Titana nanoparticles will be studied on the basis of LaMer theory and Sugimoto model.

2. Nucleation

Generally, the nucleation is considered as a first stage of embryos formation from nuclei which have only nanometer in size and possible growth to the final particles. In fact, the nucleation is the first formation step of the solid phase from other phases such as solid, liquid and gas. Nucleation occurs and continues with aggregation and clustering of molecules or ions in a supersaturated melt, solution or vapor to form a stable size of solid phase. It seems that supersaturation of solute in the main matrix is the important key to nucleation of particle. Indeed, the driving force for nucleation and growth phenomena is supersaturation. After dissolving the chemical species in a solvent, whether or not of a predetermined nature, to observe the nucleation and growth process the solution must be supersaturated (Mangin et al., 2009) (Kashchiev, 2001). Supersaturation is the difference between the chemical potential of the solute molecules in the supersaturated (μ) and saturated (μ_s) states respectively. For one molecule, difference is explained by:

$$\Delta\mu = \mu - \mu_S = k_B T \ln \beta \tag{1}$$

where k_B is the Boltzmann constant and T is the temperature. To simplify, we consider that the activities are equal to the concentrations and can be written as (Mangin, et al., 2009):

$$\beta = \frac{C_i}{C_s} \tag{2}$$

Here β is the supersaturation ratio, C_i is the concentration of the solute in solution and C_s its saturated or equilibrium concentration. Obviously, this ratio is dimensionless . Moreover, if $\beta > 1$, the nucleus grows; if $\beta < 1$, the nucleus dissolves; and if $\beta = 1$, nuclei and solution are at equilibrium. However, in a crystallized system, not only there must be a supersaturation condition for the system but also the system must have a number of minute solid bodies, embryos, nuclei or seed which are acting as a core of nucleation to develop crystals. Therefore, nucleation may occur spontaneously or it can happen artificially. Nowadays there is no a general agreement on forming of particles. The term 'primary' is explained for the system that has unknown crystalline matter. On the other hand, nuclei are often generated in the vicinity of crystals presented in a saturated system; this is referred to as 'secondary' nucleation (Kashchiev, 2001) (Mangin et al,. 2009).

2.1 Primary nucleation

Scientists did not know exactly when and how a nucleus is become stable within a homogeneous solution. Actually, primary nucleation is the initial formation of a nucleus where there are no other solid phases. If the solid phases are present in the system, they do not have any influence on the nucleation process. The number of monomers in a stable nucleus is different about ten to several thousand, i.e., ice nucleus may contain about 100 molecules. The formation of nuclei can be explained by a successive addition of monomer (units A) according to the formation scheme (Mersmann et al,. 2001):

$$A_1 + A = A_2 \qquad A_2 + A = A_3;.....;\ A_n + A \underset{Kd}{\overset{Ka}{\leftrightarrow}} A_{n+1} \tag{3}$$

From above scheme, it can be found that the addition to the stable nucleus or part from it, would lead to growth or shrink which can be described by the rate constant of K_a and Kd. $K_a$is the monomer addition rate constant and K_d is the rate of monomer desorbstion which is parted from nucleus. So, addition to the stable nucleus would result in growth of the nucleus. Similarly, ions or molecules in a solution can make the short-lived clusters, which occurs very rapidly. These clusters can be considered as very high supersaturation in local region and many of these embryos or 'sub-nuclei' fail to growth (Mullin, 2001). They dissolve easily because they are extremely unstable. The classical nucleation theory (CNT) which was pioneered by Volmer and Decker (Maris 2006) , where the first to argue that the nucleation rate should depend exponentially on the reversible work of the formation of a critical cluster. The free energy of nucleus formation G is expressed by the two competing terms. One of them is free volume energy (G_V) and the other is free surface energy (G_A). The positive change in free surface energy (G_A) can be provided by increasing the interfacial tension μ_{CL} between the surface of the nucleus and the surrounding solution. On the other hand, the change in free volume energy (ΔG_V) is negative because it is set free. The ΔG_V is proportional to the volume of the nucleus and increased with increasing in energy $RT\times lnS$, where $S=a/a^*$ or in ideal systems, $S = C/C^*$. The concentration of the elementary units (C) changes to the lower equilibrium concentration as follows (Mersmann et al,. 2001):

$$C = C^* - \Delta C \tag{4}$$

Fig 1 shows, the free energies of ΔG_A and ΔG_V and the total free energy $\Delta G = \Delta G_A + \Delta G_V$ against the nucleus radius. The nucleus free energy can expressed with the following equation which is obtained by the nucleus surface A_n, nucleus volume V_n, the degree of dissociation a, and the number of ions n:

$$\Delta G = \Delta G_A + \Delta G_V = A_n \gamma_{CL} - (1 - \alpha + un)\, V_n C_C R\, T \ln S \tag{5}$$

The change in total enthalpy (ΔG), is regarding to the nucleus radius r passes through a maximum value. This is because the free volume energy (ΔG_V) is related to the volume of a cluster and therefore(Mersmann et al,. 2001):

$$\Delta G_v \approx V_n \approx r^3 \tag{6}$$

whereas the free surface energy (ΔG_A) is proportion to the size of the nucleus with the following manner(Mersmann et al,. 2001):

$$\Delta G_A \approx A_n \approx r^2 \tag{7}$$

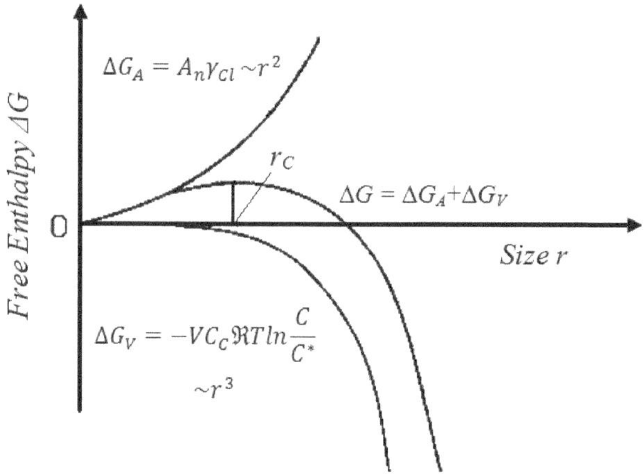

Fig. 1. Free enthalpy ΔG against nucleus size r (Mersmann et al,. 2001).

The maximum value of ΔG corresponds to the nucleus with critical radius (r_c). A thermodynamically stable nucleus exists when the radius of the nucleus reaches to r_c. Therefore, the slope of ΔG at critical radius of nucleus will be zerro(Mersmann et al,. 2001):

$$\frac{\partial \Delta G}{\partial r} = 0 \tag{8}$$

In that situation which the radius of nucleus is critical, the rate constant of addition (k_A) is equel to the desorbtion rate constant (k_D). Therefore, neither growth nor decomposition takes place. The critical radius of nucleus can be obtained by the above equations for the spherical nucleus:

$$r^*_{crit} = \frac{4\mu_{CL}}{(1-a+an)RTC_c \ln S} \tag{9}$$

By considering the Boltzmann constant as $k_B = R / N_A$ and molecular volume as:

$$V_m = \frac{1}{C_C N_A} = \frac{M}{\rho_C N_A} \tag{10}$$

the critical radius(r_{crit}) is obtained by the following equation(Mersmann et al,. 2001):

$$r^*_{crit} = \frac{4\gamma_{CL}}{k_B T \ln S} V_m \tag{11}$$

The behavior of created crystalline particles in supersaturation solution depends on their size; they can be dissolved or grow, but the free energy of the particles determines the viable of particle. The critical size r_C represents the minimum size of stable nucleus. Particles with the radius smaller than r_C will dissolve because only in this way the particle can achieve a reduction in its free energy. Similarly, particles larger than r_C will grow to the final particle. The nucleus free energy can also be expressed in terms of the number of atoms in the nucleus; one such general expression is given by(Mersmann et al,. 2001):

$$\Delta G_n = -nk_B T \ln\left(\frac{C}{C_{sat}}\right) + \sigma b n^{2/3} \quad (12)$$

where, n is the number of atoms in the nucleus, k_B is the Boltzmann constant, T is the temperature, σ is the surface tension of the cluster, b is a geometric factor and hence the critical number of atom which create a stable nucleus at the supersaturation level is :

$$n_c = \left(\frac{2\sigma b}{3k_B T \ln\left(\frac{C}{C_{sat}}\right)}\right)^3 \quad (13)$$

Alternatively, the critical nucleus can be explained in terms of the number of atoms contained within it, n_C, by differentiating Eq. (12) and finding the maximum. For nucleus, containing less than n_C atoms or ions growth is not happened and the cluster dissolves. For nucleus containing more than n_C atoms, growth is favored. The *critical nucleus* is therefore the activated complex along the reaction coordinate of nucleation(Mersmann et al,. 2001).

3. Growth

It is well known that three basic steps are involving in the formation of crystal from an initially disordered phase. These stages are considered as achievement of supersaturation or supercooling, nucleation, and growth the nuclei into single crystals of distinct phase. When the crystallized particles are formed, the particles with size greater than critical size began to grow into final size. When, the equilibrium between crystallized phase and mother phase achives the free energy is at a minimum and the growth cannot occur. The equilibrium prevents the particle growth(Dhanaraj, et al. 2010). Therefore, a change in temperature or other effective parameters on the deriving force of crystallization must be applied to start the growth process again. Actually, supersaturation or supercooling of liquid increases the driving force of crystallization with respect to component whose growth is required. Several theories such as surface energy theory, diffusion theory, adsorption layer theory, and screw dislocation theory have been proposed to analyze the concept of particle growth. These theories evaluate the mechanism and the rate of crystal growth. Although, these theories cover the growth kinetic but high level accuracy remain elusive in the study of nucleation and growth process. For this reason, scientists try to fit experimental data and adjusting the theoretical parameter accordingly. In following, a brif presentation about these theories will be given.

3.1 Crystal growth theories

3.1.1 Surface energy

In surface energy theory, the particle growth happens through the plane which has the lowest surface energy based on the thermodynamic principle. In 1878 Gibbs proposed that the growth of a crystal depends on the total free energy of a crystal in equilibrium with surrounding at constant temperature and pressure. For the stable nuclei If the free energy per unit volume is assumed to be constant in the crystal, then the surface energy must have a minimum for a given volume (Mullin, 2001; Jones, 2002):

$$\sum_{1}^{n} a_i g_i = \min. \tag{14}$$

In a crystal which is bounded by n faces, a_i is the area of the ith face and gi is the surface free energy per unit area of the ith face. Therefore, if crystal grows in supersaturated system, its shape must have an equilibrium with surrounding medium. Many scientists used this idea in their research. Wulff (Suzuki, et al. 2007) studied the relationship between growth rate at different faces and the surface free energies. He showed that the excess surface free energy $\sigma_{ijk} dA_{ijk}$ is minimum for crystal with its *{hkl}* faces when it is in equilibrium with surrounding medium. The value of σ_{hkl} could determine on the basis of the shape of crystal. For example if σ is isotropic; the form of the crystal is spherical. Also Marc and Ritzel (Dhanaraj, et al. 2010) indicated that different faces have different values of solubility. When the difference in solubility is small, the surface energy has an imposrant role on the growth of nucleus. The rate of growth can be measured by the outward rate of monemer movement in the direction perpendicular to the face. In a crystal with geometrically shape, the velocity of growth will different from face to face. The ideal case of a crystal that has geometrical shape is called 'invariant'. In this crystal, any face with equal area grow at same rate. But, the smaller faces grows faster while the smallest face grows fastest of all. A similar, but, reverse behavior would be seen in the crystal in this type in a solvent. The face with faster growth rate dissolves at a faster rate than the other faces, but the sharp outlines of the crystal are soon lost ones dissolution commences (Mullin, 2001; Jones, 2002).

3.1.2 Diffusion theory

The main work on diffusion theory backs to the researches of Noyes and Whitney (1897). They proposed that the deposition of solid monomer on the surface of crystal was essentially a diffusion process. The base of this idea has two following assumptions. At first, it can be hypothesed that a concentration gradient exists in the near layer of the growing surface. Secondly, it can be considered that the crystal growth is the reversible process (Mullin, 2001; Jones, 2002). Consequently, on the basis of these assumptions, an equation can be written between the amounts of solute that are getting to solid surface and in the bulk of the supersaturated solution, which is as follows:

$$\frac{dm}{dt} = \left(\frac{D}{\delta}\right) A \left(C - C_0\right) \tag{15}$$

where m is the mass of solid deposited in a small time t; A is the surface area of crystal; C is the solute concentration in the solution; C_0 is the equilibrium saturation concentration; D is the diffusion coefficient of the solute and δ is the length of diffusion path. The thickness δ of the torpid layer would obviously depend on the relative solid-liquid velocity, in the system without any agitation, the film thickness would be around of 150 nm but in agitation system this thickness maybe less than 100 nm. The importance of nucleation site during growth of crystals was the main consideration of Volmer. He suggested that in growth mechanism the monomers when reached to a surface, they are not attached immediately to the lattice but migrate over crystal face to find a suitable site for attachment (Mullin, 2001; Jones, 2002).

3.1.3 Adsorption layer theory

In 1939 Volmer proposed the crystal growth mechanism based on adsorption layer theory. Volmer theory or Gibbs-Volmer theory explained the growth of particles when the units of solute arrive at the crystal face they do not attract immediately into surface but they migrate over the crystal face. Consequently, there is a loosely adsorption layer of the monomers at the interface and a dynamic equilibrium is established between the layer and the bulk solution. Kossle considered that crystal is in equilibrium with the solution when it is just saturated (Dhanaraj, et al. 2010). Also the energy of attachment unit on surface is a simple function of distance only. The thickness of the adsorption layer does not exceed 10nm and may even be near 1nm. The places in the lattice surface which have the greatest attractive energy are the best place for ions, molecules or monomers to attach. After that the suitable places with less energy attract these pieces. Under the standard condition this step-wise build-up will continue till the monomers cover the surface and will create a new layer on surface completely. Before the crystal faces can continue to grow or before creating the additional layer, a center of crystallization must exist on the surface plane. Gibbs-Volmer theory suggested that there is a monolayer island on the crystal surface which is called a two-dimensional nucleus (Mullin, 2001; Jones, 2002). Kossle (Dhanaraj, et al. 2010) also showed that the attachment energy depends on the kind of crystal. If the crystal is homopolar the attachment energy is due to Van Der Wals while if the crystal is heteropolar (ionic) the energy is due to electrostatic forces. The important specific feature of new phase nucleation on substrate surfaces, as compared to homogeneous nucleation in the bulk, is the presence of various defects on the substrate. Defects are typically divided into point and linear ones; the latter include kinks, surface dislocations, and vacancies. It seems that the sites like terraces, ledge, and kinks are the suitable place to unit arriving at a crystal to attach. The attachment energy of atoms, molecules or ions is maximum when it is correlated into a kink site in a surface ledge, whilst at any point on the ledge it is greater than that for an atom attached to the flat surface. Consequently, when a monomer reaches to a crystal surface, it does not go around and integrated into the lattice immediately. Instead, it goes to find to a step and moves along it to a kink sites, where it is finally incorporated. A crystal grow fastest when have lots of kink on its face (Mullin, 2001; Jones, 2002).

4. Kinetic models

4.1 LaMer model

One of the most famous and widely used nucleation theories is the work done by LaMer in the 1950s. LaMer and Dignae (1954) described the formation of sulfur sols from the decomposition of sodium thiosulfate in hydrochloric acid. Fig. 2, shows the concentration changing profile of nucleating species (sulfur in this case) via time. The contest of the nucleation mechanism is as follows. At first, the concentration of elemental sulfur increase slowly(stage I). In this stage no appreciable nucleation occurs even if the concentration level exceed the solubility level which is called as critical concentration (Moon et al., 1998; Finney & Finke, 2008; Keshmiri & Kesler, 2006; Tovstun & Razumov, 2010). When the concentration reaches to a critical level (critical supersaturation level) the nucleation process can be started (stage II). In the second stage, because some of the solute monomer is consumed to form particles the concentration of the solute species continues to increase but in lower slope. However, as the concentration still increases, the nucleation is drastically accelerated and, finally, the concentration of the monomeric species reaches the maximum level (maximum supersaturation).

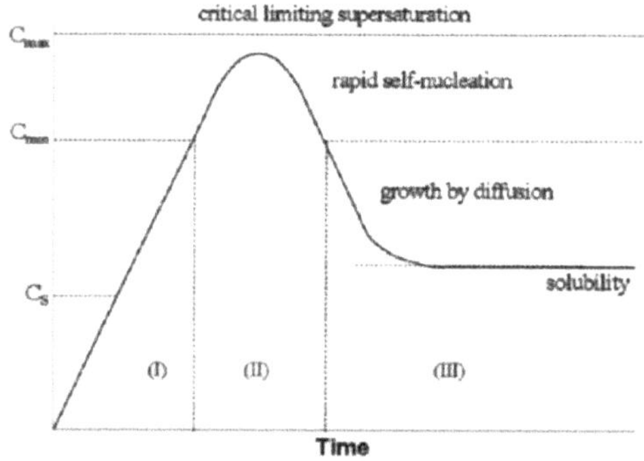

Fig. 2. The LaMer mechanism of nucleation of sulfur shows the changing pattern of sulfur concentration as a function of time(LaMer & Dignae, 1954).

At this time, the consummation of monomers to form the particles is balanced with increasing in their concentration. After that, the concentration of the monomeric species starts to be reduced. Due to the increasing in consumption of the monomeric species by growth of a number of the preformed nuclei, the concentration of the monomeric species decrease until reach to the supersaturation level. At this stage the nucleation is stopped (the end of the nucleation stage) and the remaining stable nuclei continue to grow with a supersaturation below the critical level for nucleation (Stage III). When the concentration of monomeric species becomes close to the solubility level of the bulk solid, the particle growth finishes (Finney & Finke, 2008; Keshmiri & Kesler, 2006; Tovstun & Razumov, 2010). There is another nucleation and growth theory, which is called as Turkevich model. In this model, it can be found that the particle formation followed kinetics consisting of an induction period, fast and linear increase particles number and then a rapid decrease in the rate.

4.2 Experimental models

Although, nucleation and growth process have a lot of interests to study, the highest levels of accuracy and precision have not made just by theoretical approach. It seems that, the theoritical approach would be come out by fitting experimental data with adjusting theoretical parameters simultaneously. Oxtoby (Oxtoby, 1992) has observed that the nucleation theory is one of the few areas of science in which the agreement of predicted and measured rates to within several orders of magnitude is considered a major success. Recently, some nucleation and growth studies based on experiment are developed. Becker and Doring (Toralff & Marro, 1987) evaluated and formulated the growth kinetic of particles as a steady-state process. The main goal of a theoretical description is to explain the size selection mechanism, the formation kinetic of monodisperse and the shape of the precipitated solids. There are approaches describing thermodynamic and dynamical growth mechanisms have been proposed by a lot of researchers (Privman, et al., 1999a; Privman and Mozyrsky, 1999b; Park, et al., 2001; Gorshkov & Privman, 2010; Libert, et al.

2003; Kulkarni & Zukoski 2002a; Dixit & Zukoski 2001; Dixit & Zukoski, 2002b). Dmitri V. Talapin in 2001 evaluated the evolution of an nanoparticle and presented a model allowing to calculate the progressed size of an nanoparticle under reaction, diffusion, and mixed reaction-diffusion control. The assumed model describing a place with a solution and durable concentration of monomer. The monomers can add to the nanoparticle surface and help it to growth or it can remove from the surface. It seems that the most probable for nanoparticles colloidal growth is polynuclear-layer ,so , the flux of monomer J_g^{react} toward the particles surface is described by a first order surface reaction (Talapin, et al. 2001, 2002):

$$J_g^{react} = 4\pi r^2 k_g^{flat} [M]_r \exp\left[-a\frac{2\gamma V_m}{rRT}\right] \tag{16}$$

where $[M]_r$ is the concentration of the monomer near the particle surface and k_g^{flat} is the growth rate constant for growth of a flat $(r = \infty)$ surface (Talapin, et al. 2001, 2002). If the rate of the dissolution of a solid does not depend on the monomer concentration in solution, it can be said that the flux of the monomer toward the particles is as follows:

$$J_g^{react} = -4\pi r^2 k_d = -4\pi r^2 k_d^{flat} \exp\left[\beta\frac{2\gamma V_m}{rRT}\right] \tag{17}$$

where

$$k_d^{flat} = B_d \exp\left(-\frac{\Delta\mu_d^{\infty}}{RT}\right) \tag{18}$$

is the dissolution rate constant for a flat $(r = \infty)$ interface. The equilibrium constant for the dissolution of the bulk material gained by the ratio between and :

$$K_{eq} \cong \frac{k_d^{flat}}{k_g^{flat}} = C_0^{flat} \tag{19}$$

Considering of the Fick low for the diffusion of the monomers from the bulk solution toward the particle surface and neglecting the size of the nanoparticle compare to the thickness of the diffusion layer under real condition the following equation can be written:

$$J_{diff} = 4\pi Dr\left([M]_{bulk} - [M]_r\right) \tag{20}$$

Where $[M]_{bulk}$ is the concentration of the monomer in the bulk of the solution and D is the diffusion coefficient. The number of monomers remove and add to the particles surface equal to those diffused toward the bulk of the solution, so:

$$J_g^{react} - J_d^{react} = J_{diff} \tag{21}$$

Now with substitution of the equations (18), (19), (20) and (21) the following equation for determination of the stationary concentration of the monomer near the particle surface can be expressed (Talapin et al., 2001, 2002):

$$[M]r = \frac{D[M]_{bulk} + rk_d^{flat}\exp\left[\beta\frac{2\gamma V_m}{rRT}\right]}{rk_g^{flat}\exp\left[-a\frac{2\gamma V_m}{rRT}\right] + D} \tag{22}$$

The size dependent rate of the particle radius is obtained from the combination of the Equations (20) and (22) :

$$\frac{dr}{dt} = VmDC_0^{flat}\left\{\frac{\frac{[M]_{bulk}}{C_0^{flat}} - \exp\left[\frac{2\gamma V_m}{rRT}\right]}{r + \frac{D}{k_g^{flat}}\exp\left[-a\frac{2\gamma V_m}{rRT}\right]}\right\} \tag{23}$$

If the process is the pure diffusion control i.e. $D\langle\langle k_g^{flat}$ and r^{*}>> *1*, equation (23) can be simplified to equation obtained by Sugimoto which is explained in the next part. In 1999, Tadao Sugimoto developed a model to explain the nucleation and the growth of nanoparticles which is based on LaMer model. This model described the formation of nanoparticles that their nucleation and growth comfort with LaMer diagram and controlled by diffusion (Sugimoto, 2001; Sugimoto et al., 2000a; 200b). The Sugimoto model is briefly explained as follow. When C reaches C_{max},(Fig. 2) the steady state of a mass balance between the supply rate of solute (R_S), its consumption rates for nucleation (R_N), and the growth of generated nuclei (R_G), is established, and the following relationship may be held:

$$R_N + R_G - R_S = 0 \tag{24}$$

This mass balance equation may be replaced by a corresponding differential equation for the number of the generated stable nuclei, n_∞ (Sugimoto, 2001; Sugimoto et al., 2000a; 2000b):

$$\upsilon_0\frac{dn_\infty}{dt} + \dot{\upsilon}n_\infty = QV_m \tag{25}$$

In Eq.(25) the generation of stable nuclei shows itself as the first term in the left-hand and similarly, the second term indicates the growth of the generated stable nuclei. Where υ_0 is the minimum particle volume of the stable nuclei, $\dot{\upsilon}$ is the mean volume growth rate of the stable nuclei, n is the their number density, Q is the supply rate of solute in mol per unit time, and V_m is the molar volume of the solid. The stable nuclei are sufficiently large to be grown to the final product particles, and they are defined as nuclei whose each particle volume is over the maximum particle volume of the stationary nuclei (Sugimoto, 2001; Sugimoto et al., 2000a; 200b). It's assumed that Q is constant during the short nucleation period (Q_0), and may be regarded as a constant during nucleation stage and shortly after growth. If it's assumed that there is a time-averaged mean volume growth rate of the stable nuclei during the nucleation stage, the behavior of n as a function of time by using the time-averaged growth rate is as follows:

$$n(t) = \frac{Q_0V_m}{\dot{\upsilon}}\left[1 - \exp\left(\frac{-\dot{\upsilon}t}{\upsilon_0}\right)\right] \tag{26}$$

Therefore, the number of final particles is given by:

$$n_{\infty} = \frac{Q_0 V_m}{\dot{\upsilon}} \tag{27}$$

On the other hand, it has been assumed that the growth of nanoparticles is controlled by diffusion, the linear growth rate, *dr/dt*, in the growth stage may be given by (Sugimoto, 2001; Sugimoto et al., 2000a; 200b):

$$\frac{dr}{dt} = \frac{DV_m}{r}\left[C - C_s \exp\left(\frac{2\gamma V_m}{rRT}\right)\right] \tag{28}$$

In Eq. 28, D is the diffusivity of the solute including free ions or their halide complexes, C is the molality of the solute, C_∞ is the solubility of the bulk solid in terms of total molality of solute and γ is the specific surface energy of the solid For maximum supersaturation. With having $\dot{\upsilon}$ and $S_m(=C_{max}/C_s)$, the initial particle radius of the stable nuclei during the nucleation stage r_0, the below equation is established between r_0, D, $\dot{\upsilon}$:

$$\dot{\upsilon} = 4\pi r_0 D V_m C_s\left[S_m - \exp\left(\frac{2\gamma V_m}{r_0 RT}\right)\right] \tag{29}$$

5. Nucleation and growth of TiO_2 nanoparticles

When TiO_2 powder was comerically producted in the early of twenty centery, it was used widely in a lot of applications such as membrane, sensor devices, photocatalyst. These applications can be classified in two major categories which are named as energy and enviromental. The employment of TiO_2 powders in these applications depends on the properties of the titania material which are affected by particle size, morphology, kind of polymorphe. Anatase, brookite and rutile are the main polymorphes of the TiO_2 and each of its phase has its own properties. In the nanoscale, the difference between properties of titanium is more considrelable. For example, anatase nanopowder shows advantages in applications such as photocatalytic material or dye-sensitized solar cells, and the particle size, shape and the morphology of TiO_2 nanomaterial have an important role in the performance of prepared material (Behnajady et al., 2008; Lee et al., 2009; Ni et al., 2006). On the other hand, rutile phase exhibits a higher refractive index, hiding power, good chemical stability. It is becoming a candidate material for high-temperature separation and catalysis applications (Li et al., 2006). Consequently, phase stability is very important for advanced application of TiO_2 nanomaterial. The phase transformation occurs during calcinations which is effected by the particle size and morphology. It is well known that the particle size and morphology are controlled by nucleation and growth procedure of TiO_2 nanoparticles. Therefore, it is necessary to have a detailed understanding of the nucleation and growth mechanism, as well as processes such as aggregation and coarsening. Nowadays, several methods such as flame reactor (Mehta et al., 2010), melt-phase separation technique (Zhaoa & Zhang, 2011), atomic layer deposition (King et al., 2009), combustion of $TiCl_4$ (West et al. 2007), atmospheric pressure CVS process (Rahiminezhad-Soltani et al., 2011), solvothermal (Wang & Xiao, 2006), thermal hydrolysis

(Santacesaria et al., 1985), sol-gel (Mahshid et al., 2007; 2008) have been developed to prepare TiO_2 nanoparticle with designed and controlled properties (crystalline phase, size, shape, etc.). Formation of TiO_2 nanoparticles have been studied in many papers with different approaches and aspects. For example, the influence of synthesis condition on preparation of TiO_2 by the hydrothermal method has shown that the size of particle increases from 18 to 26 nm when the synthesis temperature rise from 140 to 200°C (Ryua et al., 2008). But, the nucleation and growth kinetic study of nanaocrystalline TiO_2 formation from aqueous solutions and sol-gel methods have attracted many attentions (Courtecuisse et al., 1996; Mizukami, 1999; Baek et al., 2009; Kim et al., 1999; Gaoa et al., 2004; Jiaa et al., 2009; Nishidea & Tieng et al., 2011). Demopoulo and his co-workers (Charbonneau et al., 2009) produced the nanostructured TiO_2 particles by precipitation from forced hydrolysis of aqueous Ti(IV) chloride solution. The kinetic study of precipitation was shown that the nucleation rate is slow while growth rate is fast and follows Avrami model equation. They reported that the formation of nanostructure spherical particles is based on two steps. In the first step, the nuclation and aggregation of elementary nanocrystallite (≈ 10 - 2 nm) is acquired and in the second step the growth of elongated self -assembled nanofibers along rutile (110) atomic planes (Charbonneau et al., 2009). Oskam et al., (2003) reported that the average size of primary particles, depends on the parameters such as like coarsing time, temperature and solution chemistry. Solution chemistry can be modified by using different anions, solvent, pH or concentrations. He prepared precipitated anatase nanoparticles from the homogeneous acidified solution (pH=1) using titanium(IV) isopropoxide as precursor and using water-to-titanium mole ratio of about 200. It was shown that the secondary particles could be formed by epitaxial self-assembly of primary particles when longer time and higher temperature were used. The formation of secondary particles increases at higher temperatures because the tendency of primary particles to attend in the attachment process rises. It was indicated that the average particles radius increases linearly with time on the basis of the Lifshitz-Slyozov-Wagner model for coarsening (Oskam, et al. 2003). One of the complete research for studying of the nucleation and growth kinetic of titania nanoparticles was done by Zhang (Zhang et al., 2008). In his research titnium dioxid particles produced by the hydrolysis of a titanium tetrachloride ($TiCl_4$) precursor in aqueous solution at temperature below 100°C. They studied on the formation of TiO_2 particles in the three periods; induction, initial growth and saturated growth. Induction time depends extremely on the concentration of $TiCl_4$ and pH. Induction time was shorter with higher pH or [$TiCl_4$]. Higher supersaturation was made by increasing the solution temperature, [$TiCl_4$], and pH which results in shorter induction time for nuclation and a faster growth rate. The growth rate can be represented by the following power-low relationship:

$$d = Cs^{P} \tag{30}$$

which d is the growth rate and is equal to 0.6 at lower pH. The growth rate is equal to 3.9 nm/min at higher pH in 25°C. The particle critical size for growth was 0.44nm. In addition, the particle growth rate has the parabolic behavior relationship with the degree of supersaturation (Zhang, et al. 2008). As it has been mentioned above another appropriate technique for preparation of titanium oxide nanoparticles is sol-gel process. Sol-gel is a one the best techniques that can be applied to prepare and analyze the formation of nano titania. Some effective parameters on the nature of the sol-gel process are metal precursor, temperature, pH of solution and the presence/absence of catalyst. Proposed benefits of the sol gel process are considered as easy control of formation process (nucleation and growth), high stability, better homogeneity, and high purity. The sol-gel process has been widely employed for preparing of titanium dioxide particles because the control of hydrolysis and polycondensation reactions can be easily stablished and gotten the appropriate properties (Zhang et al., 2009). TiO_2 nanomaterial can be synthesized by sol gel process with different titanium precursors. Typically, synthesizing of TiO_2 nanoparticles via sol-gel process include an acid-catalyzed hydrolysis step of titanium (IV) alkoxide followed by condensation. The development Ti-O-Ti is favorite with low water. Present of large excess of water lead to develop polymetric Ti-O-Ti chains (Mahshida et al., 2007; 2009). The study on nucleation and growth kinetic of titana nanoparticles prepared by sol gel method shows that the rate constant for coagulation of particles increases with temperature because the velocity of monomer through the particles has high dependency of temperature. Secondary particles were formed and the growth of particles incresed when they passed critical radius and became stable (Chenl & Mao, 2007). Jean and Ring (1986) studied on the growth kinetic of titania monospheres and noticed that the data fitted with theoretical curve for growth under diffusion control (LaMer diagram). They used the curve and determined the diffusion coefficient of particles which was $\approx 10^{-9}$ cm^2/s (Jean & Ring, 1986). Keshmiri and Troczynski (2002) syntheszied the ultra-fine and spherical of TiO_2 nanoparticles with a controlled precipitation of sol-gel technique. They found that the nucleation period was limited to a short period of time and the growth process was induced by another step. On the basis of this idea it can be possible to make nucleation through supersaturation phenomena (refer to LaMer diagram), and it is possible to produce almost perfect monodispersed particles (Keshmiri & Troczynski 2002). Li et al., (2002) prepared TiO_2 nano powder by sol-gel method. They used tetra-n-butyl-titanate and dionized water as the starting materials. The grain size of powder as a function of calcination temperature was studied and they found out that the grain size increase with the increasing of calcinations temperature. The following equation was used to explain the growth rate (Li, et al. 2002):

$$u = a_0 \dot{\upsilon}_0 \left[\exp\left(\frac{-Q}{kT} \right) \right] \left[1 - \exp\left(-\frac{\Delta F_\upsilon}{kT} \right) \right] \tag{31}$$

where a_0 is the particle diameter, $\dot{\upsilon}_0$ the atomic jump frequency, Q the activation energy for an atom to leave the matrix and attach itself to the growing phase, ΔFv the molar free energy difference between two phases. For non-crystallization $\Delta Fv \gg KT$. So Eq. (31) can be reduced as (Li, et al. 2002):

$$u = a_0 \dot{\upsilon}_0 \left[\exp\left(\frac{-Q}{kT} \right) \right] \tag{32}$$

When the calcination temperature is high, the activation energy is very small, therefore, the growth rate is large. So the grain size increases very quickly as the calcination temperature increases. When the calcination temperature is low, the activation energy is very large and the growth rate becomes slow. So the grain size increases very slowly (Li, et al. 2002). A series of complete experiment on formation of TiO_2 particles have been assembled by Sugimoto

(Sugimoto, et al. 2002; Sugimoto & Zhou, 2002; Sugimoto, et al. 2003a). Using the sol-gel method for synthesis of TiO_2 nanoparticles, different sizes and shapes can be obtained by changing the reaction parameters. In his experiments, uniform anatase TiO_2 nanoparticles were formed by the gel-sol process from a condensed $Ti(OH)_4$ gel preformed by the hydrolysis of a Ti–triethanolamine (TEOA) . the titanium complex is hydrolyzed by degrees in water and completely converted to a rigid gel of $Ti(OH)_4$ after aging at 100°C for 24 hr (first aging). Then, the $Ti(OH)_4$ gel is totally transformed into anatase TiO_2 nanoparticles by additional aging at 140°C in an autoclave for 3 days (second aging) . In his experiment, the deposition process hydroxo complexes of titanium ion onto the growing TiO_2 determined the step for the formation of TiO_2 particles and not the dissolution of the $Ti(OH)_4$ gel. Consequently, the supply rate of the solute is equivalent to the dissolution rate of the gel and is increased with the consumption rate of solute and the particles of TiO_2 grow to keep the equilibrium concentration level (Sugimoto, et al. 2003b; 2007). His results showed that, however a mass balance exists between the rates of supply and consumption of solute toward the particles, the nucleation rate is related and a function of solute concentration which is equilibrium with the gel and also is independent of the consumption rate increase of solute for the particle growth. In this way, LaMer mechanism doesn't work in this situation because the nucleation of TiO_2 will not be finished until the precursor gel remains. Sogimoto pointed out that in such situation it is hard to separate the nucleation and growth process (Sugimoto et al., 2003b; Sugimoto, 2007). But, he defined a different way to separate nucleation and growth stage. Fig(3), shows the changing pattern of the time evolution of the concentration vs. time during the first aging at 100°C and the second aging at 140°C. Under the standard condition a different way of separation can be distinguished. In stage 1, the concentration of titanium ion isdecreased by a factor of about 2 orders of magnitude for the precipitation of $Ti(OH)_4$ gel. But, in second stage its order factor is one. He suggested that the factor is lowered probably because the progress of hydrogen bonding throughout the gel hydroxide groups when temperature raised to 140°C. He observed that at the earlier of the second aging, the nucleation

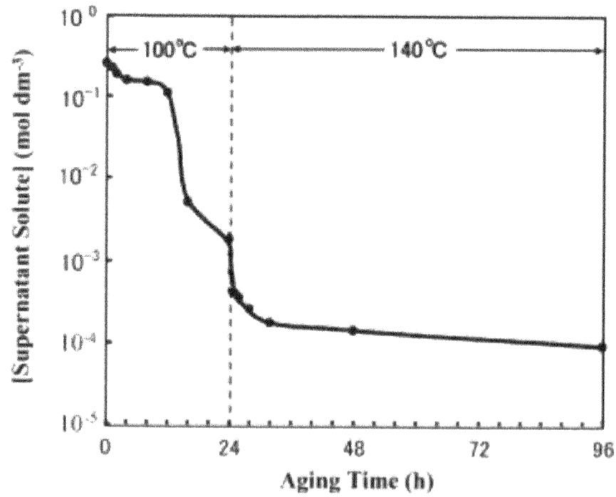

Fig. 3. Time evolution of the concentration vs. time during the first aging at 100°C and the second aging at 140°C under standard sol-gel system for TiO_2 nanoparticles (Sugimoto, 2007).

of TiO_2 particles happend, but it was completed soon with the drop of the supernatant concentration, followed by the growth of the nuclei under the supersaturation below the critical level for nucleation (Sugimoto et al., 2003b; Sugimoto, 2007).

Recently, we studied on the nucleation and growth of TiO_2 nanoparticles which were synthesized in two stages in an aqueous peroxotitanate solution by using HNO_3, TTIP, and TIPO as starting materials. The established controlled conditions for the preparation of TiO_2 nanoparticles were as follows: First, a stock solution of Ti^{+4} was prepared by mixing titanium isopropoxide with triethanol amine (TEOA) at a molar ratio of TIPO: TEOA=1 : 2 under dry air to form a stable Ti^{4+} compound against the hydrolysis reaction at room temperature. Then, 10 ml of the stock solution was mixed with the same volume of doubly distilled water. The pH was controlled by addition of $HClO_4$ or NaOH solution. The solution (pH=9.6) was placed in a screw-capped Pyrex bottle and aged at 100°C for 36 h. Finally, the resulting highly viscous gel was mixed with 80 ml (2×10^{-3} molar) nitric acid and stirred at 25°C for 3 h to dissolve the gel and so prepared a dark solution with pH=1. The solution was set in pool water at various temperatures ranging from 60 to 80°C. Solution was heated to achieve super saturation of Ti in solution. In each 4 min time interval, a sample has been taken from prepared solution and analyzed by atomic adsorption. A condenser was used to stop the evaporation of solution and each 4 min a sample has been taken throughout the process. TiO_2 particles were separated from the resulting Ti^{4+} suspension by centrifugation. This procedure has been repeated for several times and every time the concentration of Ti4+ ions was measured via atomic adsorption. Titanium alkoxide can be hydrolyzed and condensed through following steps:

$$\text{Ti-OR} + H_2O \leftrightarrow \text{Ti-OH} + \text{ROH} \quad \text{(I)}$$

$$\text{...Ti-OH} + \text{OR-Ti...} \leftrightarrow \text{...Ti-O-Ti...} + \text{ROH} \quad \text{(II)}$$

$$\text{...Ti-OH} + \text{OH-Ti...} \leftrightarrow \text{...Ti-O-Ti...} + \text{HOH} \quad \text{(III)}$$

Depending on the pH of solution, several compounds of titanium (IV), e.g. TiO^{2+}, $TiOH^{+3}$,$Ti(OH)_2^{+2}$, and $TiO(OH)^+$, can be formed. Experimental results have shown that the uniform nanoparticles can be obtained in solution with more acidic condition (pH~1) (Sugimoto & Zhou 2002; Sugimoto et al., 2002a). Therefore, it is reasonable that an acidic solution is suitable to choose for our theoretical approach. When the pH of solution is near to 1, the TiO^{+2} ions are the dominant complex in the solution and TiO_2 can be nucleated by the aforementioned equations (II and III). It can be found that the growth of particles occurs by the addition of Ti^{+4} to the growing particles. It has been shown that the sufficient number of O^- or OH^- sites are available along the TiO_2 particle planes, which are appropriate positions to create octahedral coordination for the incoming Ti^{+4} ions. Moreover, after the attachment of Ti^{4+} ions along the planes, equal number of new O^- or OH^- sites are created for further attachment of Ti^{4+} ions (Baek, et al. 2009). In Fig. 4, the concentration-changing pattern of [Ti] ions in the solution under the standard condition in 70°C can be observed, which is similar to the LaMer diagram. For better explanation, the diagram is divided in two different stages. In the first stage, the liquid in the solution evaporates slowly which leads to the increase in the concentration of titanium ions. In contrast, in the second stage, the concentration of titanium ions starts to decrease as TiO_2 particles start to precipitate. The supersaturation ratio, S, was obtained from the LaMer diagram (Fig. 4). The maximum value of supersaturation ratio, S_m, is equal to 3.19. The radius and free energy of formation of the nucleus at the supersaturation, r^*_m and ΔG^*_m respectively, are calculated from Gibbs-Thomson equation which are $r^*_m = 4.42nm$ and $\Delta G^*_m = 1.16\times10^{-16}J$ (=343K). At the supersaturation level, the high driving force is available (Fig. 4). Therefore, the nucleation process is easiest and continues until the radius of the nucleus of stable particles reaches (r_0).

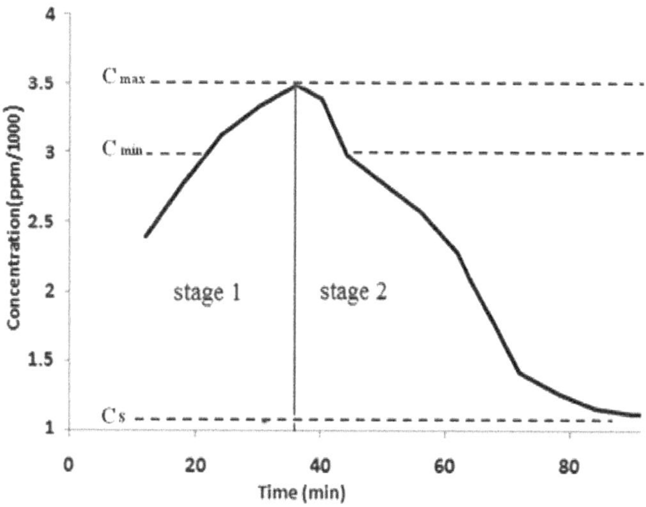

Fig. 4. Time evolutions of Ti^{4+} concentration in the standard system: stage 1;[Ti] concentration increasing. stage 2; [Ti] concentration decreasing (Mehranpour et al., 2010).

When, the stable particles are present, the nucleation stops and particles start to grow until the final product is produced. To obtain the primary radius of stable embryo, the Gibbs-Thomson equation was used and the determined radius is equal to $r_0=6.1nm$. Fig. 5 shows the TEM image of the TiO_2 particles. It is clear that the size of formed TiO_2 nanoparticles is equal to *7.1nm (d=14.2nm)* and has a good agreement with the theoretical size. The supply rate of the monomer in the nucleation period, Q_0, was obtained from the tangential slope of the *[Ti]-t* diagram (Fig. 4) at $[Ti]_{max}$ which is equal to $7.03\times10^{-7}mol.dm^{-3}.s^{-1}$. Using the supply rate of monomer and the final particle density, the amount of mean volumic growth rate of the stable nuclei are calculated from Equ. 27 as 2.98×10^2 nm^3s^{-1}. Finally, from equation 29 the diffusion coefficient of ions in solution is $D = 6.18\times10^{-8}$ cm^2s^{-1}. The effect of temperature on the formation of particles was investigated in the diffrent temperature. The results show that the nucleation and growth of TiO_2 nanoparticles is Extremely dependent on the temperature. It can be observed the increase in temperature has caused a shift in the LaMer diagram to the left side. So, the maximum supersaturation can be reached in higher temperatures. In fact, temperature has a major effect on the diffusion coefficient of monomers. According to Arrhenius equation, the rise in temperature causes an increase of the diffusion coefficient which effects on the acceleration of hydrolysis of *Ti* in solution and the supply rate of the monomer in the nucleation period, Q_0, rises.

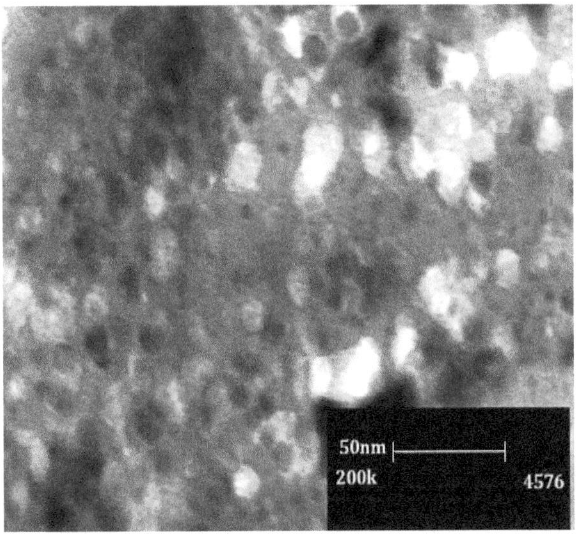

Fig. 5. TEM image of TiO_2 nanoparticles at the end of nucleation and start of growth stage. The radius is large enough to be stable and growth to the final particles ($r^* = r_0$).

Therefore, it was predictable that the nucleation and growth of particles occurred earlier in upper temperature. As we mentioned above, TiO_2 phases have their own properties and the phases transformation which is associated by the nuclation and grain growth of new phase inside the old phase has attracted a considerable attention (Delogu 2009; Gole et al., 2008; Napolitano et al., 2010). The activation energy of anatase to rutile phase transformation has been reported from 147 to 837 kJ/mol (Hsiang & Lin, 2006; 2008). Zhang and Banfield (2000) suggested that in anatase samples with denser particle packing, rutile nucleates primarily at interfaces between contacting anatase particles. But in anatase samples with less dense particle packing, rutile nucleates at both interfaces and free surfaces of anatase particles (Gribb & Banfield, 1997). To analyze, the phase transformation of synthesized titania nanoparticles, they were thermally heated at 200, 400, 600, and 800°C for 2 hours. Figure 6 shows the phase contents of rutile in samples which were treated isochronally at 2 hours within the temperature range 200 to 800°C. The results indicate that the process involves the transformation of anatase in the starting material to rutile from its initial 23% at 200°C to 100% at 800°C. Rapid reaction rate of phase transformation shows that there is high driving force for the transition (free energy change for phase transformation). It means that a lot of nucleation and growth site for transformation are available in prepared sample. The kinetics of anatase-rutile phase transformation is followed by Avrami equation as follows (Hsiang and Lin 2006):

$$X = 1 - \exp(-kt)^n \tag{33}$$

where X is the fraction of transformation, t is the time, k is the kinetic constant. From the Arrhenius plot of various kinetic constants, and with calculation of slop, activation energy is obtained as −20 kj/mol.

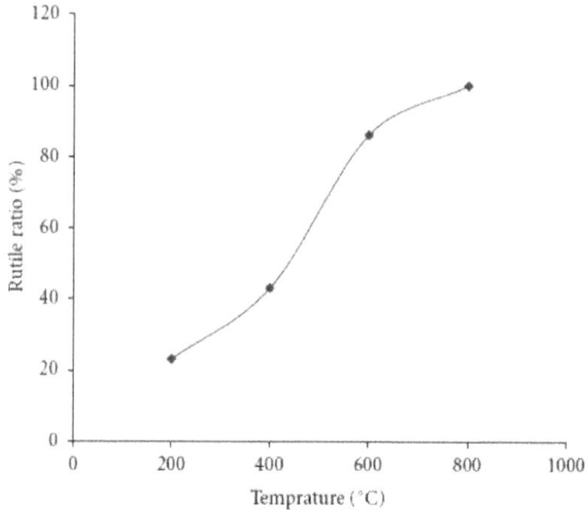

Fig. 6. Evolution of the amount of Rutile phases in the various temperatures (Mehranpour et al., 2010).

6. Conclusion

Nowadays, titanium oxide nanoparticles have the most interest due to their properties. These properties make them to use in a wide range of advance. It is well known that the size of nanoparticles plays an important role in electrical and optical characteristic (Miao et al., 2004; Oskam et al., 2003). Also the other factor effecting on properties of nano powder is the phase composition and the particle size of each phase. Hence, it is essential to have completely control on forming of particles to achieve the favorite particle size, shape, and size distribution. Therefore, a theoretical approach to understand the mechanism of TiO_2 nanocrystal formation provides a greater control over the size, shape, and composition and results an ability to tune the abovementioned properties simply by varying the crystallization conditions. In this chapter, at first, the kinetic of nucleation and growth of particles have been generally discussed and some of scientific reports have been analyzed. Then, the kinetic of nucleation and the growth mechanism of Titana nanoparticles were studied on the basis of LaMer theory. In comparison with other synthesized TiO_2 nanoparticles the kinetics of phase transformation of our sample is very fast and so proposes that the activation energy of nucleation and growth is more than other prepared nanoparticles. It is concluded that the lowering of activation energy is due to diffusion controlled synthesis process and the uniformity of size and shape of resultant material.

7. References

Bae, H., Lee, M., Kim, W., & Rhee, C. (2003). Dispersion properties of TiO_2 nano-powder synthesized by homogeneous precipitation process at low temperatures. *Colloids and Surfaces A: Physicochem. Eng. Aspects*, Vol. 220, No. 1-3, pp. 169-177, ISSN: 0927-7757

Baek, I. C., Vithal, M., Chang, J. A., Yumb, J.-H., Nazeeruddin, M., & Grätzel, M. (2009). Facile preparation of large aspect ratio ellipsoidal anatase TiO_2 nanoparticles. *Electrochemistry Communications*, Vol. 11, pp. 909–912. ISSN: 1388-2481.

Behnajady, M. A., Modirshahla, N., Shokri, M., Elham, H., & Zeininezhad, A. (2008). The effect of particle size and crystal structure of titanium dioxide nanoparticles on the photocatalytic properties. *Journal of Environmental Science and Health A*, Vol. 43, No. 5, pp. 460–467, ISSN 1093-4529.

Charbonneau, C., R. Gauvin, & Demopoulos, G. P. (2009). Nucleation and growthofself-assembled nanofibre-structured rutile(TiO_2) particles via controlled forced hydrolysis of titanium tetrachloride solution. *Journal of Crystal Growth*, Vol. 312, No.1 , pp.86–94, ISSN 0022-0248.

Chenl, X., & Mao, S. S. (2007). Titanium Dioxide Nanomaterials: Synthesis, Properties, Modifications, and Applications. *Chemical Reviews*, Vol. 38, No. 41, pp. 2891-2959, ISSN 1522-2667.

Courtecuisse, V. G., Chhor, K., Bocquet, J.-F. O., & Pommier, C. (1996). Kinetics of the Titanium Isopropoxide Decomposition in Supercritical Isopropyl Alcohol. *Industrial & Engineering Chemistry Research*, Vol. 35, No. 8, pp. 2539-2545, ISSN 1520-5045.

Delogu, F. (2009). A mechanistic study of TiO_2 anatase-to-rutile phase transformation under mechanical processing conditions. *Journal of Alloys and Compounds*, Vol. 468, No. 1-2, pp. 22-27. ISSN: 0925-8388

Dhanaraj, G., Byrappa, K., Prasad, V., & Dudley, M. (2010). Crystal Growth Techniques and Characterization: an overveiw. *Springer Handbook of Crystal Growth*. Springer, ISBN: 978-3-540-74182-4, London.

Dixit, N. M., & Zukoski, C. F. (2001). Kinetics of crystallization in hard-sphere colloidal suspensions. *Physical Review E*, Vol. 64, No. 4, pp. 041604-1, ISSN 1550-2376.

Dixit, N. M., & Zukoski, C. F. (2002). Nucleation rates and induction times during colloidal crystallization: Links between models and experiments. *Physical Review E*, Vol. 66, No. 5, pp. 051602-12, ISSN 1550-2376.

Dohsh, S., Takeuch, M., & Anpo, M. (2003). Effect of the local structure of Ti-oxide species on the photocatalytic reactivity and photo-induced super-hydrophilic properties of Ti/Si and Ti/B binary oxide thin films. *Catalysis Today*, Vol. 85, No. 2-4 , pp. 199–206, ISSN: 0920-5861.

Finney, E. E., & Finke, R. G. (2008). Nanocluster nucleation and growth kinetic and mechanistic studies: A review emphasizing transition-metal nanoclusters, *Journal of Colloid and Interface Science*, Vol. 317 , No. 3, pp. 351–374, ISSN 0021-9797.

Fokin, V. M., Yuritsyn, N. S., & Zanotto, E. D. (2005). Nucleation and Crystallization Kinetics in Silicate Glasses:Theory and Experiment, J. W. Schmelzer, *Nucleation Theory and Applications*, WILEY-VCH Verlag GmbH & Co., ISBN 978-3-527-40469-8, Weinheim.

Gaoa, Y., Masudaa, Y., Seo, W.-S., Ohta, H., & Koumoto, K. (2004). TiO_2 nanoparticles prepared using an aqueous peroxotitanate solution. *Ceramics International*, Vol. 30, No. 7, pp. 1365–1368, ISSN 0272-8842.

Gasser, U., Weeks, E. R., Schopeld, A., & Pusey, P. N. (2001). Real-Space Imaging of Nucleation and Growth in Colloidal Crystallization. *Science*, Vol. 292, No. 5515, pp. 258-261, ISSN 0036-8075.

Gole, J. L., Prokes, S. M., & Glembocki, O. J. (2008). Efficient Room-Temperature Conversion of Anatase to Rutile TiO_2 Induced by High-Spin Ion Doping. *Journal of Physical Chemistry C*, Vol. 112, No. 6, pp. 1782–1788, ISSN 1932-7447.

Gorshkov, V., & Privman, V. (2010). Models of synthesis of uniform colloids and nanocrystals. *Physica E*, Vol. 43, No. 1, pp. 1–12, ISSN 1386-9477.

Gribb, A. A., & Banfield, J. F. (1997). Particle size effects on transformation kinetics and phase stability in nanocrystalline TiO_2. *American Mineralogist*, Vo. 82, No. 7-8 , pp. 717–728, ISSN 1945-3027.

Hsiang, H.-I., & Lin, S.-C. (2008). Effects of aging on nanocrystalline anatase-to-rutile phase transformation kinetics. *Ceramics International*, Vol. 34, No. 3, pp. 557–561, ISSN 0272-8842.

Hsiang, H.-I., & Lin, S.-C. (2006). Effects of aging on the kinetics of nanocrystalline anatase crystallite growth. *Materials Chemistry and Physics*, Vol. 95, No. 2-3, pp. 275–279, ISSN 0254-0584.

Huang, Y., & Pemberton, J. E. (2010). Synthesis of uniform, spherical sub-100nm silica particles using a conceptual modification of the classic LaMer model. *Colloids and Surfaces A: Physicochem. Eng. Aspects*, Vol. 360, pp. 175–183, ISSN 0927-7757.

Jean, H. J., & Ring, A. T. (1986). Nucleation and growth of monosized TiO_2 powders from alcohol solution. *Langmuir*, Vol. 2, No. 2, pp. 251-255, ISSN 0743-7463.

Jiaa, H., Zhenga, Z., Zhaoa, H., Zhangb, L., & Zou, Z. (2009). Nonaqueous sol–gel synthesis and growth mechanism of single crystalline TiO_2 nanorods with high photocatalytic activity. *Materials Research Bulletin*, Vol. 44, No. 6, pp. 1312-1316, ISSN 0025-5408.

Jones, A. G. (2002). Crystal agglomeration and disruption. In *Crystallization Process System*, Butterworth Heinemann, ISBN 0750655208, Oxford.

Kashchiev, D. (2001). Cluster Properties. In *Nucleation basic theory and application*. Butterworth-Heinemann, ISSN 0750646829, Burlington .

Keshmiri, M., & Kesler, O. (2006). Colloidal formation of monodisperse YSZ spheres: Kinetics of nucleation and growth. *Acta Materialia*, Vol. 54, No. 2, pp. 4149–4157, ISSN 1359-6454 .

Keshmiri, M., & Troczynski, T. (2002). Synthesis of narrow size distribution sub-micron TiO_2 spheres. *Journal of Non-Crystalline Solids*, Vol. 311, No. 1, pp. 89-92, ISSN 0022-3093.

Kim, S.-J., Park, S.-D., & Jeong, Y. H. (1999). Homogeneous Precipitation of TiO_2 Ultrafine Powders from Aqueous $TiOCl_2$ Solution. *Journal of the American Ceramic Society*, Vol. 82, No. 4, pp. 927–932, ISSN 1551-2916.

King, D. M., Zhou, Y., Hakim, L. F., Liang, X., Li, P., & Weimer, A. W. (2009). In situ synthesis of TiO_2-functionalized metal nanoparticles. *Industrial & Engineering Chemistry Research*, Vol. 48, No. 1, pp. 352–360, ISSN 0888-5885.

Kulkarni, A. M., & Zukoski, C. F. (2002). Nanoparticle Crystal Nucleation: Influence of Solution Conditions. *Langmuir*, Vol. 18, No. 8, pp. 3090-3099, ISSN 0743-7463.

Kumar, K.-N. P., Keizer, K., & Burggraaf, A. J. (1994). Stabilization of the porous texture of nanostructured titania by avoiding a phase transformation, *Journal of Materials Science Letters*, Vol. 13, No. 1, pp. 59–61, ISSN 1573-4811.

Lee, K.-M., Suryanarayanan, V., & Ho, K. (2009). Influences of different TiO_2 morphologies and solvents on the photovoltaic performance of dye-sensitized solar cells. *Journal of Power Sources*, Vol. 183, No. 2, pp. 635–641, ISSN 0378-7753.

Li, B., Wang, X., Yan, M., & Li, L. (2002). Preparation and characterization of nano-TiO2 powder. *Materials Chemistry and Physics,* Vol. 78, No. 1, pp 184–188, ISSN 0254-0584.

Li, Y., Liu, J., & Jia, Z. (2006). Morphological control and photodegradation behavior of rutile TiO_2 prepared by a lowtemperature process, *Materials Letters,* Vol. 60, No. 13-14, pp. 1753–1757, ISSN: 0167-577X.

Libert, S., Gorshkov, V., Privman, V., Goia, D., & Matijevic, E. (2003). Formation of monodispersed cadmium sulfide particles by aggregation of nanosize precursors. *Advances in Colloid and Interface Science* , Vol. 100, pp. 169–183, ISSN 0001-8686.

Mahshid, S., Askari, M., Ghamsari, M. S., Afshar, N., & Lahuti, S. (2009). Mixed-phase TiO_2 nanoparticles preparation using sol-gel method, *Journal of Alloys and Compounds,* Vol. 478 No.1-2, pp. 586-589, ISSN 0925-8388.

Mahshida, S., Askaria, M., & Sasani Ghamsar, M. (2007). Synthesis of TiO2 nanoparticles by hydrolysis and peptization of titanium isopropoxide solution. *Journal of Materials Processing Technology,* Vol. 189, No. 1-3, pp. 296-300, ISSN 0924-0136.

Mangin, D., Puel, F., & Veesler, S. (2009). Polymorphism in Processes of Crystallization in Solution: A Practical Review. *Organic Process Research & Development,* Vol. 13, No. 6, pp. 1241–1253, ISSN 1083-6160.

Maris, H. J. (2006). Introduction to the physics of nucleation. *Comptes Rendus Physique,* Vol. 7, No. 9-10, pp. 946–958, ISSN 1631-0705.

Mehranpour, H., Askari, M., Sasani Ghamsari, M., & Farzalibeik, H. (2010). Study on the Phase Transformation Kinetics of Sol-Gel Drived TiO_2 Nanoparticles. *Journal of Nanomaterials,* Vo. 2010, pp 1-5, ISSN 1687-4129.

Mehranpour, H., Askari, M., Sasani Ghamsari, M., & Farzalibeik, H. (2010). Application of Sugimoto model on particle size prediction of colloidal TiO_2 nanoparticles. Nanotech 2010 Vol. 1, Chapter 3: Nanoparticle Synthesis & Applications,

Mehta, M., Sung, Y., Raman, V., & Fox, R. O. (2010). Multiscale Modeling of TiO2 Nanoparticle Production in Flame Reactors: Effect of Chemical Mechanism. *Industrial & Engineering Chemistry Research,* Vol. 49,No. 21, pp.10663–10673, ISSN 0888-5885.

Mersmann, A., Heyer, C., & Eble, A. (2001). Activated Nucleation. In A. Mersmann, *Crystallization Technology Handbook.* Taylor & Francis Group, ISBN: 0-8247-0528-9, Basel.

Miao, L., Tanemuraa, S., Tohb, S., Kanekob, K., & Tanemura, M. (2004). Fabrication, characterization and Raman study of anatase-TiO_2 nanorods by a heating-sol-gel template process. *Journal of Crystal Growth,* Vol 264, No. 1-3, pp.246–252, ISSN 0022-0248.

Moon, J., Carasso, M. L., Krarup, H. G., Kerchner, J. A., & Adair, J. H. (1998). Particle-shape control and formation mechanisms of hydrothermally derived lead titanate. *Journal of Materials Research,* Vol. 14, No. 3, pp. 866-875, ISSN 2044-5326.

Mullin, J. W. (2001). Crystal Growth. In *Crystallization,* Butterworth Heinemann, ISBN 0750648333, Oxford, pp. 216-250.

Nakade, S., Saito, Y., Kubo, W., Kanzak, T., Kitamura, T., Wada, Y., et al. (2003). Enhancement of electron transport in nano-porous TiO_2 electrodes. *Electrochemistry Communications* , Vol. 5, No. 9, pp.804–808, ISSN 1388-2481.

Napolitano, E., Mulas, G., Enzoa, S., & Delogu, F. (2010). Kinetics of mechanically induced anatase-to-rutile phase transformations under inelastic impact conditions. *Acta Materialia,* Vol. 58, No. 10, pp. 3798-3804, ISSN: 1359-6454.

Ni, M., Leung, M. K., Leung, D. Y., & Sumathy, K. (2006). analytical study of the porosity effect on dye-sensitized solar cell performance. *Solar Energy Materials and Solar Cells,* Vol. 90, No. 9, pp.1331–1344, ISSN 0927-0248.

Nishidea, T., & Mizukami, F. (1999). Effect of ligands on crystal structures and optical properties of TiO_2 prepared by sol-gel processes. *Thin Solid Films,* Vol. 353, No. 1, pp. 67-71, ISSN: 0040-6090.

Oskam, G., Nellore, A., Penn, R. L., & Searson, P. C. (2003). The Growth Kinetics of TiO_2 Nanoparticles from Titanium(IV) Alkoxide at High Water. *Journal of Physical Chemistry B,* Vol. 108, No. 8, pp.1734-1738, ISSN 1520-5207.

Oxtoby, D. W. (1992). Homogeneous nucleation: theory and experiment. *Journal of Physics: Condensed Matter,* Vol. 4 , No. 4, pp. 7627-7650, ISSN 1361-648X

Park, J., Privman, V., & Matijevic, E. (2001). Model of Formation of Monodispersed Colloids. *Journal of Physcal Chemistry B , Vol. 105,* No.1, pp. 11630-11635, ISSN 1520-5207.

Privman, V., & Mozyrsky, D. (1999b). Diffusional growth of colloids, *Journal of Chemical Physics,* Vol. 110, No. 18, pp. 9254-9258, ISSN 0021-9606 .

Privman, V., Goia, D. V., Park, J., & Matijevic, E. (1999a). Mechanism of Formation of Monodispersed Colloids by Aggregation of Nanosize Precursors. *Journal of Colloid and Interface Science,* Vol. 213, No. 1, pp. 36–45, ISSN 0021-9797.

Rahiminezhad-Soltani, M., Saberyan, K., Shahri, F., & Simchi, A. (2011). Formation mechanism of TiO_2 nanoparticles in H_2O-assisted atmospheric pressure CVS process. *Powder Technology,* Vol. 209,No. 1-3, pp.15–24, ISSN: 0032-5910.

Ruiz, A. M., Cornet, A., & Morante, J. R. (2004). Study of La and Cu influence on the growth inhibition and phase transformation of nano-TiO_2 used for gas sensors. *Sensors and Actuators B,* Vol. 100, No.1-2, pp.256–260, ISSN 0925-4005.

Ryua, Y. B., Jung, W. Y., Leea, M. S., Jeongb, E. D., Kimb, H. G., Park, S.-S., et al. (2008). Effect of synthesis conditions on the preparation of titanium dioxides from peroxotitanate solution and their photocatalytic activity. *React. Kinet. Catal. Lett.,* Vol. 93, No. 2, pp. 333–341, ISSN 1588-2837.

Santacesaria, E., Tonello, M., Storti, G., Pace, R. C., & Carra, S. (1985). Kinetics of Titanium Dioxide Precipitation by Thermal Hydrolysis. *Journal of Colloid and Interface Scienc,* Vol. 111. No. 1, pp.44-53, ISSN 0021-9797.

Sugimoto, T., (2001), *Monodispersed particles.* Elsevier Science B.V., ISBN 0444895698, Amsterdam.

Sugimoto, T. (2007). Underlying mechanisms in size control of uniform nanoparticles. *Journal of Colloid and Interface Science,* Vol. 309, No. 1, pp. 106–118, ISSN 0021-9797.

Sugimoto, T., & Shiba, F. (2000a). Spontaneous nucleation of monodisperse silver halide particles from homogeneous gelatin solution II: silver bromide. *Colloids and Surfaces A: Physicochemical and Engineering Aspects,* Vol. 164, No. 2-3, pp. 205–215, ISSN 0927-7757.

Sugimoto, T., Zhou, X. (2002). Synthesis of Uniform Anatase TiO_2 Nanoparticles by the Gel–Sol Method 2. Adsorption of OH− Ions to $Ti(OH)_4$ Gel and TiO_2 Particles. *Journal of Colloid and Interface Science,* Vo. 252, No. 2, pp. 347–353, ISSN 0021-9797.

Sugimoto, T., Shiba, F., Sekiguchi, T., & Itoh, H. (2000b). Spontaneous nucleation of monodisperse silver halide particles from homogeneous gelatin solution I: silver chloride. *Colloids and Surfaces A: Physicochemical and Engineering Aspects, Vol. 164,* No. 2-3, pp. 183–203, ISSN 0927-7757.

Sugimoto, T., Zhou, X., & Muramatsu, A. (2002). Synthesis of Uniform Anatase TiO_2 Nanoparticles by Gel-Sol Method 1. Solution Chemistry of Ti(OH)(4−n)+ n Complexes. *Journal of Colloid and Interface Science,* Vol. 252, No. 2, pp 339–346, ISSN 0021-9797.

Sugimoto, T., Zhou, X., & Muramatsu, A. (2003b). Synthesis of uniform anatase TiO_2 nanoparticles by gel–sol method 3. Formation process and size control. *Journal of Colloid and Interface Science,* Vol. 259, No. 1, pp. 43–52, ISSN 0021-9797.

Sugimoto, T., Zhou, X., & Muramatsu, A. (2003a). Synthesis of uniform anatase TiO_2 nanoparticles by gel–sol method 4. Shape control. *Journal of Colloid and Interface Science,* Vol. 259, No. 1, pp. 53–61, ISSN 0021-9797.

Suzuki, T., Iguchia, E., Teshima, K., & Oishia, S. (2007). Wulff's relationship of ruby single crystals grown from molybdenum trioxide flux studied by contact angles of liquid droplets. Chemical Physics Letters , Vol. 438, No. 1-3, pp. 127-131, ISSN 0009-2614.

Talapin, D. V., Rogach, A. L., Haase, M., & Weller, H. (2001). Evolution of an Ensemble of Nanoparticles in a Colloidal Solution: Theoretical Study. *Journal of Physical Chemistry B,* Vol. 105 No. 49, pp. 12278-12285, ISSN 1932-7447.

Talapin, D. V., Rogach, A. L., Shevchenko, E. V., Kornowski, A., Haase, M., & Weller, H. (2002). Dynamic Distribution of Growth Rates within the Ensembles of Colloidal II-VI and III-V Semiconductor Nanocrystals as a Factor Governing Their Photoluminescence Efficiency. *Journal of the American Chemical Society,* Vol. 124, No. 20, pp. 5782-5790, ISSN 0002-7863.

Tieng, S., Azouani, R., Chhor, K., & Kanaev, A. (2011). Nucleation−Growth of TiO2 Nanoparticles Doped with Iron Acetylacetonate. *Journal of Physical Chemistry C,* Vol. 115, No. 13, pp. 5244–5250, ISSN 1932-7447.

Toralff, R., & Marro, J. (1987). Cluster kinetics in the lattice gas model: the Becker-Doring type of equations. *Journal of Physics C: Solid State Physics,* Vol. 20, No. 17, pp. 2491-2500, ISSN: 0022-3719.

Tovstun, S. A., & Razumov, V. F. (2010). Theoretical analysis of methods for the colloidal synthesis of monodisperse nanoparticles. *High Energy Chemistry,* Vol. 44, No. 3, pp. 196-203, ISSN 1608-3148.

Wang, X. M., & Xiao, P. (2006). Morphology tuning in nontemplated solvothermal synthesis of titania nanoparticles. *Journal of Materials Research , Vol. 21,* No. 5, pp.1190-1203.

West, R. H., Celnik, M. S., Inderwild, O. R., Kraft, M., Beran, G. J., & Green, W. H. (2007). Toward a Comprehensive Model of the Synthesis of TiO_2 Particles from $TiCl_4$. *Industrial & Engineering Chemistry Research,* Vol. 46, No. 19, pp.6147-6156, ISSN: 1022-6680.

Zhang, G., Roy, B. K., Allard, L. F., & Cho, J. (2008). Titanium Oxide Nanoparticles Precipitated from Low-Temperature Aqueous Solutions: I. Nucleation, Growth, and Aggregation. *Journal of American Ceramic ,* Vol 91, No. 12, pp. 3875–3882, ISSN 0888-5885.

Zhang, J., Xu, Q., L, M., Feng, Z., & Li, C. (2009). UV Raman Spectroscopic Study on TiO_2. II. Effect of Nanoparticle Size on the Outer/Inner Phase Transformations. *The Journal of Physical Chemistry C*, Vol. 113, No. 5, pp 1698–1704, ISSN 1932-7447.

Zhanga, H., & Banfield, J. F. (2000). Phase transformation of nanocrystalline anatase-to-rutile via combined interface and surface nucleation. *Journal of Materials Research*,Vol. 15, pp. 437-448, ISSN 1022-6680.

Zhaoa, Y., & Zhang, W. (2011). Preparation Principle and Crystal Growth Kinetics of Loaded Nanocrystal TiO_2. *Advanced Materials Research*, Vol. 194-196, pp.585-588, ISSN 1022-6680.

Synthesis, Characterization and Biological Applications of Water-Soluble ZnO Quantum Dots

Raphaël Schneider[1], Lavinia Balan[2] and Fadi Aldeek[1]
[1]*Laboratoire Réactions et Génie des Procédés (LRGP), Nancy-University, CNRS, Nancy,*
[2]*Institut de Science des Matériaux de Mulhouse (IS2M), Mulhouse, France*

1. Introduction

Quantum dots (QDs) or semiconductor nanocrystals are of great interest to fundamental studies but have also potential applications as biological probes (Medintz et al., 2005), fluorescent biosensor (Costa-Fernandez et al., 2006), light-emitting diodes (LEDs) (Lim et al. 2007), and solar cells (Robel et al., 2006). Owing to the effect of quantum confinement, QDs show exceptional physical and chemical properties such as sharp and symmetrical emission spectra, high quantum yield (QY), good photo- and chemical stability, and size-dependent emission-wavelength tunability (Bruchez et al., 1998; Chan et al., 1998). For biological labelling, the most studied QDs are the nanocrystals of CdSe and CdTe (Aldeek et al., 2008) and the corresponding core/shell structured QDs (such as CdSe/ZnS, CdTe/ZnS or CdTe/ZnTe) that are more robust against chemical degradation or photooxidation than the parent cores (Law et al., 2009). Recent findings have highlighted the acute toxicity of II-VI semiconductor QDs without an external layer of a nontoxic material on biological systems (Schneider et al., 2009; Dumas et al. 2010). This toxicity results mainly from the decomposition and release of heavy metal ions and formation of reactive oxygen species. The toxicity of cadmium is a concern that will also limit the use of these visible or near IR emitting nanocrystals, especially for applications directly related to human health. Synthesis of low toxicity QDs and especially Cd-free QDs is the most challenging aspect of working with these materials in biological and medical fields. A promising member of the Cd-free QD is ZnO. However, ZnO nanoparticles are not stable in water. This instability is related to their surface luminescent mechanisms. Water molecules are able to attack the luminescent centers on the ZnO surface and destroy them rapidly. This chapter describes the strategies that have been developed over the last years to transfer ZnO QDs in water.

2. ZnO photoluminescence

For the past two decades, study of ZnO nanoparticles has become a leading edge in nanotechnology. With the reduction of size, novel electrical, chemical and optical properties are introduced, which are largely believed to be the result of surface and quantum confinement effects (Ahmad et al., 2011).

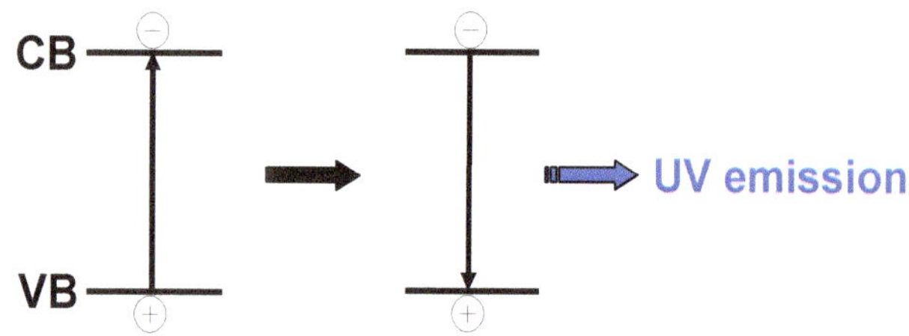

Fig. 1. Exciton emission of ZnO quantum dots.

At room temperature, wurtzite-ZnO QDs exhibit dual emission bands in the near UV and visible range (Djurisic et al., 2006; Zhang et al., 2010; Xiong, 2010). The UV emission at about 370 nm (band gap = 3.37 eV at room temperature) is associated with the radiative recombination of an electron from the conduction band with a hole from the valence band (band-edge emission) (Figure 1) or in a trap near the valence band (near-band-edge emission). Because the recombination of the photo-generated electron with a hole in the valence band is directly associated with the band gap, it is a size-dependent emission due to quantum confinement.

The origin of the visible luminescence (also called deep-level emission) is more complicated, its mechanism is controversial and not fully understood. Numerous hypotheses have been proposed to explain the visible luminescence such as oxygen vacancies (V_O), zinc vacancies (V_{Zn}), oxygen intersticials (O_i), zinc intersticials (Zn_i), and antisite oxygen (O_{Zn}) (Zhang et al., 2010). Among them, two mechanisms involving oxygen vacancies are the most cited. The first one is the recombination of a shallowly trapped electron with a hole in a deep trap (Figure 2a). The second one is the recombination of an electron in a singly occupied oxygen vacancy with a photo-generated hole in the valence band (Figure 2b) (Xiong, 2010). Additionnally, copper and alkaline metals impurities, surface-bonded hydroxides, and donor-acceptor complexes have also been proposed to be responsible of the visible photoluminescence.

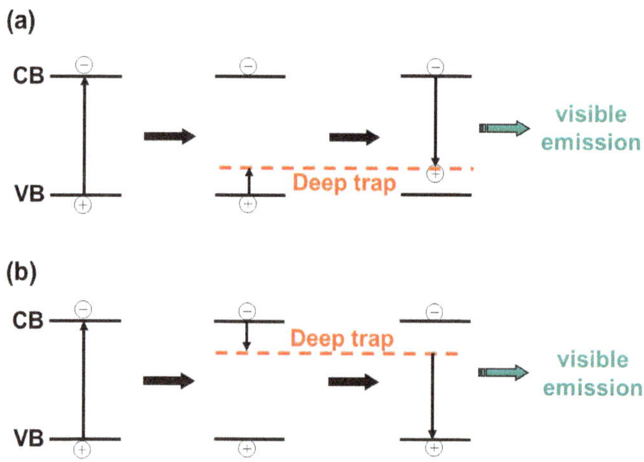

Fig. 2. Defect emission of ZnO quantum dots.

3. Synthesis and biological applications of ZnO quantum dots

Because the final materials are highly cristallised and of high purity, physical methods such as chemical vapour deposition (An et al., 2006), pulsed laser deposition (Choi et al., 2001), and chemical vapour transport and condensation processes (Ding et al., 2004) are the most popular methods to produce ZnO QDs with fine UV emission (Djurisic et al., 2006). However, these routes are not free of disadvantages and harsh conditions are often required preventing these methods from being used for large-scale production and their wide application.

Moreover, visible emission is the most-suited for biomedical and bio-imaging applications because most cells and tissues appear blue under UV light.

ZnO QDs with high defect concentrations are therefore the most appropriate for such applications. Because the sol-gel technique syntheses have many advantages such as process simplicity and high homogeneity, and that they offer the possibility of preparing large quantities of small-sized ZnO QDs at low cost and at low temperature, they constitute one of the best choice to obtain ZnO QDs with highly visible emission. Decomposition of organozinc derivatives and electrolysis constitute new and alternative methods for the production of green-yellow emitting ZnO QDs and will be described at the end of this paragraph.

3.1 The sol-gel synthesis of ZnO quantum dots

The sol-gel process can briefly be defined as the conversion of a precursor solution into an inorganic solid by chemical means. The sol-gel route for the production of ZnO QDs involves generally the basic hydrolysis of zinc acetate $Zn(OAc)_2$ mediated by an alkaline hydroxide or tetramethylammonium hydroxide (TMAH) in an alcohol (Figure 3) (Bahnemann et al., 1987; Spanhel et al., 1991).

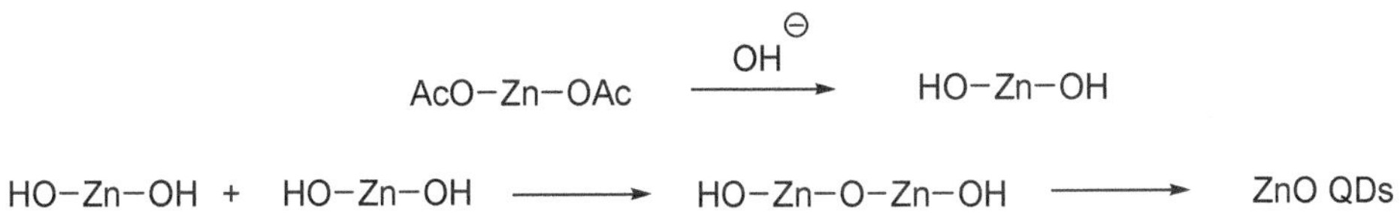

Fig. 3. The sol-gel synthesis of ZnO nanoparticles.

After the nucleation stage, the ZnO photoluminescence (PL) exhibits a continuous red-shift and its quantum yield (QY) decreases gradually (van Dijken et al., 2001). Protective organic ligands must therefore be added on the ZnO surface to inhibit further crystal growth and aggregation. Polymers, alkoxysilanes, oleic acid and triethanolamine are the most commonly stabilisers used to coordinate with Zn atoms on the surface of ZnO QDs to hinder the formation of bulk ZnO.

3.1.1 Polymer-capped ZnO QDs

The conventional sol-gel technique of ZnO synthesis was first modified in 2005 to prepare ZnO nanocrystals stabilized by poly(ethylene glycol)methyl ether (PEGME). Zinc salts $[CH_3(CH_2CH_2O)nCH_2COO]_2Zn$ were hydrolysed with LiOH in ethanol to get ZnO(PEGME) QDs with diameters ranging from 1 to 4 nm and PL tunable from blue to yellow depending on the [LiOH]/[Zn] ratio used for the synthesis (Xiong et al., 2005; Xiong et al., 2005).

Although stabilized by hydrophilic poly(ethyleneglycol) moieties, the water-dispersability of these ZnO QDs has not been studied.

Poly(ethyleneimine) (PEI) can also be used to disperse ZnO QDs in water with preservation of their optical properties (Joshi et al., 2009). ZnO QDs, produced by hydrolysis of $Zn(OAc)_2$ with LiOH, were first stabilized by trisodium citrate through hydrogen bonding and then treated by PEI to get stable dispersions of the dots in water (Figure 4).

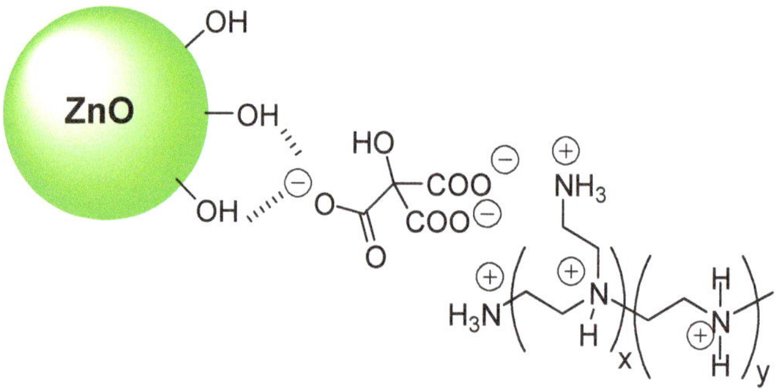

Fig. 4. Proposed mechanism for the stabilization of ZnO QDs using sodium citrate and PEI.

ZnO QDs produced by this method are ca. 5 nm in diameter, exhibit a strong green-yellow fluorescence at 555 nm upon excitation at 360 nm and were found to be stable in aqueous solution for at least 45 days.

Coating ZnO QDs with a polymer shell is an efficient method for stabilizing the photophysical properties of the cores. Xiong *et al.* were the first to report the preparation of core/shell ZnO/polymer QDs by initiating the polymerization with the inherent free radicals pre-existing on ZnO QDs surfaces (Xiong et al., 2006). For that purpose, zinc methacrylate was hydrolysed with LiOH to obtain blue-emitting methacrylate-coated ZnO QDs. Copolymerisation of these nanoparticles with methyl methacrylate for instance affords uniform and monodisperse ZnO core QDs with an inorganic core diameter of 2.1 nm. The thickness of the polymer shell is about 1 nm. Once dispersed in organic solvents, the PL QY of ZnO@polymer QDs reached 80%.

A similar strategy was employed by the authors to prepare water-stable ZnO QDs (Xiong et al., 2007). Poly(methacrylic acid)(PMMA)-capped ZnO QDs were co-polymerised with methacrylic acid to produce ZnO@PMMA microspheres of about 150 nm in diameter that exhibit blue fluorescence (PL QY = 22%) once dispersed in water. From the viewpoint of practical applications, such large nanoparticles are not suitable for biological and medical fields, especially for labelling functional subcellular or proteins, as larger particles with diameters above 25 nm often affect their biological function and are more likely to be recognized and cleared by the phagocyte (Medintz et al., 2005).

The first successful labelling of living cells with ZnO QDs was only described in 2008 (Xiong et al., 2008). Green- or yellow-emitting ZnO QDs were prepared by basic hydrolysis of zinc methacrylate using LiOH or NaOH, respectively, and concomitant polymerisation with poly(ethylene glycol)methyl ether methacrylate. Final ZnO QDs have a copolymer shell with the internal hydrophobic polymethacrylate layer protecting the nanocrystals and the external hydrophilic poly(ethylene glycol)methyl ether groups allowing their transfer in aqueous medium (Figure 5).

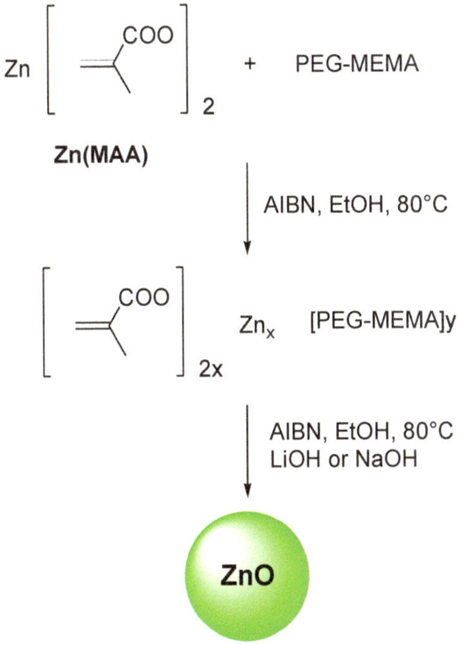

Fig. 5. Synthesis of ZnO@poly(MMA-co-PEG-MEMA) quantum dots.

The obtained nanocrystals have diameters of ca. 3-4 nm and exhibit a stable photoluminescence in water (up to 50% PL QY for QDs prepared with LiOH). Efficient labelling of the cytoplasm of human hepatoma cells was achieved using aqueous dispersions of these dots at a concentration of 0.1 mg/mL.

Small-sized polymer-coated ZnO QDs were also recently prepared by hydrolysis of zinc methacrylate by LiOH and concomitant AIBN-initiated polymerisation with 2-(dimethylaminoethyl) ethylmethacrylate (DMAEMA) (Figure 6) (Zhang et al., 2010).

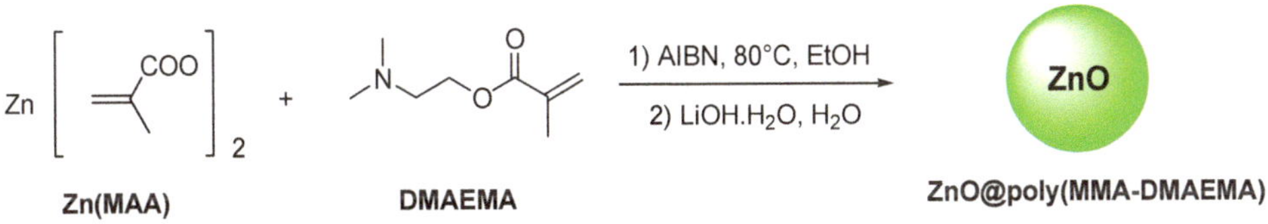

Fig. 6. Synthesis of ZnO@poly(MMA-DMAEMA) quantum dots.

The obtained QDs are ca. 4 nm in diameter, exhibit a strong yellow luminescence and have a PL QY of 21%. Positively-charged ZnO@poly(MMA-DMAEMA) QDs were found to associate with negatively charged plasmid DNA (p-DNA) to form spherical particles around 200 nm in diameter. Using the Luciferase reporter gene, it was found that QD/p-DNA complexes could efficiently transfect COS-7 cells and allowed real-time imaging of gene transfection in live cells (Figure 7).

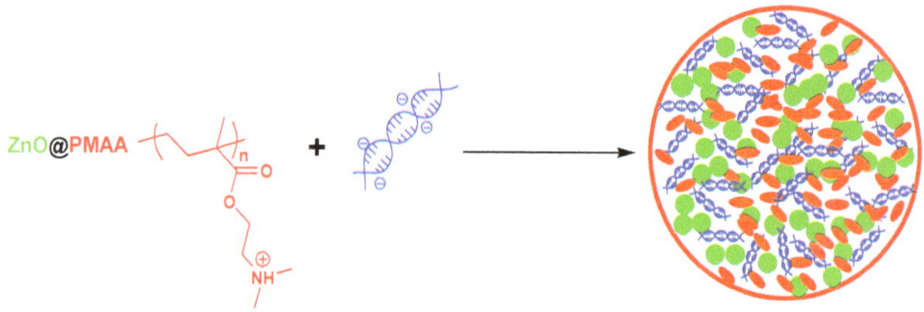

Fig. 7. Schematic of QD/pDNA complex formation.

ZnO QDs have also successfully been integrated into polymer (chitosan - $M_w \approx 200$ kDa - and acrylic acid) nanospheres (Yan et al., 2011). In that one-pot process, chitosan, $Zn(NO_3)_2$ and acrylic acid were first heated at 80°C and ZnO QDs formed by hydroxide ions generation from chitosan. Polymerisation was initiated by addition of potassium persulfate $K_2S_2O_8$ and chitosan finally cross-linked using glutaraldehyde. The obtained chitosan-ZnO QDs nanospheres have an average diameter of ca. 150 nm and ZnO QDs were found to be homogeneously distributed in the shell of the nanospheres. A strong blue-emission centered at 440 nm could only be observed for these nanospheres using high amounts of $Zn(NO_3)_2$ relative to chitosan.

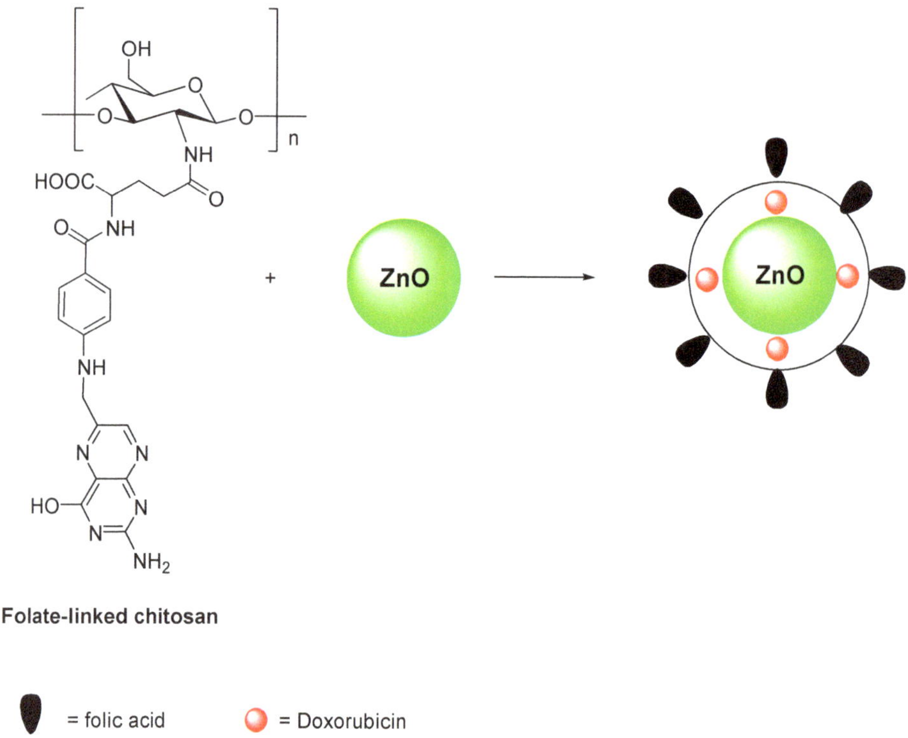

Fig. 8. Preparation of ZnO quantum dots encapsulated by folate-conjugated chitosan.

The natural and biodegradable polymer chitosan was also used to encapsulate and stabilize ZnO QDs through electrostatic interactions. Chitosan ($M_w \approx 50$ kDa) was first conjugated to folic acid and then associated to ZnO QDs produced by hydrolysis of $Zn(OAc)_2$ with an aqueous solution of diethanolamine in the presence of oleic acid used as stabilizer (Yuan et al., 2010). The obtained ZnO QDs have an average diameter of 3 nm, exhibit strong blue fluorescence at 440 nm, and were found to be stable in water for at least one year. The "ZnO-chitosan-folate" platform can associate Doxorubicin, an antineoplastic agent used in tumor treatments, through physical and chemical interactions and thus be used as nanocarrier (Figure 8). The drug-release responses at pH = 5.3 or 7.4 were characterized by an initial rapid drug release (ca. 85% within 36 h) followed by a controlled release.

3.1.2 Silanization of ZnO QDs

An alternative and interesting methodology for surface-functionalization and stabilization of ZnO nanocrystals is based on organosilanes chemistry to inhibit decomposition in aqueous media. Since the first report of ZnO nanoparticles surface modification with silane, this strategy has successfully been used for the stabilisation of ZnO QDs in water. Trialkoxysilanes have a strong affinity with hydroxylated surfaces, form covalent bonds with them, thereby creating a shielding barrier of cross-linked polysiloxanes that protects the nanocrystal at the core (Bruce et al., 2005; Soares et al., 2008).

ZnO nanoparticles with diameters of ca. 20 nm (hydrodynamic diameter of 60 nm) can simply be prepared by heating of $Zn(OAc)_2.2H_2O$ in methanol at 68°C for 7 h. These QDs were immediately capped by 3-aminopropyltrimethoxysilane (APTES) to inhibit the particle growth at the end of the synthesis. After excitation at 350 nm, ZnO@APTES nanocrystals show two emission bands at ca. 380 and 560 nm. A second capping (SiO_2 or TiO_2) was introduced through the sol-gel process at the periphery of ZnO@APTES nanocrystals to improve the PL intensity and to red-shift the fluorescence emission peaks. Optimal SiO_2 or TiO_2 shell thickness in both cases was found to be 0.5 nm (Wu et al., 2007). Once a SiO_2 shell introduced at the periphery of ZnO@APTES QDs, the visible green-yellow emission was found to be markedly increased. When ZnO@APTES QDs were capped with TiO_2, the PL emission was slightly red-shifted to 520-590 nm. A high PL QY was also obtained for core/shell ZnO/TiO_2 nanocrystals (79% in MeOH). When mung beens seeds were germinated in the presence of these ZnO colloidal dispersions, a strong fluorescence in the cell walls of the vascular cylinder was observed, indicating a good uptake of the nanocrystals and their good potential in bio-imaging studies.

Another report of silazination to get small-sized, water-dispersible and stable ZnO QDs was made in 2007 (Jana et al., 2007). Oleate-stabilized ZnO QDs produced by hydrolysis of $Zn(OAc)_2$ with tetramethylammonium hydroxide (TMAH) were treated by *N*-(2-aminoethyl)aminopropyltrimethoxysilane (AEAPS) and TMAH in a two-stage silanization process to prepare dots with an average diameter of ca. 5 nm (PL QY = 16-20% in water) (Figure 9). This two-step silanization was developed to minimize polysiloxane by-products formation during the surface-capping of ZnO QDs.

QDs with a mixed shell containing amine and phosphonate groups at their periphery were also prepared by this method to vary the surface charge and thus the isoelectric points of the nanocrystals. The primary amine groups of the AEAPS ligand were also found to contribute to the photostability of ZnO QDs, probably through interaction of the nitrogen atoms with surface zinc atoms. This property was used for the detection of various aldehydes, including the biologically important glucose and acetaldehyde, by photoluminescence. Imine functions formed upon reaction of peripheral NH_2 functions with aldehydes induced a marked photostability decrease (up to 71% with *o*-phthaldehyde).

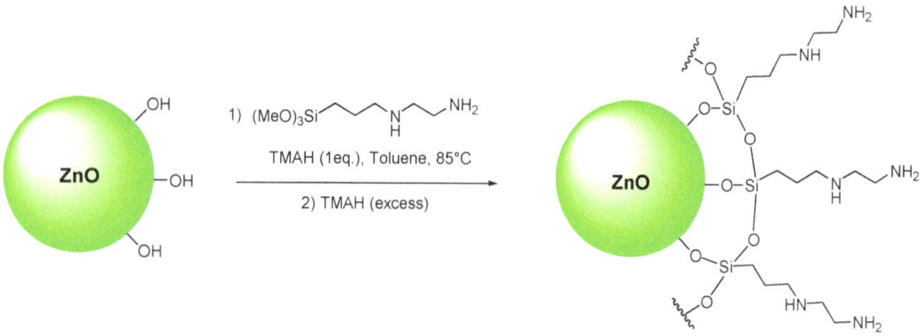

Fig. 9. Synthesis of ZnO@AEAPS quantum dots.

Our group used as similar strategy to anchor 2-[ethoxy(polyethyleneglycol)propyl] triethoxysilane at the surface of hydrophobic oleate-capped ZnO QDs (Figure 10) (Moussodia et al., 2008).

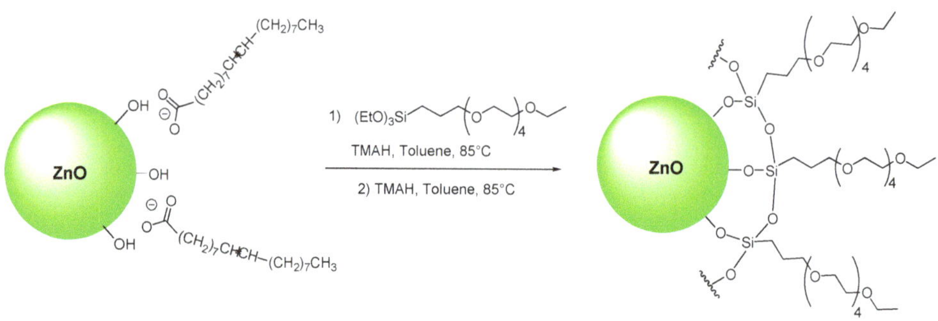

Fig. 10. Synthesis of PEG-siloxane capped ZnO quantum dots.

Well-dispersed green-emitting ZnO QDs of ca. 6.0 nm in diameter were obtained after surface functionalization. Their PL QY was found to be 8.0% in water. ZnO nanocrystals with mixed functional groups (PEG and amine) could also be prepared using a mixture of the PEG-silane and 3-aminopropyltriethoxysilane in a 4:1 molar ratio during the silanisation step. The amine groups at the periphery of the dots can be used for the covalent attachment of biomolecules allowing these ZnO QDs to be used as fluorescent labels for ultrasensitive detection and imaging.

To increase the ZnO materials' flexibility in bio-related applications, we also developed the synthesis of poly(amidoamine) (PAMAM) dendrons bearing a reactive siloxane group at the focal point (Moussodia et al., 2010). These new dendrons were prepared by divergent synthesis through repetitive Michael addition using methyl acrylate and amidation with ethylene diamine starting from 3-aminopropyltrimethoxysilane. Covalent anchorage of the silane function at the surface of ca. 5 nm diameter ZnO QDs was achieved by treatment of first and second generation dendrons (respectively noted G1 and G2) by an excess of TMAH

(Figure 11). Due to the presence of polar amine and amide groups and of the high affinity of PAMAM for aqueous media, ZnO@G1 and ZnO@G2 QDs could easily be dispersed in water or phosphate-buffered saline (PBS). ZnO@G1 and ZnO@G2 QDs typically exhibit the green-yellow fluorescence centered at ca. 550 nm (their PL QYs measured immediately after redispersion in water were found to be 19 and 17%, respectively). Interestingly, their PL intensities were found to increase upon storage of the dots in water. The PL QY of ZnO@G1 reached 59% after 20 days, probably due to a surface-ordening of the siloxane-capping, which results in ZnO QDs with more efficiently protected luminescent centers from water attack.

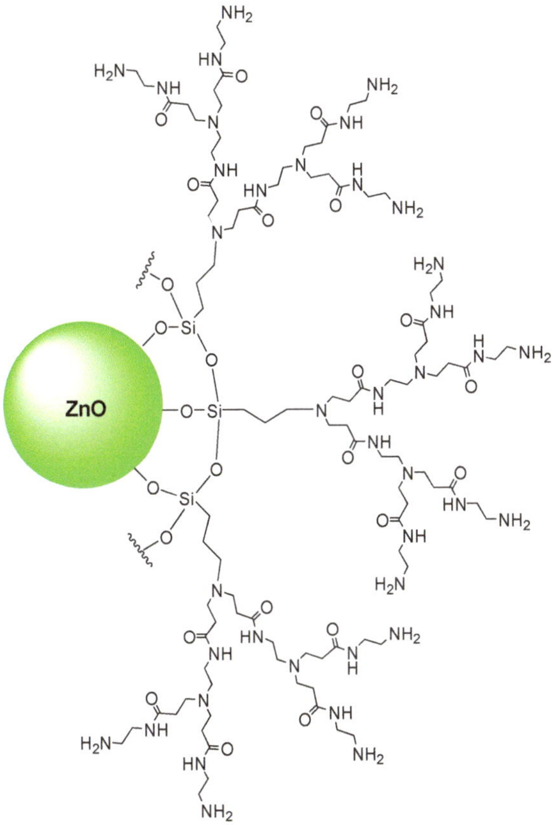

Fig. 11. Structure of ZnO@G2 quantum dots.

By varying the concentration of the LiOH base and thus the pH value of the solution during the sol-gel process, Tang *et al.* have recently prepared color tunable emission ZnO QDs. Smaller particles (average diameter of 3.3 nm) were formed at pH = 12 than at lower pH (4.5 nm at pH = 10, 5.5 nm at pH = 8, and 6.5 nm at pH = 6) because ZnO nuclei were formed faster at higher pH resulting in competition growth. Blue (λ_{em} = 455 nm), green (λ_{em} = 494 nm), yellow (λ_{em} = 570 nm), and orange (λ_{em} = 590 nm)-emitting ZnO QDs were formed when precipitation values were 12, 10, 8, and 6, respectively (Tang et al., 2010). All these nanocrystals exhibit good quantum yields in ethanol (> 20%) and were stabilized in water by encapsulation into silica nanospheres of ca. 50 nm in diameter (ca. 20 ZnO QDs per nanosphere). The surface of these nanospheres was further modified with *N*-(2-aminoethyl)aminopropyltrimethoxysilane to get positively-charged nanoparticles stable in water from pH = 4.5 to 9. These nanoparticles were successfully used to label NIH/3T3 cells and were found to be attached at the cell surface through electrostatic interactions.

The sol-gel method has also successfully been used for the synthesis of Gd^{3+}-doped ZnO QDs, dual nanoprobes integrating fluorescence and magnetic resonance imaging functionalities into a single nanomaterial (Liu et al., 2011). $Gd(OAc)_3$ and $Zn(OAc)_2$, with Gd^{3+}/Zn^{2+} molar ratios varying between 0.02 and 0.3, were both hydrolysed with TMAH in ethanol in the presence of oleic acid to afford Gd-doped ZnO QDs with diameters ranging from 3.5 to 5.0 nm depending on the percentage of dopant used (high doping concentration affords the smallest nanocrystals). These Gd-doped ZnO QDs typically exhibit the green-yellow fluorescence related to oxygen vacancies and the highest PL intensity (PL QY = 34%) was attained for a Gd/Zn molar ratio of 0.08. These nanocrystals were successfully transferred in aqueous medium through surface functionalization with *N*-(2-aminoethyl)aminopropyltrimethoxysilane. MTT cell proliferation assay indicated that Gd-doped ZnO QDs were not toxic at concentrations up to 1 mM. Gd-doped ZnO QDs were successfully used for the fluorescence imaging of HeLa cells. In MRI studies, they exert strong positive contrast effect with a large longitudinal relaxivity (r_1) of water proton of 16 $mM^{-1}s^{-1}$.

Luminomagnetic Fe-doped ZnO nanoparticles with diameters ranging from 6 to 9 nm were also recently synthesized by treatment of $Zn(NO_3)_2$ and $Fe(NO_3)_3$ by potassium hydroxide. The green fluorescent ZnO nanocrystals doped with ca. 10% iron exhibit also a paramagnetic behaviour at room temperature (coercivity = 5.1 milliTesla and magnetization of remanence = $7.6.10^{-3}$ emu/g) (Dutta et al., 2010). ZnO:Fe QDs were further dispersed in water through surface functionalization with *N*-(2-aminoethyl)aminopropyltrimethoxysilane and next conjugated with folic acid using EDC and *N*-hydroxysuccinimide. Because folate receptors are over-expressed on numerous human cells (breast, ovaries, lungs, kidneys, ...), the folic acid-conjugated ZnO:Fe luminomagnets have great potential for various biomedical applications including bio-labelling, bio-separation, magnetic resonance imaging and targeted drug delivery.

ZnO QDs with an average diameter of 5 nm produced by the sol-gel method were also recently used as sensors for the carbohydrate antigen 19-9 (CA 19-9), the preferred label for pancreatic cancer (Gu et al., 2011). A functionalized silicon substrate was first prepared by reaction with 3-aminopropyltrimethoxysilane followed by treatment with glutaraldehyde (Figure 12). An antibody (Ab) monolayer was next assembled on the aldehyde-activated surface and CA 19-9 antigens linked with their antibodies through immunoreaction (samples with various CA 19-9 concentrations from 0.1 to 180 U/mL were prepared). In a parallel step, CA 19-9 Abs were immobilized on ZnO QDs through electrostatic interactions in benefit of the high isoelectric point (IEP = 9.5) of ZnO. The sandwich-immunosensor was finally obtained by the ZnO-Ab association with the substrate. Detection of CA 19-9 can be performed by fluorescence through measurement of the band-edge emission of ZnO QDs at 372 nm (linear increase of PL intensity with increase of CA 19-9 concentration), but also through electrochemical measurements, the latter method being more sensitive than photoluminescence.

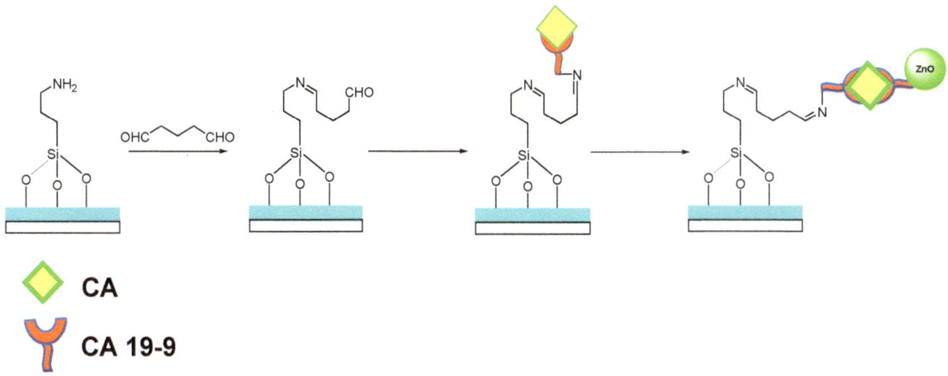

Fig. 12. Schematic protocol of sandwich-immunosensor.

3.1.3 Cyclodextrin-capped ZnO QDs.

Another recent example of using ZnO QDs as sensors was reported by Rakshit et al. Carboxymethyl β-cyclodextrin (CMCD) was used to disperse yellow-emitting core-shell ZnO/MgO QDs in aqueous solution (carboxymethyl groups coordinate to the ZnO surface as carboxylate groups of fatty acids). The surface-anchored cyclodextrins were found to retain their host capabilities for inclusion of small hydrophobic molecules, like the organic dye Nile Red. ZnO/MgO QDs were shown to be excellent donors in fluorescence energy transfer (FRET) to Nile Red. The 640 nm-fluorescence of Nile Red was observed upon excitation of ZnO/MgO QDs in the UV (Figure 13). The Nile Red emission following resonance energy transfer exhibits a pronounced thermochromic shift (linear blue shift with increasing temperature) and thus suggest the possible use of the CMCD-capped ZnO/MgO QDs-Nile Red assemblies as thermometers in aqueous solution (Rakshit et al., 2008).

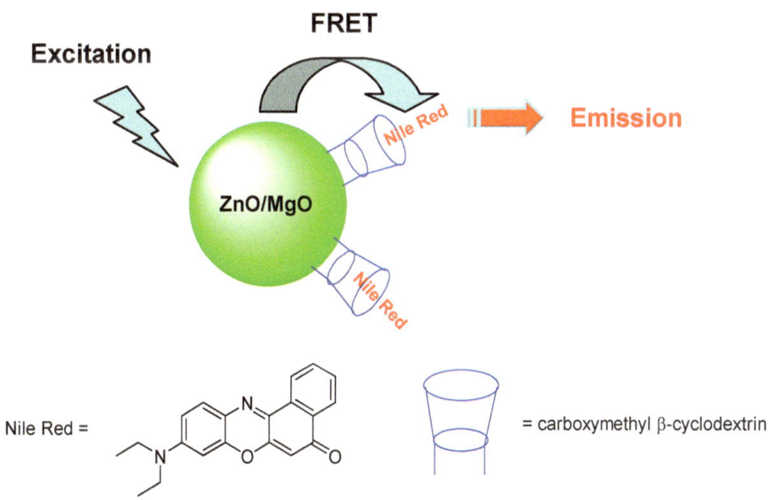

Fig. 13. FRET between the yellow-emitting ZnO/MgO QDs and Nile Red included in the cavities of carboxymethyl β-cyclodextrins.

3.1.4 Encapsulation of ZnO QDs into phospholipid micelles.

ZnO nanocrystals with a noncentrosymmetric structure produced via non-hydrolytic sol-gel route (Joo et al., 2005) based on the ester-elimination between zinc acetate and 1,12-dodecanediol can be used for high contrast nonresonant nonlinear bio-imaging applications (Kachynski et al., 2008). These nanoparticles were dispersed in water using folic acid-conjugated phospholipid micelles as stabilizer. Sum frequency generation imaging of human KB cells over-expressing the folate receptor showed a strong internalisation of these ZnO nanocrystals which have great potential not only as non linear optical probes for diagnostic, but also for the generation of photochemical reactions inside the cells.

3.2 Decomposition of dialkylzinc for the synthesis of ZnO QDs

Decomposition/oxidation of organozinc precursors like Et_2Zn or Cy_2Zn in the air in the presence of alkylamines or amphiphilic hyperbranched polymers like stearate-modified polyether polyols allows the preparation of ZnO QDs with control over nanoparticles diameters and photophysical properties (Glaria et al., 2008; Richter et al., 2009). The hydrophobic nanomaterials prepared can however only be dispersed in organic solvents (THF, toluene) and are therefore not suitable for biological applications.

Oxidation of Cy_2Zn in the presence of hyperbranched polymers (HYPAM) can however give access to water-dispersible ZnO QDs (Saliba et al., 2010). HYPAM were prepared by reaction of a hexaester with a tris-amine and surface functionalized with amphiphilic building blocks formed by alkyldiacids (up to C_{18}) connected to monomethyl poly(ethyleneglycol) ($mPEG_{750}$) (Figure 14).

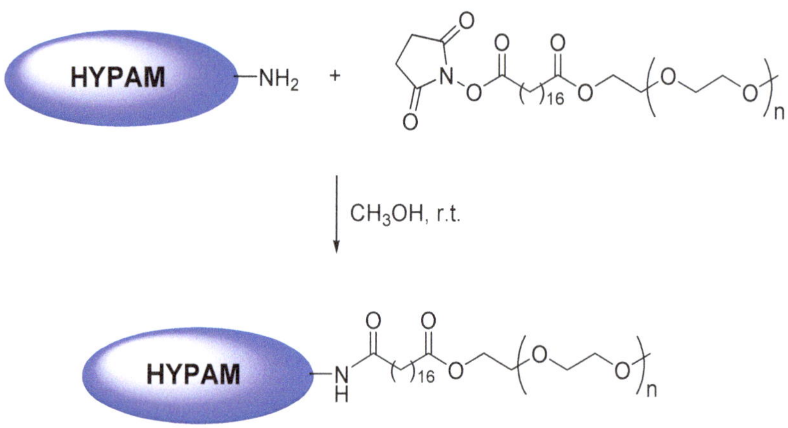

Fig. 14. Preparation of HYPAM polymer with a double-shell structure.

The HYPAM4-C_{18}-PEG_{750} polymer has hydrodynamic diameters of ca. 9 and 13 nm in dilute aqueous solution and in THF solution, respectively. The multiple amine groups of the HYPAM core favour interactions with Cy_2Zn, the inner hydrophobic shell acts as a protective layer for the ZnO QDs produced by hydrolysis, while the outer PEG shell assures their dispersibility in aqueous solution. Depending on the polymer/ Cy_2Zn weight ratio used for the synthesis, ZnO QDs with diameters ranging from 3.7 to 4.5 nm can be produced (the hydrodynamic diameter is ca. 30 nm). These dots exhibit a broad fluorescence emission centered at 575 nm but their PL QY is modest (5%).

3.3 Electrolysis for the production of ZnO QDs

ZnO nanoparticles of ca. 5 nm diameter can be prepared by electrochemical deposition under oxidizing conditions using zinc metal as anode. These QDs were first surface-functionalized with 3-mercaptopropyltrimethoxysilane (MPTS) and then treated with iodoacetyl-PEG2-biotin. A similar strategy was used for the functionalization of ZnO films (Selegard et al., 2010). Due to the high affinity of the biotin ligand with avidin, these materials have great potential for specific targeting in bio-imaging or for recognition studies in biosensing applications (Figure 15).

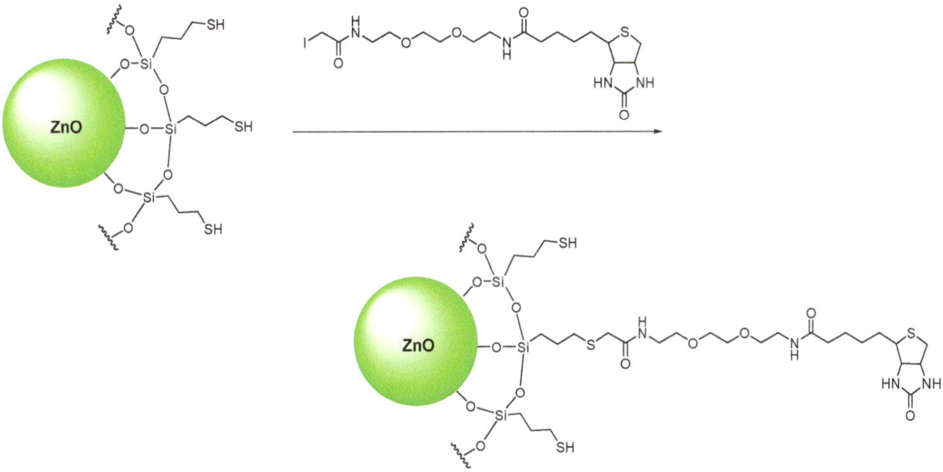

Fig. 15. Surface-functionalization of ZnO@MPTS quantum dots with iodoacetyl-PEG2-biotin.

4. Conclusions

In recent years, the luminescent properties of ZnO nanoparticles have attracted considerable attention for numerous applications like ultraviolet light emitting devices, flat panel displays as low voltage phosphor, photovoltaic solar cells, and biosensing devices.

With the developments in fabrication techniques of nanomater-sized, water-dispersible and stable nanocrystals, many efforts have been made to stabilize ZnO QDs in aqueous medium for biological applications, including employing organic ligands, coating nanoparticles with inorganic shells, and capping nanoparticles surfaces with polymers. In addition, the low toxicity of ZnO and its high natural abundance make it a good alternative to cadmium-based II-VI semiconductors, which cause toxicity via photoinduced reactive oxygen species (ROS) generation.

This review highlights recent advances in ZnO QDs synthesis and surface capping for bio-applications and shows that simple and low cost methods have been developed for the production of water stable ZnO QDs with tunable photoluminescence and high quantum yields. These nanocrystals have successfully been used as fluorescent probes for numerous bio-imaging applications, gene delivery, sensing, and as carrier for targeted drug delivery. All these studies prove that ZnO QDs, as a type of safe and cheap luminescent labels, have an inspiring prospect in biological and medical fields.

5. References

Ahmad, M.; Zhu, J. (2011). ZnO based advanced functional nanostructures: synthesis, properties and applications. *J. Mater. Chem.*, 21, 599-614.

Aldeek, F.; Balan, L.; Lambert, J.; Schneider, R. (2008). The influence of capping thioalkyl acid on the growth and photoluminescence efficiency of CdTe and CdSe quantum dots. *Nanotechnology*, 19, 475401.

An, S.J.; Park, W.I.; Yi, G.-C.; Kim, Y.-J.; Kang, H.-B.; Kim, M. (2004). Heteroepitaxal fabrication and structural characterizations of ultrafine GaN/ZnO coaxial nanorod heterostructures. *Appl. Phys. Lett.*, 84, 3612.

Bahnemann, D.W.; Kromann, C.; Hoffmann, M.R. (1987). Preparation and characterization of quantum size zinc oxide: a detailed spectroscopic study. *J. Phys. Chem.*, 91, 3789-3798.

Bruce, I.J.; Sen, T. (2005). Surface modification of magnetic nanoparticles with alkoxysilanes and their application in magnetic bioseparations. *Langmuir*, 21, 7029-7035.

Bruchez, Jr. M.; Moronne, M.; Gin, P.; Weiss, S.; Alivisatos, A.P. (1998). Semiconductor nanocrystals as fluorescent biological labels. *Science*, 281, 2013-2016.

Chan, W.C.W.; Nie, S. (1998). Quantum dot bioconjugates for ultrasensitive nonisotopic detection. *Science*, 281, 2016-2018.

Choi, J.H.; Tabata, H.; Kawai, T. (2001). Initial preferred growth in zinc oxide thin films on Si and amorphous substrates by a pulsed laser deposition. *J. Cryst. Growth*, 226, 493-500.

Costa-Fernandez, J.M.; Pereiro, R.; Sanz-Medel, A. (2006). The use of luminescent quantum dots for optical sensing. *Trends Anal. Chem.*, 25, 207-218.

Ding, Y.; Gao, P.X.; Wang, Z.L. (2004). Catalyst−nanostructure interfacial lattice mismatch in determining the shape of VLS grown nanowires and nanobelts: a case of Sn/ZnO. *J. Am. Chem. Soc.*, 126, 2066-2072.

Djurisic, A.B.; Leung, Y.H. (2006). Optical properties of ZnO nanostructures. *Small*, 2, 944-961.

Dumas, E.; Gao, C.; Suffern, D.; Bradforth, S.E.; Dimitrijevic, N.M.; Nadeau, J.L. (2010). Interfacial charge transfer between CdTe quantum dots and Gram negative vs Gram positive bacteria. *Environ. Sci. Technol.*, 44, 1464-1470.

Dutta, R.K.; Sharma, P.K.; Pandey, A.C. (2010). Design and surface modification of potential luminomagnetic nanocarriers for biomedical applications. *J. Nanopart. Res.*, 12, 1211-1219.

Glaria, A.; Kahn, M.L.; Cardinal, T.; Senocq, F.; Jubera, V.; Chaudret, B. (2008). Lithium ion as growth-controlling agent of ZnO nanoparticles prepared by organometallic synthesis. *New J. Chem.*, 32, 662-669.

Gu, B.; Xu, C.; Yang, C.; Liu, S.; Wang, M. (2011). ZnO quantum dot labeled immunosensor for carbohydrate antigen 19-9. *Biosens. Bioelectron.*, 26, 2720-2723.

Jana, N.R.; Yu, H.-h.; Ali, E.M.; Zheng, Y.; Ying, J.Y. (2007). Controlled photostability of luminescent nanocrystalline ZnO solution for selective detection of aldehydes. *Chem. Commun.*, 1406-1408.

Joo, J.; Kwon, S.G.; Yu, J.H.; Hyeon, T. (2005). Synthesis of ZnO nanocrystals with cone, hexagonal cone, and rod shapes via non-hydrolytic ester elimination sol-gel reactions. *Adv. Mater.*, 17, 1873-1877.

Joshi, P.; Ansari, Z.A.; Singh, S.P.; Shankar, V. (2009). Synthesis and characterization of highly fluorescent water dispersible ZnO quantum dots. *Adv. Sci. Lett.*, 2, 360-363.

Kachynski, A.V.; Kuzmin, A.N.; Nyk, M.; Roy, I.; Prasad, P.N. (2008). Zinc oxide nanocrystals for nonresonant nonlinear optical microscopy in biology and medicine. *J. Phys. Chem. C*, 112, 10721-10724.

Law, W.-C.; Yong, K.-T.; Roy, I.; Ding, H.; Hu, R.; Zhao, W.; Prasad, P.N. (2009). Aqueous-phase synthesis of highly luminescent CdTe/ZnTe core/shell quantum dots optimized for targeted bioimaging. *Small*, *5*, 1302-1310.

Lim, J.; Jun, S.; Jane, E.; Baik, H.; Kim, H.; Cho, J. (2007). Preparation of highly luminescent nanocrystals and their application to light-emitting diodes. *Adv. Mater.*, *19*, 1927-1932.

Liu, Y.; Ai, K.; Yuan, Q.; Lu, L. (2011). Fluorescence-enhanced gadolinium-doped zinc oxide quantum dots for magnetic resonance and fluorescence imaging. *Biomaterials*, 32, 1185-1192.

Moussodia, R.-O.; Balan, L.; Schneider, R. (2008). Synthesis and characterization of water-soluble ZnO quantum dots prepared through PEG-siloxane coating. *New J. Chem.*, 32, 1388-1393.

Moussodia, R.-O.; Balan, L.; Merlin, C.; Mustin, C.; Schneider, R. (2010). Biocompatible and stable ZnO quantum dots generated by functionalization with siloxane-core PAMAM dendrons. *J. Mater. Chem.*, 20, 1147-1155.

Medintz, I.L.; Uyeda H.T.; Goldman E.R.; Mattoussi, H. (2005). Quantum dot bioconjugates for imaging, labelling and sensing. *Nat. Mater.*, 4, 435-446.

Rakshit, S.; Vasudevan, S. (2008). Resonance energy transfer from β-cyclodextrin-capped ZnO:MgO nanocrystals to included Nile Red guest molecules in aqueous media. *ASC Nano*, 2, 1473-1479.

Richter, T.V.; Schüler, F.; Thomann, R.; Mülhaupt, R.; Ludwigs, S. (2009). Nanocomposites of size-tunable ZnO-nanoparticles and amphiphilic hyperbranched polymers. *Macromol. Rapid Commun.*, 30, 579-583.

Robel, I.; Subtamanian, V.; Kuno, M.; Kamat, P.V. Quantum dot solar cells. Harvesting light energy with CdSe nanocrystals molecularly linked to mesoscopic TiO_2 films. *J. Am. Chem. Soc.* 2006, *128*, 2385-2393.

Saliba, S.; Serrano, C.V.; Keilitz, J.; Kahn, M.L.; Mingotaud, C.; Haag, R.; Marty, J.-D. (2010). Hyperbranched polymers for the formation and stabilization of ZnO nanoparticles. *Chem. Mater.*, 22, 6301-6309.

Schneider, R.; Wolpert, C.; Guilloteau, H.; Balan, L.; Lambert, J.; Merlin, C. (2009). The exposure of bacteria to CdTe-core quantum dots: The importance of surface chemistry on cytotoxicity. *Nanotechnology*, 20, 225101.

Selagard, L.; Khranovskyy, V.; Söderling, F.; Vahlberg, C.; Ahren, M.; Käll, P.-O.; Yakimova, R.; Uvdal, K. (2010). Biotinylation of ZnO nanoparticles and thin films: a two-step surface functionalization study. *ACS Appl. Mater. Interfaces*, 2, 2128-2135.

Spanhel, L.; Anderson, M.A. (1991). Semiconductor clusters in the sol-gel process: quantized aggregation, gelation, and crystal growth in concentrated zinc oxide colloids. *J. Am. Chem. Soc.*, 113, 2826-2833.

van Dijken, A.; Makkinje, J.; Meijerink, A. (2001). The influence of particle size on the luminescence quantum efficiency of nanocrystalline ZnO particles. *J. Lumin.* 92, 323-328.

Soares, J.W.; Whitten, J.E.; Oblas, D.W.; Steeves, D.M. (2008). Novel photoluminescence properties of surface-modified nanocrystalline zinc oxide: toward a reactive scaffold. *Langmuir*, 24, 371-374.

Tang, X.; Guang Choo, E.S.; Li, L.; Ding, J.; Xue, J. (2010). Synthesis of ZnO nanoparticles with tunable emission colors and their cell labeling applications. *Chem. Mater.*, 22, 3383-3388.

Wu, Y.L.; Lim, C.S.; Fu, S.; Tok, A.I.K. (2007). Surface modifications of ZnO quantum dots for bio-imaging. *Nanotechnology*, 18, 215604.

Xiong, H.-M.; Wang, Z.-D.; Liu, D.-P.; Chen, J.-S.; Wang, Y.-G.; Xia, Y.-Y. (2005). Bonding polyether onto ZnO nanoparticles: an effective method for preparing polymer nanocomposites with tunable luminescence and stable conductivity. *Adv. Funct. Mater.*, 15, 1751-1756.

Xiong, H.-M.; Liu, D.-P.; Xia, Y.-Y.; Chen, J.-S. (2005). Polyether-grafted ZnO nanoparticles with tunable and stable photoluminescence at room temperature. *Chem. Mater.*, 17, 3062-3064.

Xiong, H.-M.; Wang, Z.-D.; Xia, Y.Y. (2006). Polymerisation initiated by inherent free radicals on nanoparticle surfaces: a simple method of obtaining ultrastable (ZnO)polymer core-shell nanoparticles with strong blue fluorescence. *Adv. Mater.*, 18, 748-751.

Xiong, H.-M.; Xie, D.-P.; Guan, X.-Y. ; Tan, Y.-J.; Xia, Y.-Y. (2007). Water-stable blue-emitting ZnO@polymer core-shell microspheres. *J. Mater. Chem.*, 17, 2490-2496.

Xiong, H.-M.; Xu, Y.; Ren, Q.-G.; Xia, Y.-Y. (2008). Stable aqueous ZnO@polymer core-shell nanoparticles with tunable photoluminescence and their application in cell imaging. *J. Am. Chem. Soc.*, 130, 7522-7523.

Xiong, H.-M. (2010). Photoluminescent ZnO nanoparticles modified by polymers. *J. Mater. Chem.* 20, 4251-4262.

Yan, E.; Wang, C.; Wang, S.; Sun, L.; Wang, Y.; Fan, L.; Zhang, D. (2011). Synthesis and characterization of fluorescent chitosan-ZnO hybrid nanospheres. *Mater. Sci. Engineer. B*, 176, 458-461.

Yuan, Q.; Hein, S.; Misra, R.D.K. (2010). New generation of chitosan-encapsulated ZnO quantum dots loaded with drug: synthesis, characterization and in vitro drug delivery response. *Acta Biomater.*, 6, 2732-2739.

Zhang, L.; Yin, L.; Wang, C.; Lun, N.; Qi, Y.; Xiang, D. (2010). Origin of visible photoluminescence of ZnO quantum dots: defect-dependent and size-dependent. *J. Phys. Chem. C*, 114, 9651-9658.

Zhang, P.; Liu, W. (2010). ZnO QD@PMMA-co-PDMAEMA nonviral vector for plasmid DNA delivery and bioimaging. *Biomaterials*, 31, 3087-3094.

Part 2

State-of-the-Art Nano-Composites

4

Nanocomposite Materials with Oriented Functionalized Structure

Petrişor Zamora Iordache, Nicoleta Petrea,
Rodica Mihaela Lungu, Răzvan Petre, Ciprian Său and Ioan Safta
Scientific Research Center for CBRN Defense and Ecology
Romania

1. Introduction

Functionalized materials are material structures able to meet certain functions and tasks in well established interaction conditions. The functionality of a nanostructured material is completely conditioned by the well-defined relation holding between the functionality of the host structure (carrier support - CS) and the structure upon which it is carried out (target structure - TS). The functionality and the functions of a material are carried out by means of its physical, chemical or other characteristics, and by means of which it interacts with neighbour material systems, in order to fulfil the assigned functions. The most well-known examples of functionalities are those of: chemical reticulation, detection, controlled stimulation, chemical separation, etc. (fig.1). Most of the natural material structures are found in the form of nano or micrometrically structured composites, and dependency and interaction relations are established between their structural elements, depending on the internal structure and the physical and chemical characteristics of the neighbouring material structures. Composite materials are mixtures of material phases, either amorphous, or structurally well delimited, varied from a morphostructural and morphochemical point of view. The basic structure of composite materials is represented by nanostructured material phases, as material superior form of organization of atomic and molecular structures. As a result of physical and chemical interactions, the structural components of natural composites establish complex physical and chemical equilibrium states, characterised by the fact that the functional structure of this type of composites is masked or cancelled.

The functionality concept is a relatively defined concept, taking as a reference point the investigation and order needs of the surrounding material structures. Basically, the function of a functionalized material defines the way in which the functional structure affects a certain process or a certain material structure, not necessarily neighbouring it. All material structures are carriers of certain types of inherent functionalities, by means of which they specifically interact with the exterior, but which do not have selectivity for certain target structures. Most modern scientific and technological applications need nanostructured or nanocomposite materials, having an oriented functional structure. These constrains impose that natural composite materials (CM) should be submitted to morphostructural and morphochemical modelling processes, in order to use them as oriented functionality materials. From the point of view of the sense of functionality orientation, the materials with oriented functional structure (M_{OF}) can be defined as materials whose structure and characteristics model and transform in sequences that are phenomenologically ordered in space and time, the structural evolution and the properties of the target structures (TS).

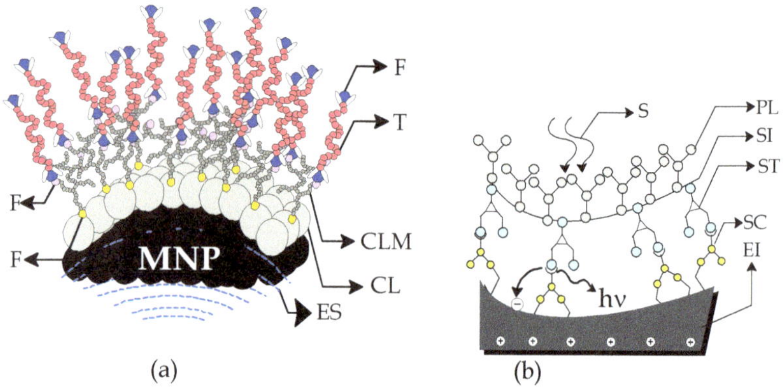

Fig. 1. Some typical examples of nanostructures wih orientated functionality (a) Polyfunctional structure with macromolecular separation tasks (ES - external magnetic field stimulus; CL - coated layer; MNP - magnetic nanoparticles; T - target macromolecules; CLM - reticulation molecules; F - functionalities) (b) Functional structure of a interfaces with sensitization tasks (EI - electric interface; SC - signal carrier; ST - signal transducer; SI - structural interface; PL - protective layer; S - external stimulus)

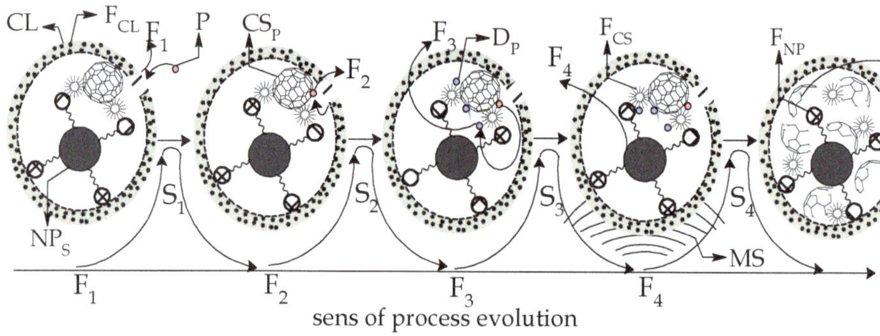

Fig. 2. Typical diagram of an orientated process for controled separation of organic

Figure 2 presents the functional diagram of a M_{OF} intended for the reticulation, degradation and controlled separation of the organic pollutants in the surface waters. The material obtained by means of controlled functionalising processes was modelled so that it could carry out the nano- and microfiltering (F_1), reticulation (F_2), degradation (F_3), magnetic separation (F_4) and biodegradation (F_5) functions of the organic pollutants in the surface waters, containing chemical functionalities of the aminated, hydroxilated, carboxylated, phenolated, carbonylated, thiolated and halogenated types, derivatives or chemical structures having mixed functionality. In phase S_1, the material comes in contact with the pollutants (P), reticulating them at the level of the functional structures of the CL and CS_P structural components of the composite. CL carries out the compositing functions of the CS_P

and NP_S structural elements (fig.2) and the filtering functions at molecular level of the contaminated water. In order to carry out the reticulation function of the pollutants, the CL, CS_P and NP_S structural components have been chemically functionalized with glutaraldehyde (F_{CS}) and epyclorhidrine (F_{CL}), so that, we might obtain, by means of a compositing process, a material having a nanometric porous structure and usable in the molecular filtering processes (fig.2). The pollutants' reticulation (F_2) at the level of the three structural components takes place in S_2 phase, each of them separating a certain category and type of target pollutants. Following the reticulation processes, it is likely that the reticulated pollutants should be partially or totally degraded, as a result of the chemical crosslinking interactions holding between the functional groups and the pollutants. The degradation function of the pollutants (D_P) is carried out at the level of all the structural elements of M_{OF}, and is due to the polyfunctional structure and to the nano- and micro-fiberlike morphostructure of the composite. In order to degrade the pollutants (F_3), M_{OF} was functionalized and composited, so that, due to the density of functional groups and due to the space distribution of its structural components, nano and microchannel structures may be formed, which may degrade and encapsulate the pollutants and the resulting degradation products, without offering the possibility for them to escape in the filtered environment. The controlled separation process of the depolluting products is done by means of the F_4 function, carried out by the NP_S nanostructural component.

NP_S contains Fe_3O_4 nanoparticles functionalized with glutaraldehyde, which assures the separation of the composite filled with pollutants. These nanoparticles can be stimulated, in order to be separated in a controlled manner, by means of an external magnetic field (MS) (fig.2). The NP_S structural component can reticulate organic pollutants that have chemical aminated functions, by means of the F_{NP} functional structure.

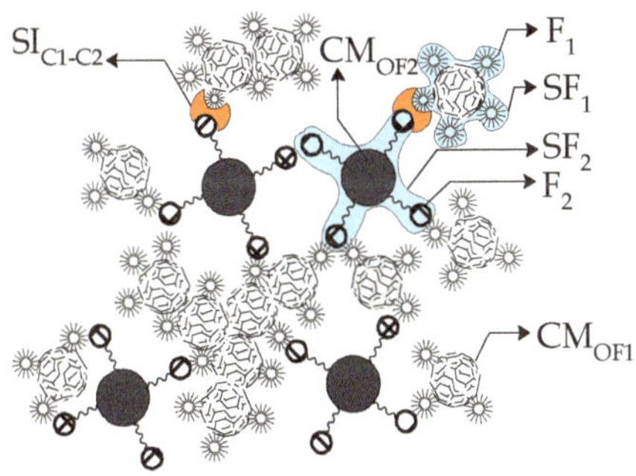

Fig. 3. Typical structural and interfaceing diagram of a composite material (CM_{OF1}, M_{OF2} - component material phase of M_{OF}; F_1, F_2 - functional elements of CM_{OF}; SF_1, SF_2 - microsurfaces distributions of functionalities; $SI_{C1\text{-}C2}$ - compositing interfaces of CM_{OF})

The biodegradation function (F_5) is the result of the appropriate modelling of the morphological and morphochemical structure of M_{OF}, so that, when the composite charged with pollutants is exposed to the environment factors, it might be degraded and assimilated, without modifying the structure and the functionality of the environment. The CL structural component encapsulates the reticulated pollutants and the degradation products, so that the environment may assimilate more easily the M_{OF} charged with pollutants, without generating short or long term toxic effects.

There is not a well defined limit between the intelligent materials and the functionalized materials, as in both cases material structures which are "programmed" to execute pre-settled tasks are involved. The polyvalent functionalization of M_{OF} involves the discretization of the obtaining processes, the carrier material phases being integrated in stages and by different processes. The most sensitive stage in the obtaining process of M_{OF} involves the preserving and quantification of the functional structure and of the carried out functions, so that they are not lost, masked or modified, following the compositing processes. The preservation and quantification of the functionality need the interposition of material phase connection, which should link the material structures that enter the composites structure (CM_{OF}) (fig.3). These connecting structures are called interface structures (IS) and have the role to protect and connect each structural component, as well as to preserve their functions. IS (fig.1) establish transition microregions between structural components (fig.3), the most widespread being the chemical interfaces.

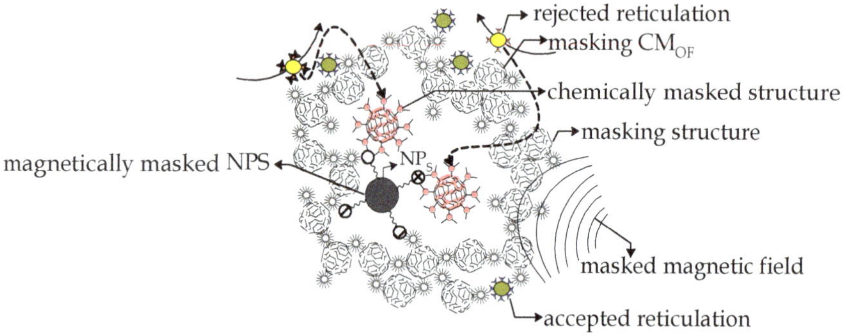

Fig. 4. General mechanism for functional structure masking of polyfunctionalized M_{OF}

Chemical interfaces favour the interposition between two homogenous material phases, different in point of structure and physicochemical properties of the functional groups, coated layers or molecular phases. As it is difficult to find the structural interface which may realise, at the same time, the connecting and preserving functions of the functionality, in most cases, between the nanomaterial composites and its structural components, molecular layers interpose to meet one of the two imposed conditions. Depending on the application and the type of functionalised composite, IS has an important role in stabilizing the morphochemical and morphostructural functionality by establishing complex physicochemical equilibria, which influence the dielectric structure, chemical and thermical stability, electric conductivity, magnetic moment, radiative stimulation, molecular diffusion, etc. The dimensions, the morphological structure, the morphochemical structure and the manner of distribution of IS are strictly conditioned by the chemical structure of the interfaced material phases and by the functionality we seek to obtain. For instance, in the case of the magnetically stimulable nanostructures, the thickness of the coating layer may affect, by screening, the magnetical separation processes (fig.4). In order to obtain structural interfaces that would not mask or cancel the functionality of the composite, each material structure of the composite is obtained and functionalised separately, following that the compositing processes of the different structural components would be carried out in a controlled manner and in distinct stages. The functionalised nanocomposite materials have become indispensable tools in most scientific and technological fields, due to the flexibility of the structural and functional modelling mechanisms, as well as due to the available manipulation and control possibilities. Environment science and protection (water depollution, waste control dispersion, water purification, water filtration, etc.), biochemical sensing, detection and nanomedicine are the most promising fields in which this type of materials can be applied.

2. Mono- and polyvalently functionalized nanocomposite materials

By definition, mono- or polyvalently functionalised nanocomposite materials (CM) are the carriers of one ore more functionalities. The structural and functional modelling of composites depends on the type of application and functions that the composite must carry out. In depolluting applications, the polyvalently functionalised composites have the advantage that they may be structurally and functionally modelled in an oriented manner, so that they might reticulate, encapsulate and separate a large number of organic pollutants.

2.1 Functional structure modeling of orientated functionalized materials

Chemical functionalization compounds ar those chemical compounds able to establish stable polydirectional chemical bonds, not only with the carrier support, but also with the target molecular structure. Some applications (analytical applications, molecular separations, etc.) constrain the use of M_{OF}, imposing that the reticulation processes should not degrade the crosslinked structure. Compounds with high chemical aggressiveness are not proper to be used in the analytical processes of sensitization, reticulation, or macromolecular separation, which are specific to the biological compounds, organic pollutants, nanoprobes, etc. Highly aggressive reticulation chemical compounds can degrade, alter and disperse the morphochemical structure of reticulated target structures.
The most used classes of chemical functionalization compounds are triazines (fig. 5), sulphochlorides (fig.6), periodates (fig.7), cyanate compounds and their derivates (BrCN, ClCN, hydrazine - fig.8), aldehydes (glutaraldehyde, N-hydroxysuccinimide, benzoquinone, etc. – fig. 9) and the epoxidic compounds (epichlorhydrine - fig. 10).
The glutaraldehyde is the most proper compound to be used in macromolecular functionalizing processes, as it can easily form stable functional structures, which do not degrade reticulated structures. The rest of the functionalization chemical compounds mentioned in the literature (antibodies, biological markers, functionalization compounds for the noble metals or for those with low reactivity, etc.) is characterized by the fact that they are specific to a certain carrier support layer and reticulate a strictly limited spectre of target structures. As functionalizable carrier supports, different types of polymeric nanostructures may be used (cellulose, dextrans, elastomers, glicols, amines, acetamides, etc.), nanostructured elements (oxidic structures, metals, fulerenes, quantum dots, etc.), other nanostructured materials having their surface and depth structure enriched with chemical groups (-H, -OH, -SH, H_2O, etc.) that may establish chemical reticulation reactions. The compositing process of the structural components of composites may take place in one or more stages, depending on the functionality and the assigned functions, in relation to the affinity and chemical stability of each functional component, as well as depending on the morphological and morphochemical structure of the structural components and of the composite we seek to obtain.

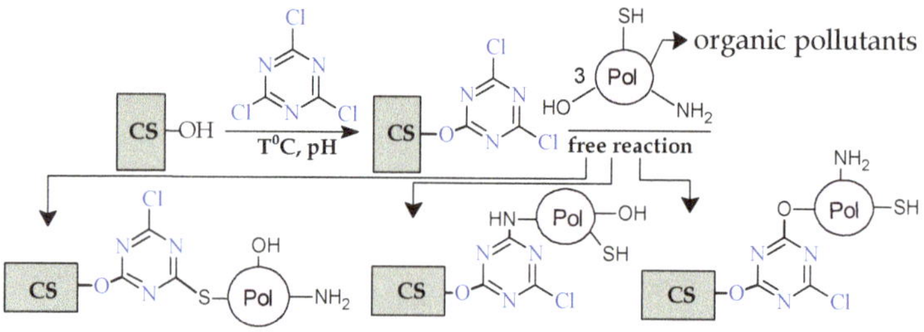

Fig. 5. General mechanisms of the macromolecular reticulation of triazinic-functionalized carrier supports (hidroxy-enriched carrier support)

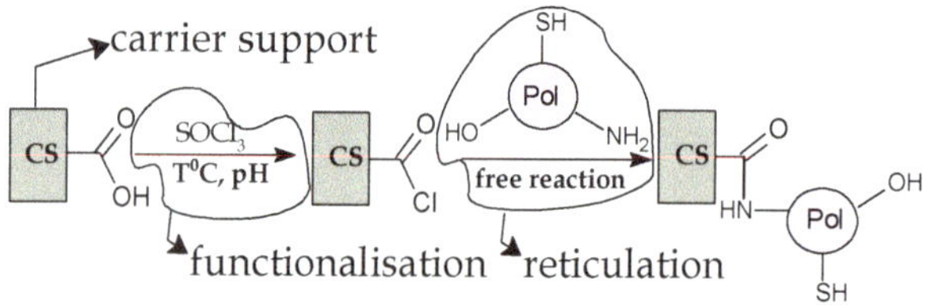

Fig. 6. General mechanisms of the macromolecular reticulation of halidic-functionalized carrier supports (carboxy-enriched carrier support)

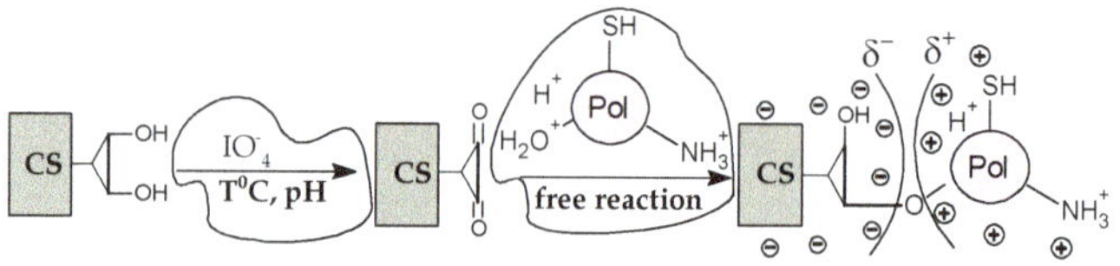

Fig. 7. General mechanisms of the macromolecular reticulation of IO-4-functionalized carrier supports (poliol-enriched carrier support)

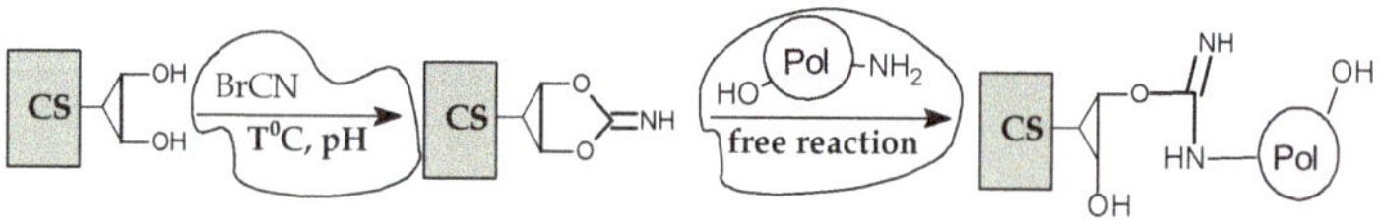

Fig. 8. General mechanisms of the macromolecular reticulation of cyan-functionalized carrier supports (hydroxy-enriched carrier support)

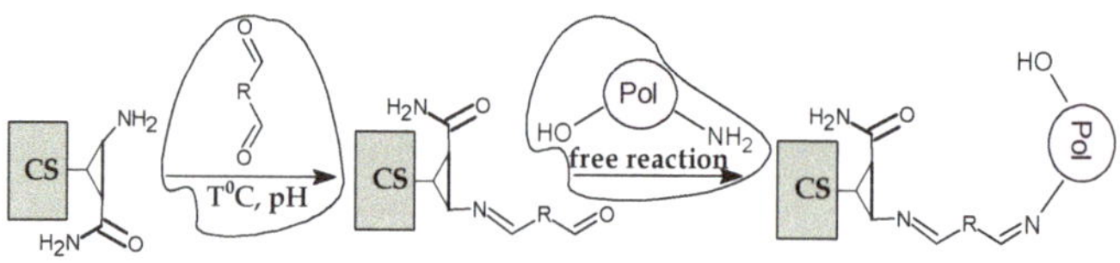

Fig. 9. General mechanisms of the macromolecular reticulation of aldehydic-functionalized carrier supports (amino-enriched carrier support)

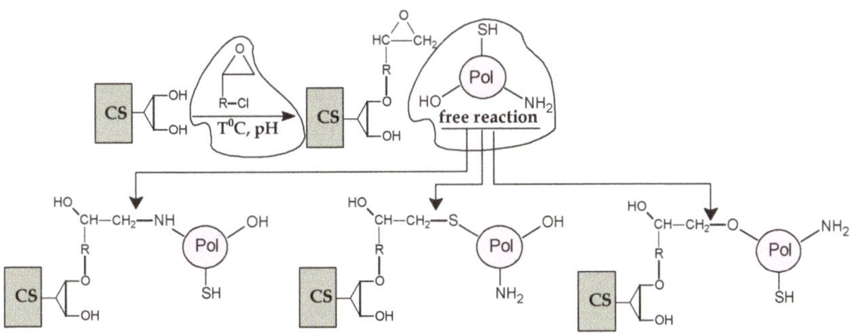

Fig. 10. General mechanisms of the macromolecular reticulation of epoxy-functionalized carrier supports (poliol-enriched carrier support)

2.2 Morphological and morphochemical structure of nanostructured materials

The surface and depth morphological and morphochemical structure of the orientedly functionalized composite materials is radically modelled by the reticulation and self reticulation processes, and by the local physical and chemical processes, established between the structural elements of the composite (self reticulation, induced mechanical tensions, magnetic and electric dipole moments, hydrogen bonds, molecular or atomic reticulation sites) (fig.11).

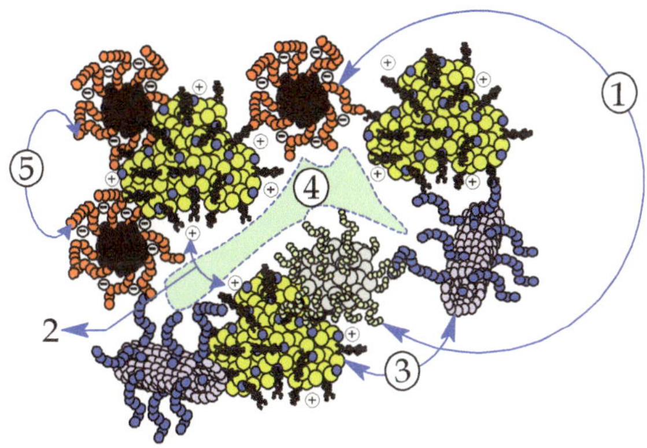

Fig. 11. General mechanisms of the morphological and morphochemical modelling of CM_{OF} 1 - functional constraints; 2 - charge constraints; 3 - mechanical and morphotopological constraints; 4 - free space; 5 - functional self-constraints

The agglomeration processes are specific both for the nanostructured materials, and for the functionalized nanocomposite materials, due to self reticulation processes, surface impurities, the electrostatic and micromagnetic polarization, superficial tensions between solvents and the dispersed functionalized material. Self reticulation is due to functionalization imperfections, which lead to the formation of ununiformly functionalized surfaces that favour the occurrence of atomic and molecular sites, able to initiate and establish reticulation bonds with the chemical functions of the neighbouring morphostructures. M_{OF} morphology is modelled by the morphology of its structural component, by the compositing techniques and conditions, by the expression mechanisms of the functionality and by the molecular association processes at a nanotopological level. The functionality and the morphostructure of the structural components are conditioned by the physical and chemical equilibria established between the structural elements of the composite. The morphological structure of M_{OF} may be modelled by choosing the adequate compositing methods (reverse micelle, photolysis, hydrolisis, etc.) and by the strict control of conditions and of the compositing phases (temperature, solvation effects, rate of agitation, etc.). The morphological and morphochemical depth structure of the composites cannot be modified during the compositing process, and the external modelling factors may lead to loosing of the functionality of the composite or of its strutural components. Reverse micelle method is the most convenient technique of obtaining nanostructured materials and functionalized composites.

3. Nanostructured material interfaces

Material interfaces (MI) are material structures that favour the stable binding of two or more material structures having different chemical and physical properties. The surface chemical structure of most of the inorganic structures does not bear direct chemical functionalization processes, and the growth of material interfaces is needed (functionalization molecule layers, coating layers) which should favour stable chemical bonds between the involved material structures (fig.12). Impurities, microsurface defects and the reduced reactivity of microsurfaces are the main elements that condition the functionality of this type of nanostructures. In order to favour the physical and chemical interaction processes between the organic and inorganic material phases, in most cases, a series of processes and chemical compounds are used to modify their surfaces energetically and chemically: silans, acid or basic chemical treatments, chemical passivations, etc.

In the case of high functionalized oriented structure composites, each structural component of the composite needs the deposition of one interfaceable chemical structures having a certain specificity in relation to the rest of the composited structures. Preliminary chemical treatments due to eliminate impurities and to modify the treated microsurfaces energetically aim to favour chemical reticulation reactions and to obtain uniformly distributed functional interfaces. The elimination of surface impurities is indicated in the case of oxidic nanomaterials and amorphous material structures, whose surface is likely to be impurified with traces of water, hydroxyl groups, metals, ions, etc. The most efficient method to eliminate these impurities from the surface of oxidic structures or of amorphous material structures consists of treating the interest surfaces with acid mixtures (HNO_3, H_2SO_4, H_3PO_4) (Lucas et al., 2007). As a result of the acid treatments, part of the treated support structure is partially degraded, and the metals and metal ions are removed. After having carried out the acid treatment, the composite nanoparticles are washed with solvents, which should remove the resulted degradation compounds. In most cases, acid treatments are followed by chemical treatments intended to passivate the modified nanostructures electrostatically. Organometallic salts are usually used (citrates, oxalates, etc.) in order to eliminate the electric charge density induced in the surface structure (Campos et al., 2002).

Prevention of the agglomeration and flocculation processes can be made by introducing some additional repulsion forces, so that the support phase, chemically and energetically modified, should be positioned in a chemical, electrostatic and micromagnetostatic equilibria with the host fluid. The spreading of the ionic phases in aqueous host medium is influenced by the presence of the residual ions (Na^+, $(CH_3)NH_3^+$, NH_4^+, NO_3^-, ClO_4, etc.) and by the dielectrical intrinsic structure of the cristaline structure of the dispersed phase, establishing electrostatic Van-der-Waals bonds with the amphoteric hydroxyl groups (Campos et al., 2002).

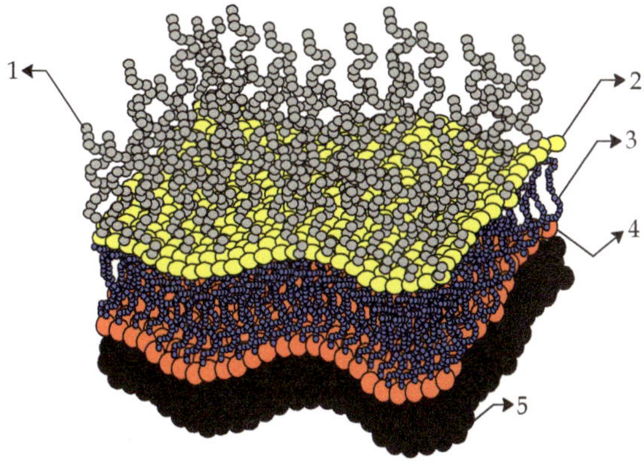

Fig. 12. General morphological and morphochemical structure of material interfaces
1 - crosslinking molecules; 2 - reticulation interface; 3 - organic laid-down coated layer; 4 - energetically and chemically modified surfaces structures; 5 - inorganic substrate

The favouring of dispersion may be done by inducing some negative (dispersion in acid environments) or positive (dispersion in alkaline medium) electric charges at the level of dispersed phase surface (Massart et al., 1995). The dispersion of the inorganic phase in organic host medium (cyclic hydrocarbons, paraffine oils, liniar caten hydrocarbons, etc.) involves coating them with an amphiphilic molecule layer, in order to initiate the steric repulsion processes. Molecular reticulation prevention is due to the thermic agitation movement of the terminal hydrocarbons that lead to the formation of specific cristaline structures. The most used surfactants are sodium bis (2-ethylhexyl) sulphosuccinate (AOT), cetyl trimethyl ammonium bromide (CTAB), lipids (lecithin, phosphatidyl glycerol dihexadecyl phosphate, etc.), oleic acid, stearic acid, etc.

Silanes and cyanides are the most indicated chemical agents to modify the nanostructured surfaces chemically and energetically, so that they should establish easily stable chemical bonds with OH-, H-, -O-, M-O and M (M - metal atom) groups found on most of the material structure surfaces. Cyanates compounds (BrCN, ClCN) favorise the forming of a large spectrum of reactive chemical groups on oxidic surfaces, of the type: nitrile: (-CN), isonitrilic, cianamidic ($-(R)_2N-CN$), azinic ($(C=N-)_2$, isocianic (-N=C=O), carbodiimidic (N=C=N), aminated (primar, secondary, tertiary), compounds having nitrogen and mixed functionality. This type of reactive chemical groups have reticulation properties which are similar to those mentioned in figures 5-10. The organofunctional silanes reaction with -OH groups surfaces is a substitution reaction at the silicium atom, resulting silylated surfaces in which the Si atoms fix on the surface by covalent chemical bonds, via oxygen (fig. 13).

These bonds may be established directly or in the presence of water, through of an intermediary silanol. The reactivity of the hydroxilated surfaces with the silans decreases in the following order: $Si-NR_2>SiCl>Si-NH-Si>Si-O_2CCH_3>Si-OCH_3>Si-OCH_2CH_3$. Si-O-microsurface bonds are the critical factor that sustains the reaction, not only in aprotic conditions, but also in non aprotic ones.

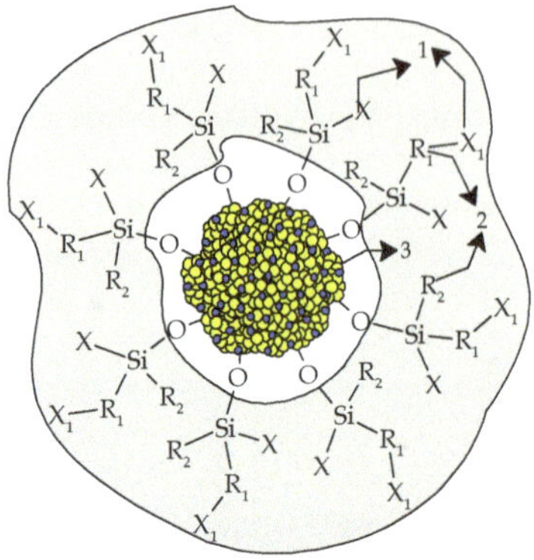

Fig. 13. General structure of an organosilane organic-inorganic interface (1 - hidrolisable functionalities: -OH, NH2, O-O, etc.; 2 - space linkers; 3 - inorganic modified substrate)

Methoxysilanes are able to sustain reactions with the surfaces from an energetic and chemical modification point of view, not only in aprotic and non aprotic conditions, having lower reactivity compared to the ethoxylans and needing catalytic conditions for sustaining. An important control factor of the physical and chemical characteristics of the chemically modified microsurfaces is given by the length of the chemical linkers interposed between the organic functionalities and the Si atom fixed on the modified surface, imposing a series of specific chemical reactivity constrains (Jall et al., 2004; Chechik et al., 2000). The organic surface imposes a series of steric limitations on the accessibility of the functional groups which are placed in their vicinity, and the space linkers with appreciable lenght confer an increased mobility in order to functionalize the entire target surface. Aluminium and titanium form stable surface oxides by hydrolitic processes. Their oxidized surfaces have enough -OH groups to allow the coupling of the organofunctional silanes in the same conditions with those applied to the silicides layers.

Copper, zinc and iron form surface oxides, which are instable from a mechanic and hydrolitic point of view. The reticulation process of the silanes on these types of surfaces involves two or more silans, one of which is a chelate agent (polyamines, polycarboxylic acids). The second silane is chosen depending on the chemical reactivity of the organic component and on its capacity to initiate condensing reactions with the first silan.

In the case of weak reactive metals (Ni, Au, Pt, etc.), the reticulation of silanes on their surface involves the formation of coordinative bonds (phosphines, mercapto, amino-functionalized silanes). In order to do this, a second silan must be used, which may present reactivity towards the organic component. Titan, zirconium and nickel form stable hydrides. The ability of several metals to form amorphous alloys with hydrogen is exploited in order to adsorb and coordinate the silicon hydrides on the metal surface. The functionalization hydride of silanes and the treated surface of the metal realease hydrogen in the presence of bases or noble metals.

4. Physical and chemical properties of functionalized nanosurfaces

Nanocomposites and nanostructured functionalised materials are those materials which carry their own chemical and physical properties, showing them in their next vicinity. The physical and chemical characteristics of structural components and of their composites depend completly on how they have been obtained: precursors, solvents, physical and chemical properties of the structural interfaces, temperature, pH, etc. The main processes responsible for the morphostructural, morphochemical and morphofunctional modification of depth material interfaces are: a. chemical or electrostatic adsorbtion sites of: neighbouring molecular radicals, metalic ions, molecules with high diffusion degree and small molecular mass (fig.14); b. hydrogen bonds or other similar chemical bonds (fig.14); c. the type, density and lenght of molecular space-linkers (fig.14); d. the type of chemical bonds established by the material interfaces with the organic and/or inorganic interfaced structures.

In the case of organosilanic interfaces, the chemical and electrostatic sites are due to functional groups of silane which carry out functions in the reticulation of the target structures and in the molecular or atomic sites which possess electric dipole moment. The structural modelling process is a dynamic one, taking place in the phase of nucleation of the organosilane sites and at the level of the growing surface of the laid-down functionalization interface. As a direct consequence, micro- and nanostructures empty sites (space) are generated in the depth structure, favouring the diffusion of charged ions and of the structures with lower molecular mass (fig.14). The impurities reticulated in the depth structure come from the reaction medium, as most silanization processes are catalyzed processes (alcohols, NH_4OH, etc.), that may generate in the reaction medium different molecular structures ($R_nSi(OH)_{4-n}$, $N^{j+}H_n(OH)_m$ (j+n+m=5), H^+, $-OH^-$, H_3O^+, etc.) that may diffuse and reticulate in the depth of the laid-down functionalized interface. Self assembled monolayer (SAM) material interfaces have the advantage that they suppress morphochemical and morphostructural processes, but they do not present the same variety and functional afinity by comparison with layered material interfaces. The obtaining of layered material interfaces with varied and stable functionality may be an advantage for a series of applications in detoxification, depollution and decontamination, as they may reticulate, encapsulate and host a wide spectrum of toxic compounds in their structure.

Interfaces with high content of hydrogen bonds and chemical bonds with hydrophobic character can reticulate and host large amounts of aminated, carboxylated, hydroxylated, hydration water, active hydrogen sites and molecular radicals. Using silanes of the NH_2R_n-$Si(-O-)_m$ (n+m = 4) type in processes of deposition of functionalized interfaces leads to the formation of polymeric silanic structures of the $n[SiO_{1.5\gamma}\text{-}R(NH_2)](NH_2)_{n\delta}$ type, enriched with -OH, amino, free water and reticulated H_3O^+ functional forms (γ characterizes the degree of cohydrolisis of $Si(-O-)_m$; δ characterizes the amount of aminated groups) (Iordache et al., 2009). The thickness of the laid-down organosilane layer and the amount of reticulated aminated groups depends on the type of the used silane, the cohydrolisis time and the type of the used catalyzer. In the case of F_3O_4 nanoparticles coated with organosilane polymer by cohydrolisis of the (3-aminopropyl)trietoxysilane, functionalized interfaces of the F_3O_4-$n[SiO_{1.5\gamma}\text{-}(CH_2)_3(NH_2)](NH_2)_{n\delta}$ type have been obtained. On the basis of the analyis of the aquired experimental data by energy-dispersive X-ray spectroscopy (tab.1), the growth rate of the coated layer has been determined, as being modelled by the analythic relation (1). The graph representation of the analythic relation (1) is presented in figure 16.

Nano and micro-sites established in the depth of the laid-down organosilane coated layer favour the diffusion of the lower mass molecular compounds, due to the type of space-linkers and to the flexibility of Si-O-Si ($120^o<\theta<180^o$) bonds (Tolstoy et al., 2003). This fact is due to the unhydrolizable character of molecular space-linkers, which orientate the neighbouring hydrolizable structures depending on the electric dipole moment, the chemical interaction potential and the mechanic tensions established between them and their neighbouring macromolecules.

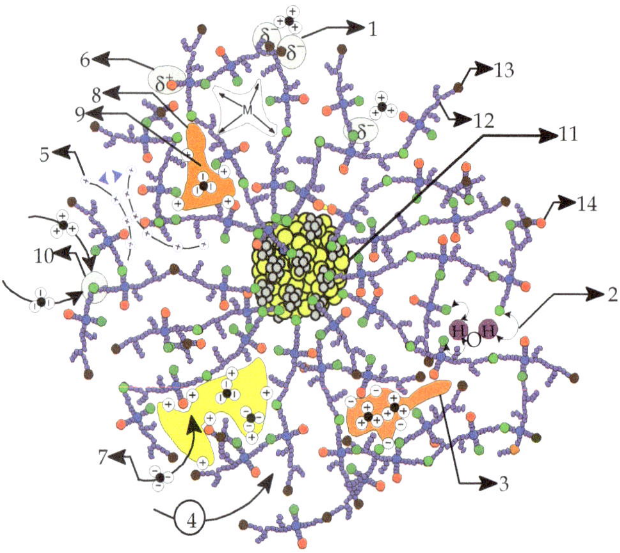

Fig. 14. General physical and chemical mecanisms who model the morphochemical and morphological structure of functionalized nanosurfaces (1 - positively charged molecular radicals; 2 - hydrogen bonds; 3 - free space sites filed with positive charged molecular radicals; 4 - acces patway for smoll molecules; 5 - positively charged nanosurfaces; 6 - molecules with strong electric dipole; 7 - negatively charged molecular radicals; 8 - positively charged topological nanosites; 9 - free space for negatively charged molecular radical retaining; 10 - bonds susceptible for breaking; 11 - inorganic interfaced nanostructure; 12 - space linkers; 13 - organic functionality; 14 - space linkers)

The size of the free micro-sites established in the depth of the coated layer has the magnitude of dimensions comparable with that of molecular space-linkers, modelling the mechanic tensions and the charge distribution. Functional organosilane groups (fig.14) could be inserted on the surface and in the depth of interfaced structures, using as silanization agent: silanols, amino silanes, siloxanes, epoxy silanes, carbinolic-functionalized silanes, acrylate/metacylate-functionalized silanes, carboxy silanes, mercapto silanes or anhydrides-functionalized silanes (Denmark & Sweis, 2002). Functional organosilanic groups may reticulate nucleation macromolecular structures of the $n[SiO_{1.5\gamma}\text{-}RX]Y_{n\delta}$ type (X functional silanic group: $-NH_2$, -SH, OH, -COOH, etc.) or the sites or molecular radicals (Y) in the reaction volume, modifying the morphochemical and morphofunctional distribution, as well as the electric charge density established on surfaces of the deep micro- or nanosites of material interfaces. In the case of the aminosilanes catalyzed by NH_4OH, the following bonds Si-O-Si, Si-NH-Si, $Si\text{-}O(NH^{+}{}_3)\text{-}Si$, $SiNH^{+}{}_3$, $[M(NH_3)_n]^{m+}$, result during the process of the $n[SiO_{1.5\gamma}\text{-}R(NH_2)](NH_2)_{n\delta}$ laid-down coated layer. These bonds can be easily broken by the molecular radicals present in the reaction medium, resulting new types of functional group: Si-OH, $Si\text{-}NH_2$, etc.

Crt.	N (Wt%)	O (Wt%)	Si(Wt %)	Fe(Wt%)
P1	0.00	60.45	17.62	21.93
P2	1.97	42.10	22.02	33.91
P3	3.93	60.79	8.56	26.72
P4	2.49	51.40	19.08	24.20
P5	2.90	53.09	26.98	13.02
P6	4.06	53.22	14.05	17.74
P7	1.04	55.87	18.47	22.30
P8	3.50	44.06	20.93	13.76

Table 1. Contents of Fe, Si, N and O found on SEM investigated microsurfaces

$$\ln\{([N]+[Fe])t^2\} - \frac{t}{[Si][Fe]}\frac{1}{\ln\{([N]+[Fe])t^2\}} = f(t) \qquad (1)$$

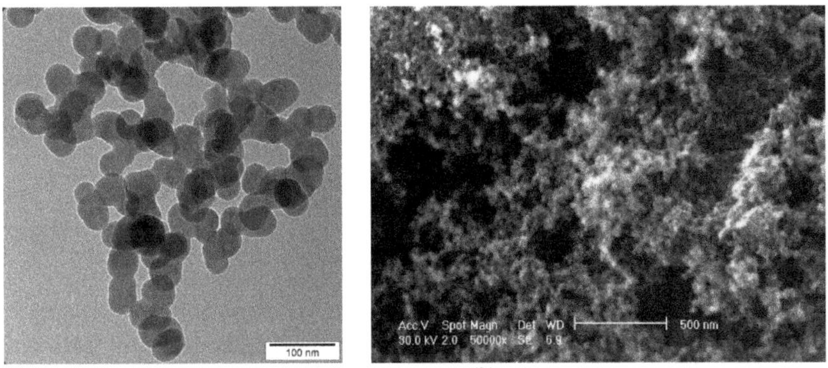

(a) (b)

Fig. 15. F_3O_4-n[$SiO_{1.5\gamma}$-$(CH_2)_3(NH_2)$]$(NH_2)_{n\delta}$ nanoparticles
(a) TEM image (180.000x) (b) SEM image

In the case of Fe_3O_4 nanoparticles coated with n[$SiO_{1.5\gamma}$-$(CH_2)_3(NH_2)$]$(NH_2)_{n\delta}$, the experimental data we have obtained by electronic transmission microscopy (ETM), electronic scanning microscopy (ESM) and EDX prove that the morphostructural and morphochemical modelling takes place at the level of the laid-down organosilane layer and at the level of the n[$SiO_{1.5\gamma}$-$(CH_2)_3(NH_2)$]$(NH_2)_{n\delta}$ nucleation sites. This fact entails that chemical stabilization and passivation of the laid-down interfaces take place directly during the coating process, by microlocal passivation of nucleation sites and at the level of the growth surface, in the presence of the reaction medium content.

For biochemical reticulation applications, unstable chemical reticulated impurities on the surface and in the depth of coating layers can be removed by washing them with H_2O, NaCl or organomethalic salts (citrates), having the role to stabilize the suspension nanoparticles from a chemical and isoelectrical point of view. The obtained suspensions (4%) are used with the pH adjusted to 7÷8 by proper dilution.

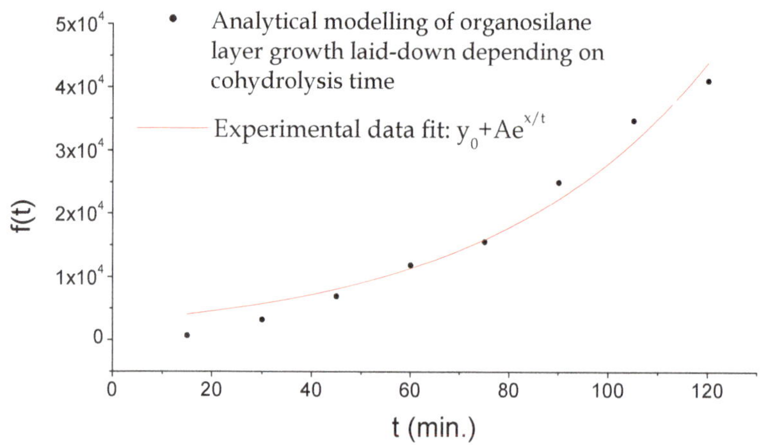

Fig. 16. f(t) function graph

5. External stimulable nanocomposite materials

Externally stimulable nanocomposites represent one of the latest priorities and needs in nanoscience and nanotechnology field, having applications in various fields, such as: chemical and biochemical detection (sensors, differentiated and undifferentiated prelevation of biochemical analytes, etc.) (Festag et al., 2005), nanomedicine (Murcia & Naumann; Bekyarova Haddon & Parpura; Gao; Zheng & Huang; Watanabe & al.; Meziani & al.; Lévy & Doty; Hattory & Maitani; Lellouche; Kommareddy, Shenoy & Amiji; 2005), environment science and protection (depollution, decontamination), chemistry (molecular and macromolecular separation). The stimulation concept designates that way by which a material structure registers a change in its energetic or structural structure under the influence of an external stimulation factor. In this sens, electromagnetic stimulation processes are very well know (fluorescence, phosphorescence, photocatalysis, semiconductor stimulation), magnetic (RMN, RES), radiative (elementary particles, nuclear and molecular radiation). By implementation of some control functions at the level of a stimulable material nanostructure, oriented controlled stimulable nanostructures can be obtained in order to measure, quantify and control the stimulable induced states. Polyvalently functionalized nanocomposite structures are hard to obtain, especially in reticulation applications with targets coming from systems with complex morphochemical and morphostructural, biochemical and biological organisation. Kinematic and dynamic control methods of the stimulable material structures placed in carrier host medium with complex molecular structure are relatively limited and reduced to control by magnetic, electric and electromagnetic stimulation. These nanoparticles have been functionalized so that they should reticulate undifferentiatedly and undestructively the target biological structures, which vary from a morphostructural and morphofunctional point of view.

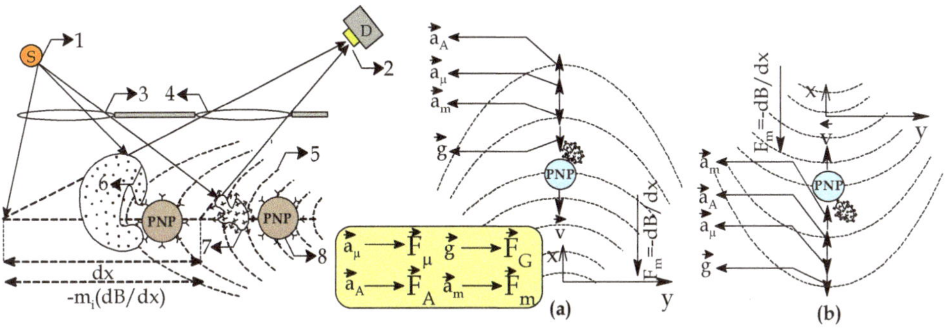

Fig. 17. Diagram of the configuration of constant metric biological detection technique 1 - scan source; 2 - detection sensor; 3 - scan gain; 4 - detector gain; 5 - magneti field stimulus; 6 - reticulated microorganism; 7 - other type of reticulated microorganism; 8 - functionalised magnetic stimulable nanostructure (PNP); M_i - mass of NPB

Analytical techniques with constant metrics use only a stimulation factor for the sensitization of a large spectrum of analytes. Existent analytical methods for biological sensitization (Bhunia, 2008; Jasson, 2010) are variant in respect of a specific required stimulation factor, being unfriendly in respect of sensitization of a large analite spectrum. The analytical method proposed in figure 17 starts from the assumption that the set of values formed by the magnetic moment attached to biochemical structures (m) by reticulation processes and the mass (M) attached to reticulation macrocomplexes are uniquely determined for each type of biochemical structure. As the prelevation and magnetic discrimination processes of the analytes take place in a host carrier fluid, we started from the assumption that the friction force (F_μ), weight force (F_G), archimedic force (F_A) and micromagnetical force (F_m) (fig.17) are the only ones that can model the kinematics and the dynamics of the magnetic dicrimination. In order to reduce the number of the observables and degrees of freedom that affect the kinematics and the dynamics of the Fe_3O_4 nanoparticles reticulated on the surface of the targeted biostructures (NPB), a force field configuration was proposed (fig.17a, fig. 17b), in which NPB movement is orientated, as follows: a. parallel with the gravitational acceleration vector (fig.17.b); b. opposed to the gravitational acceleration vector (fig.17.a). The observables attached to the configured biochemical sensitization process are the specific speeds (v_i) of NPB movement in magnetic field and the specific surface fluorescence of the reticulated biochemical structures (τ_i), so that an univoque analytical relation can be established between the discriminated biostructure and its specific surface macromolecular fluorescence mark. From an analytical point of view, the set of observables (v_i, τ_i) can be correlated by software, so that, for each v_{1i} (a), v_{2i} (b) of sensitized NPB_i, according to the configurations proposed in figure 17, one may identify those sequences that present the same fluorescence mark (τ_i), according to relations (4) and (5). The index „i" designates the type of the sensitized biochemical structure and is defined on the entire reticulation and sensitivization-detection spectrum. Analytical equations (4) and (5) model and describe the proposed constant metric analytical model, which is deducted from fundamental equations (2) and (3), which describe the NPB movement in configurations 17a and 17b.

$$F_{ARH} + F_F = F_{EM\text{-}1} + Mg \tag{2}$$

$$F_{EM-2} + F_{ARH} = Mg + F_F \tag{3}$$

where: F_{ARH} is the archimedic force, $F_{EM-1/2}$ are the magnetic field gradients, F_F is the friction force opposed to NPB movement in the host fluid, $F_F = -kx$, M is the mass of NPB.

$$M_{NPB}(i) = 2m(\frac{dB_1}{dx_1} - \frac{dB_2}{dx_2}) / \sqrt{2c^5 v_f \left(\frac{1}{x_2} - \frac{1}{2x_1}\right)} \tag{4}$$

$$M_{NPB}(i) = m(\frac{dB_1}{dx_1} - \frac{dB_2}{dx_2}) / \{\frac{\sqrt{2v_2^5 v_f}}{2x_2} + \frac{\sqrt{2v_1^3 v_f}}{2x_1} \ln \frac{v_1 + (v_2 - v_1)e^{\sqrt{\frac{2v_2}{v_f}}}}{v_2} - \frac{\sqrt{2v_1^5 v_f}}{4x_1}\} \tag{5}$$

where: dB_1/dx_1 şi dB_2/dx_2 are the gradients of the magnetic field realized on the sensitization distance, x_1 and x_2 are the sensitization distances, v_1 and v_2 are the characteristic speeds of each NPB, v_f is the speed of the carrier fluid which hosts the NPB.

In the above mentioned conditions, equation (4) describes those NPB that, following the magnetic acceleration process, achieve a constant speed (c), at the moment of the sensitization of discrimination and for a certain value of the field gradient (dB/dx). Equation (5) is specific for those NPB that, after the magnetic acceleration process, did not achieve a constant limit speed (c). According to equations (4) and (5) it results that analytical methods with invariant metric may be configured for biological detection and monitoring by selective magnetic discrimination. Also, according to these equations, it results that the sensitization domain may be improved by modifying the magnetic field parameters (dB/dx) of the external stimulation factors. The magnetic field represents a convenient method for the undestructive stimulation of magnetic dipole carrier functionalized material structures, which, having a considerable action range, as compared to the rest of the known stimulation factors. Due to the fact that most organic compounds are diamagnetic, the magnetic field substantiates new undestructive selective techniques for controlled stimulation, as the stimulable material nanostructures are not masked by their neighbouring material structures.

6. Structure and functionality modelling of nanocomposite materials designed for undestructive reticulation of biological structures

The functionalization of nanoparticle and nanocomposite structures for biochemical reticulation requires a set of conditions involving: the specificity and the recognition of the reticulated structure, the chemical and functional stability, avoidance of the biochemical interaction with the host carrier, avoidance of self reticulation and agglomeration processes. Most of the applications, especially the analytical one, require undestructive biological and biochemical reticulation in respect to the targeted biochemical structure. Glutaraldehyde is the chemical functionalization and reticulation compound frequently used in the analytical biological process (fig.10), as it presents a weighted chemical aggressiveness and it preserves the internal morphochemical struture of the reticulated structures. Glutaraldehyde is a dialdehyde whose functional molecular residue (CHO) can reticulate undifferentiatedly the hydrogen active, amino- and tiol- sites, present on the surface and in the depth structure of most of biochemical structures. Polyvalent functionalizations require a series of cautions intended for the chemical stabilization of the functionalized nanostructures, so that different implemented functional groups may not mask or cancel each other's functionality and functions. In order to investigate this aspect, Fe_3O_4 nanoparticles have been coated with $n[SiO_{1.5\gamma}-(CH_2)_3(NH_2)](NH_2)_{n\delta}$ and functionalized monovalently (with glutaraldehyde - GL) and polyvalently (with glutaraldehyde and epichlorhydrine - GL+ECH) (Iordache et al., 2009). For their morphochemical structure, the following structural relations have been proposed Fe_3O_4-{$n[SiO_{1.5\gamma}-(CH_2)_3(NH_2)](NH_2)_{n\delta}-(GL)_{n\varepsilon}$} (np-$(GL)_{n\varepsilon}$), $(ECH)_{n\beta}$-[Fe_3O_4-$n[SiO_{1.5\gamma}-(CH_2)_3(NH_2)](NH_2)_{n\delta}$]-$(GL)_{n\varepsilon}$ (($(ECH)_{n\beta}$-(np-$(GL)_{n\varepsilon}$) (ε and β - modelling coefficients for functional fractions CHO, respectively Cl-C and -CH(O)CH-) (Iordache et al., 2009). For the functionalization of nanoparticles, 4 grams of chemically stabilized coated nanoparticles have been used (4% suspension), in the presence of 7 ml of glutaraldehyde (25%) ((6)). Polyvalently functionalized particles have been obtained by the functionalization of np-$(GL)_{n\varepsilon}$ with an epichlorhidrine solution, previously prepared (NaOH, H_2O, acetone and 0.5 ml epiclorhidrine) (fig.9).

$$Fe_3O_4 - n\left[SiO_{1.5\gamma} - (CH_2)_3(NH_2)\right](NH_2)_{n\delta} + nC_5H_8O_2(GL) \xrightarrow{NaOH} Fe_3O_4 - n\left[SiO_{1.5\gamma} - (CH_2)_3(NH_2)\right](NH_2)_{n\delta} - (GL)_{n\varepsilon} \quad (6)$$

Glutaraldehyde functionalized nanoparticles have been exposed to air in order to investigate the undifferentiated reticulation potential of the saprophyte microorganisms, as the prelevated samples have been investigated by TEM (fig. 18). The same type of investigations has been carried out on B. Cereus, St. Aureus, E. Coli, Ps Aeruginosa, and the obtained results have been presented in figure 19. Also, the biochemical reticulation potential of np-$(GL)_{n\varepsilon}$, and $(ECH)_{n\beta}$-(np-$(GL)_{n\varepsilon}$ has been investigated, by ricin reticulation (Petrea et al., 2009). The samples have been investigated by scanning electron microscopy (fig. 20) and by confocal microscopy (fig.23). Before the investigations were performed by confocal microscopy, the functionalized nanoparticles were marked fluorochromically with rhodamine B, right in the phase of the laying down of the coated layer (cyan color in fig.23). The ricin was marked fluorochromically with fluorescein isothiocyanate (FITC) (green colour in fig.23). The results of the biological and biochemical rericulation investigations have proved that the obtained functionalized nanoparticles present a well defined functionality, and that the morphochemical structure of the reticulated structures has not been degraded (fig.18, fig.19).

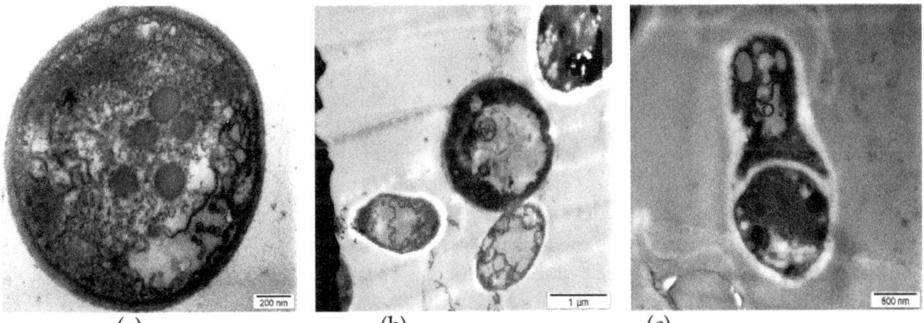

(a) (b) (c)

Fig. 18. Investigation images of the np-$(GL)_{n\varepsilon}$ biochemical reticulation potential on saprophyte microorganisms (images acquired by Philips S208 TEM microscope)

It was observed that np-$(GL)_{n\varepsilon}$ are specifically distributed on the reticulated surfaces, according to the surface morphochemical structure of microorganisms. Most likely, distribution maps of the reticulated nanoparticles follow the patterns of the morphochemical distribution of aminated, thiolated, carboxylated, and carbonilated functional groups placed on the surface of microorganisms. The obtained data emphasize the np-$(GL)_{n\varepsilon}$ potential in the field of specific biochemical reticulation, as nanoprobes, or in the field of undestructive biochemical investigation (fig.18).

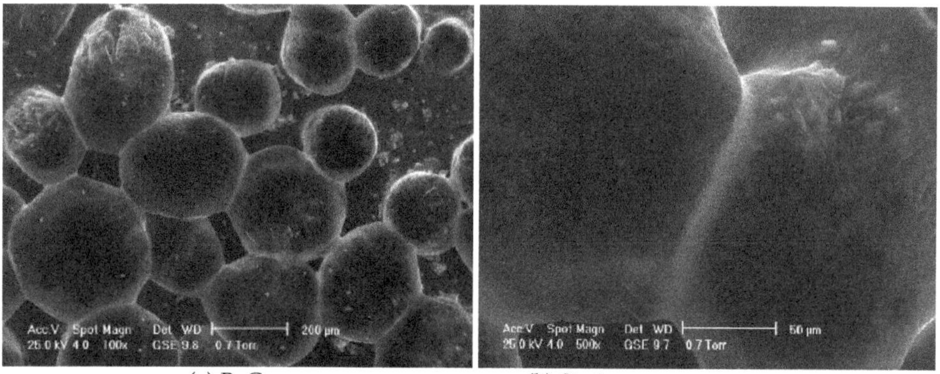

(a) B. Cereus (b) St. Aureus

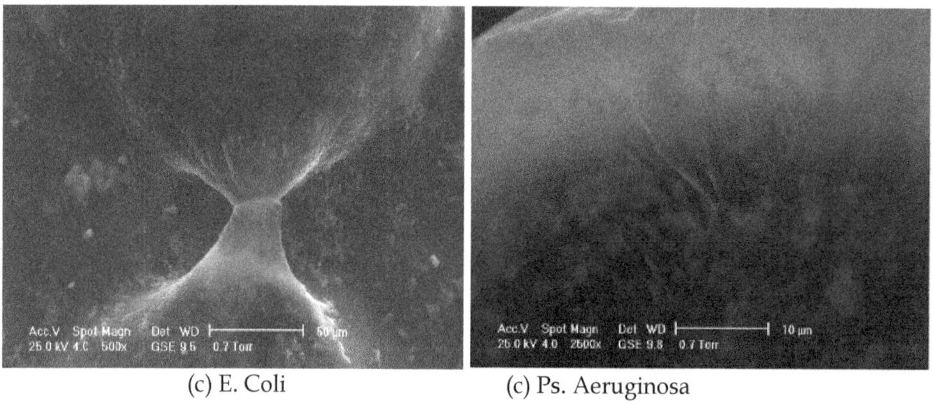

(c) E. Coli (c) Ps. Aeruginosa

Fig. 19. Investigation images of np-$(GL)_{n\varepsilon}$ biochemical reticulation potential on B. Cereus, St. Aureus, E. Coli, Ps. Aeruginosa (images acquired by ESEM XL30 SEM microscope)

Using the elementary morphological segmentation method (Iordache et al., 2009), in order to determine the elementary morphostructural parameters, it was proved that the main agglomeration mechanism is micromagnetic. Also, it was proved that the functional structure of the obtained functionalized nanoparticule suspension is stable, and it does not initiate self-reticulation bonds. In the case of np-$(GL)_{n\varepsilon}$ and $(ECH)_{n\beta}$-np-$(GL)_{n\varepsilon}$, which include Fe_3O_4 nanoparticles obtained by coprecipitation, the agglomeration domains present a spherical form (fig.20a, fig.20d, fig.21b). In the case of functionalized particles containing Fe_3O_4 obtained by reverse micelle technique, the agglomeration domains present an acicular form (fig.20b, fig.20c, fig.21a). These results prove the polyvalent functional structure of

ECH)$_{n\beta}$-np-(GL)$_{n\varepsilon}$, according to the specific functionalization mechanisms proposed in figure 22b.

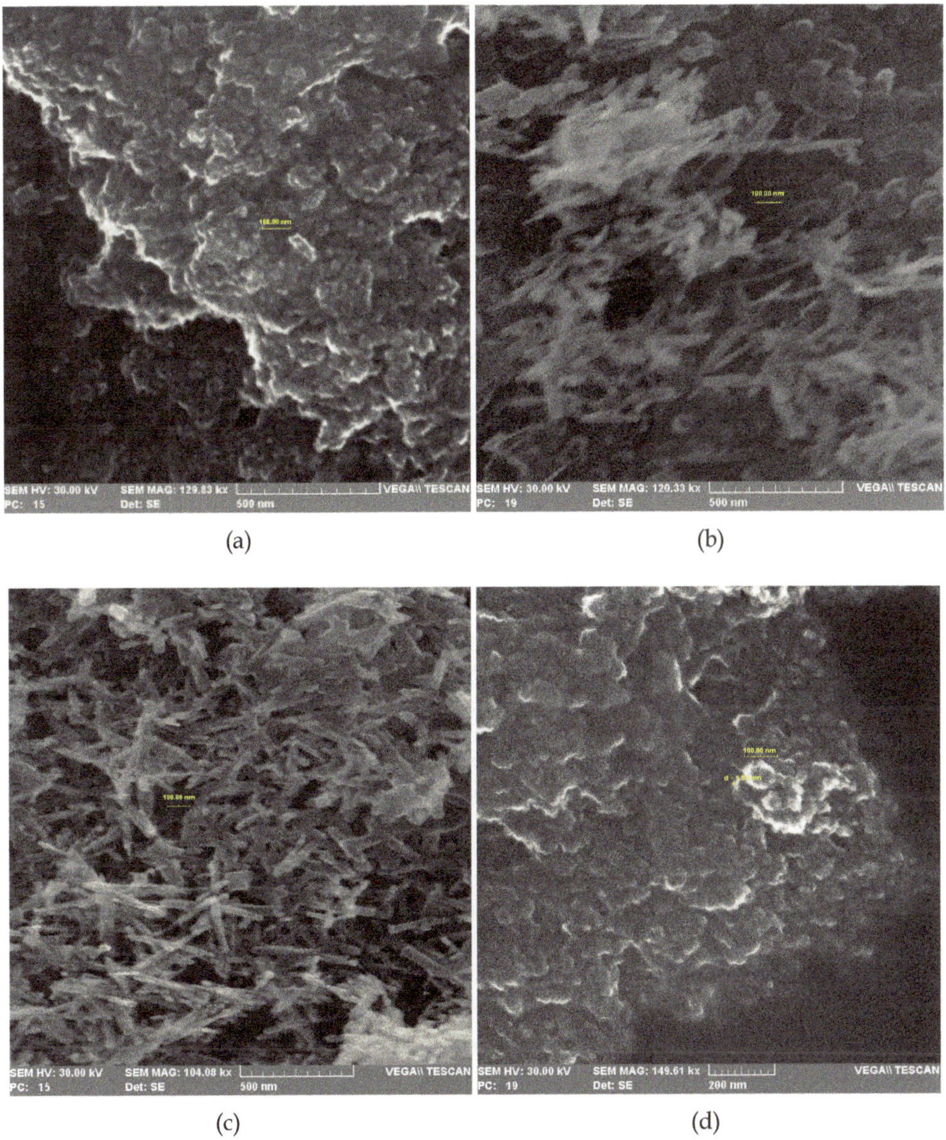

Fig. 20. Investigation images of (ECH)$_{n\beta}$-npa-(GL)$_{n\varepsilon}$ and npa-(GL)$_{n\varepsilon}$ biochemical reticulation potential on ricin (images acquired by VEGA II LMU SEM microscope)
(a) np-(GL)$_{n\varepsilon}$ (Fe_3O_4 obtained by coprecipitation) (b) np-(GL)$_{n\varepsilon}$ (Fe_3O_4 obtained by reverse micelle) (c) (ECH)$_{n\beta}$-np-(GL)$_{n\varepsilon}$ (Fe_3O_4 obtained by reverse micelle) (d) (ECH)$_{n\beta}$-np-(GL)$_{n\varepsilon}$ (Fe_3O_4 obtained by coprecipitation)

Reticulation investigations carried out on ricin have proved that there is no significant difference between the reticulation mechanisms of microorganisms and those of ricin. Ricin reticulation tests have proved that the reticulation processes specific to toxins are degenerated, and a single functionalized nanoparticle crosslinks more macromolecules on ricin by contiguity.

The results demonstrate that the morphological and morphochemical structure of macromolecular biological structures play an important role in the evolution of reticulation processes, conditioning the mechanisms and their specificity, as well as the morphological and morphochemical structure, physicochemical properties and the stability of the resulted agglomeration domains. Morphochemical structure relations and the proposed functionalization mechanisms (fig.22) have been deduced by taking into account the obtaining conditions of the structure functionalities and the experimental data acquired during the chemical and biochemical reticulation investigation processes. Chemical and functional stability of the functionalized nanoparticle suspension, as well as the uniformity and the specificity of the biochemical reticulation processes, can be explained only if we accept the fact that the layer of functionalizing molecules, laid down on the coated layer, is continuous and uniform.

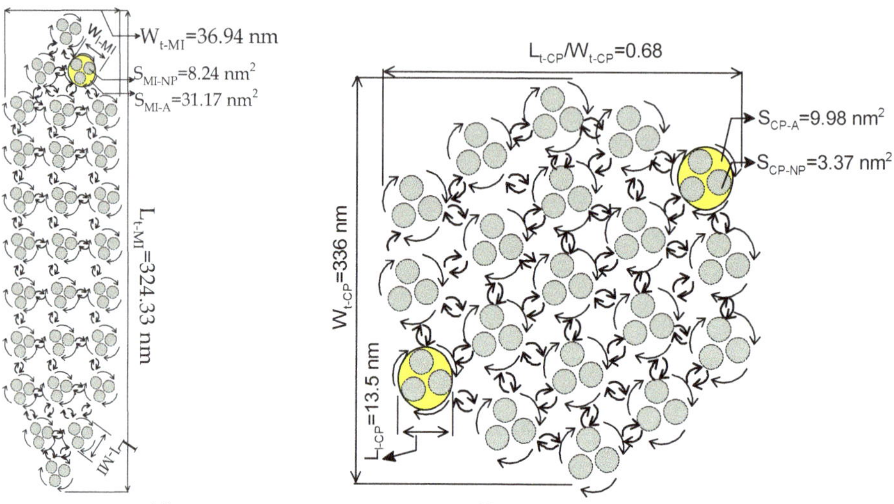

Fig. 21. The adopted elementary and bulk morphostructure domains for (a) $(ECH)_{n\beta}$-np-$(GL)_{n\varepsilon}$ ($Fe_3O_4 \rightarrow$MI) şi (b) $(ECH)_{n\beta}$-np-$(GL)_{n\varepsilon}$ ($Fe_3O_4 \rightarrow$CP) reticulated nanoparticles on ricin surface (W_t, W_l, L_t, L_l = geometrical dimension of elementary and bulk domains

The discontinuities on the functional structure of surfaces involve the existence of more different biochemical reticulation sites, favouring self-reticulation processes of nanoparticles in suspension. The acquired experimental data disprove the existence of masking, neutralisation or repeal processes between the CHO, CCl or CH(O)CH functional groups. Most likely, a stable chemical equilibrium is established between the CHO, CCl and CH(O)CH functional groups, as they are distributed on the surface of functionalized nanoparticles, according to a well determined morphochemical distribution map.

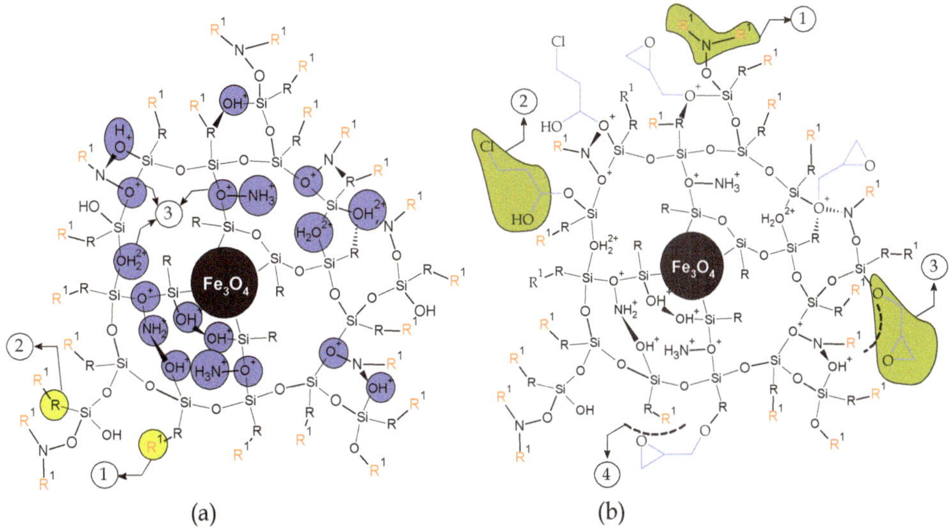

Fig. 22. Propesed morphochemical and morphofunctional structure for (a) npa-$(GL)_{n\varepsilon}$ and (b) $(ECH)_{n\beta}$-np-$(GL)_{n\varepsilon}$ nanoparticles ((a): 1 - GL; 2 - propylamine chemical residues; 3 - the most probable morphochemical distribution of chemical functionalities in the depth of coated layer; (b): 1 - CHO functionalities; 2 - Cl functionalities; 3 - epoxy functionalities; 4 - masked CHO functionalities)

7. Structure and functionality modelling of nanocomposites intended for ecological depollution

Research in the waste water and solid waste management usually encounters three main challenging problems: the huge quantities to be processed, the large varieties of contaminants and impurities, and the conversion of solid residuals coming from specific treatments into chemically and biologically inert materials for their safe disposal. Additionally, the useful compounds recycling are preferential design target for most of researches, but in many cases the safe disposal of inactivated residuals is prevailing. In spite of the many advances in waste water treatment, the problem of removal the organic pollutants is still pending around the elaborated technologies developed under the tertiary waste waters treatment concept. Thus, oxidative technologies are expensive and degradation products are still polluting (Gogate & Pandit, 2004). Flocculation and coagulation (Cheremisinoff, 2002) as well as ionic exchange technologies (Robinson & al., 2001), are limited in their efficiency for the removal of organic compounds. Chlorination is often use in the treatment of organic polluted waters, but its drawback is well known, when a large number of pollutants are converted into other compounds with the same toxicity. Microbiological technologies (Tedder & Pohland, 1990) and photocatalytic technologies (Hashimoto & al., 2005) are very selective ones and totally inappropriate for a broad spectrum of organic pollutants in the treated waste waters. Membrane materials (Cassano & al., 2001), particulate nanomaterials and functionalized hybrid nanomaterials as well as functionalized composite material adsorbents seems to be the best solutions for treatment of waste waters highly contaminated with a large number of organic compounds from different classes. Only the prices of these materials are restrictive for their application at larger scale. The main factors that restrict depollution methods and technologies of surface waters charged with organic pollutants are due to: a. a great number (spectrum) of organic pollutants (OP) contained by contamined waters; b. morphostructural and morphofunctional varieties of pollutants; c. dispersion mechanisms of pollutants in the environment; d. lack of controlled separation possibilities of pollutants from the source of pollution; e. lack of control upon degrading products, resulted during depollution processes; f. collecting, stocking, storage and management depollution waste.

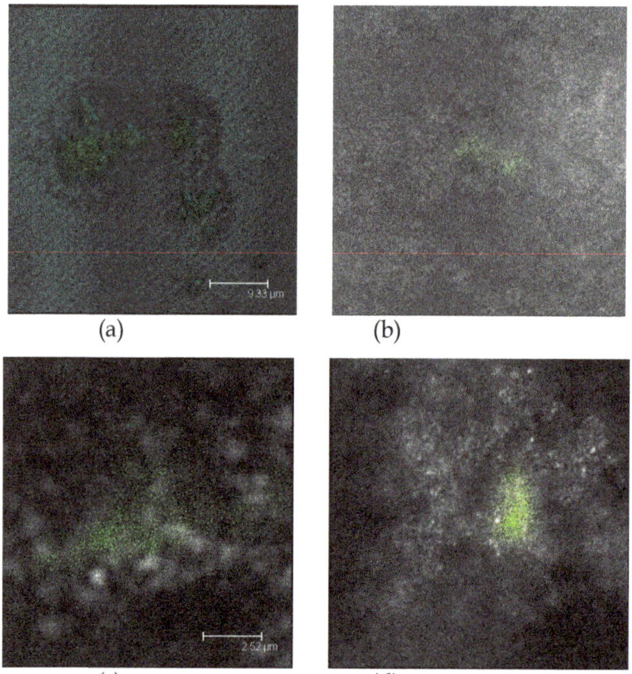

Fig. 23. Investigation images of $(ECH)_{n\beta}$-np-$(GL)_{n\varepsilon}$ and np-$(GL)_{n\varepsilon}$ biochemical reticulation potential on ricin (confocal investigation using a multifoton broadband Leica TCS SP 2 microscope) (a) np-$(GL)_{n\varepsilon}$ (Fe_3O_4 obtained by coprecipitation) (b) $(ECH)_{n\beta}$-np-$(GL)_{n\varepsilon}$ (Fe_3O_4 obtained by coprecipitation) (c) np-$(GL)_{n\varepsilon}$ (Fe_3O_4 obtained by reverse micelle) (d) $(ECH)_{n\beta}$-npa-$(GL)_{n\varepsilon}$ (Fe_3O_4 obtained by reverse micelle)

The spectrum and the morphostructural and morphofunctional variety (SVM) specific to the content of organic pollutants are due to the variety and diversity of the materials used in industrial and domestic activities. SVM limitation, control and determination is a complex issue, which cannot be solved by conventional procedures, as it is conditioned by the exigence, necessities and material needs of the society. From the point of view of the morphochemical type and variety, surface contaminated waters contain inorganic pollutants (salts and metal derivates, mineral acid salts, mixed composition suspensions and amorphous material structures, simple metallic oxides, mixed metallic oxides, etc.), carrier vectors of radioactive elements, organic pollutants (hydrocarbons, alcohols, phenols, dervivates of benzen, organic acids and their derivates, carbonylic compounds and their derivates, amines, organic halogens and their derviates, aromatic derivates, organometallic compounds, detergents, etc.), biological pollutans (bacteria, viruses) and biochemical pollutants (toxines, proteins, peptines, sugar, etc.) (Evanghelou, 1998). Most of the organic pollutants degradate in time, generating other polluting compounds that may present high toxicity. Not only the organic pollutants, but also their degradation products, may affect on long and medium term human health and the functionality of the environment.

The dispersion and the diffusion of pollutants is constrained by their physical and chemical properties. Once introduced in the external water circuit, pollutants become uncontrollably, randomly and irreversibly dispersed on wide surfaces. Due to the complexity of the above mentioned factors, the separation processes of the organic pollutants in surface waters and industrial contaminated flux water will be carried out incrementally, using complex technologies and solutions. In most cases, separation processes are only realised partially, as used tehnologies and materials are oriented towards the removal of certain pollutants or of some limited classes of organic pollutants.

Oxidic materials with oriented functional structure (M_{OF}) offer promising development and new implementing solutions for the development of new methods to control depollution of contaminated waters, due to the modelling possibilities of functional structure, and to extend the spectrum of separate pollutants.

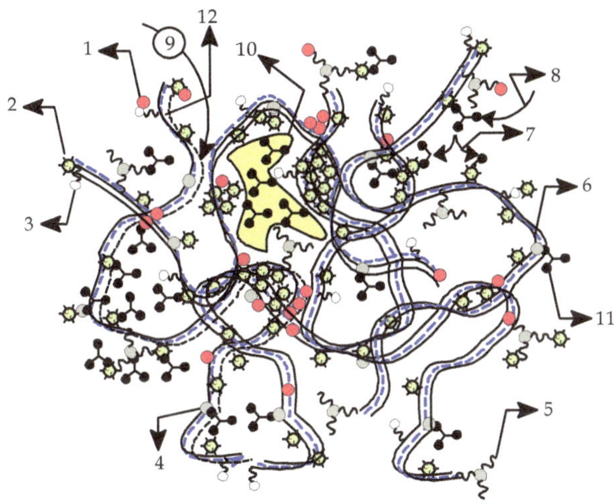

Fig. 24. General mechanisms for the retention, degradation and encapsulation of pollutants 1 - Su; 2 - functionalised magnetite; 3 - OH functional groups; 4 - amino-functionalities; 5 - epoxy-functionalities; 6 – fiberlike cellulose; 7- ways of pollutant degradation; 8 – retention of pollutants; 9 – paths for pollutant diffusion in the depth of composites; 10 – free space for encapsulation; 11 – pollutants; 12 – paths of pollutant host carrier in the depth of composite

M_{OF} separated pollutants spectrum depends directly on the type, polyvalency and stability of the obtained functional structure. Natural oxidic materials, with mixed nanostructures are the ideal candidates in order to obtain polyvalently functionalized materials, due to the complexity of the surface chemical structure, rich in reactive molecular fractions: -O-M, M-OH, H, -O-O-, hydrolizable molecular residue, electrically charged sites, embedded organic compounds, etc. Moreover, for the applications involving depollution, as well as for the obtaining and functionalization of M_{OF}, there is no need for special chemical and technological conditions, as their usability is constrained by the enrichment and the stabilization of the functionality structure. The stability of this type of materials involves the morphochemical and morphofunctional stability, in order to preserve the morphostructural, morphochemical and morphofunctional characteristics, as well as to prevent the degradation of the support material, carrier for the reticulated pollutants.

In most depollution situations, the chemical and morphostructural stabilization of the functionalized composites is done under water, so that the depolluting applications are conditioned by the possibility of the composite to reticulate the pollutants in free chemical conditions (fig.5÷10). According to the above proposed morphostructural and morphofunctional modelling mechanisms, a polyvalently functionalized composite has been obtained. This composite integrates in its structure a mixture of natural oxides (Su), functionalized with BrCN (~100 grams, about 10 grams of np-$(GL)_{n\varepsilon}$ nanoparticles and 100 grams of cellulose acetate functionalized with epichlorhidrine ((7)). The chemical composition of Su was determined by Wavelength Dispersive X-ray Fluorescence (WDXRF), and it was found that Su contains the following main fractions: SiO_2 (70.13%), Al_2O_3 (14.33%), Fe_2O_3 (4.36%), K_2O (2.78%), Na_2O (0.46%), MgO (0.87%), P_2O_5 (0.26%), TiO_2 (0.58%), CaO (0.87%).

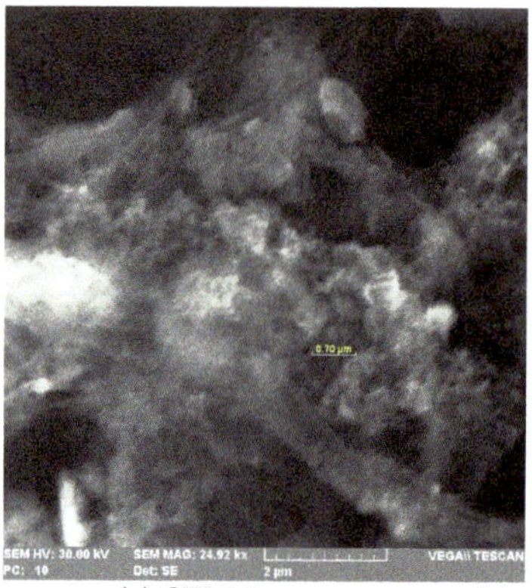

(a) SEM image

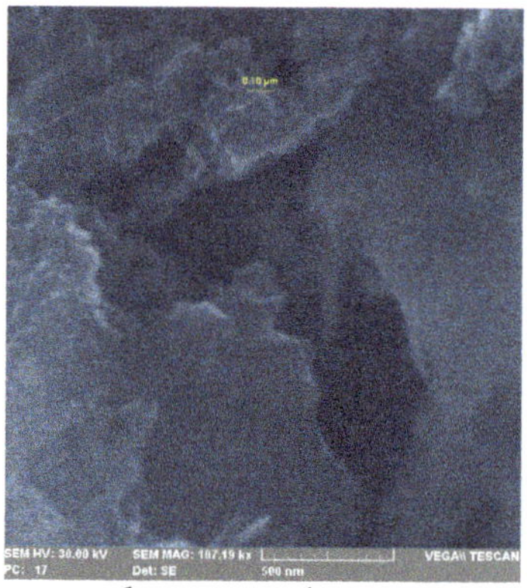

(b) elementar morphostructural segmentation

(c) EDX image

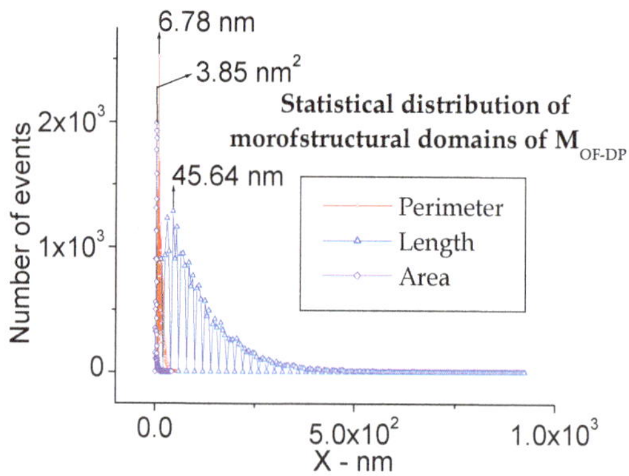

(d) morphostructure parameters of M_{OF-DP}

Fig. 25. Morphological and morphochemical structure of M_{OF-DP}

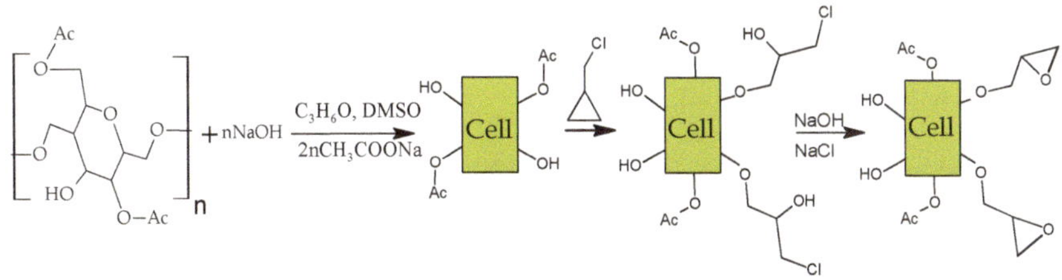

(7)

The composite presents a nano- fiberlike character (fig.24, fig.25a), due to the nano- and microstructural character of the water regenerated cellulose. Functionalized cellulose acetate performs the reticulation, compositing and spatial distribution functions of the Su and functionalized magnetite, as structural components. The structural elements of the composite are spatially distributed, on the surface and in the depth of cellulose microfibers, so that they lead to the formation of nano- and micro-membranary structures. Using the elementary segmentation method (fig.25b), the mean distribution values of the elementary morphostructural parameters have been determined: perimeter (6.78 nm), lenght (17 ÷ 924 nm), area (3.85 nm^2) (fig.25d). The morphostructural character of the nanocomposite allows that the contaminated water may be filtered, at molecular level, from a mechanic, chemical, bichemical and biological point of view, according to the reticulation, degradation, and encapsulation mechanisms proposed in figure 24. It was noticed that the composite allows the development of the saprophyte bacteria on its surface (fig.26), and points out the stable character of the material functionality and the biodegradability of its composite structure. Moreover, the biodegradable nature of the composite is also given by the nature of its composite structure, as its structural elements can be assimilated without inducing toxic effects in relation to the structure and functionality of the environment.

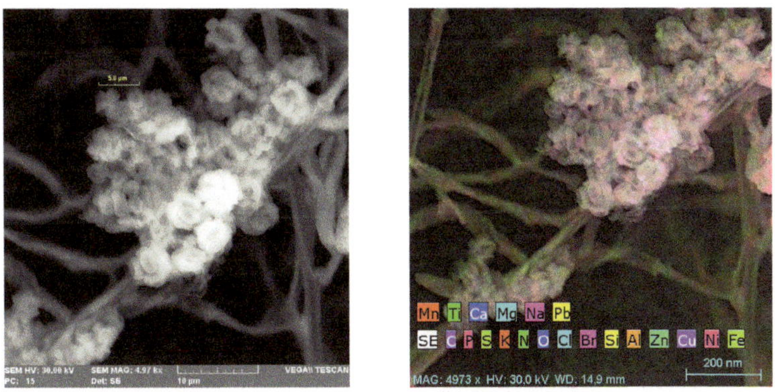

(a) SEM image (b) EDX image

Fig. 26. Growth and develop of saprophite microorganisms on surface of $M_{OF\text{-}DP}$

According to the proposed functional mechanisms and to the experimental data acquired by FTIR, the resulted functional structure of the composite derives from the functionalisation of its structural elements with BrCN, glutaraldehyde, and epichlorhydrine, respectively. FTIR acquired data reveal that the composite contains four main classes of functional groups (fig.27): amines and compounds with nitrogen, metal oxides, metal-organic and mixed functional structures. Amines and compounds with nitrogen functionalities are owed to the functionalization processes of surface oxide components with BrCN. Most likely, in the first stage BrCN hydrolyses, leading to the formation of HCN, which attacks metal ions, anions and electric charged microsites.

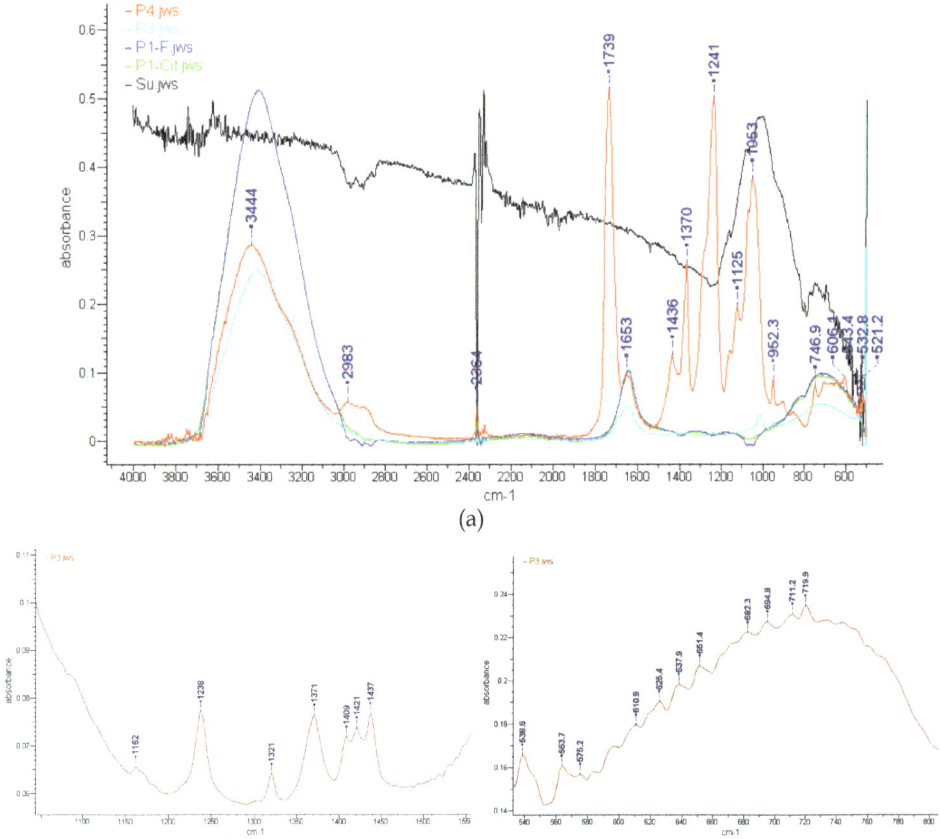

Fig. 27. IR spectra of composite and their structural components
(a) IR spectra of: citrate passivated Fe_3O_4 nanoparticle (P1-Cit), Fe_3O_4-n[$SiO_{1.5\gamma}$-$(CH_2)_3(NH_2)$]$(NH_2)_{n\delta}$-$(GL)_{n\varepsilon}$ (P1-F), functionalised cellulose (P4), Su and $M_{OF\text{-}DP}$ (P3) (b) 1000-1550 cm^{-1} IR spectra of composite (c) 500-800 cm^{-1} IR spectra of composite

The CN^- anion interaction with surface oxides leads to the formation of a wide variety of functional groups, such as: $NX_mY_n \rightarrow H$ (hydrogen bonds), $N^{+i}X_m \rightarrow OH^{+n}{}_j$ ((i = 0,1, j = 1÷3, n = 0÷2), water-hydrogen bonds), $NH^+{}_3$, CN, C=NH, =C-NX_mY_n, NX_mY_n, $C^{m+}N^{n+}X_o$ (n+m=5; X, Y = H, -OH, halogens, other metalic ions). In return, aminated functionalities interact with metal ions in the structure of oxidic components, resulting new types of functional mixed structures, such as: (FeOFe, $FBrO_2$, $OClF_4$) (506 cm^{-1}), (Fe-O_2) (503 cm^{-1}), peroxides (920-750 cm^{-1}), O=CClF (501 cm^{-1}), O=$SiCl_2$ (501.1 cm^{-1}), C-Cl (501.9 cm^{-1}), (ClHBr, $P_2O^{6-}{}_4$)(508 cm^{-1}), (FNCl, $Al(OH)^{-4}$, CdO, ClNNN, VF_5,) (720 cm^{-1}), (($OReI_4)^-$, $(Si_2O_7)^{6-}$) (508 cm^{-1}), (HCN)(721 cm^{-1}), (LiNC)(722.9 cm^{-1}) (Nakamoto, 2009, 2009). Most likely, the shape of the absorption band 950-550 cm^{-1} is due to the significant contribution of the absorbing oxide forms, of the following types: MO^{-4}: $(SiO_4)^{4-}$ (819, 596 cm^{-1}), $(PO_4)^{3-}$ (938, 567 cm^{-1}), $(SeO_4)^{2-}$ (822, 856 cm^{-1}), $(ClO_4)^{-}$ (928 cm^{-1}), $(BrO_4)^{-}$ (878, 801 cm^{-1}), $(TiO_4)^{4-}$ (770, 761 cm^{-1}), $(ZrO_4)^{4-}$ (846, 792 cm^{-1}), $(HfO_4)^{4-}$ (800, 796 cm^{-1}), $(VO_4)^{4-}$ (818, 780 cm^{-1}), $(VO_4)^{3-}$ (826, 804 cm^{-1}), $(CrO_4)^{2-}$ (863, 833 cm^{-1}), $(CrO_4)^{3-}$ (844, 775.8 cm^{-1}), $(MoO_4)^{2-}$ (897, 837 cm^{-1}), $(WO_4)^{2-}$ (931, 838 cm^{-1}), $(SiO_4)^{4-}$ (819, 596 cm^{-1}), $(MnO_4)^{2-}$ (902, 834, 820, 821 cm^{-1}), $(MnO_4)^{3-}$ (789, 778 cm^{-1}), $(TcO_4)^{-}$ (921 cm^{-1}), $(ReO_4)^{-}$ (971, 920 cm^{-1}), $(FeO_4)^{2-}$ (832, 790 cm^{-1}), $(RuO_4)^{-}$ (845, 830 cm^{-1}), $(RuO_4)^{2-}$ (840, 804 cm^{-1}), OsO_4 (965.2, 960.1 cm^{-1}), $(CoO_4)^{4-}$ (670, 633 cm^{-1}) (Nakamoto, 2009). The absorption peaks placed at 1645, 1710, 3335 cm-1, can be assigned to C = O functional groups, which are characteristic for polyurethane polymeric structures, resultated from the reticulation of -OH with aminated residues, placed on the surface of oxides. The absence of the absorbtion peaks of C=O, sugest that glutaraldehide and functionalized Fe3O4 nanoparticle were entirely reticulated by the functionalities containing nitrogen during material composition processes. Chemical reticulation processes have favoured the compositing process, by embedment of functionalized nanoparticles of magnetite and Su on surface and in the deep structure of cellulose.

8. Conclusions

This chapter presents the main physicochemical and functional mechanisms substantiating the understanding and proper modelling of the process of obtaining composite materials with functional oriented structure. These mechanisms were substantiated on the theoretical and experimental data obtained by the team of authors and their collaborators inside applicative projects for the sensitising and detection of biochemical structure and for the separation of organic pollutants in contaminated waters.

The chapter proposes a new analytical method with invariant metrics for sensitization and biochemical monitoring, based on using magnetically stimulable functionalized nanostructures, able to prelevate a great number of biochemical analytes, undifferentiatedly and undestructively. Biochemical invariance sensitization-detection is provided by imposing analytical and practical constraints at the level of analytical acquisition and processing systems, according to the configuration of the magnetic field of the discriminator, molecular weight, specific biochemical fluorescence spectrum. We have presented the analytical structure and the elements that model the process for obtaining magnetically stimulated functionalised composites, orientated towards the detection and undestructive biological reticulation.

The second part of this chapter presents the modelling and the obtaining of new polyvalently functionalized types of materials oriented to the reticulation, encapsulation and separation of organic pollutants in contaminated waters. The obtained material has a strong, biodegradable character, being modelled morphostructurally and morphofunctionally, so as to separate by reticulation a large number of organic pollutants, toxins and microorganisms. The chapter describes the main physicochemical mechanisms that model the conditions of separation, degradation and encapsulation of pollutants.

9. Acknowledgment

The authors are grateful for the financial support granted by the CNMP (National Research, Development and Innovation Plan; Programme 4) in projects 31-001/2007, 81-002/2007 and

32-165/2008. We would also like to thank the Scientifical Research Center for CBRN Defence and Ecology for the support we benefited from. Also, for their contributions to the obtaining of results, we are indebted to Dionesie Bojin, Lucia Mutihac, Viorel Ordeanu and Andrada Pretorian. For contributions to text editing and translating we are grateful to Loredana Iordache.

10. References

Lucas, I. T.; Durand-Vidal, S.; Dubois, E.; J. Chevalet & P. Turq. (2007). Surface charge density of maghemite nanoparticles: Role of electrostatics in the proton exchange. *Journal of Physical Chemistry C*, Vol.111, No.50, pp. 18568–18576, ISSN 1932-7447

Campos, A. F. C.; Tourinho, F. A.; T. Cotta A. P. G. & J. Depeyrot. (2002). A theoretical approach of succimer complexation at the surface of ferrite nanoparticles in order to improve the synthesis of biocompatible magnetic fluids. *European Cells and Materials*, Vol. 3, Suppl. 2, pp. 151-153, ISSN 1473-2262

Massart, R.; Roger, J. & Cabuil, V. (June 1995). New Trends in Chemistry of Magnetic Colloids: Polar and Non Polar Magnetic Fluids, Emulsions, Capsules and Vesicles. *Brazilian Journal of Physics*, Vol. 25, No. 2, pp.135-141, ISSN 0103-9733

Jal, P. K.; Patel, S. & Mishra, B. K. (2004). Chemical modification of silica surface by immobilization of functional groups for extractive concentration of metal ions. *Talanta*, No. 62, pp.1005-1028, ISSN 0039-9140

Chechik, V.; Crooks, R. M. C. & Stirling, J. M. (2000). Reactions and Reactivity in Self-Assembled Monolayers, *Adv. Mater.*. Vol. 12, No. 16, pp.1161-1171, ISSN 0035-9648.

Iordache, P. Z.; Şomoghi, V.; Petrea, N.; Petre, R.; Dionezie, B.; Ordeanu, V.; Hotăranu, A. & Mutihac, L. (2009). The obtaining of nanocomposite slurries with practical aplication in real time biological fixing. *J. Optoelectron. Adv. M.*, Vol. 11, No. 5, pp. 736-743, ISSN 1454-4164

Tolstoy, V. P.; Chernyshova, I. V. & Skryshevsky, V. A. (2003). *Handbook of Infrared Spectroscopy of Ultrathin Films*, Wiley & Sons, ISBN 0-471-35404-x (alk. paper), Hoboken, New Jersey

Denmark, S. E. & Sweis R. F. (2002). Cross-Coupling Reactions of Organosilicon Compounds: New Concepts and Recent Advances, *Chem. Pharm. Bull.*, Vol. 50, No. 12, pp.1531-1541, ISSN 0009-2363

Festag, G.; Klenz, U.; Henkel, T.; Csa´ki, A.& Fritzsche, W. (2005). *Biofunctionalization of Metallic Nanoparticles and Microarrays for Biomolecular Detection, In: Biofunctionalisation of Nanomaterials*, Challa S. S. R. Kumar, (Ed.), 150-182, Wiley WCH, ISBN 3-527-31381-8, Weinheim

Murcia M. J. & Naumann C. A. (2005). *Biofunctionalization of Fluorescent Nanoparticles, In: Biofunctionalisation of Nanomaterials*, Challa S. S. R. Kumar, (Ed.), 1-40, Wiley WCH, ISBN 3-527-31381-8, Weinheim

Bekyarova, E.; Haddon, R. C. & Parpura V. (2005). *Biofunctionalization of Carbon Nanotubes, In: Biofunctionalisation of Nanomaterials*, Challa S. S. R. Kumar, (Ed.), 41-71, Wiley WCH, ISBN 3-527-31381-8, Weinheim

Gao, Y. (2005). *Biofunctionalization of Magnetic Nanoparticles, In: Biofunctionalisation of Nanomaterials*, Challa S. S. R. Kumar, (Ed.), 72-98, Wiley WCH, ISBN 3-527-31381-8, Weinheim

Zheng, M & Huang, X. (2005). *Biofunctionalization of Gold Nanoparticles, In: Biofunctionalisation of Nanomaterials,* Challa S. S. R. Kumar, (Ed.), 99-124, Wiley WCH, ISBN 3-527-31381-8, Weinheim

Watanabe, J; Park, J.; Ito, T.; Takai M. & Ishihara K. (2005). *Biofunctionalization of Phospholipid Polymer Nanoparticles, In: Biofunctionalisation of Nanomaterials,* Challa S. S. R. Kumar, (Ed.), 125-149, Wiley WCH, ISBN 3-527-31381-8, Weinheim

Meziani, M. J.; Lin, Y. & Sun, Y. P. (2005). *Conjugation of Nanomaterials with Proteins, In: Biofunctionalisation of Nanomaterials,* Challa S. S. R. Kumar, (Ed.), 183-234, Wiley WCH, ISBN 3-527-31381-8, Weinheim

Le´vy R. & Doty R. C. (2005). *Stabilization and Functionalization of Metallic Nanoparticles: the Peptide, In: Biofunctionalisation of Nanomaterials,* Challa S. S. R. Kumar, (Ed.), 235-269, Wiley WCH, ISBN 3-527-31381-8, Weinheim

Hattori Y. & Maitani, Y. (2005). *Folate-linked Lipid-based Nanoparticles for Tumor-targeted Gene Therapy: the Peptide, In: Biofunctionalisation of Nanomaterials,* Challa S. S. R. Kumar, (Ed.), 270-298, Wiley WCH, ISBN 3-527-31381-8, Weinheim

Lellouche, J. P. (2005). *Magnetic Core Conducting Polymer Shell Nanocomposites for DNA Attachment and Hybridization, In: Biofunctionalisation of Nanomaterials,* Challa S. S. R. Kumar, (Ed.), 299-329, Wiley WCH, ISBN 3-527-31381-8, Weinheim

Kommareddy, S.; Shenoy, D. B. & Amiji, M. M. (2005). *Gelatin Nanoparticles and Their Biofunctionalization, In: Biofunctionalisation of Nanomaterials,* Challa S. S. R. Kumar, (Ed.), 330-352, Wiley WCH, ISBN 3-527-31381-8, Weinheim

Bhunia, A. K. (2008). Biosensors and Bio-Based Methods for the Separation and Detection of Foodborne Pathogens. *Advances in Food and Nutrition Research,* Vol.54, pp. 1–44, ISSN 1043-4526

Jasson, V.; Jacxsens, L.; Luning, P.; Rajkovic, A. & Uyttendaele M. (2010). Alternative microbial methods: An overview and selection criteria. *Food Microbiology,* No.27, pp. 710–730, ISSN 0740-0020

Iordache, P. Z.; Şomoghi, V.; Savu, I.; Petrea, N.; Mitru, G.; Petre, R.; Dionezie, B.; Ordeanu, V.; Hotăranu, A. & Mutihac, L. (2009). Biochemical Fixing and Activation Mechanisms of Amino Groups Deposited on the Surface and in the Depth of Polymeric Layers of the n[$SiO_{1.5}$-$(CH_2)_3NH_2$] Type Deposited on the Fe_3O_4 Nanoparticle Surface. *Revista de chimie - Materiale Plastice,* Vol. 46, No. 2, pp. 162–168, ISSN 0034-7744

Petrea, N.; Iordache, P. Z.; Şomoghi, V.; Savu, I.; Mureşan, M.; Petre, R.; Rece, L.; Lungu, R.; Pretorian, A.; Mitru, G.; Dionezie, B.; Savu, B.; Mutihac, L.; Kim, L. & Ordeanu, V. (2009). The Obtaining of High Sensitivity and Specificity Transducer Media for Sensors Specialized in Biological Toxin Detection, *Digest Journal of Nanomaterials and Biostructures,* Vol. 4, No. 4, pp.699-712, ISSN 1842-3582

Iordache, P. Z.; Lungu, R. M.; Epure, G.; Mureşan, M.; Petre, R.; Petrea, N.; Pretorian, A.; Dionezie, B.; Mutihac, L. & Ordeanu, V. (2011). The determination of the nanostructurated materials' morphology, by applying the statistics of the structural element maps, *Journal of Optoelectronics and Advanced Materials,* Vol. 13, No. 5, pp. 550-559, ISSN 1454-4164

Gogate, P.R.; Pandit AB.; (2004). A review of imperative technologies for wastewater treatment I: oxidation technologies at ambient conditions, *Advances in Environmental Research,* Vol. 8, No. 3-4, pp. 501-551, ISSN 1093-0191

Cheremisinoff, N.P. (2002). *Handbook of water and wastewater treatment technologies*, Butterworth-Heinemann, ISBN 0-7506-7498-9, Boston

Robinson, T.; McMullan, G.; Marchant, R. & Nigam, P. (2001). Remediation of dyes in textile effluent: a critical review on current treatment technologies with a proposed alternative, *Bioresource Technology*, Vol. 77, No. 3, pp. 247-255, ISSN 0960-8524

Tedder, D.W. & Pohland, F.G. (1990). *Emerging Technologies in Hazardous Waste Management*, ACS Symposium Series, ISBN 0-8412-1747-5, Washington, DC

Hashimoto, K.; Irie, H. & Fujishima A. (2005). TiO_2 Photocatalysis: A Historical Overview and Future Prospects, *Japanese Journal of Applied Physics*, Vol. 44, No. 12, pp. 8269–8285, ISSN 0021-4922

Cassano, A.; Molinari, R.; Romano, M. & Drioli, E. (2001). Treatment of aqueous effluents of the leather industry by membrane processes: A review, *Journal of Membrane Science*, Vol. 181, No. 1, pp. 111-126, ISSN 0376-7388

Evanghelou, V., P. (1998). *Environmental soil and water chemistry - Principles and applications*, John Wiley & Sons, ISBN 0-471-16515-8, Hoboken, New York

Nakamoto, K. (2009). *Infrared and Raman Spectra of Inorganic and Coordination Compounds - Part A: Theory and Applications in Inorganic Chemistry*, 5rd ed., Wiley, ISBN 978-0-471-74339-2, Hoboken, New Jersey

Nakamoto, K. (2009). *Infrared and Raman Spectra of Inorganic and Coordination Compounds - Part B: Applications in Coordination, Organometallic, and Bioinorganic Chemistry*, 5rd ed., Wiley, ISBN 978-0-471-74493-1, Hoboken, New Jersey

5

Synthesis of Carbon Nanomaterials Using High-Voltage Electric Discharge Techniques

A. D. Rud, N. I. Kuskova, L. I. Ivaschuk,
L. Z. Boguslavskii and A. E. Perekos
G. V. Kurdyumov Institute for Metal Physics of NASU,
Institute of Pulse Research and Engineering of NASU,
Ukraine

1. Introduction

New spatial forms of carbon - fullerenes, nanotubes, graphene, etc., attract significant interest since the time of their discovery due to their unique physicochemical and mechanical properties (Afanas'ev et al., 2001; Gogotsi, 2006; Guozhong Cao, 2004; Eletskii & Smirnov, 1995; Shenderova et al., 2002). A lot of investigations have been carried out in this field for the recent years. Hence, the problems of the development of effective synthesis, separation and purification methods for carbon nanomaterials (CNM) remain importance. Therefore, it is of special interest to clarify the possibility of application of electric discharge techniques, such as the electric wires explosion (EWE) (Kuskova, 2005; Kuskova et al., 2010; Rud et al., 2007, 2011), the spark erosion of materials (Bulgakov et al., 2009; Rud et al., 2007), developed for manufacturing metallic nanopowders, and the electric breakdown of dielectric liquids (EBOL) (Rud et al., 2011) for synthesis of CNM and investigation of structural and physical properties of synthesized materials.

The physical basis of electric discharge technology of synthesis of different forms of CNM consists in the injection into working medium– a source of carbon, energy need for its heating, evaporation and destruction through a passage of powerful (up to 1 MA) current pulses with frequencies of 0.1-10 Hz. As a result, structural and phase transformations of carbon or destruction of molecules of organic liquids on individual fragments take place with their subsequent ultrafast cooling and synthesis of different types of CNM. Thus, it is possible to control effectively the structural and phase state of the synthesized CNM by the followings:

- variation of the injected into working medium energy by changing the energy deposited in the capacitor bank and the number of current pulses;
- selection of the working medium - a source of carbon (using graphite conductors of different geometry or hydrocarbons with different chemical nature).

Electric explosions of graphite conductors were performed in different cooling media (ethanol, toluene, hexane) in the interval of deposited energies of capacitor bank W_0 from 0.1 to 45 kJ and the inductance of the discharge circuit L from 1 to 50 μH. Charging voltage U_0 and capacitance C of the capacitor bank was varied from 10 to 50 kV and from 1 to 36 μF, respectively. The electrical circuit of the experimental setup is shown in Fig. 1. The use of graphite conductors with different diameters and lengths enables to vary the value of injected specific energy into the substance (per unit mass of graphite) from 1 to 300 kJ/g. There are two kinds of electric explosion (EE) modes which depend on the value of the specific energy w injected into the conductor: low-energy if $w<w_s$, and high-energy if $w> w_s$, where w_s is graphite sublimation specific energy (62 kJ/g).

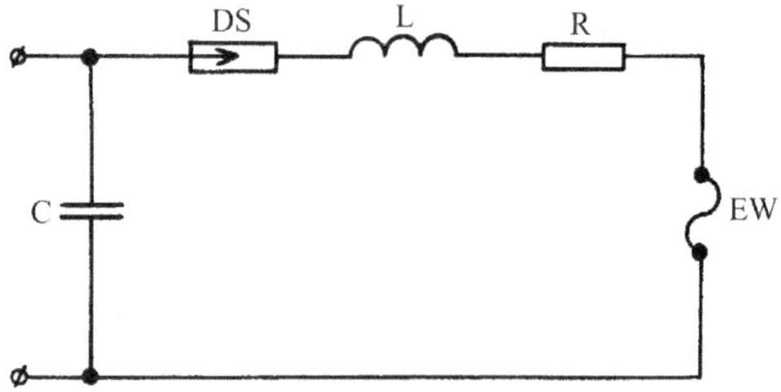

Fig. 1. RLC- electrical circuit with an exploding conductor:
C - capacitor bank,
DS - discharge switch,
L - inductance,
R - resistance,
EW - exploding wire (conductor).

The method of electrical breakdown of organic liquids has been implemented on the electrical equipment (Fig. 2). The electric-discharge reactor is filled out with liquids, differ by degree of hybridization of carbon atoms in the molecule, but with the same number of carbon atoms contained in the ring:

- benzene C_6H_6 - belongs to arenes with sp^2-hybridisation of carbon atoms in planar ring molecule;
- cyclohexane C_6H_{12} - belongs to alkanes with sp^3-hybridisation of carbon atoms in nonplanar ring molecule.

The indicated organic liquids are non-polar dielectrics with a zero dipole moment (Speight, 2004).

The working liquid with colloidal solution of carbon nanoparticles is decanted from the reactor and centrifugated during 0.3-2 hours after series of breakdowns (up to 50,000 pulses). The produced material is exsiccated at the sparing temperatures (up to 500 K) with a purpose to form dry powder.

X-Ray data were collected using diffractometer HZG-4 with filtered Cu Kα or Co Kα radiations and standard powder diffractometer with monochromatic Mo Kα radiation (primer pirographite monochromator). Electron microscopy investigations were performed on the high resolution microscopes JEOL JEM-2100F and LEO SUPRA 50VP. Raman spectroscopy analysis was conducted at room temperature on Horiba Jobin-Yvon T64000 spectrometer (λ=514 nm), mass spectra - on the high resolution mass spectrometer MX-1320 with field ion source. The magnetic properties were measured by means of the ballistic method in the magnetic field ranged from 0 to 800 kA/m at temperature interval of 77 to 673 K.

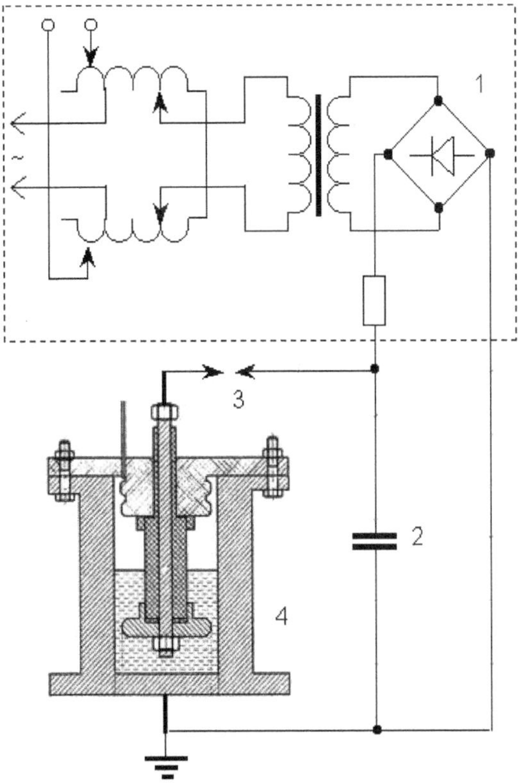

Fig. 2. Block-scheme of the high-voltage electric-discharge facility and reactor:
1 - charging unit,
2 – pulse energy storage system (capacitor bank),
3 - discharge switch ,
4 - discharge reactor with organic liquid.

2. Phase composition of carbon nanomaterials produced by the method of electrical explosion of graphite conductors

2.1 Physical aspects of synthesis of new carbon allotropes by an electric explosion technique

The electrical wires explosion is one of the most powerful techniques for pulse action on a substance that results of rapid transformation of the electric field energy into internal energy of conductor matter. It is a wide used technology for production of metallic nanopowders. Exploding material is subjected to the combined impact of high temperature and pressure ($\sim10^5$ K and $\sim5\cdot10^2$ MPa, respectively), electrical and magnetic fields. It was proposed (Kuskova, 2005; Kuskova et al., 2010; Rud et al., 2007, 2011) to utilize EWE technique for synthesis of carbon nanomaterials in consequence of structural transformations graphite → diamond, graphite → nanotubes or graphite → fullerenes due to heating of graphite rods by powerful current pulse.

The analytical time profiles T(t) and P(t) and phase trajectories at carbon phase diagram should be calculated for different electroexplosion modes of graphite rods to predict the phase compositions of the obtained products. Let us consider the behavior of conductor (graphite rod) at the EE process. It starts from homogeneous heating of solid conductor to melt point. A specific electric conductivity of material changes during melting process and results the change of current flowing through the conductor. Liquid conductor continues to heat holding its shape because the time of the progress of hydrodynamic instabilities exceeds a task time scale ($t > 10^{-8}$ c). A pressure inside of conductor has magnetic and gas-kinetic components.

The system of equations, that simulates the process of heating of a solid cylindrical conductor that expands or liquid conductor which does not undergo phase transformations in an electric circuit with capacity C and inductance L, has the following form (Kuskova, 1997):

$$\frac{\partial \rho}{\partial t}+\frac{1}{r}\frac{\partial(\rho r v)}{\partial r}=0 \qquad (1)$$

$$\rho\left(\frac{\partial v}{\partial t}+v\frac{\partial v}{\partial r}\right)=-\frac{\partial P}{\partial r}-\frac{1}{2\mu r^2}\frac{\partial\left(r^2 B_\varphi^2\right)}{\partial r}, \qquad (2)$$

$$\rho\left(\frac{\partial \varepsilon}{\partial t}+v\frac{\partial \varepsilon}{\partial r}\right)=-P\frac{1}{r}\frac{\partial(rv)}{\partial r}+\frac{j^2}{\sigma} \qquad (3)$$

$$\frac{\partial B_\varphi}{\partial t}+\frac{\partial(vB_\varphi)}{\partial r}=\frac{1}{\mu\sigma}\frac{\partial}{\partial r}\left(\frac{1}{r}\frac{\partial(rB_\varphi)}{\partial r}\right) \qquad (4)$$

$$\frac{d^2(LI)}{dt^2}+\frac{d(RI)}{dt}+\frac{I}{C}=0 \qquad (5)$$

$$\rho=\rho_0\left[1-\alpha(T-T_0)\right] \qquad (6)$$

$$\sigma=\frac{\sigma_0}{1+\beta(T-T_0)}\left(\frac{\rho}{\rho_0}\right)^{\gamma} \qquad (7)$$

where ρ - density, P - pressure, ε - specific internal energy, T - temperature, α - thermal expansion coefficient, ρ_0 - density of conductor's matter at the initial temperature T_0, σ_0 - specific electroconductivity of conductor's matter at the temperature T_0, β - temperature coefficient of electroconductivity, γ - coefficient, μ - magnetic permeability, B_φ - magnetic induction component.

The obtained solutions for the system of equations (1) - (7) of solid and liquid phases allow (after exclusion of the time from dependences P(t) and T(t)) to find the phase trajectory of matter in the process of homogenous heating of conductor by powerful current pulse (P, T - terms):

$$P(T) \approx \frac{\mu}{4\pi^2 a_0^{\ 2}} \left(\frac{m c_0 U_0 (T - T_0)}{R_0 L} \right)^{2/3} \left(1 - \frac{r^2}{a_0^{\ 2}} \right) \text{ for } T < T_m \tag{8}$$

$$P^0 \approx \frac{k_m^2 \mu}{4\pi^2 a_0^{\ 2}} \left(\frac{m c_0 U_0 (T_m - T_0)}{R_0 L} \right)^{2/3} \left(1 - \frac{r^2}{a_0^{\ 2}} \right) \text{ for } T = T_m \tag{9}$$

$$P(T) \approx P^0 \approx \text{const for } T_m < T < T_b, \tag{10}$$

where k_m - coefficient, $1<k_m<3$; a_0 - initial radius of conductor; m –mass of conductor; c_0 – specific heat capacity at the initial temperature T_0; R_0 – resistance at the T_0; T_m – melting temperature; T_b – boiling temperature; P^0 – maximum pressure.

Transition to specific characteristics of the matter enables to obtain the following expression of the maximum magnetic pressure:

$$P^0 = \frac{k_m^2 \mu}{4} \left(\frac{\sigma_0 \rho_0 c_0 U_0 (T_m - T_0) a_0}{\pi L} \right)^{2/3} \left(1 - \frac{r^2}{a_0^{\ 2}} \right) \text{for } T_m<T<T_b \tag{11}$$

Further heating of liquid conductor by current pulse leads to its evaporation in the form of explosion. A secondary discharge initiation in the electroexplosion products results the formation of the plasma channel and shock waves generation. Fast expansion of gas-vapor cavity, condensation of electroexplosion products and synthesis of nanosized particles of initial material or new materials takes place after current termination depending on physicochemical properties of the medium and time-energy parameters of electroexplosion. Therefore, the electrical parameters of the experimental apparatus (Fig.1) and explosion conductor required to achieve P, T-conditions of phase transformation can be calculated using the expressions (8)-(11). It is possible to control an explosion process through these parameters and properties of environment media in reactor to produce nanosized particles in different structural state.

2.2 Phase trajectories of carbon in process of electroexplosion

At the initial stage of the fast (~1-5 μs endurance) elecroexplosion, magnetic pressure in a conductor exceed gas-kinetic one considerably and rise their top values. For so called low-energy modes of EE (injected into explosion material energy is less than sublimation energy), the value of the maximum pressure in centre of graphite conductor, calculated by expressions (8)-(11) or experimentally measured, does not exceed 20 MPa. According to the phase diagram of graphite, liquid phase does not appear at these P, T conditions but only the partial sublimation of graphite in the form of carbon clusters or graphene sheets occurs. Hence, a low-energy mode of EE with a current break because of the formation of the nonconductive gas-vapor carbon cave should be realized for synthesis fullerenes or carbon nanotubes. The mentioned synthesis occurs after breakdown of hot ionized gas of carbon clusters.

For high-energy modes of EE (injected into explosion material energy essentially exceed sublimation energy), the value of the maximum pressure in the centre of the conductor is about 10 GPa that is possible to expense gas-kinetic or reactive pressure. The reactive pressure comes from the evaporation of the wave spreading from the surface into the conductor body and shock waves spreading from the discharge channel at the secondary breakdown in an electroexplosion process. Condensation of liquid carbon at these P, T conditions can result in diamond synthesis.

The phase trajectories of carbon at phase diagrams for different kinds of graphite conductor electroexplosion (low- and high-energy) can be calculated using expressions (8)-(11) and estimating thermodynamic conditions in gas-vapor cave (Fig.3). The phase trajectories for central layers of the graphite rod of the high-energy electroexplosion experiments correspond to phase transition graphite → diamond and they are plotted in the Fig. 3 (curves 1, 2). The trajectory 1 matches to immediate phase transition graphite → diamond, what is impossible at conditions this work owing to deficit of capacitor deposited energy injected into graphite rod. The curve 2 corresponds to conditions of the high-energy mode of these experiments when phase trajectory of carbon can be described as sequence of graphite→liquid carbon→diamond transitions. For comparison, the phase trajectory of carbon for the low-energy mode, when electric breakdown of carbon gas realizes in EE process without melting of graphite conductor, is plotted in the Fig. 3 (curve 3).

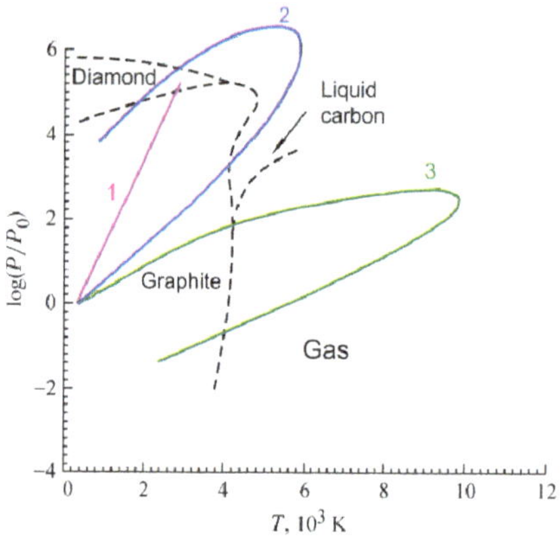

Fig. 3. Carbon phase diagram (dashes - lines of phase equilibrium) and phase trajectories of carbon at graphite electroexplosion (1,2 - transition graphite → diamond, 3 - graphite → fullerenes).

It can be seen from Fig. 3 that extreme conditions occuring in a process of low- or high-energy modes of graphite electroexplosion match to diapason of thermodynamic parameters required for high-probability spontaneous formation of new phases of carbon. The synthesized phases at these conditions can possess amorphous, nano- or microcrystalline structure. A size of individual particles strongly depends on parameters of EE process. The produced nanostructures are susceptible to the relaxation in the case of quenching conditions of nonobservance.

2.3 Technological conditions of nanocarbon allotropes synthesis during graphite electroexplosion

It makes possible to find a correlation between electrical parameters of experimental apparatus, size and properties of an exploded graphite rod and required thermodynamic parameters for synthesis of new allotropes of carbon. The mechanism for synthesis of diamond and fullerenes is considered below in order to clarify the conditions of their manufacture by EWE technique.

The first stage of the synthesis of fullerenes and nanotubes in EE process is a destruction of graphite layered structure on separated graphene sheets containing various defects including broken covalent bonds between carbon atoms which results from fast Joule heating. It leads to the formation of pentagonal cycles in the structure of the graphene sheet required for its folding in fullerene-like cage. Therefore, the injected into graphite conductor specific energy w should be several times lower than sublimation energy of graphite w_S. The conditions of partial sublimation of graphite to the graphene sheets and clusters from thermodynamic point of view ($P < 10^7$ Pa, $T < 5 \cdot 10^3$ K, $(dP/dT)_1 < 2 \cdot 10^3$ Pa/K) can be realized at a low-energy mode of EE. At the second stage of EE, at the conditions of $P \approx 10^3$ Pa and $T \approx 2 \cdot 10^3$ K in the fast expanded gas-vapor cave, an assemblage of various clusters into fullerene molecule or graphite sheet folding into nanotube occurs.

The synthesis of diamond proceeds in conditions of a high-energy mode of EE and it associates with two stages. The first stage is fast compression of graphite due to pinch-effect (the compression of a plasma filament by the magnetic forces), melting the central part of the conductor and the formation of liquid carbon ($P > 10^{10}$ Pa, $T > 5 \cdot 10^3$ K, $(dP/dT)_2 > 2 \cdot 10^6$ Pa/K). Injected energy W into a graphite rod should be greater than total sublimation energy of the whole sample W_S. At the second stage, a fast cooling results in crystallization of liquid carbon in diamond.

Given limitations (from above and from below) for the rate of the change of the pressure dP/dT in the central part of the conductor during EE process allow to find a correlation between the phase composition of the CNM synthesized, velocity of the current increase and current density in the conductor. A range at the phase diagram where the exploding substance falls is characterized by both the temperature increased owing to Joule heating and magnetic pressure. It can be estimated for the low-expanding cylinder conductor from the equation (Kuskova, 2005):

$$\frac{\partial P}{\partial r} = -\frac{1}{2\mu r^2}\frac{\partial}{\partial r}(r^2 B_\varphi{}^2) \tag{12}$$

For homogeneous heating of a conductor the magnetic induction $B\varphi$ is:

$$B_\varphi = \frac{\mu I}{2\pi a^2} r = \frac{\mu j r}{2} \tag{13}$$

where I - current, a - conductor radius, j - current density, μ - magnetic permeability.

The solution of equation (12) is of the form:

$$P(r) = P(a) + \frac{1}{4}\mu j^2 (a^2 - r^2) \tag{14}$$

where P(a) – pressure at the conductor surface.

The rates of change of top pressure (in the centre of conductor at r=a) and temperature in a solid cylinder conductor can be evaluated from the expressions:

$$\frac{\partial P}{\partial t} \approx \frac{1}{2} a^2 \mu j \frac{\partial j}{\partial t} \approx a^2 \mu \frac{j^2}{2\tau}, \tag{15}$$

$$\frac{\partial T}{\partial t} \approx \frac{j^2}{\sigma\rho c} \tag{16}$$

where τ - current rise time to the top magnitude, σ – specific conductivity, c – specific heat capacity, ρ – density.
From expressions (15) and (16) we get

$$\frac{\partial P}{\partial T} = \frac{\mu\sigma\rho c}{2\pi j}\frac{dI}{dt} \tag{17}$$

Based on the pointed out limitations for $(dP/dT)_{1,2}$ and the expression (17), the relations between current density and its rate of the rise need to synthesis of diamond (18) and fullerenes (19) can be written as:

$$j < k_1\sigma\rho c\mu(dI/dt) \tag{18}$$

$$j > k_2\sigma\rho c\mu(dI/dt) \tag{19}$$

where $k_{1,2} = \left[2\pi(dP/dT)_{1,2}\right]^{-1}$.
As the synthesis of diamond from graphite phase is stipulated by high pressure, the rate of the current rise in the conductor is a crucial controlling factor as it is evident from (18). When the current density increases (in a thin conductor), the intensity of heating of the conductor rises as well that results in the sublimation of graphite at a low pressure. Substituting the graphite characteristic values into (18) and (19), the following conditions are obtained for synthesis of:

$$\text{diamond} - j < k_3(dI/dt), \text{ where } k_3 = 10^{-1}\ s/m^2 \tag{20}$$

$$\text{fullerenes} - j > k_4(dI/dt), \text{ where } k_4 = 10^{2}\ s/m^2 \tag{21}$$

From expressions (18) and (19) approximate relations between the conductor radius a and the capacity of capacitor batteries C can be obtained. Using the sinusoidal dependence of current, which is valid for beginning of phase transformation, allows to find j, dI/dt and their ratio:

$$\frac{j}{dI/dt} = \frac{\sqrt{LC}}{\pi a^2}\,tg\frac{t}{\sqrt{LC}} \tag{22}$$

Then the synthesis conditions of new carbon phases which are similar to (20) and (21) can be given as:

$$\text{for diamond - } \sqrt{LC} < k_3\pi a^2 ctg(t/\sqrt{LC}) \tag{23}$$

$$\text{for fullerenes - } \sqrt{LC} > k_4\pi a^2 ctg(t/\sqrt{LC}) \tag{24}$$

As modes of energy injection are usually chosen such that the phase transformations (melting or sublimation) occur in the interval of $\pi/4 < t/\sqrt{LC} < \pi/3$, then

$ctg(t/\sqrt{LC}) \approx 1$. Taking into account a ratio of injected energy w into the graphite conductor and its sublimation one w_S, the synthesis conditions can be received:

$$\text{for diamond - } \sqrt{LC} < k_3 \pi a^2 \text{, } w > w_S \quad (25)$$

$$\text{for fullerenes - } \sqrt{LC} < k_4 \pi a^2 \text{, } w < w_S \quad (26)$$

Using expressions (25) and (26) enables to forecast the phase composition of synthesized carbon nanomaterials without oscillographic testing of electroexplosion process.

2.4 Structure state and magnetic properties of carbon nanomaterials produced by the method of electrical explosion of graphite conductors

It was theoretically shown in 2.1-2.3 that extreme conditions arising at the electroexplosion of the graphite conductor conform to the thermodynamic parameters of the new carbon allotropes formation. The interaction of powerful current pulses with solid matter occurs at two stages. The first stage is heating, melting and evaporation of the matter in the equilibrium state. The second stage is nonequilibrium processes of ionization and condensation of EE products. Synthesis of new carbon forms can happen at the both stages. To verify the expressions (25) and (26) obtained in the part 2.3, the EE experiments of high-energy and low-energy modes were performed at the experimental setup (Fig. 1) with the oscillographic records of current I and voltage U pulses (Fig. 4).

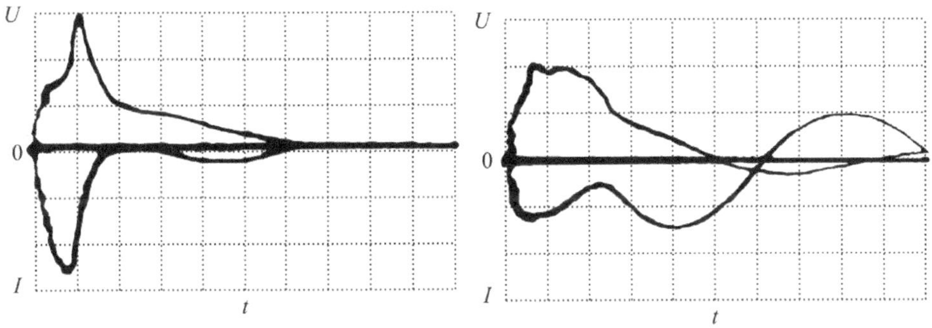

(a) (b)

Fig. 4. Typical oscillograms of voltage (top) and current (down) for different modes of electrical explosion of graphite conductors:
a) U_0 = 30.5 kV, a = 0.15 mm; m_t=2 μs/cell, m_u= 10 kV/cell, m_i = 2.2 kA/cell,
b) U_0 = 25 kV, a = 1 mm; m_t=2 μs/cell, m_u=10 kV/cell, m_i = 8.7 kA/cell.

The typical oscillogram for a low-energy mode of graphite electroexplosion with the secondary breakdown of evaporated carbon gas after current pause is shown in Fig. 4a. The high-energy EE mode with melting of the central part of the graphite rod and subsequent electrical breakdown of the evaporated surface layers is given at Fig. 4b. The values of dI/dt and j are estimated at the point of the phase transitions when current has a maximum value.

The experimental results of the synthesis products of graphite electroexplosion are presented in Fig. 5-8. The synthesis of a diamond-like phase occurs during electrical explosion of graphite at a high-energy mode realized in hexane with the following parameters found from the oscillogram (Fig. 4b): the injected specific energy w into the conductor >100 kJ/g; the current density j is about 10^9 A/m^2 and the rate of the current rise dI/dt is about 10^{10} A/s, and it satisfies to (20). The typical XRD pattern is shown in Fig. 5. It is clearly seen, that the phase composition of the produced CNM consists of the ordinary graphite and the cubic diamond (space group Fd3m). Further decreasing injected energy to w≈62 kJ/g leads to synthesis of a mixture of ordinary graphite and lonsdalite (hexagonal modification of diamond with space group P63/mmc).

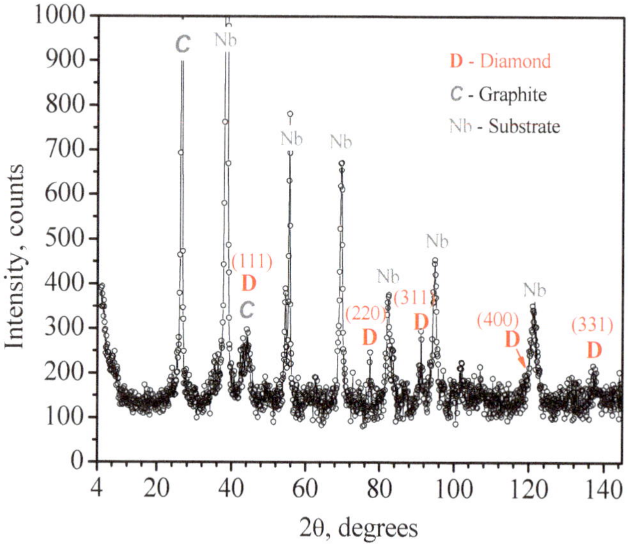

Fig. 5. X-Ray diagram of product of electrical explosion of graphite rod at w≈100 kJ/g in hexane (Cu Kα radiation).

According to the condition (26), the further decrease of specific energy injected into the graphite conductor (realization of the low-energy mode, Fig. 4a) enables to reach the thermodynamic parameters which are necessary for the synthesis of fullerenes and carbon nanotubes during EE process (P~10^3 Pa, T~ $2{\cdot}10^3$ K). The building of fullerene molecules from carbon cluster or folding of graphene sheets in the nanotubes occurs in the second phase of the electrical explosion in a fast expanding carbon evaporated cave and it depends on the size of the cluster (a higher w reflects to a smaller cluster size).

An electron microscopy investigation of CNM produced at low-energy parameters (8 kJ/g < w < 10 kJ/g) shows the presence of nanographite and carbon nanotubes (Fig. 6). Carbon nanotubes are aggregated at bunches with the diameters up to 20 nm and the length up to 5μm. It is important to note that the separate nanotubes have composite structure and hierarchical organization. In the products of electric explosion of graphite in the hexane, fragments of graphene sheets are present (Fig. 6b).

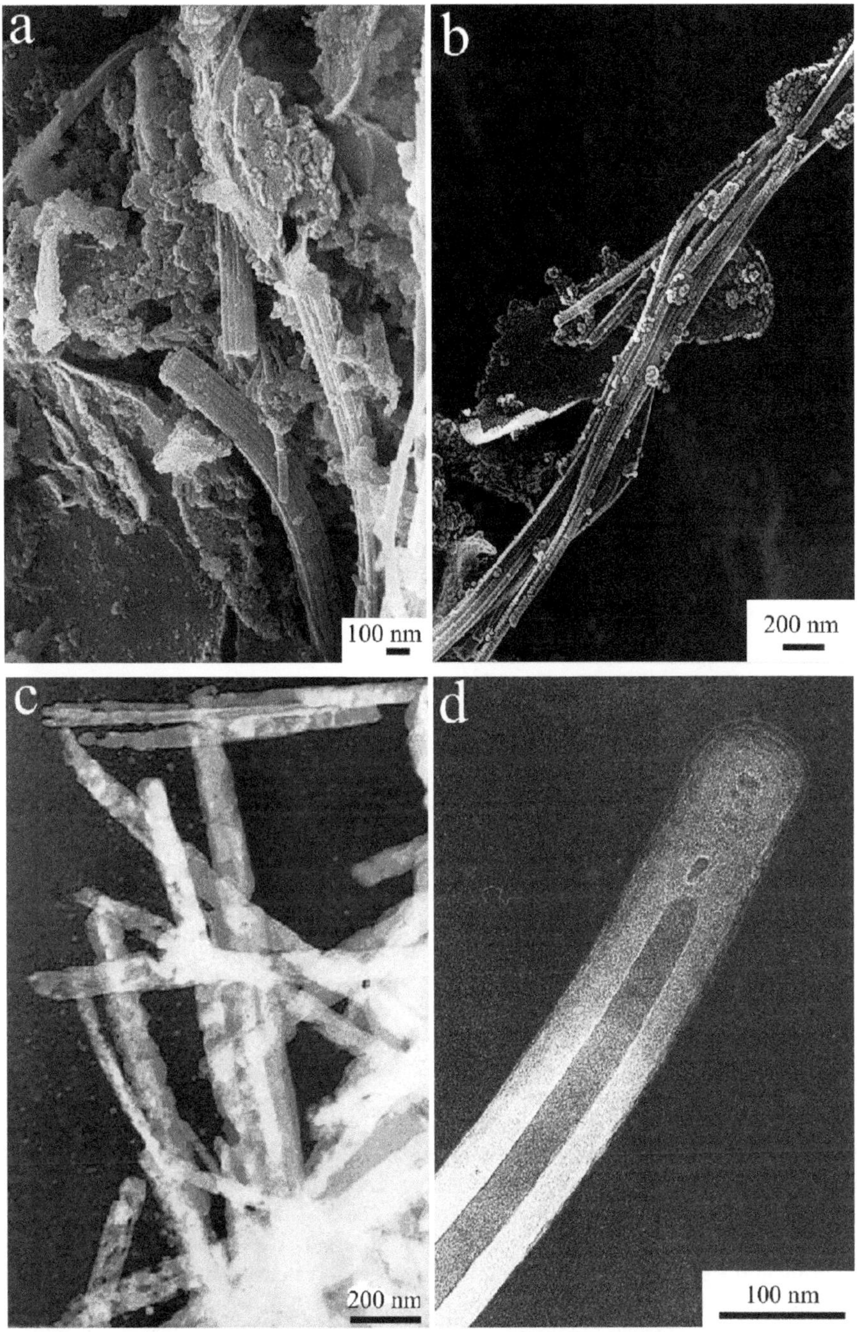

Fig. 6. Electron microscope images of the products of electrical explosion of graphite in hexane (a, b) and toluene (b, c) with injected energy 8 kJ/g <w< 10 kJ/g.

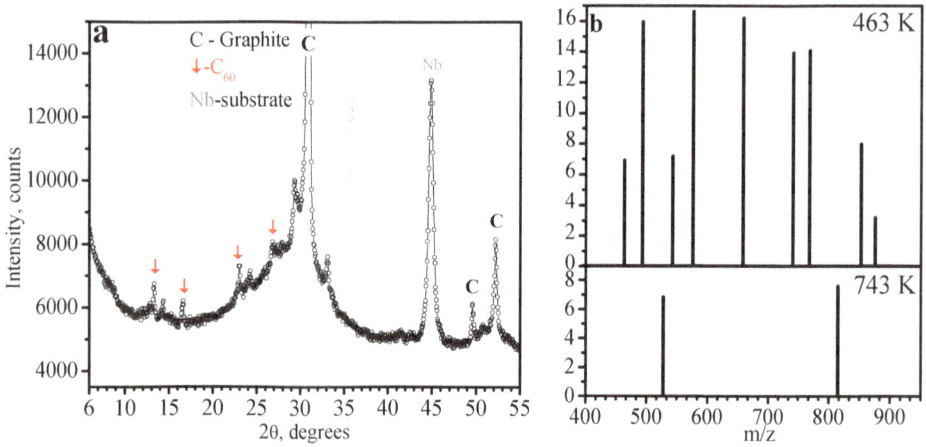

Fig. 7. Products of electrical explosion of graphite in ethanol (a) and toluene (b) with injected energy 10 kJ/g <w <20 kJ/g:
a) fragment of the XRD diagram, Co Kα radiation,
б) mass spectra at different temperatures of emitter.

It was found that CNM synthesized by electric explosion of graphite have strong ferromagnetic properties, typical to ferromagnetic materials (type of nickel). The value of specific saturation magnetization is ~57 A·m^2/kg at the temperature of liquid nitrogen and 20 - 30 A·m^2/kg at room temperature (Fig. 8a). The Curie temperature is 425 K (Fig. 8b). According to the chemical analysis, CNM have less than 2 wt.% (0.5 at.%) impurity of iron due to the erosion of the steel electrodes and the body of reactor during electroexplosion. This amount of iron can provide not more than 10% of the value of specific saturation magnetization which demonstrate CNM. In addition, the Curie temperature of iron T_c is 1043 K. Therefore, we can say with reasonable confidence that the magnetic properties of CNM are associated with ferromagnetic condition of carbon atoms but not iron impurities. It should be pointed out that the value of the specific saturation magnetization of this powder decreases in 2 order after 3 year storage on the air.

The information about existence of magnetic ordering in CNM, but with a much smaller degree of specific magnetization (on a 2 orders), have recently appeared in literature (Makarova & Palacio, 2006). At the present moment, there is no general theory that would explain the magnetic properties of CNM. At the extreme conditions of electrical explosion under impact of high temperatures and pressures, the specific electronic state of carbon atoms can result in a ferromagnetic state appearing.

In recent years a number of reports of observation of the magnetic ordering in different allotropic forms of carbon, such as in the fullerenes, fullerene hydrides, carbon nanotubes, nanocrystalline graphite and graphene (Makarova, 2004); Belavin et al. ,2004; Alexandre, 2008; Yazyev, 2008) are presented. Several possible mechanisms explaining this phenomenon are considering at the present: exchange magnetism perfect crystalline structure containing sp^2-sp^3 hybridized atoms; induced magnetism; the magnetism caused by structural imperfections in the various forms of carbon (Makarova & Palacio, 2006; Ugeda, 2010). It is considered that under extreme conditions (such as high temperature and pressure, irradiation by fast particles, laser heating) special types of ordering of carbon atoms or defects in structure of CNM, favorable for the formation of the spontaneous ferromagnetic state, can appear. Such extreme conditions are typical for electroexplosion synthesis methods which can lead to a high degree of defectiveness of the crystal structure and the formation of specific electronic states of the synthesized products that make them ferromagnetic.

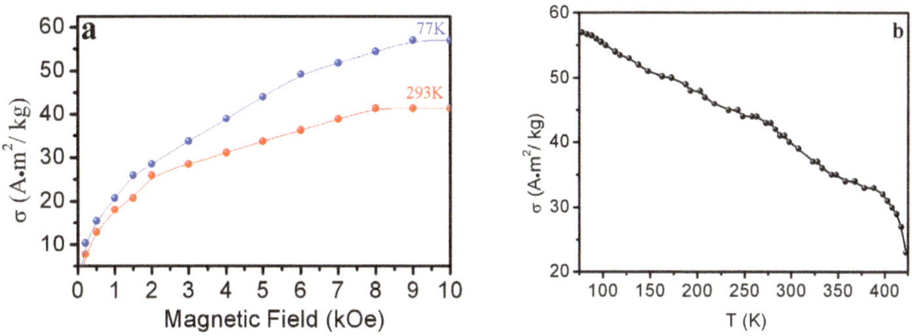

Fig. 8. The dependence of specific magnetization (σ) of the graphite explosion products in toluene: a) vs applied magnetic field, b) vs temperature (saturation magnetic field H=10 kOe).

3. Structure of amorphous carbon synthesized by an electric breakdown of organic liquids technology

Impact of powerful current pulses on organic liquids results in initiation of a zone of high temperatures and pressures, in which a destruction of C-H bonds in hydrocarbon molecules occurs with the formation of nanocarbon in solid state (Rud et al., 2011). The electric discharge treatment of hydrocarbons was conducted using standard high-voltage equipment (Fig. 2) in oscillation regime to realize an electrical breakdown of liquid dielectrics (the injected energies in the range 0.1 - 2 MJ/l). The benzene and cyclohexane with the equal amount of carbon atoms and a different degree of their hybridization in the molecule were chosen as working media. The typical oscillograms of voltage and current in a discharge circuit are plotted at the Fig. 9.

The investigations of the products of EBOL synthesis completed by X-Ray diffractometry and Raman spectroscopy testify that they are typical amorphous carbon (Robertson, 2002; Boukalov et al. 2006; Casiraghi, 2005). The Raman spectra of the amorphous carbon produced from different types of organic liquids are characterized by intensive broad G- and D- bands. The standard positions of D (1350 cm^{-1})- and G (1580 cm^{-1}) -bands for graphite are indicated in Fig. 10. It is well known, that AC is characterized by very strong structural disorder. This structural parameter can be calculated from the ratio of intensities of D-and G-bands - I_D/I_G. It is clearly seen that the value of I_D/I_G is very large for the both powders that proves a strong structural disorder (Table 1). Tuinstra and Koenig (Tuinstra & Koening, 1970; Robertson, 2002) have shown that the I_D/I_G value is inversely proportional to the block size L_a along the direction *a*:

$$\frac{I_D}{I_G} = \frac{C(\lambda)}{L_a} \tag{27}$$

where C(λ) is a wavelength-dependent pre-factor.

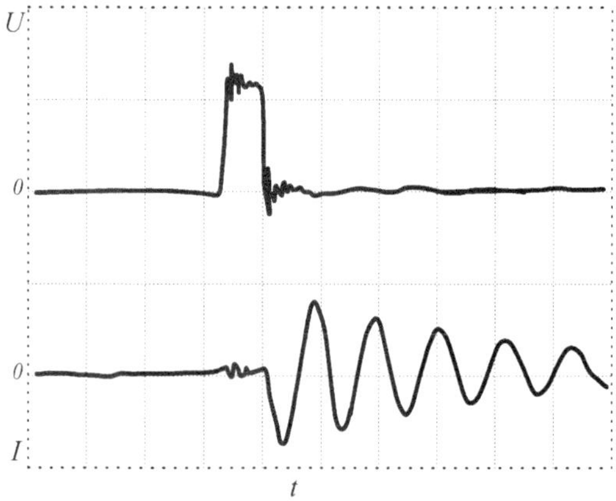

Fig. 9. Typical oscillograms of voltage (top) and current (bottom).
m_t=5 μs/cell, m_U= 25 kV/cell, m_I= 9.52 kA/cell.

The sizes of blocks L_a calculated from (27) are presented in Table 1. It is clearly seen that AC is characterized by a small blocks size, which practically does not depend on the type of working liquid.

Working liquid	G-band		D-band		I_D/I_G	L_a, Å
	Center, cm^{-1}	FWHM, cm^{-1}	Center, cm^{-1}	FWHM, cm^{-1}		
C_6H_6	1595	75	1353	101	1	43
C_6H_{12}	1585	69	1348	59	0.96	45

Table 1. Data from Raman spectra of amorphous carbon synthesized in hydrocarbon liquids with different chemical nature (Center - the position of maximum, FWHM - full width at half height, I_D/I_G - the ratio of the integrated intensities of D-and G-bands, L_a - the size of the particles along the graphene plane)

The two additive broad bands at 1236 cm^{-1} and 1489 cm^{-1} are present in the Raman spectrum of the AC synthesized in cyclohexane (Fig.10) that indicates the presence of sp^2-sp^3 bonds and CH groups (Speight, 2004) in the material. There are only low-intensity bands at 1000 - 1100 cm^{-1} corresponding to the C-H vibrations in an aromatic ring in the spectrum of the AC produced by electrical discharge treatment of benzene (Fig.10). It is associated with the use of benzene as a working medium for synthesis (Robertson, 2002; Speight, 2004). The data of Raman spectroscopy indicate the amorphous structure of the synthesized powders.

The traditional method of the structural analysis of amorphous materials is the method of radial distribution function (RDF) analysis (Elliot, 1984; Egami & Billinge, 2003; Petkov, 1989). It is based on the relationship between the function of radial distribution of the atomic density ρ(r) and the intensity of coherent X-ray scattering I_{ce} which is expressed in the form of the integral equations:

$$I_{ce} = NF^2 + NF^2 \int_0^{\infty} 4\pi r^2 \left[\rho(r) - \rho_0\right] \frac{\sin(sr)}{sr} dr \quad (28)$$

$$4\pi r^2\rho(r) = 4\pi r^2\rho_0 + \frac{2r}{\pi}\int_0^\infty s\left[i(s)-1\right]\sin(sr)ds \tag{29}$$

where N - is the number of the scattering atoms, F^2 - the atomic factor, $i(s)=I_{ce}/F^2$ the structure factor, I_{ce} - the coherent-scattered intensity in electron units, s - the diffraction vector, 2θ - the scattering angle, $4\pi r^2\rho(r)$ - the RDF function of atom density.

Calculated structure factors (SF) and RDF of AC synthesized by the EBOL technology in benzene and cyclohexane are shown on the Fig. 11. It is seen that AC produced in the benzene possesses clearly defined graphite-like type of short range order. The position and intensity of the second peak on SF for AC produced in cyclohexane (Fig. 11, left, bottom) testify to a possible presence of diamond-like type of short range order in its atomic structure.

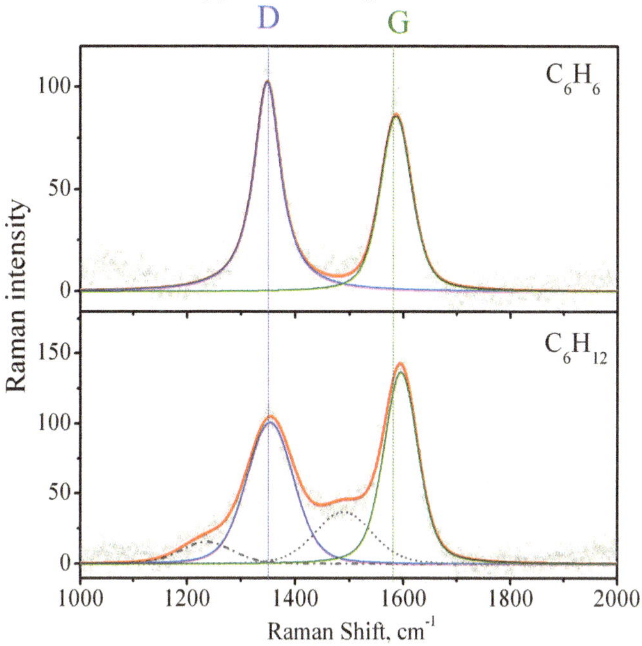

Fig. 10. Raman spectra (λ = 514 nm) of the products of electrical breakdown of benzene (top) and cyclohexane (down). Standard position for D (1350 cm^{-1}) and G (1581 cm $^{-1}$) bands of graphite are indicated by dotted lines.

The electron microscopic studies were carried out for more detailed characterization of produced materials. The products of electrical discharge processing of different hydrocarbon liquids are agglomerated powders with an extended surface (S_{BET} ~ 150 m^2/g) (Fig. 12a). High-resolution electron microscopy has shown that the individual particles have complex morphology which depends on the chemical nature of working liquid, that is, a carbon source. Thus, the micrographs of the AC, produced by electrical discharge treatment of benzene, clearly show a layered structure with a characteristic distance of ~ 0.35 nm between the individual layers (Fig. 12b). However, in the case of cyclohexane, the morphology of the particles is more complicated. Fig. 12c shows visible core up to 5 nm in size surrounded by a shell of about 5 layers with an interlayer distance of ~ 0.36 -0. 37 nm. This "core-shell" structure is typical onion-like carbon (Mykhaylyk et al., 2005; Shenderova et al., 2002; Shenderova & Gruen, 2006;). It is worthy of note that separated onions are united into agglomerates with external common graphite-like shell.

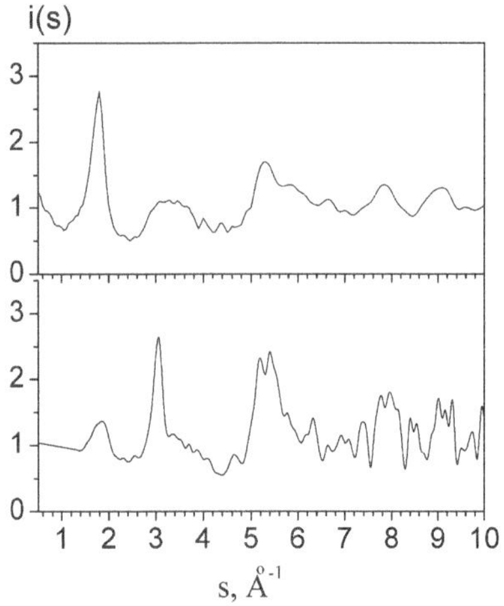

Fig. 11. SF (left) and RDF (right) of AC synthesized by the EBOL technology in benzene (top) and cyclohexane (bottom).

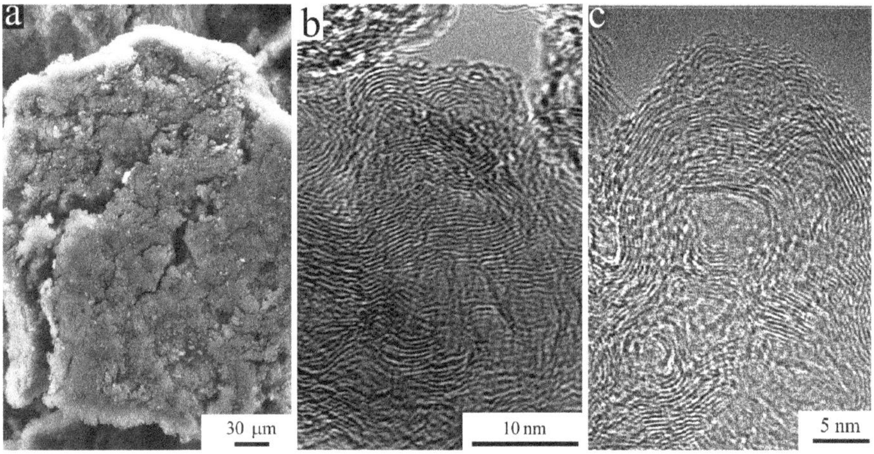

Fig. 12.Typical micrographs of products of electric discharge treatment of hydrocarbon liquids: (a) particle-agglomerate with a developed surface;
(b,c) morphology of particles produced in benzene (b) and cyclohexane (c).

4. Conclusion

The new electric discharge technologies using the methods of electrical wire explosion of graphite conductors and electric breakdown of organic liquids are developed to produce carbon nanomaterials, containing fullerene-like clusters of the C_{60}-C_{70}, nanotubes, nanodiamonds and amorphous carbon. The phase composition of CNM obtained could be effectively controlled by variation of the energy parameters of the synthesis process and chemical nature of the surrounding medium.

A model of formation of different phases of CNM during electroexplosion of graphite is proposed. Carbon nanotubes and fullerenes are formed from fragments of graphene, obtained from explosive fractured graphite, if the injected specific energy into graphite is substantially smaller than the graphite sublimation energy. Nanodiamonds are formed by phase transformation graphite-liquid-diamond, if the injected specific energy into sample is substantially greater than the graphite sublimation energy.

It was discovered that the synthesis products have strong ferromagnetic properties. The magnetization value of the synthesized carbon nanomaterials (about the 57 $A \cdot m^2/kg$) amounts the value close to that of pure nickel.

The structure of amorphous carbon synthesized by electric breakdown of organic liquids technique directly depends from chemical nature of the working medium - source of carbon. The fact of onion-like carbon synthesis by electric discharge treatment of cyclohexane is established. This technology enables to produce onion-like carbon in a large scale at normal conditions.

5. Acknowledgment

This work was partially supported by the joint projects of NASU-STCU (# 4951).

6. References

Afanas'ev D.V., Baranov G.A., Belyaev A.A., Dyuzhev G.A. & Zinchenko A.K. (2001). Producing fullerenes by evaporation of graphite stationary CO_2-laser. *Technical Physics Letters*. Vol. 27. No10, (May 2001), C. 31 - 36, ISSN: 1063-7850, (in Russ.)

Alexandre S.S., Mazzoni M.S. & Chacham H. (2008). Edge States and magnetism in carbon nanotubes with line defects. *Physical Review Letters*. Vol. 100, No. 14, (April 2008), pp. 146801 [4 pages], ISSN: 0031-9007

Belavin V.V., Bulusheva L.G., Okotrub A.V., & Makarova T.L. (2004). Magnetic ordering in C60 polymers with partially broken intermolecular bonds. *Physical Review B*. Vol. 70, (October 2004), pp.155402 (5 pages), ISSN:1098-0121

Boukalov S.S., Mihaltsyn L.A., Zubavichus Y.V., Leytets L.A. & Novikov Y.N. (2006). Investigation of the structure of graphite and other sp2 hydrocarbon materials by the methods of micro-Raman spectroscopy and X-ray diffraction. *Rossijskij himicheskij zhurnal*, Vol.1. No 1, pp. 83-91. ISSN 0373-0247, (in Russ.)

Bulgakov A.V., Bulgakova N.M., Burakov I.M. et al. (2009). *Nanosized Material Synthesis by Action of High-Power Energy Fluxes on Matter*, Kutateladze Institute of Thermophysics, ISBN: 978-5-89017-010-1, Novosibirsk, (in Russ.).

Casiraghi C., Ferrari A. C. & Robertson J. (2005). Raman spectroscopy of hydrogenated amorphous carbon. *Physical Review B*, Vol. 72. No.8, (August 2005), pp. 085401 [14 pages]. ISSN: 1098-0121

Egami T. & Billinge S.J.L. (Eds.). (2003). *Underneath the Bragg Peaks*. Pergamon, ISBN: 080426980, UK

Eletskii A.V. & Smirnov B.M. (1995). Fullerenes and carbon structures. *Physics-Uspekhi*, Vol. 38, (September, 1995), pp. 935-964, ISSN: 0320 - 0116

Elliot S.R. (1984). *Physics of amorphous materials*, Longman, ISBN 0-582-44636-8, USA.

Gogotsi Y. (Ed.). (2006). *Nanomaterials Handbook*, Taylor & Francis Group, ISBN: 0-8493-2308-8, Boca Raton

Guozhong Cao. (2004). *Nanostructures & Nanomaterials: Synthesis, Properties & Applications*, Imperial College Press, ISBN: 1860944159, London

Kuskova N.I., Tkachenko S.I., Koval S.V. (1997). Investigation of liquid metallic wire heating dynamics. *Journal of Physics: Condensed Matter*. Vol. 9, No. 29, (September 1996), pp. 6175 - 6184, ISSN: 0953-8984

Kuskova N.I. (2005). Phase Transformations of Carbon Heated at High-Power Current Pulse. *Technical Physics Letters*. Vol. 31, No 17, (February 2005), pp. 28-32. ISSN 1063-7842

Kuskova N.I., Rud A.D., Baklar V.U. & Ivaschuk L.I. (2010). Physical aspects of the formation of various allotropic forms of nano-sized carbon in the process of electrical explosion. *Technical Physics*. Vol. 80, No. 9, (December 2009), pp. 57 - 62, ISSN: 0044-4642

Makarova T.L. (2004). Magnetic properties of carbon nanostructures. The Overview. *Fizika i Tekhnika Poluprovodnikov*. Vol 38, No. 6, (December 2003), pp. 641 - 664, ISSN: 0015-3222 (in Russ.)

Makarova T. & Palacio F. (Eds.). (2006). *Carbon Based Magnetism: An Overview of the Magnetism of Metal Free Carbon-based Compounds and Materials*, Elsevier Science, ISBN: 0444519475

Mykhaylyk O.O., Solonin Y.M., Batchelder D.N. & Brydson R. (2005). Transformation of nanodiamond into carbon onions: A comparative study by high-resolution transmission electron microscopy, electron energy-loss spectroscopy, x-ray diffraction, small-angle x-ray scattering, and ultraviolet Raman spectroscopy. *J. Appl. Phys*. Vol. 97, No. 7, (January 2005), pp. 074302 (16 pages). ISSN: 0021-8979

Petkov V. (1989). RAD, a program for analysis of X-ray diffraction data from amorphous materials for personal computers. *Journal of Applied Crystallography*. Vol. 22, No. 4, (August 1989), pp. 387-389, ISBN: 0021-8898

Rud A.D., Perekos A.E., Ogenko V.M., Shpak A.P., Uvarov V.N., Chuistov K.V., Laknhik A.M., Voinash V.Z. & Ivaschuk L.I. (2007). Different state of carbon produced by high-energy plasmochemistry synthesis. *Journal of Non-Crystalline Solids*. Vol. 353, No. 32 - 40, (October 2007), pp. 3650 - 3654, ISSN: 0022-3093

Rud A.D., Kuskova N.I., Ivaschuk L.I., Zelinskaya G.M. & Biliy N.M. (2011). Structural State of carbon nanomaterials Produced by High-Energy Electric Discharge Techniques. *Fullerenes, Nanotubes, and Carbon Nanostructures*, Vol. 19, No. 1, (January 2011), pp. 120 - 126, ISSN 1536 -383X

Robertson J. (2002). Diamond-Like carbon. *Materials Science and Engineering: R: Reports*. Vol. 37, No. 4 - 6, (May, 2002) pp.129 - 281. ISSN: 0927-796X

Shenderova O.A & Gruen D.M. (Eds.). (2006). *Ultrananocrystalline Diamond: Synthesis, Properties and Applications*. William Andrew Publishing, ISBN 0815515243 . Norwich, NY.

Shenderova O.A., Zhirnov V.V. & Brenner D.W. (2002). Carbon nanostructures. *Critical Reviews in Solid State and Materials Sciences*. Vol. 27, No.3/4, pp. 227-356. ISSN: 1040-8436

Speight J. G. (Ed.). (2004). *Lange`s Handbook of chemistry. 16th Edition*, McGraw-Hill Professional, ISBN 0-07-143220-5, New York.

Tuinstra F. & Koening J.L. (1970). Raman Spectrum of Graphite. *Journal of Chemical Physics.*, Vol. 53, (November 1969), pp.1126-1130. ISSN 0021-9606

Ugeda M. M., Brihuega I., Guinea F., & Gómez-Rodríguez J. M. (2010). Missing Atom as a Source of Carbon Magnetism. *Physical Review Letters*. Vol. 104, No. 9, (March 2010), pp. 096804 [4 pages], ISSN 0031-9007

Yazyev O.V. (2008). Magnetism in Disordered Graphene and Irradiated Graphite. Physical Review Letters. Vol. 101, No. 3, (July 2008), pp. 037203 [4 pages], ISSN 0031-9007

6

Review of Nanocomposite Thin Films and Coatings Deposited by PVD and CVD Technology

Krzysztof Lukaszkowicz
Silesian University of Technology
Poland

1. Introduction

Commonly it is assumed that the nanocrystalline materials are composed of elements like grains, crystallites, layers, e.g., of a size of ca. 100 nm. (more typically less than 50 nm; often less than 10 nm - in the case of superhard nanocomposite, materials for optoelectronic applications, etc.) at least in one direction. The definition give above limits the size of the structure elements, however it has to be seen only as a theoretical value and doesn't have any physical importance.

Thin films and coatings are applied to structural bulk materials in order to improve the desired properties of the surface, such as corrosion resistance, wear resistance, hardness, friction or required colour, e.g., golden, black or a polished brass-like. The research issues concerning the production of coatings are one of the more important directions of surface engineering development, ensuring the obtainment of coatings of high utility properties in the scope of mechanical characteristics and wear resistance. Giving new utility characteristics to commonly known materials is frequently obtained by laying simple monolayer, multilayer or gradient coatings using PVD methods (Dobrzanski et al., 2005; Lukaszkowicz & Dobrzanski, 2008). While selecting the coating material, we encounter a barrier caused by the fact that numerous properties expected from an ideal coating are impossible to be obtained simultaneously. The application of the nanostructure coatings is seen as the solution of this issue.

Nanostructure and particularly nanocomposite coatings deposited by physical vapour deposition or chemical vapour deposition, have gained considerable attention due to their unique physical and chemical properties, e.g. extremely high indentation hardness (40-80 GPa) (Veprek et al., 2006, 2000; Zou et al., 2010), corrosion resistance (Audronis et al., 2008; Lukaszkowicz et al., 2010), excellent high temperature oxidization resistance (Vaz et al., 2000; Voevodin & Zabinski, 2005), as well high abrasion and erosion resistance (Cheng et al., 2010; Polychronopoulou et al., 2009; Veprek & Veprek-Heijman, 2008).

In the present work, the emphasis is put on current practices and future trends for nanocomposite thin films and coatings deposited by physical vapour deposition (PVD) and chemical vapour deposition (CVD) techniques. This review will not be so exhaustive as to cover all aspects of such coatings, but the main objective is to give a general sense of what has so far been accomplished and where the field is going.

2. Synthesis of nanocomposite films and coatings

In conventional bulk materials, refining grain size is one of the possibilities for hardness increase. The same is true for nanocomposite films or coatings. Looking for coating materials with high hardness, higher then the hardness of traditional polycrystalline coatings, there are founding materials with unique nanometric structure properties. With a decrease in grain size, the multiplication and mobility of the dislocations are hindered, and the hardness of materials increase according to the Hall-Petch relationship. This effect is especially prominent for grain size down to tens of nanometers. However, dislocation movement, which determines the hardness in conventional materials, has little effect when the grain size is less than approximately 10 nm. There is a critical value of the grain size at which a maximum hardness can be achieved (Fig. 1). For grains with diameters less than the critical value the material should be "softened" due to the action of a new deformation mechanism. i.e., the Hall-Petch relation is inverted (Koch et al., 2007; Voevodin et al., 2005; Zhang et al., 2007).

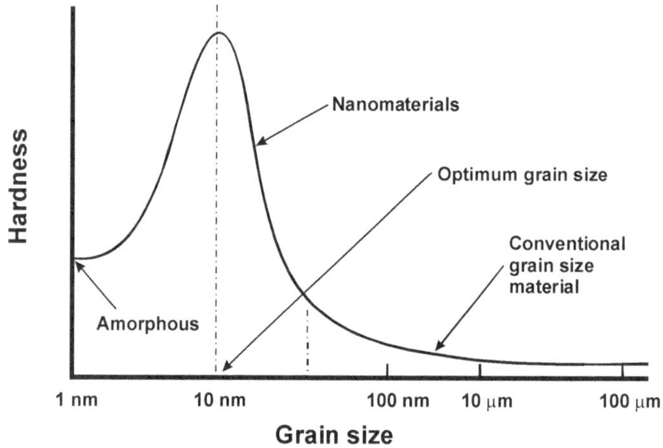

Fig. 1. Schematic illustration of materials hardness versus grain size (Zhang et al., 2007).

According to the Hall-Petch equation, the strength properties of the material rise along with the reduction of the grain size. In case of the coatings deposited by the PVD and CVD processes, the structures obtained, with grain size ~10 nm results with the maximum mechanical properties. Coatings of such structure present very high hardness >40 GPa, ductility, stability at high temperatures, etc.(Tjong et al., 2004; Veprek et al., 2006).

The known dependency between the hardness and abrasion resistance became the basis for the development of harder and harder coating materials. The progress in the field of producing coatings by the physical vapour deposition and/or chemical vapour deposition process enables the obtainment of coatings of nanocrystal structure presenting high mechanical and usable properties. The coatings of such structure are able to maintain a low friction coefficient (self-lubricating coatings) in numerous working environments, maintaining high hardness and increased resistance (Donnet & Erdemir, 2004; Voevodin & Zabinski, 2005).

The main concept for achieving of high hardness of nanostructure coatings and good mechanical properties and high strength related to it, particularly in case of nanocomposite coatings is the restriction of the origination and the movement of dislocations. High hardness and strength of the nanocomposite coatings are due to the fact that the movement of dislocations is suppressed at small grains and in the spaces between them, which causes the appearance of incoherent deformations. When the grain size is reduced to that of nanometers, the activity of dislocations as the source of the material ductility is restricted. This type of coatings is also characterized with a large number grain boundaries with a crystalline/amorphous transition across grain-matrix interfaces, restricting the forming and development of cracks. Such mechanism explains the resistance to fragile cracking of nanocomposite coatings. Simultaneously, the equiaxial grain shapes, high angle grain boundaries, low surface energy and the presence of the amorphous boundary phase facilitating the slide along the grain boundaries cause high plasticity of the nanocomposite coatings (Rafaja et al., 2006; Voevodin et al., 2005).

According to the model proposed by Veprek (Veprek & Reiprech, 1995;Veprek, 1997, Veprek et al., 1998) nanocomposite coatings comprise at least two phases, a nanocrystalline phase and a matrix phase, where the matrix can be either nanocrystalline or amorphous phase. In this design, the nanocrystalline grains must be 3÷10 nm in size and separated by 1÷2 nm within an amorphous phase as shown in Fig. 2.

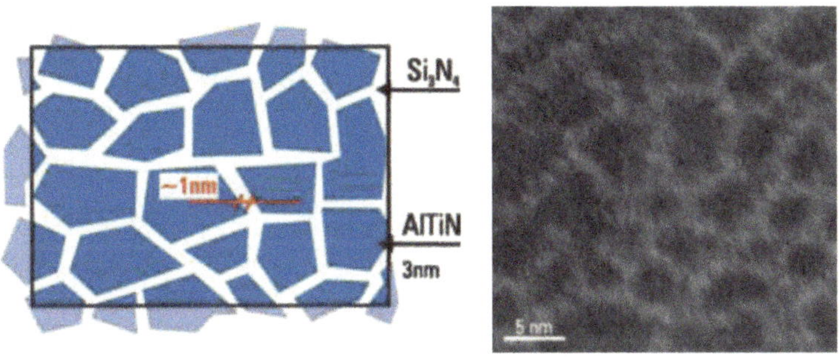

Fig. 2. Model and TEM image of nanocomposite structure (Source: PLATIT AG).

Nanocomposite coatings are usually formed from ternary or quaternary system with nanocrystalline (nc-) grains of hard transition metal-nitrides (e.g. TiN, TiAlN, CrN, BN, VN, ZrN, W_2N, etc.), carbides (e.g. TiC, VC, WC, etc.), borides (e.g. TiB_2, TiB, CrB_2, etc.), oxides (e.g. Al_2O_3, TiO_2, B_2O_3, SiO_2, etc.), or silicides (e.g. $TiSi_2$, $CrSi_2$, $ZrSi_2$, etc.) surrounded by amorphous (a-) matrices (e.g. Si_3N_4, BN, C, etc.) (Moore et al., 2007). There are many different nanocrystalline grain/amorphous matrix (nc-/a-) systems, such as: nc-TiN/a-Si_3N_4, nc-TiN/a-Si_3N_4/a-&nc-$TiSi_2$, nc-CrN/a-Si_3N_4, nc-W_2N/a-Si_3N_4, nc-TiAlN/a-Si_3N_4, nc-TiN/a-BN, nc-TiC/a-C, nc-WC/a-C (Moore et al., 2007, Voevodin et al., 2005; Zhang et al., 2007). The synthesis of such nanocomposite coatings depends on the ability to co-deposit both phases.

Various analyses revealed that the synthesized TiAlSiN coatings exhibited nanostructured composite microstructures consisting of solid-solution (Ti,Al,Si)N crystallites and amorphous Si_3N_4. The Si addition caused the grain refinement of (Ti,Al,Si)N crystallites and its uniform distribution with percolation phenomenon of amorphous silicon nitride. The (Ti,Al)N crystallites became finer with a uniform distribution as the Si content was increased. Si addition changes microstructure from columnar to isotropic (Fig. 3). Grain size decreases with an increase in Si content (Park et al., 2003; Yu et al., 2009).

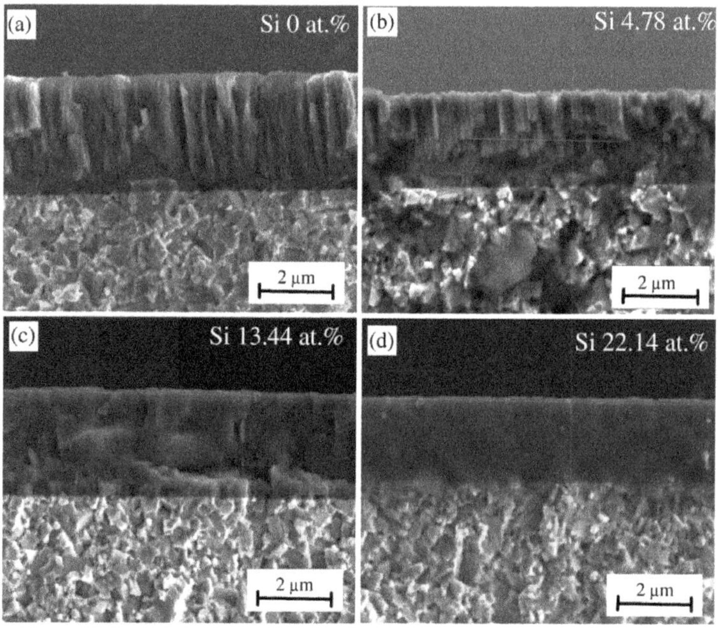

Fig. 3. Cross sectional morphology of TiAlSiN coatings with different Si content (Yu et al., 2009).

Phase size decrease occurred in the nanocomposite coatings to a value of few nanometers, offered an unique possibility to design smart tribological coatings called „chameleon", for their ability to decrease wear and friction by changing surface chemical composition and microstructure according to the applied load and environmental changes - like a chameleon changes the colour in case of emergence (Voevodin & Zabinski, 2000).

Nanocomposite coatings make it possible to achieve a combination of several layers with opposite properties, such as: high hardness, strength, plasticity or low friction coefficient in the single layer. First „chameleon" coatings were produced using mixtures of oxides and disulphides (PbO/MoS_2, Sb_2O_3/MoS_2, ZnO/WS_2), which were applicable in a wide range of temperatures (Zabinski et al., 1992; Zabinski et al., 1993).

It was the next steep to create the advanced multilayer coatings, where the previously achieved nanocomposite layers were put together with layers in form of diffusion barrier and as a result a self adapting surface was achieved, resistant against cyclic temperature changes. In the following new materials were developed, which are wear resistant, they were developed as a combination of nanocrystall based carbides (TiC, WC), ceramics (YSZ i AlON), bisulphites (MoS_2 i WS_2) and amorphous DLC for achieving the nanocomposite structure of the coating (Voevodin et al., 2001; 2002). An example of „chameleon" coatings is the composite coating produced within the W-C-S system, which consist of 1÷2 nm WC and 5÷10 nm WS_2 grains embedded in the amorphous DLC matrix. The $WC/DLC/WS_2$ nanocomposite exhibits self-adaptation to tribological conditions that occur in aerospace system. This adaptation was found in crystallization and reorientation of initially nanocrystalline and randomly oriented WS_2 grains, graphitization of the initially amorphous DLC matrix, reversible regulation of the composition of the transfer film between WS_2 and graphite with environmental cycling from dry to humid air, and a possible DLC/WS_2 synergistic effect, providing friction reduction in oxidizing environments (Voevodin & Zabinski, 2000).

A scheme of a „chameleon" coating of the YSZ/Au/DLC/MoS_2 type as well the WC/DLC/WS_2 type, where the amorphous matrix and hard nanocrystalline phases (YSZ or WC) were used for achieving optimal mechanical properties are presented in Fig. 4.

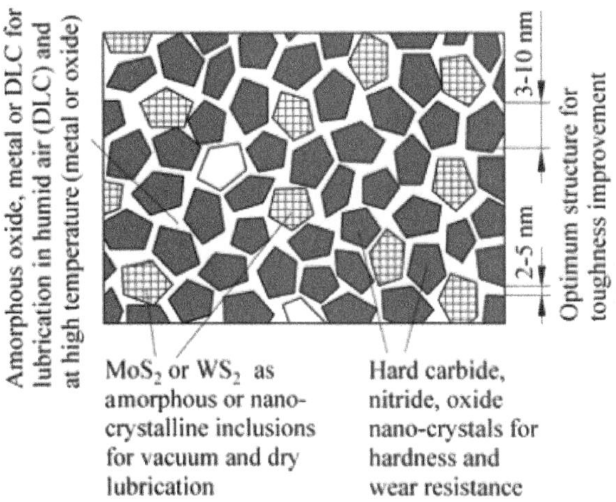

Fig. 4. Schematic conceptual design of a nanocomposite tribological coating with chameleon-like surface adaptive behaviour (Voevodin et al., 2005).

Carrying out investigations over nanocomposite coatings it was found, that the creation mechanisms of such structures have a form of spinodal transformation. During deposition of nanocomposite coatings using physical vapor deposition methods a spinodal character of the segregation of the crystallized phase (e.g. TiN) and amorphous (np. Si_3N_4) was observed during production of nanocomposite coating of the nc-TiN/a-Si_3N_4 type. Such layer is composed of nanocrystalline TiN structure of 3÷4 nm in diameter, positioned in amorphous silicon nitride Si_3N_4.

During deposition of nanocomposite coatings in the PVD process the spinodal decomposition occurs when following conditions are fulfilled (Zhang & Veprek, 2006):

- appropriate temperature occurs making possible a complete segregation of phases during the deposition process of coatings by controlled diffusion;
- sufficiently high nitrogen pressure, for achieving of stechiometric nitride phases by a given coating deposition rate.

Different techniques are available for the preparation of nanocomposite coatings. The most promising methods are plasma assisted chemical vapour deposition (PACVD), thermally-induced chemical vapour deposition (TCVD), magnetron sputtering, vacuum arc evaporation, laser ablation and hybrid techniques consisting of a combination of these. Other techniques, such as ion beam deposition, electrodeposition are also used (Table 1).

Group	Sub-group	Methods
Physical vapour deposition (PVD)	Thermal evaporation (TE)	Pulsed laser deposition (PLD)
		Electron beam deposition (EB-PVD)
	Sputter deposition	Magnetron sputtering
		Ion beam sputtering
	Arc vapour deposition	Vacuum arc deposition
		Filtered arc deposition
	Ion implantation	Ion beam deposition (IBD)
Chemical vapour deposition (CVD)		Plasma enhanced CVD (PECVD)
		Plasma assistant CVD (PACVD)
		Electron cyclotron resonance CVD ((ECR-CVD)

Table 1. Main preparation methods for nanocomposite thin films (Zhang et al., 2007).

Uniform deposition for complicated geometries and high deposition rate are the advantages of the CVD method. However, the main concern for the CVD process is that the precursor gases, $TiCl_4$, $SiCl_4$ or SiH_4, may pose problems in deposition because they are a fire hazard and corrosive. In conventional thermally activated CVD heating provides sufficient high temperature for the dissociation of gaseous species. In case of $TiCl_4$ at low temperatures the chemical equilibrium is one the side of the chlorides, high temperatures (~1000°C) are needed to shift the equilibrium towards TiN and HCl.

$$TiCl_4 + 2H_2 + 0.5N_2 \rightarrow TiN + 4HCl \quad (1)$$

For most applications (e.g. high speed steel, hot work tool steel, etc.) the low deposition temperature is required to prevent substrate distortion and loss of mechanical properties. It precludes the use of substrates having melting points or tempering temperature much lower than the reaction temperature. Plasma assisted CVD exhibit a distinct advantages over thermal CVD owing to its lower deposition temperature (Koch et al., 2007; Tjong, 2007).

Vacuum arc evaporation technology has high ionization (≥90 %) but limited material. Magnetron sputtering is a low temperature technology, which can deposit various materials, but its ionization is low. In magnetron sputtering process the energetic ion (several hundred to a few thousand electronovolts) bombardment is used to vaporize the target. During magnetron sputtering process, the sputtered atoms carry more energy than the evaporated atoms, therefore sputter-grown films usually have higher density. Moreover, ion bombardment of the growing film can restrict the grain growth and permit the formation of nanocrystalline. In arc evaporation process atoms are removed from source by thermal or electron means. Thermal evaporation has a limitation in multicomponent materials since one of the metallic elements typically evaporates before the other. The advantages of deposition of coatings by means of vacuum arc are the high deposition rate and fully ionized plasma. The disadvantages are the emission of droplets of molten metal from the cathodic spot and the uncontrolled movement of the cathodic spot over the surface. Nowadays one of the commonly used coating equipment for large-scale industrial production is the LARC® Technology (Fig. 5). The most important advantages on the LARC® Technology come from the rotating cathodes and their lateral position. The cathodes are in permanent rotation; the magnetic field is generated by coils and permanent magnets controlled both vertically and radially (Cselle et al., 2002; Koch et al., 2007).

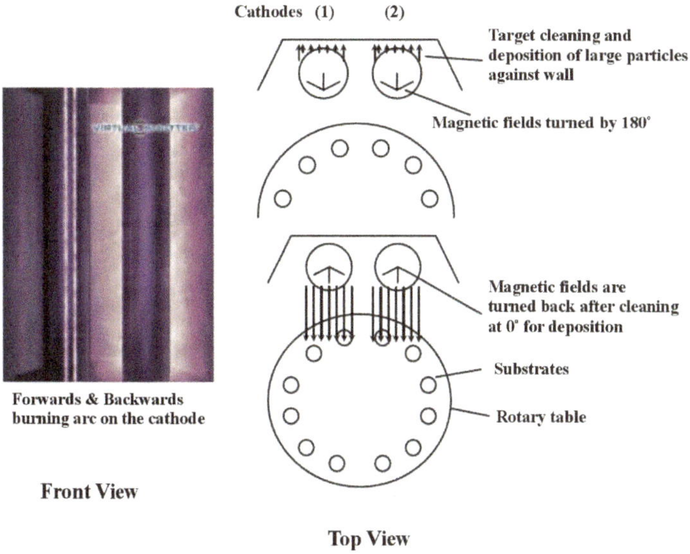

Fig. 5. Coating equipment with the LARC® Technology (Lateral Rotating Arc-Cathodes) (Cselle et al., 2002).

To enhance the strong points of above mentioned technologies and avoid their weakness the hybrid deposition techniques are used. Yu and co-workers have used hybrid physical vapour deposition system, which combine ion plating and sputter plating in order to modify the properties of coatings, and to enhance controllability of the ion energy and the manageable coatings component (Yu et al., 2009). Voevodin and co-workers used the pulsed laser ablation (PLD) of graphite in combination with magnetron sputtering of titanium and filtered cathodic arc deposition (FCAD) of graphite with magnetron sputtering of titanium (Fig. 6) in order to prepare the unhydrogenated amorphous diamond-like carbon (DLC) and fullerene-like carbon nitride (CNx) (Voevodin et al., 2005).

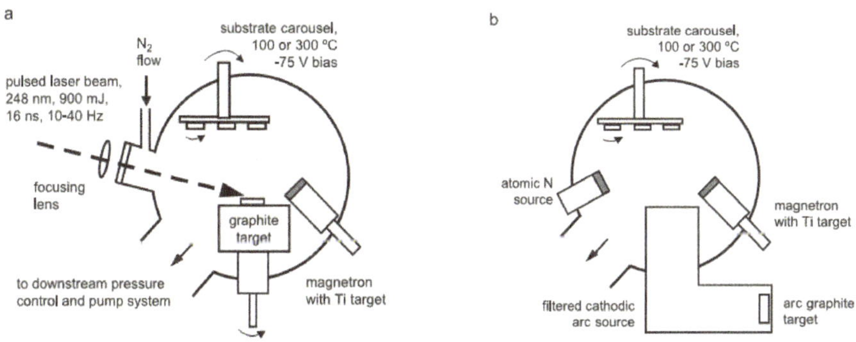

Fig. 6. Schematic of deposition arrangements to produce DLC and CNx coatings by (a) pulsed laser deposition and (b) filtered cathodic arc deposition. A Ti magnetron was used to grow a functionally graded Ti-TiC-C interface by either a hybrid of laser ablation and magnetron sputtering (a) or a hybrid of filtered cathodic arc and magnetron sputtering (b) (Voevodin et al., 2005).

3. Characterization of nanocomposite films

Film characterization is the inevitable and important step for ensuring of high-quality film for the intended application. Different characterization techniques can be used to identify nanocomposite coatings.

3.1 High resolution transmission electron microscope (HRTEM)

Since nanoscale irregularities have the enormous influence on bulk behavior, the high resolution transmission electron microscope has become a powerful and indispensable tool for characterizing nanostructured materials. HRTEM is a powerful technique due to its high imaging spatial resolution (~0.1 nm). Using correct operating conditions and well-prepared samples (specimen materials must be specially prepared to thickness which allow electrons to be transmitted through), high-resolution image characteristics are interpretable directly in terms of projections of individual atomic positions. As the wavelength of electrons is much smaller than that of light, the optimal resolution of transmission electron microscope images is many orders of magnitude better than that of a light microscope. Imaging with the HRTEM enables individual atomic columns to be resolved in most inorganic materials, making it possible to determine the atomic-scale microstructure of lattice defects and other inhomogeneities. Structural features of interest include planar faults such as grain boundaries, interfaces and crystallographic shear planes, linear faults such as dislocations and nanowires, as well as point defects, nanosized particles and local surface morphology. Additional information can be obtained from high-resolution studies, including unique insights into the controlling influence of structural discontinuities on a range of physical and chemical processes such as phase transformations, oxidation reactions, epitaxial growth and catalysis (Materials Evaluation and Engineering, Inc., 2010; Smith, 2005).

3.2 Scanning electron microscope (SEM)

Scanning electron microscopy is a very useful technique in the characterization of thin films and undoubtedly the most widely used of all electron beam instruments. SEM is a method for high-resolution imaging of surfaces. Scanning electron microscopy provides magnified images by using electrons instead of light waves (in case of conventional light microscopes). The advantages of SEM over light microscopy include much higher magnification (> 1,000,000×) and greater depth of field up to 100 times. The resolution of modern scanning electron microscopes is typically <10 nm (image resolution of about 0.5 nm can be now achieved in the most recent generation field-emission-gun SEM), so in principle SEM could be used to measure nanocrystalline grain size. The problem is obtaining the appropriate surface contrast for grain sizes in the nanoscale regime. To create the SEM image, the incident electron beam is scanned in a raster pattern across the sample's surface. When an electron beam interacts with a bulk specimen, a variety of signals can be generated. Figure 7 shows the various types of electrons and electromagnetic radiation produced when a high energy electron beam interacts with a bulk specimen (Liu, 2005).

The popularity of the SEM can be attributed to many factors: the versatility of its various modes of imaging, the excellent spatial resolution, the very modest regarding sample preparation, the relatively straightforward interpretation of the acquired images, the accessibility of associated spectroscopy and diffraction techniques. Most electron microscopes have extra attachments to allow them to generate and analyse X-rays or generate and analyse the energy spectra of the secondary emission electrons. These attachments create a more powerful tool for the study of qualitative and quantitative chemical analysis, for example X-ray photoelectron spectroscopy (XPS), Auger electron spectroscopy (AES), energy dispersive X-ray emission (EDX), wavelength dispersive X-ray emission (WDX), electron back-scattered diffraction (EBSD) (Smith, 2005).

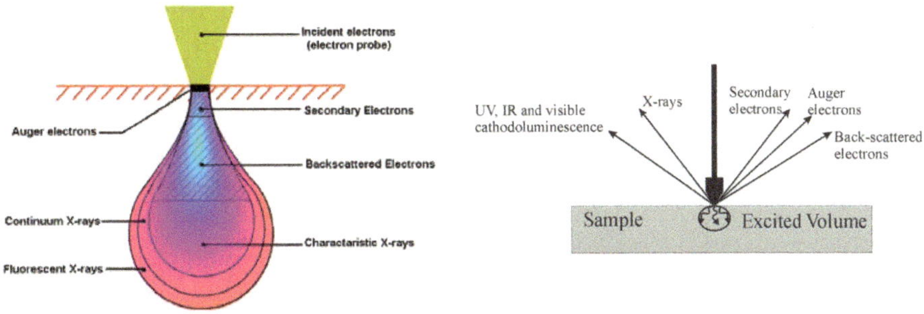

Fig. 7. Schematic view of the different signals generated after an electron beam interacts with a bulk specimen (Materials Evaluation and Engineering, Inc., 2010).

3.3 X-ray photoelectron spectroscopy (XPS)

X-ray photoelectron spectroscopy is the most useful analysis technique used to obtain chemical information about different elements at sample surfaces. In XPS the sample is placed in a high vacuum environment and the beam of low-energy X-rays (or electron) irradiates the sample and produce photo-ionisation. The energy of the ejected photoelectrons is a function of its binding energy (which is the energy required to remove the electron from its atom) and is characteristic of the element from which it was emitted. The equation which describes the energy of the photo-ionisation process is:

$$E_B = h\upsilon - E_K - W \quad (2)$$

where E_B is binding energy of the electron from a particular energy level, $h\upsilon$ is the incident X-ray photon (or electron) energy, E_K is the kinetic energy of the ejected photoelectron, W is the spectrometer work function (Clarke & Eberhardt, 2002).

From the binding energy we can obtain some important information about the samples (Fig. 8) under investigation: the relative quantity of each element, the elements from which it is made, the chemical state of the elements present, depth distribution (Zhang, 2007).

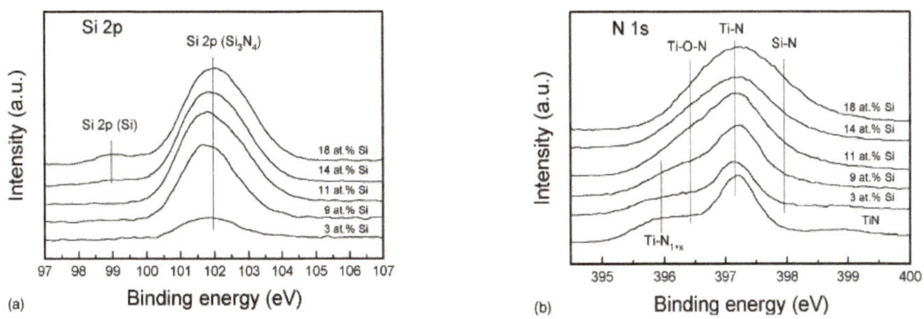

Fig. 8. Typical XPS core-level spectra of (a) Si 2p and (b) N 1s from several Ti–Si–N films (Jiang et al., 2004).

3.4 Auger electron spectroscopy (AES)

Auger electron spectroscopy provides information about the chemical composition of the outermost material comprising a solid surface or interface. The principal advantages of AES over other surface analysis methods are excellent spatial resolution, surface sensitivity (~20 Å), and detection of light elements (Materials Evaluation and Engineering, Inc., 2010). Different penetrations and accuracies of selected compositions analysis method presents Table 2.

Analysis method	Elemental range	Detection limits (at. %)	Spatial resolution	Penetration
XPS	Li-U	0.1÷1	100 μm	1.5 nm
AES	Li-U	0.5	10 nm	0.5÷7.5 nm
RBS	Li-U	1.0	1÷4 mm	2÷30 nm
EDX	Be-U	0.1	0.5÷2.0 μm	1÷3 μm

Table 2. Composition analysis methods (Zhang, 2007).

3.5 Raman spectroscopy

Raman spectroscopy is a method of chemical analysis, based on inelastic scattering of monochromatic light, usually from a laser source. Photons of the laser light are absorbed by the sample and then reemitted. Frequency of the reemitted photons is shifted up or down in comparison with original monochromatic frequency, which is called the Raman effect. This shift provides information about vibrational, rotational and other low frequency transitions in molecules. The sample is illuminated with a laser and scattered light is collected. The wavelengths and intensities of the scattered light can be used to identify functional groups in a molecule (Fig. 9) (Xu & Kumar, 2007).

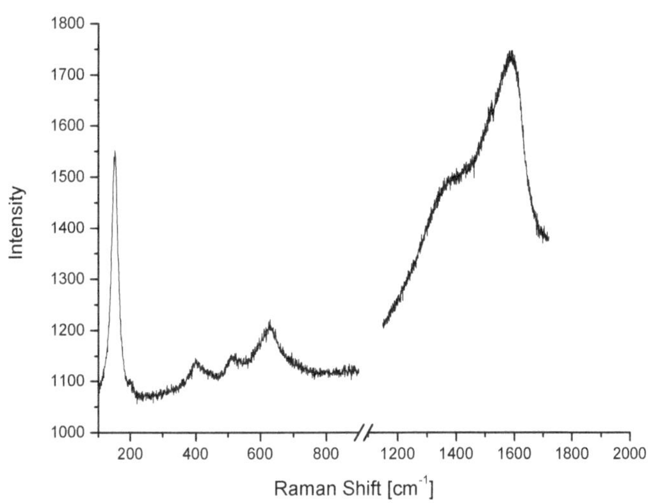

Fig. 9. Raman spectra of the AlCrSiN/DLC coating.

3.6 Grain size

In most studies of the grain size of nanocrystalline materials X-ray line-broadening analysis or direct measurements by transmission electron microscopy are performed. To estimate the grain size, the well-known Scherrer formula is used by measuring the full width at half maximum (FWHM) of the XRD diffraction peak:

$$d = \frac{K\lambda}{\beta \cos\theta} \tag{3}$$

where d is crystallite size, K is the shape factor, λ is the wavelength of the X-rays , β is the full width at half-maximum XRD peak in radians, and θ is the position of the peak maximum.

The Scherrer equation is limited to nano-scale particles. It is not applicable to grains larger than about 100 nm. Otherwise, other methods, in particular transmission electron microscopy, must be used (Koch et al., 2007).

3.7 Atomic force microscopy (AFM)

Atomic force microscopy has become a powerful tool to detect structural changes at the nanoscale and the importance of AFM, as characterization technique, is further increasing with recent developments in nanoscience and nanotechnology. The scope of AFM applications includes high-resolution examination of surface topography (Fig. 10), compositional mapping of heterogeneous samples and studies of local mechanical, electric, magnetic and thermal properties. These measurements can be performed on scales from hundreds of microns down to nanometers. Areas as large as about 100 μm square to less than 100 nm square can be imaged (Magonov & Yerina, 2005; Roa et al., 2011).

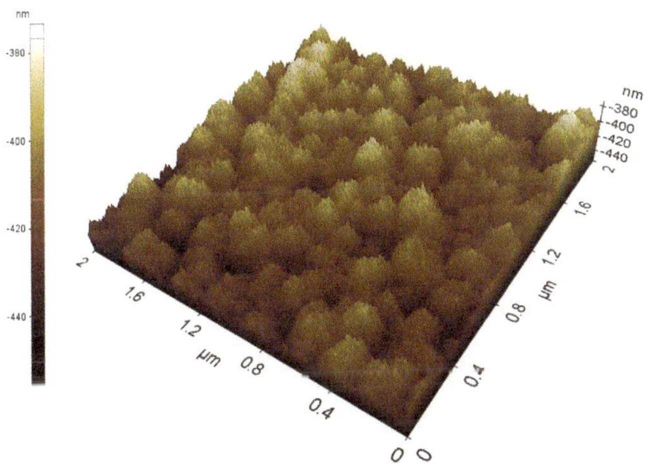

Fig. 10. AFM topography of the CrN/DLC coating.

In conjunction with the atomic force microscopy nanoindenting is performed. Nanoindenting is a new method to characterize material mechanical properties on a very small scale. Features less than 100 nm across, as well as thin films less than 5 nm thick, can be evaluated. Test methods include indentation for comparative and quantitative hardness determination and scratching for evaluation of wear resistance and thin film adhesion. Indentation forces ranging from 1 μN to 100 mN can be made to measure material hardness (Materials Evaluation and Engineering, Inc., 2010).

4. Applications of nanocomposite films

Since the introduction in 1993 of nc-$Ti_{1-x}Al_xN$/a-Si_3N_4 nanocomposite coatings for machining applications, such as steel turning, drilling and milling (Holubar et al., 2000), many new nanocomposite coatings have been developed in order to increase the machining speed, the lifetime of the coated tools and improve the quality of the machined surface.
Due to high hardness and wear resistance nanocomposite coatings show excellent results even at normal cutting conditions. The high hardness of the nanocomposite coatings can be maintained at high temperatures. The enormous hardness will be extremely important for dry cutting and for machining with minimum quantity lubrication.
Modern nanostructured coatings for structural and functional applications, are used mainly for wear protection of machining tools and for the reduction of friction in sliding parts. The nanocomposite coatings find increasing number of applications also on tools for forming, e.g. extrusion, stamping, etc. The demands on the tool material of an extrusion die are high hardness, creep resistance, yield strength, toughness at elevated temperatures, wear and corrosion resistance. It should be suitable for surface coatings. Hard coating, such as nc-AlCrN/a-Si_3N_4, nc-AlTiN/a-Si_3N_4 deposited by physical vapour deposition on extrusion dies have been used in order to increase their life time and properties (Fig. 11) (Faga et al., 2007; Veprek & Veprek-Heijman, 2008).

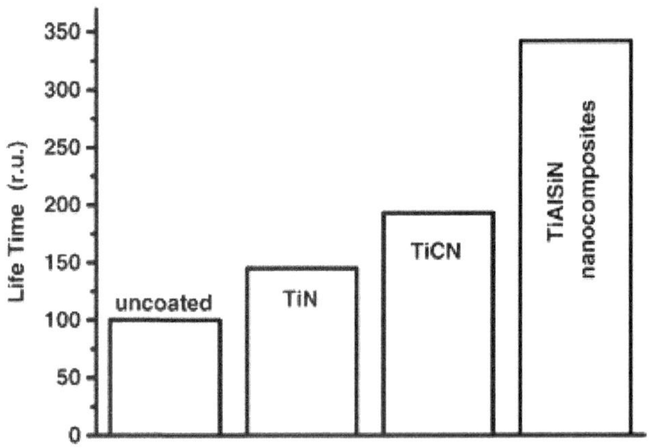

Fig. 11. The performance of uncoated and coated tool for folding boxes for cigarettes (Veprek & Veprek-Heijman, 2008).

An important advantage of the nc-TiN/a-Si_3N_4, nc-AlTiN/ a-Si_3N_4, and nc-AlCrN/ a-Si_3N_4 nanocomposites coatings is their high thermal stability and high oxidation resistance up to temperatures of more than 800°C (Veprek & Veprek-Heijman, 2008).

Nanocomposite coatings can be used for biomedical applications. Implant biomaterials interact with the body through their surface. Consequently, the properties of the material surfaces are critically important in determining biocompatibility, biological responses and bone-implants interface reaction. Surface coatings like traditional amorphous diamond-like carbon modified by doping with silicon and chromium biocompatible coatings are one of the ways of improving both the mechanical, physical and chemical properties of implants in direct contact with the blood (Okpalugo et al., 2007). In orthopaedic and dental implant application, surface coatings have been used to impart bioactivity to the metal substrates to enhance and accelerate skeletal fixation.

Nanocomposite coatings are widely used in solar energy conversion and optoelectronic applications. Nanocomposite thin films or coatings for solar energy conversion should have a high absorption of solar radiation and low thermal emission at their operating temperatures. There are many different thin films solar cells system, such as: a-Si:H/a-SiGe:H/nc-Si:H, a-Si:H/nc-Si:H/nc-Si:H, a-Si:H/nc-Si:H. Other promising thin films solar cells may include $CuInS_2$, CuInGaSe, CdTe, TiO_2, CdS (Yin, 2007). Hydrogenated amorphous silicon (a-Si:H) or nanocrystalline silicon (nc-Si) is currently used in thin film transistor liquid displays (TFT-LCDs).

5. Future trends

The nanocomposite coatings are still at the beginning of their development. Further research activity in the field of nanocomposite films and coatings should be concentrated on the following problems (Musil, 2007; Voevodin et al., 2005):

- development of films with controlled grains size,
- development of new technology system for production of nanocomposite coatings in new physical and/or chemical conditions,
- increased utilization of coatings in many types of applications including those in the aerospace and automotive industries,
- development of protective coatings with oxidation resistance exceeding 2000°C,
- development of hybrid coatings with nanophase biomaterials for biomedical applications,
- nanocrystallisation of amorphous materials,
- development and scale-up of deposition processes,
- development of high fracture toughness coatings,
- development of new advanced coatings with unique physical and functional properties.

6. Summary

The interest, caused by Feynman in the sixties of the last century, in processes occurring in nanometric scale has resulted in a very dynamical development of investigation projects in nanotechnology, nanomechanics and nanomaterials. Nanocomposite thin films and coatings are also the part of these activities.

Nanocomposite thin films and coatings, that characterised with nanocrystalline grains of transition-metal nitride or carbide embedded in an amorphous phase, are a promising instrument for improving the characteristics of various functional surfaces and are rapidly trend of surface engineering. To date, several hard coatings based on nanocomposites have been successfully developed and commercialized for engineering applications. There are many possible future applications of nanocomposite thin films and coatings. Further improvement of the presently available and newly developed nanocomposite coatings will push the technology towards new horizons. Several techniques, in particular PVD and CVD are now available for the deposition of nanocomposite films on various kinds of substrates.

Despite the big number of scientific investigations in the field of nanocomposite thin films and coatings, with results published in professional and competent literature, still some lack and deficiency of verified material knowledge occurs.

7. Acknowledgment

The research was partially financed within the framework of the Polish State Committee for Scientific Research Project No N N507 550738 headed by Dr Krzysztof Lukaszkowicz.

8. References

Audronis M.; Leyland A.; Kelly P.J. & Mathews A. (2008). Composition and structure-property relationships of chromium-diboride / molybdenum-disulhipe PVD nanocomposite hard coatings deposited by pulsed magnetron sputtering. *Applied Physics A*, vol. 91, (April 2008), pp. 77-86, ISSN 0947-8396.

Cheng, Y.C.; Browne, T.; Heckerman, B. & Meletis, E.I. (2010). Mechanical and tribological properties of nanocomposite TiSiN coatings. *Surface and Coatings Technology*, vol. 204, (April 2010), pp. 2123-2129, ISSN 0257-8972.

Clarke, A.R. & Eberhardt, C.N (2002). *Microscopy techniques for materials science*, Woodhead Publishing Limited, ISBN 1-85573-587-3, Cambridge, England.

Cselle, T.; Morstein, M.; Holubar, P.; Jilek, M. & Karimi A. (2002). Nanostructured coatings and processes on an industrial scale, *Proceedings of the Gorham 2002 Conference – Advanced Coatings and Surface Systems for Cutting Tools and Wear Parts*, pp. 1-17, Atlanta, Georgia, USA, November 18-20, 2002.

Dobrzanski, L.A.; Lukaszkowicz, K.; Zarychta, A. & Cunha, L. (2005). Corrosion resistance of multilayer coatings deposited by PVD techniques onto the brass substrate. *Journal of Materials Processing Technology*, vol. 164-165, (May 2005), pp. 816-821, ISSN 0924-0136.

Donnet, C. & Erdemir, A. (2004). Solid lubricant coatings: recent developments and future trends. *Tribology Letters*, vol. 17, No. 3, (October 2004), pp. 389-397, ISSN 1023-8883.

Faga, M.G.; Gauiter, G.; Calzavarini, R.; Perucca, M.; Aimo Boot, E.; Cartasegna, F. & Settineri, L. (2007). Advantages of nanocomposite coatings deposited by high power pulse magnetron sputtering technology. *Wear*, vol. 263 (September 2007), pp. 1306-1314, ISSN 0043-1648.

Holubar, P.; Jilek, M. & Sima, M. (2000). Present and possible future applications of superhard nanocomposite coatings. *Surface and Coatings Technology*, vol. 133-134, (November 2000), pp. 145-151, ISSN 0257-8972.

Jiang, N., Shen, Y.G., Mai, Y.W., Chan, T. & Tung, S.C. (2004). Nanocomposite Ti-Si-N films deposited by reactive unbalanced magnetron sputtering at room temperature. *Materials Science and Engineering B,* vol. 106, (January 2004), pp. 163-171, ISSN 0921-5107.

Koch, C.C.; Ovid`ko, I.A.; Seal, S. & Veprek, S. (2007). *Structural Nanocrystalline Materials. Fundamental and Applications,* Cambridge University Press, ISBN-13 978-521-85565, Cambridge, UK.

Liu, J. (2005). High-Resolution Scanning Electron Microscopy, In: *Handbook of microscopy for nanotechnology,* N. Yao & Z.L. Wang (EDS.), 325-360, Kluwer Academic Publisher, ISBN 1-4020-8003-4, Boston/Dordrecht/New York/London.

Lukaszkowicz, K. & Dobrzanski, L.A. (2008). Structure and mechanical properties of gradient coatings deposited by PVD technology onto X40CrMoV5-1 steel substrate. *Journal of Materials Science,* vol. 43, (May 2008), pp. 3400-3407, ISSN 0022-2461.

Lukaszkowicz, K.; Sondor, J.; Kriz, A. & Pancielejko, M. (2010). Structure, mechanical properties and corrosion resistance of nanocomposite coatings deposited by PVD technology onto the X6CrNiMoTi17-12-2 and X40CrMoV5-1 steel substrates. *Journal of Materials Science,* vol. 45, (March 2010), pp. 1629-1637, ISSN 0022-2461.

Magonov, S.N. & Yerina, N.A. (2005). Visualization of Nanostructures with Atomic Force Microscopy, In: *Handbook of microscopy for nanotechnology,* N. Yao & Z.L. Wang (EDS.), 427-454, Kluwer Academic Publisher, ISBN 1-4020-8003-4, Boston/Dordrecht/New York/London.

Materials Evaluation and Engineering, Inc. (2010). Scanning Electron Microscopy, In: *Handbook of analytical Methods for Materials,* Plymouth, MN 55441-5447, USA, Available from http://www.mee-inc.com

Moore, J.J.; Park, I.-W. & Lin, J. (2007). Nanostructured, multifunctional tribological coatings, In: *Nanocomposite Thin Films and Coatings,* S. Zhang & N. Ali (Eds.), 329-379, Imperial College Press, ISBN-13 978-1-86094-784-1, London, UK.

Musil, J. (2007). Properties of hard nanocomposite thin films, In: *Nanocomposite Thin Films and Coatings,* S. Zhang & N. Ali (Eds.), 281-328, Imperial College Press, ISBN-13 978-1-86094-784-1, London, UK.

Okpalugo, T.I.T.; Ali, N.; Ogwu, A.A.; Kousar, Y. & Ahmed, W. (2007). Amorphous and nanocomposite diamond-like carbon coatings for biomedical applications, In: *Nanocomposite Thin Films and Coatings,* S. Zhang & N. Ali (Eds.), 513-572, Imperial College Press, ISBN-13 978-1-86094-784-1, London, UK.

Park, I.W.; Choi, S.R.; Lee, M.H. & Kim, K.H. (2003). Effects of Si addition on the microstructural evolution and hardness of Ti-Al-Si-N films prepared by the hybrid system ion plating and sputtering techniques. *Journal of Vacuum Science & Technology A: Vacuum, Surface, and Films,* vol. 21, No. 4, (July 2003), pp 895-899, ISSN 0734-2101.

Polychronopoulou, K.; Baker, M.A.; Rebholz, C.; Neidhardt, J.; O`Sullivan, M.; Reiter, A.E.; Kanakis, K.; Leyland, A.; Matthews, A. & Mitterer, C. (2009). The nanostructure, wear and corrosion performance of arc-evaporated CrB_xN_y nanocomposite coatings. *Surface and Coatings Technology*, vol. 204, (October 2009), pp. 246-255, ISSN 0257-8972.

Rafaja, D.; Poklad, A.; Klemm, V.; Schreiber, G.; Heger, D.; Sima, M. & Dopita, M. (2006). Some consequence of the partial crysrallographic coherence between nanocrystalline domains in Ti-Al-N and Ti-Al-Si-N coatings. *Thin Solid Films*, vol. 514, (August 2006), pp. 240-249, ISSN 0040-6090.

Roa, J.J.; Oncins, G.; Diaz, J.; Capdevila, X.G.; Sanz, F. & Segarra, M. (2011). Study of the friction, adhesion and mechanical properties of single crystals, ceramics and ceramic coatings by AFM. *Journal of the European Ceramic Society*, vol. 31, (April 2011), pp. 429-449, ISSN 0955-2219.

Smith, D.J. (2005). High-Resolution Transmission Electron Microscopy, In: *Handbook of microscopy for nanotechnology*, N. Yao & Z.L. Wang (EDS.), 427-454, Kluwer Academic Publisher, ISBN 1-4020-8003-4, Boston/Dordrecht/New York/London.

Tjong, S.C. (2007). Properties of chemical vapor deposited nanocrystalline diamond and nanodiamond/amorphous carbon composite films, In: *Nanocomposite Thin Films and Coatings*, S. Zhang & N. Ali (Eds.), 329-379, Imperial College Press, ISBN-13 978-1-86094-784-1, London, UK.

Tjong, S.S. & Chen, H. (2004). Nanocrystalline materials and coatings. *Materials Science and Engineering R*, vol. 45, (September 2004), pp. 1-88, ISSN 0927-796X.

Vaz, F.; Rebouta, L.; Goudeau, P.; Pacaud, J.; Garem, H.; Riviere, J.P.; Cavaleiro, A. & Alves, E. (2000). Characterisation of $Ti_{1-x}Si_xN_y$ nanocomposite coatings. *Surface and Coatings Technology*, vol. 133-134, (November 2000), pp. 307-313, ISSN 0257-9872.

Veprek, S. & Reiprich, S. (1995). A concept for the design of novel superhard coatings. *Thin Solid Films*, vol. 268, (November 1995), pp. 64-71, ISSN 0040-6090.

Veprek, S. & Veprek-Heijman, M.J.G. (2008). Industrial applications of superhard nanocomposite coatings. *Surface and Coatings Technology*, vol. 202, (July 2008), pp. 5063-5073, ISSN 0257-8972.

Veprek, S. (1997). Conventional and new approaches towards the design of novel superhard materials. *Surface and Coatings Technology*, vol. 97, (December 1997), pp. 15-22, ISNN 0257-8972.

Veprek, S.; Mannling, H.D.; Karvankova, P. & Prochazka, J. (2006). The issue of the reproducibility of deposition of superhard nanocomposites with hardness of ≥ 50 GPa. *Surface and Coatings Technology*, vol. 201, (March 2006), pp. 6064-6070, ISSN 0257-8972.

Veprek, S.; Nesladek, P.; Niederhofer, A.; Glatz, F.; Jilek, M. & Sima, M. (1998). Recent progress in the superhard nanocrystalline composites: towards their industralization and understanding of the origin of the superhardness. *Surface and Coatings Technology*, vol. 108-109, (October 1998), pp. 138-147, ISSN 0257-8972.

Veprek, S.; Niederhofer, A.; Moto, K.; Bolom, T.; Mannling, H.D.; Nesladek, P.; Dollinger, G. & Bergmaier, A. (2000). Composition, nanostructure and origin of the ultrahardness in nc-Tin a-Si_3N_4 a- and nc-$TiSi_2$ nanocomposites with H_v 80 to 105 GPa. *Surface and Coatings Technology*, vol. 133-134, (November 2000), pp. 152-159, ISSN 0257-8972.

Voevodin, A.A. & Zabinski, J.S. (2000). Supertough wear-resistant coatings with 'chameleon' surface adaptation. *Thin Solid Films,* vol. 370, (July 2000), pp. 223-231, ISSN 0040-6090.

Voevodin, A.A. & Zabinski, J.S. (2005). Nanocomposite and nanostructured tribological materials for space applications. *Composites Science and Technology,* vol. 65, (April 2005), pp. 741-748, ISSN 0266-3538.

Voevodin, A.A.; Hu, J.J. & Zabinski, J.S. (2001). Tribological properties of adaptive nanocomposite coatings made of yttria stabilized zirconia and gold. *Surface and Coatings Technology,* vol. 146-147, (September-October 2001), pp. 351-356, ISSN 0257-8972.

Voevodin, A.A.; Jones, J.G.; Back, T.C.; Zabinski, J.S.; Strel`nitzki, V.E. & Aksenov, I.I. (2005). Comparative study of wear-resistant DLC and fullerene-like CN_x coatings produced by pulsed laser and filtered cathodic arc deposition. *Surface and Coatings Technology,* vol. 197, (July 2005), pp. 116-125, ISSN 0257-8972.

Voevodin, A.A.; O`Neil, J.P. & Zabinski, J.S. (2002). Tribological performance and tribochemistry of nanocrystalline WC/amorphous diamond-like carbon composites. *Journal of Vacuum Science & Technology A,* vol. 20, (November-December 2002), pp. 1434-1444, ISSN 0734-2101.

Voevodin, A.A.; Zabinski, J.S. & Muratore, C. (2005). Recent Advances in Hard, Tough, and Low Friction Nanocomposite Coatings. *Tsinghua Science and Technology,* vol. 10, No. 6, (December 2005), pp. 665-679, ISSN 1007-0214.

Xu, Z. & Kumar, A. (2007). Synthesis, Characterization and Applications of Nanocrystalline Diamond Films, In: *Nanocomposite Thin Films and Coatings,* S. Zhang & N. Ali (Eds.), 207-279, Imperial College Press, ISBN-13 978-1-86094-784-1, London, UK.

Yin, Y. (2007). Nanocomposite thin films for solar energy conversion, In: *Nanocomposite Thin Films and Coatings,* S. Zhang & N. Ali (Eds.), 381-417, Imperial College Press, ISBN-13 978-1-86094-784-1, London, UK.

Yu, D.; Wang, C.; Cheng, X. & Zhang, F. (2009). Microstructure and properties of TiAlSiN coatings prepared by hybrid PVD technology. *Thin Solid Films,* vol. 517, (July 2009), pp. 4950-4955, ISSN 0040-6090.

Zabinski, J.S.; Donley, M.S. & McDevit, N.T. (1993). Mechanistic study of the synergism between Sb_2O_3 and MoS_2 lubricant system using Raman spectroscopy. *Wear,* vol. 165, (January 1993), pp. 103-108, ISSN 0043-1648.

Zabinski, J.S.; Donley, M.S.; Dyhouse, V.J. & McDevit, N.T. (1992). Chemical and tribological characterization of PbO-MoS_2 films grown by pulsed laser deposition. *Thin Solid Films,* vol. 214, (March-April 1992), pp. 156-163, ISSN 0040-6090.

Zhang, R.F.; Veprek, S. (2006). On the spinodal nature of the phase segregation and formation of stable nanostructure in the Ti-Si-N system. *Materials Science and Engineering A,* vol. 424 (May 2006), pp. 128-137, ISSN 0921-5093.

Zhang, S.; Sun, D. & Bui, X.L. (2007). Magnetron Sputtered Hard and Yet Tough Nanocomposite Coatings with Case Studies: Nanocrystalline TiN Embedded in Amorphous SiN_X, In: *Nanocomposite Thin Films and Coatings,* S. Zhang & N. Ali (Eds.), 1-110, Imperial College Press, ISBN-13 978-1-86094-784-1, London, UK.

Zou, C.W.; Wang, H.J.; Li, M.; Yu, J.F.; Liu, C.S.; Guo, L.P. & Fu, D.J. (2010). Characterization and properties of TiN-containing amorphous Ti-Si-N nanocomposite coatings prepared by arc assisted middle frequency magnetron sputtering. *Vacuum*, vol. 84, (February 2010), pp. 817-822, ISSN 0042-207X.

New Methods and New Types of Functionalised Nanocomposites Intended for the Ecological Depollution of Waters

Nicoleta Petrea, Petrişor Zamora Iordache,
Rodica Mihaela Lungu, Ioan Safta, Razvan Petre and Andrada Pretorian
Scientific Research Center for CBRN Defense and Ecology
Romania

1. Introduction

Nanotechnology as techniques for the removal of organic pollutants implies the use of certain types of nanostructures or nanostructured materials as a carrier host for the reticulation, encapsulation and degradation of pollutants. The main mechanisms through which nanocomposite structures separate or destroy organic pollutants are those of the catalytic oxidation, reduction, sorption or reticulation types (Lu, Zhao & Wan, 2010, Choi et al., 2009). Most materials and methods which are based on nanotechnologies have high selectivity regarding specific organic pollutants, and their applicability rely on the economic factor. The limited spectrum of separated or degraded pollutants, the decontamination time and the lack of control over degradation products are key factors which limit the widespread use of most nanostructured materials in organic decontamination processes.

Nanotechnologies offer promising solutions to the depollution field, thanks to their remarkable properties: depollution process control ability, structural and functional modelling flexibility, with a view to improving depollution parameters, high specific surface, etc. In addition, nanostructured materials offer solutions for the obtaining of new ecological depollution biodegradable materials. Microstructured and nanomembranary materials, as well as functionalized nanostructured materials have the most promising applicative potential. Functionalized nanomaterials are carriers of chemical functions capable of reticulating, incorporating and removing organic pollutants. Functionalized composite materials have depollution properties which are similar to those of functionalized materials. As compared to functionalized nanomaterials, composites materials have several possibilities of structural and functional modelling, bearing direct effect on the yield, efficiency and spectrum of reticulated pollutants. Ion exchangers or natural adsorbents (zeolites, synthetic resins, functionalized polymers, etc.) (Cheremisinoff, 2002; Evanghelou 1998) have organic depollution properties, which are similar to those of functionalized composites. Degradation residues of organic pollutants, resulting from depollution processes, as well as from secondary pollution processes caused by the latter are one of the decisive factors that limit current depollution technologies (Rosenfeld & Feng, 2011; Bayliss & Langley, 2003). Most depollution waste fractions have lower molecular weights, as compared to the initial pollutants, which can induce carcinogenic effects on humans or changes on the structural and functional properties of the biodiversity and the environment. Membrane nanostructured materials and functionalized composites have pronounced limiting effects over the dispersion effects of secondary degradation pollutants by their retention, encapsulation and separation. As a consequence of the major social and economic implications (environmental protection, natural resource conservation, medicine, nanomedicine), organic depollution has imposed itself as a priority research field, as a source which generates solutions and implementable technologies. This chapter presents the experimental results acquired as a result of the obtaining and testing of new biodegradable functionalized composite materials, intended for the removal of a wide range of organic pollutants. The composites we obtained were tested on four categories of polluted water, coming from milk processing industry, medicine industry, the obtaining and processing of polymers and from the sewage waters in the city of Bucharest.

2. Functionality relations between environmental elements and pollutants

The environment can be defined as the manifestation of dependence and functionality relations established between physicochemical and biological structures characteristics of soil, water and air, as its fundamental structural macroentities (Tolgyessy, 1993). Overall, the environment represents a complex and dynamic structure, on the evolutionary process, due to the way and type of interaction established between its structural elements, and material and energy mediation vectors (Fig. 1). Ecosystems represent manifestations of the local distribution manner of basic structural entities of the environment. Ecosystems represent organized systems, well defined in relation to their physical, chemical and biological structure of their structural elements, among which are established structural and functional relations of their own (Socolow et al., 1994). Each set of functions is characteristic for a particular ecosystem, reflecting its biochemical structure, mediation and transport vectors, the dynamics of internal evolution, as well as the rate, role and functions of each structural component. Any changes to global environmental factors or local macrostructural distribution of an ecosystem may irreversibly affect the structural and functional evolution. From this perspective, ecosystems can be defined as stable phases of local equilibrium, established between the biochemical structure, compositional structure and functional structure of their macro- and microcomponents.

Transport and mediation vectors (TMV) influence local and global dynamics of transformation and transport, established between the local structural components of the environment and its ecosystems. The main TMV, which form the evolution and transformation of the environment and of its ecosystems are air, water and climatic factors. The distribution and local physical and chemical characteristics of TMV model the distribution and evolution of biodiversity models, as well as the distribution and evolution of the elements which form the composition of the soil, water, rocks, sediments and minerals. The modelling process leads to the establishment of relationships and dynamic biochemical balances, always subject to the evolutionary processes of transformation and structural and functional remodelling. The development and the establishment of equilibrium states within and between the local structures of the environment, reflect on going relationship of mass and energy transfer established at the level of microstructured elements, with TMV as mediating factor. The morphochemical and morphofunctional structuring manner of the soil influence pollutant absorption, modelling and conditioning applicative technologies and depollution resources (Reddy, 2010; Lal & Shukla, 2005).

Organic fractions resulting from degradation processes of biodiversity (humic and fulvinic acids, organic derivatives and their macrocomplex, carbon, nitrogen, oxygen, water, etc.) constitute the foundation of chemical and biochemical transformation processes that model host ecosystems, and are factors which catalyse and support chemical and biochemical functionalities (Evanghelou, 1998; Haider & Schäffer, 2009).

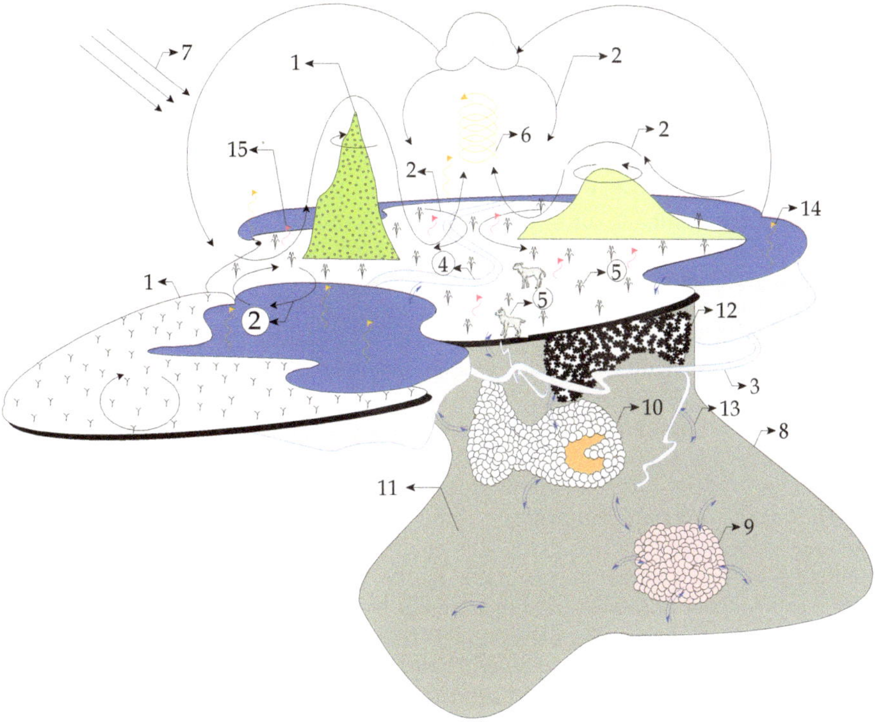

Fig. 1. Ecosystems and their structural and functional organization
(1 - surface form relief; 2 - flow of air and water vapour currents; 3 - underground water flows; 4 - surface water flows; 5 - biodiversity; 6 - energy fluxes; 7 - light; 8 - underground; 9, 10, 12 - underground mineral and rock distributions; 11 - other underground fine particulated matter; 13 - flow of matter and energy carried by fluids; 14 - sea water vapour; 15 - land water vapour)

Material and chemical structures resulting from natural processes of evolution are not a threat to biodiversity and the environment, as they support and model local exchange mechanisms and energy and material balance mechanisms. In terms of chemical functionality, the material structures resulting from natural processes of degradation can be considered inert structures, as they do not have the ability of modelling the morphochemical structure and the functions of flora, fauna and microorganisms of ecosystems. Pollutants can be defined as factors that induce sudden steps in the structural and functional evolution of an ecosystem, the biodiversity being the fastest element of the environment that responds and adapts to changes. The emergence, suppression or evolution of certain species or biological structures is conditioned by the relations established between the functionality structure of the host ecosystem and the genetic profile specific to the targeted biostructures, as a structural conservation factor. Most dependence relations established between the functionality of ecosystems and the genetic profile of their biostructures are of the evolving type, in which the ecosystem, as a material and energy source, ensures the transition of available resources in higher biological forms of organization. The suppression of material and energy resources is equivalent to the suppression of the development and evolution of biodiversity. The factors that may suppress the dependency relations established between the functionality and the structure of the environment and the biodiversity hosted by it can be factors of organic and inorganic pollution, as well as natural equilibrium factors (air, light, temperature, humidity, etc.). These factors model and provide energetic and catalytic support, as well as the dynamics and complexity of structural and material organization forms of the soil, water and air. The dynamics and complexity of the structural and material organization forms of the soil, water and air constitute the structure that supports the entire evolution and the material and functional structuring of biodiversity. Most artificial chemical compounds are material structures with high functional and chemical reactivity, small molecular weight and high diffusion coefficients, which affect the morphochemical structure and the functionality of the environment biodiversity. For this reason, artificial chemical compounds are considered and act as poisons upon microorganisms or biochemical structures with nonexplicit biological functionality (amino acids, proteins etc.). The morphofunctional uniformity and the broad spectrum of natural, synthetic organic pollutants, or those resulting from industrial and domestic activities, contained in the surface waters are the main reasons that restrain the methods, materials and current organic depollution techniques. The constraints bearing upon functionalized nanostructured materials or nanostructured composites are due to the non-selective charging with chemical compounds, which are not toxic for the environment. The limitation of the type of functionality of nanostructured materials is the second major cause that limits the depollution processes, by separately constraining the range of pollutants. Considering the above mentioned statements, the most suitable retention methods of organic pollutants are the mixed ones, which use as active principles of pollutant separation: the mechanical filtration, chemical reticulation, reverse osmosis, degradation, encapsulation and controlled extraction of depollution products.

3. Urgent needs in the field of depollution and decontamination

Inorganic depollution involves relatively low complexity processes, related to organic depollution processes. In the field of inorganic depollution there are thoroughly-studied applicative depollution methods and technologies dedicated to each class of pollutants. Inorganic depollution can be considered as a controlled process, as pollutants may be eliminated in a controlled manner, directly in the stage of generation, limiting the dispersion and secondary contamination processes. In comparison with the organic pollution, the inorganic pollution has a much simpler complexity of the spectrum of pollutants contained in surface water, extending the possibilities of separation, storage, management, treatment and reassessment. Inorganic pollution represents a priority problem of depollution, which has a high complexity level, and whose resolution implies the utilization of some partial, limited applicative solutions. Both organic pollutants and the pollutants having a mixed morphochemical structure (of the organic – inorganic type) are compounds with high and average, slightly degradable reactivity. Most organometallic pollutants and of the compounds having a mixed morphochemical structure (adsorbent inorganic matrices, host matrices, fibres, macro polymers, active suspensions, functionalized inorganic mixtures, etc.) come from food industry, medicine industry, polymer industry, wood processing industry, etc. The slightly degradable character of organic pollutants changes the organic pollution into a degenerative process that generates in time new types of degradation compounds. Most degradation compounds are toxic in relation to the environment, being able to influence, on long and short terms, the functionality and structure of the component elements of the environment. The accumulation and degradation processes are factors that support and amplify the secondary processes of pollution and biological contamination, by favouring the development and adaptation of some new biological structures, extraneous to the host ecosystems. Also, the biological pollution can influence, on long and short terms, the human health and chemical, biological and functional balance, established at the biodiversity level of the ecosystems, by favouring the uncontrollable evolution of the structure and functionality of the local and global environmental factors. By the nature of the sources, due to the complexity of industrial and domestic activities, as well as the needs and exigencies of the consumer society, huge quantities of residue result that require suitable management for collection, storage, reassessment and disposal. The identification of optimum solutions to residue management represents a priority field of depollution as a preventive and control factor. The main function of the waste management is that of controlling and limiting the pollution, by isolating and treating the resulted residue. Due to the exponential increment of the collected residue quantity, as well as to the financial and environmental implications, the depollution methods and technologies have been oriented in the direction of finding some alternative solutions of unitary depollution, disposal, reassessment and ecological reutilization of the residue (Asano et al., 2007). The aim of alternative solutions is that of minimizing the collected quantities of residue and reducing their impact upon the environment. Residue disposal technologies are limited by the implementation costs, the complexity of the type of contained contaminants, as well as the large quantity of collected residue (Cheremisinoff, 2003; Cheremisinoff & Cheremisinoff, 2005; Harrison & Hester, 2002). This type of technologies provides partial solutions, being dedicated especially to the disposal of certain classes of pollutants with high toxicity level or easily separable contaminants (paper, plastics, metals, wood, etc). Reassessment and reutilization of the collected residue have the highest applicative potential, favouring the preservation and regeneration of the natural resources and environmental factors (Baud, Post & Furedy, 2004). Research in the field of waste recovery follows more directions, among them: reassessment -reutilization (plastics, metals, precious metals, etc.), production of new biodegradable depollution materials, the recovery of wastes having organic substrates (building materials, semi-artificial fertilizers, etc.). Waste recovery has some direct active functions within depollution processes, having as direct objects: the simplification of management procedures, the reduction of pollution factors, minimizing the effects of residue and pollutants upon the environment, the preservation and regeneration of natural energy resources. Finding viable and applicable solutions of management and waste recovery, which might have a low impact upon the environment and which might be financially sustainable represents a top priority and a necessity of environment science.

The soil represents the fundamental structural element of the environment, highly exposed to pollutants, acting like an absorbent, due to its own morphological structure and due to the morphochemical and compositional structure of its component elements. The soil presents in its structure an amorphous mineral component (95 - 99 %) (acid, basic and neutral rocks and particles containing: O 46.6%, Si 27.7%, Al 8.1%, Fe 5.0%, Ca 3.7%, Na 2.8%, K 2.6%, Mg 2.1%, etc.; mechanical, chemical, biogenical sediments or their mixtures) a fluid component (characterised by: osmotic pressure, pH, oxido-reduction potential, colloidal structure, buffering capacity, absorbing capacity), a gaseous component (CO_2 - up to 1%, O_2 - 10-20%, N_2, NH_3, water vapours, H_2S, H_2, CH_4, SO_2, etc), a living matter component (bacteria, fungi, algae, protozoa, insects, arachnoides, molluscs, earthworms, etc.) and non-living matter components (humic acids, fulminic acids, humins, simple saccharides, fatty acids, alcohols, esters, starch, proteins, complex proteins, pectins, hemicellulose, cellulose, lignin, wax, bitumen, etc) (Frank & Tolgyessy, 1993). The chemical and morphochemical structure of the soil shows the complexity of its morphofunctional structure. The morphofunctional complexity is expressed by the active chemical functionality variation, capable of absorbing, encapsulating and reticulating the pollutants. Excepting the persistent compounds, in most cases of diffusion of the inorganic pollutants in the soil, the local morphochemical composition is irreversibly modified (Yu & Wang, 1997).

Organic pollutants (especially the surfactants, fatty acids, gasoline, petroleum and its derivatives) diffuse in the surface and depth structure of the structural elements of the soil, deactivating, masking or isolating the structures having functions in establishing and adjusting the chemical and biological balance mechanisms (minerals, organic matter, micro-organisms, colloids, etc.) (Perk, 2007; Evanghelou, 1998). Organic pollution has a deep persistent character and in most situations of this kind, the morphochemical, functional and biochemical soil structure is irreversibly altered. Biotechnologies remain the only solutions that have potential for organic depollution and remediation of the soil. Soil depollution and remediation biotechnologies depend on the type of removed pollutants, they affect on short-term the local biochemical balance of the ecosystems, the action mechanisms taking place in time and being noninvasive (Evanghelou, 1998). The substantiation and development of new technologies of soil depollution and remediation is a fundamental priority of environmental science, with major implications to: human health, environmental factors, human habitat distribution, redistribution of natural resources, economy.

Radiological pollution differs from the other types of pollution sources through its generating mechanisms, methods and decontamination-depollution technologies and through the effects that it induces to biodiversity and environmental factors (Bayliss & Langley, 2003). The only known radiological depollution methods are the preventive measures, aiming at the complete isolation of radiological materials and compounds and at preventing their dispersion in the structure of environmental factors. Radiological pollution generators are made up of different radioactive chemicals that have the ability to emit radiation (charged radiation, γ, neutrinos, etc.), independent of the chemical structure that hosts the unstable nucleus. The carrier support of radioactive elements in the environment occurs through external factors (aerosols, encapsulating matrices, suspensions, oxides, halides, contaminated materials, etc.) called carrier vectors. In the case of nuclear accidents, the separation of carrier vectors is a delicate problem, due to the risk of staff contamination and due to their low chemical reactivity, which does not allow the separation and encapsulation processes. Considering the physical and chemical characteristics of the carrier vectors, the only solutions with applicative potential in the separation of radiological contaminants are those based on the use of externally stimulable highly functionalized materials, in the form of sorbents or nanostructured suspensions with high specific surface, that will be able to reticulate, encapsulate, separate and store the contaminants safely, without exposing human staff (Ojovan & Lee, 2005). Radiological pollution and its effects model radically and on long term the structure and the overall functionality of the environment, human health and factors influencing macroeconomy. This type of pollution has an unpredictable nature, its implications on the environment and biodiversity, and the evolution orientation towards local and global equilibrium factors being difficult to estimate. Radiological depollution represents another prioritary research direction, due its deep implications on energy resources and on the needs of the consumer society (medicine, research, material science).

4. New types of functionalised materials and new applicative solutions concerning the ecological organic depollution of wastewaters

This paper presents the way a new class of functionalized nanocomposite materials, intended for the organic depollution of surface water from industrial and domestic activities, was obtained, investigated and tested. The means of obtaining the composite has already been presented in the previous chapter, "Nanocomposite materials with oriented functionalized structure - The modelling of structure and functionality of nanocomposites intended for ecological depollution". Considering the limitations of current methods and techniques for the separation of pollutants, research and structural and functional modelling investigations have been oriented with a view to obtaining a nanocomposite with an oriented functionality, able to: a. undifferentiatedly reticulate a large number of organic pollutants; b. encapsulate and degrade pollutants and their degradation products; c. separate and extract pollutants and depollution waste under controlled stimulability. Amorphous natural metal-oxide structures are the most suitable class of materials to be used in processes of chemical modelling and functionalization, intended for the removal of organic pollutants. Their functional, morphological and morphochemical structure, resulting from functionalizing processes, is compact and able to fully saturate its functionalized surface, with different pollutants which have been collected from the depolluted environment. Following the processes of chemical modelling and functionalization, the deep structure of the chemically modified metal-oxide components remains unchanged, giving the material a pronounced ecological and biodegradable character. In the first stage, the pollutants, reticulated on the surface and in the depth of the composite, are degraded and encapsulated, resulting in degradation products with modified functionality and toxicity. It is likely that the toxicity of the resulting degradation products might not be substantially modified, but it is very important that, following the reticulation process, the pollutants and the degradation products should remain reticulated on the surface and in the depth of the material, allowing a slow selective biodegradation of the pollutant layer collected on its surface. Equally, the stable reticulation of the pollutants and of the degradation products limits the effects of secondary pollution and contamination by dispersing the resulting degradation compounds. Such depollution materials have an increased application potential in relation to the applications of depollution under the dynamic conditions of polluted surface water flows (rivers, lakes, etc.), as they minimize the dispersion processes and favour local biodegradation processes by local sedimentation of pollutant-charged functionalized suspensions in bottom water. Due to the macromolecular character and the micrometric structure of depollution suspensions, nanostructured materials cannot be assimilated at cell level by the biodiversity they encounter, as it has already been demonstrated in the previous chapter that the surface and the depth of the composite are able to host biological structures, without affecting their morphochemical structure and metabolism.

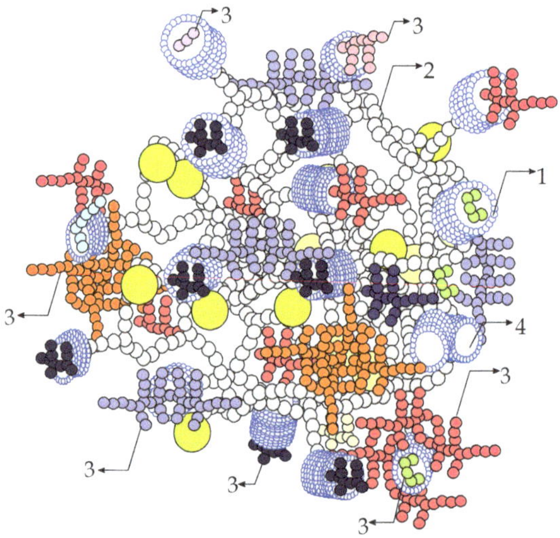

Fig. 2. (1 - mineral metal oxide; 2 - internal retaining metal-oxidic cage - mixed oxidic morphostructure; 3 - different types of organic and inorganic pollutants; 4 - external retaining metal-oxidic cage - specific and mixed oxidic component morphostructure)

The metal-oxide mixture (Su) used as a structural component of compositing of the depollution material (M_{OF-DP}) has a natural origin and has some remarkable morphostructural (fig. 4), morphochemical and functional (fig.3) features. Using the basic segmentation method proposed by Iordache et al. (Iordache et al., 2010), the main morpho-structural parameters of Su were determined and quantified (fig.4b): area (11.41 nm^2) compactness (1.29), perimeter (16.48 nm). Investigations carried out by X-rays dispersion show that Su contains, apart from oxidic fractions of the following elements Na, Mg, Al, Si, P, S, K, Ca, Ti, V, Cr, Mn, Fe, Ni, Cu, Zn , As, Rb, Sr, Y, Zr, Ba, Pb, B, O (45.76%), oxidic fractions of other elements (Sn, Ge etc. - Fig.3), C (~ 4%), halogens (Cl (~ 0.019%), Br) and nitrogen. All metal-oxide structures identified in the Su structure form nanostructured elements having a compact amorphous structure (Fig.5), whose topological distribution is most likely of the type modelled in figure 2. This was confirmed by the numerical data obtained by elementary morphological segmentation (fig. 4b) for the compactness parameters. The compactness range of values (Round) has values in the interval (1.29 ÷ 9), representing a measure of deviation from the sphericity of agglomeration domains and of the nanostructured elements which enter their structure (1 corresponds to the perfect spherical structures) (Jain, 1989). The perimeter (Round = $P^2/4\pi S$) has values on the domain (~ 10.3-107.12 nm), a measure of the size of nanoparticles and their agglomeration domains.

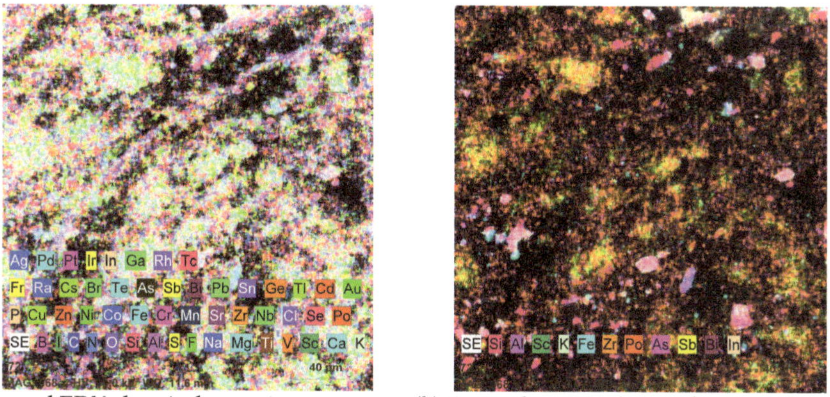

(a) General EDX chemical mapping (b) Particular EDX chemical mapping

Fig. 3. Su EDX mapping (Su laid-down on ordinary paper)

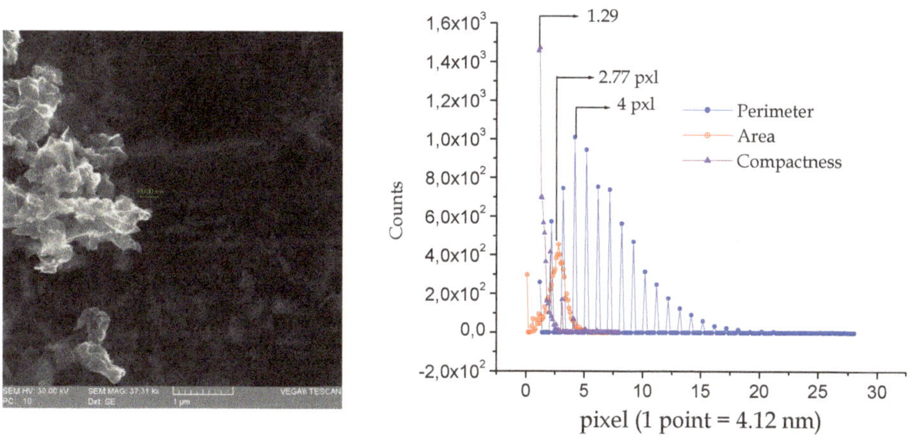

(a) SEM images of Su (b) Morphostructural parameters of Su

Fig. 4. SEM morphotopological investigation of Su (VEGA II LMU microscope)

The identification of chlorine and bromine in the Su structure implies that the metal-oxide mixture has a complex chemical functionality, which includes metal salts (MX, X - Cl, Br, etc.), halide oxides (MOX), metal-oxide-halogenated mixtures $((M_i\text{-}O\text{-}X)_n(M_j\text{-}O\text{-}X)_m$; i, j – different types of metals; m, n - compositing ratio), H and -OH chemical reactive groups, electrical charged microsites, etc. $M_{OF\text{-}DP}$ was obtained by the regeneration of the compositing mixture (functionalised magnetite, functionalised Su and solubilised celulose) in distilled water. As an applicative and testing form, $M_{OF\text{-}DP}$ was obtained as a powder by mild thermic dehydration at 80^{0}C (fig.4A.a), and as a membrane (fig.4A.b) concentrated suspension.

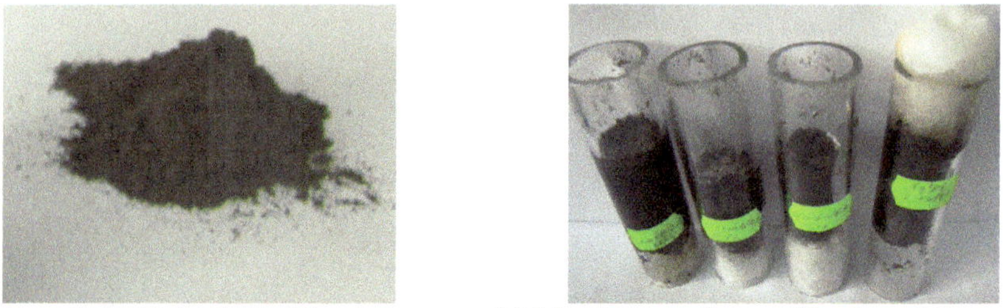

(a) Fine powder of dehydrated Su (b) Membranary layer of hydrated Su

Fig. 4A. Applicative forms of Su

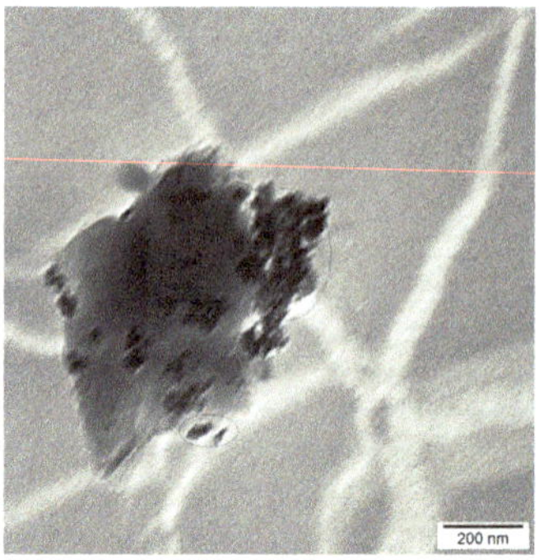

Fig. 5. TEM images of Su embedded in the structure of M_{OF-DP} –uncharged pollutants

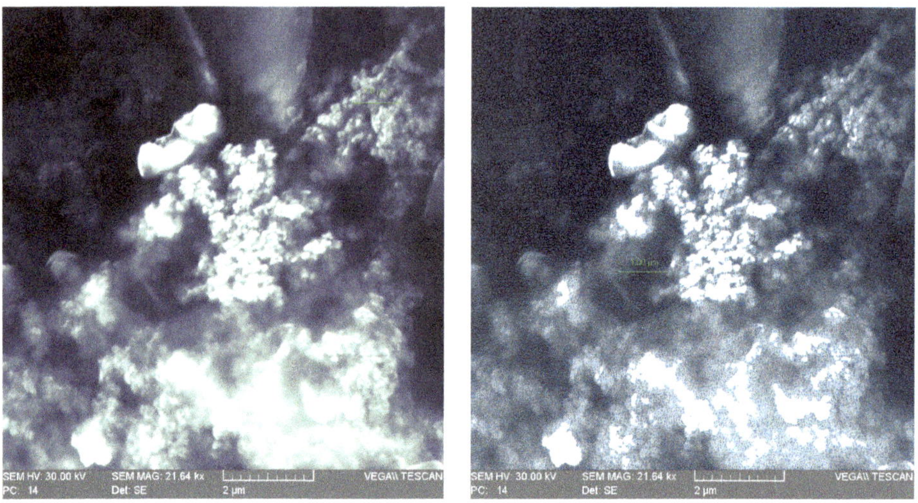

(a) SEM image of dehydrated Su (b) Elementary segmentation of the image attached in fig.6a

Fig. 6. SEM investigation of dehydrated M_{OF-DP} (VEGA II LMU microscope)

Morphostructural investigations carried out by scanning electron microscopy (SEM) show that dehydrated M_{OF-DP} has a smaller and more irregular specific surface, as compared to the hydrated M_{OF-DP}. The distributions of the agglomeration domain surfaces of the dehydrated M_{OF-DP} has values between 194 nm^2 and 12 μm^2 (fig.8). From the investigation of the distribution of the domain compactness of the agglomerated structures and of the nanoparticles whithin their structure, taking into account the analytical relationship Round $= P^2/4\pi S$, it follows that the dehydrated material suffered internal degradation processes that led to its severe discretization and to the increasing of agglomeration domains. It is likely that the increasing of the agglomeration domains may be due to secondary internal morphochemical modelling processes under the influence of temperature, as a result of which, uncontrolled reticulation processes might be established between the structural elements of the composite. It is very likely that, following the dehydration processes, M_{OF-DP} may lose and partially modify its functionality.

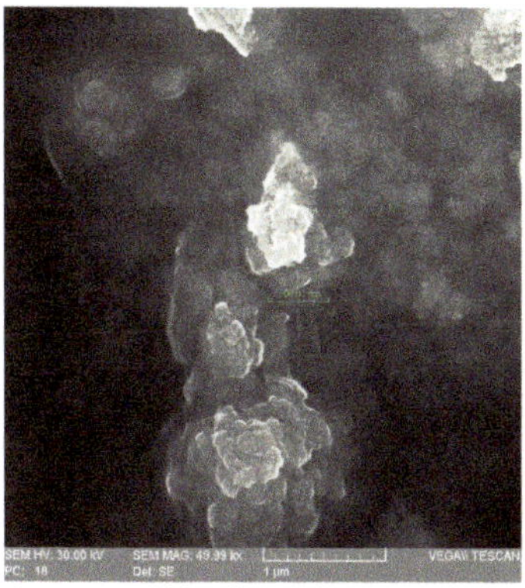

(a) SEM image of hydrated M_{OF-DP}

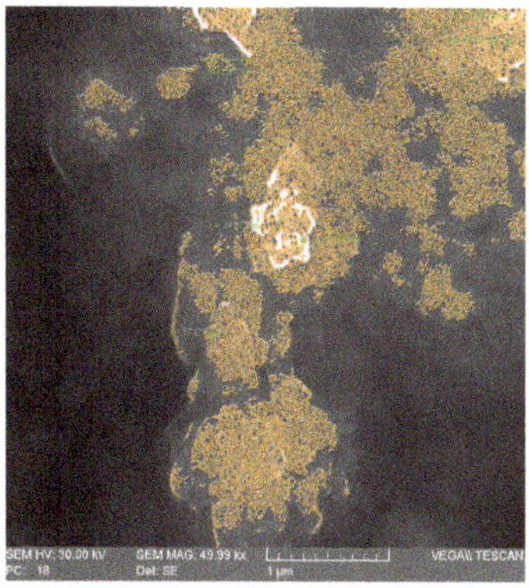

(b) Elementary segmentation of the image attached in fig.7a

Fig. 7. SEM investigation of hydrated M_{OF-DP} (acquired with VEGA II LMU microscope)

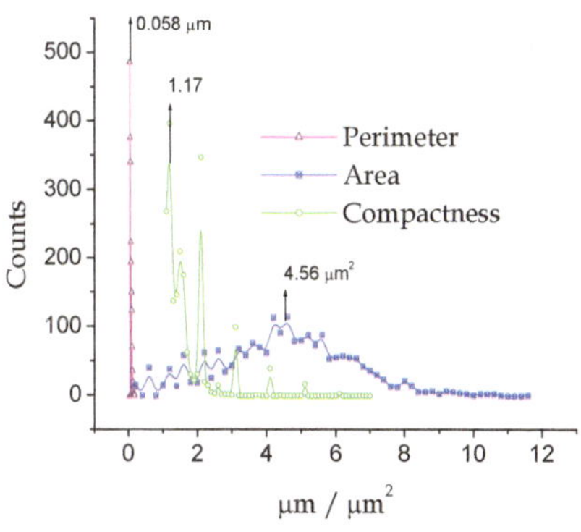

(a) Dehydrated M_{OF-DP} (attached to fig.6b)

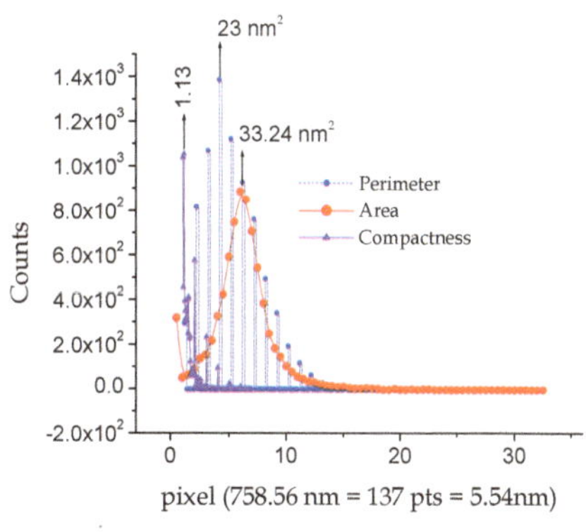

(b) Hydrated M_{OF-DP} (attached to fig.7b)

Fig. 8. Elementary morphostructural parameters of M_{OF-DP}

4.1 Depollution of sewage waters using functionalised M_{OF-DP}

This type of investigations had as main objective the determination of depollution parameters (yield and depollution efficiency) for the type of investigated water, as well as for the morphostructural and morphochemical stability of M_{OF-DP}. Investigations have been configured in accordance to the diagram in figure 9. Samples were investigated by gas chromatography coupled with mass spectrometry (GC-MS), using a cromatograph of the GC Focus type (AI300 autosampler, split / splitless injector, chromatograph column of the TR5MS - 30 m x type with Φ_{ext} = 0,25 mm and 0.25 mm stationary film thickness) and a mass spectrometer of the DSQII type. Previously, samples were concentrated by solvent extraction and evaporated in nitrogen flow. To check the saturation point of M_{OF-DP} with pollutants, 50 ml of water charged with pollutants were passed through the filtering layer (fig. 9), in four steps, the (P1, P2, P3, P4) samples were properly prepared and investigated. To determine the analytical correspondence holding between the separated pollutant quantity and the real pollutant quantity in the contaminated water (R_p), the quantity of pollutants present in the non-filtered water was determined before starting the investigating processes. The separated pollutant quantity was estimated by determining the pollutant fractions found in P_1, P_2, P_3 and P_4, after filtering (c% (pollutants) + c% (separated) = 1). The investigated samples were taken from the sewerage system of Bucharest municipality. With a view to determining the morphochemical and morphostructural stability of M_{OF-DP}, the composite charged with pollutants was investigated by scanning electron microscopy (SEM), energy-dispersive X-ray spectroscopy (EDX) and infrared spectroscopy (IR).

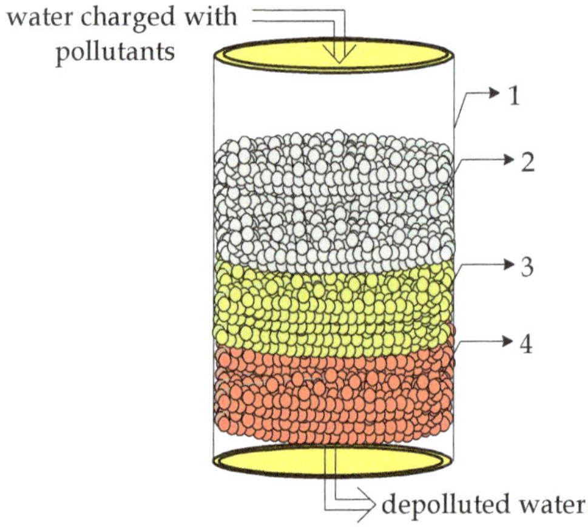

Fig. 9. Experiment configuring for the investigation of water depollution yield and efficiency of M_{OF-DP} (1 – fixing column; 2 – polluted water; 3 - M_{OF-DP} layer; 4 – fibre glass)

The results of the investigations carried out by GC-MS are presented in table 1 and they demonstrate the fact that the tested water had a high level of organic pollution, containing a large range of organic pollutants (n-tridecane, propanoic acid, 2-methyl-,1-(2-hydroxy-1-methylethyl)-2,2-dimethylpropyl ester propanoic acid, 2-methyl-,3-hydroxy-2,4,4-trimethylpentyl ester, diethyl benzene-1,2-dicarboxylate, phthalic acid, isobuthyl-2-penthylester, butyl octyl phthalate, dibutyl phthalate, ((Z)-9-octadecenenitrile).

Pollutant/ undecelable peaks	A_r-R_P	A_r-P1	A_r-P3	A_r-P4
TS_P	2848678635	15047488950	10533922491	8526288052
8.33	5094541	0	2272703	7299050
PL1	977320	0	0	0
10,75	5268711	0	0	5547832
PL2	2465265	0	0	0
PL3	1405799	0	0	0
12,49	1740771	0	0	0
12,93	2828747	0	0	3710438
14,85	3347278	0	1841992	2755159
PL4	1680227	0	0	0
16,13	523552	0	0	0
16,51	1704027	0	0	0
17,23	1278456	0	0	0
17,98	901907	0	0	1129834
PL5	132278	1730996	0	0
18,27	267140	0	0	0
PL6	507889	0	0	0
PL7	10830818	8652426	0	0
PL8	1859244	0	0	0
24,00	5434244	2712903	4682044	2903562
26,56	11791893	8461410	6699962	4400335

Table 1. The relative abundance (A_r) of decelable and partially decelable pollutants in filtered water - case of sewage waters (where: PL1 - n-tridecane; PL2 - propanoic acid,2-methyl-,1-(2-hydroxy-1-methylethyl)-2,2-dimethylpropyl ester; PL3 - propanoic acid,2-methyl-, 3-hydroxy-2,4,4-trimethylpentyl ester; PL4 - diethyl benzene-1,2-dicarboxylate; PL5 - phthalic acid, isobuthyl-2-penthylester; PL6 - butyl octyl phthalate; PL7 - dibutyl phthalate; PL8 - (9Z)-9-octadecenenitrile)

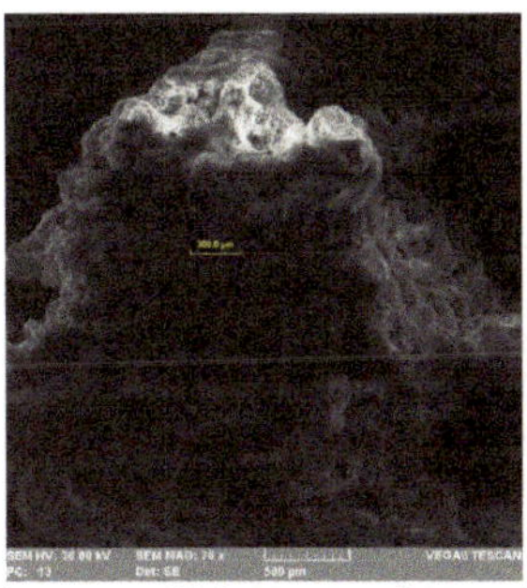

(a) SEM images of $M_{OF\text{-}DP}$ uncharged with pollutants

(b) EDX images of $M_{OF\text{-}DP}$ charged with pollutants (attached to fig.10a)

Fig. 10. SEM investigation of $M_{OF\text{-}DP}$ charged and uncharged with pollutants - case of sewage waters (images acquired with VEGA II LMU SEM microscope)

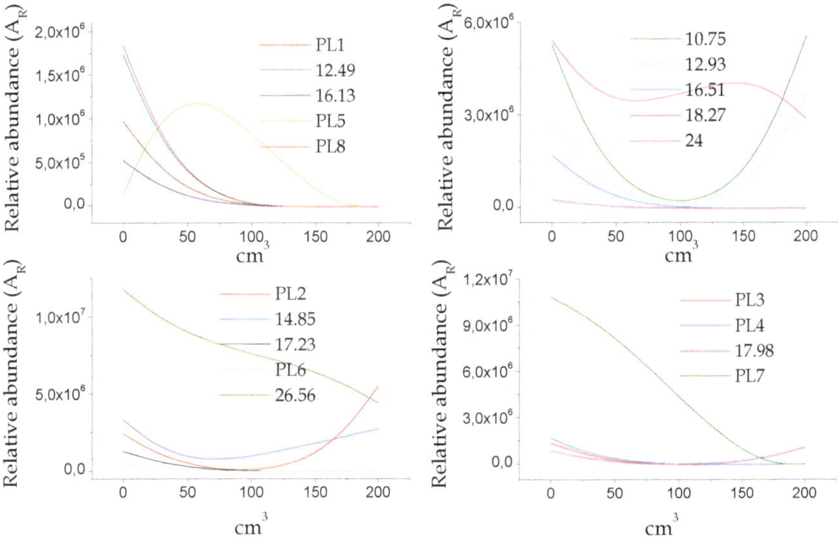

Fig. 11. Relative yield of decelable and partially decelable pollutants retained by 71.43 grams of M_{OF-DP} (according to tab.1)

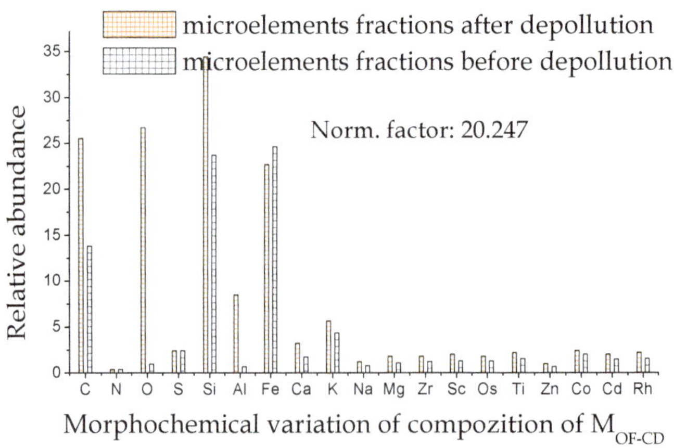

Fig. 12. Relative variation of microelement fractions in the depth and on the surface of M_{OF-DP} charged with pollutants

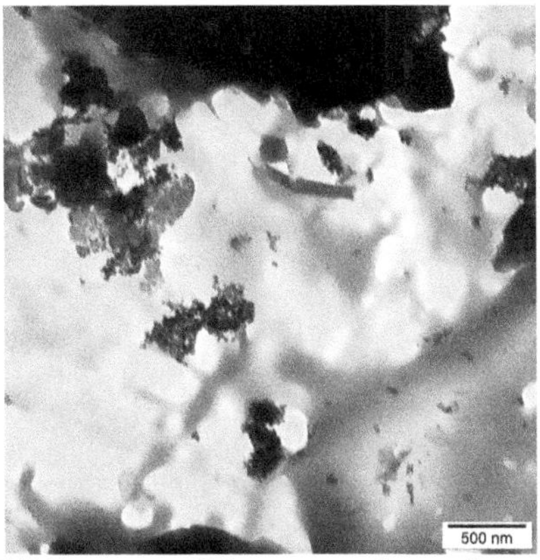

(a) TEM images of $M_{OF\text{-}DP}$ - uncharged pollutants

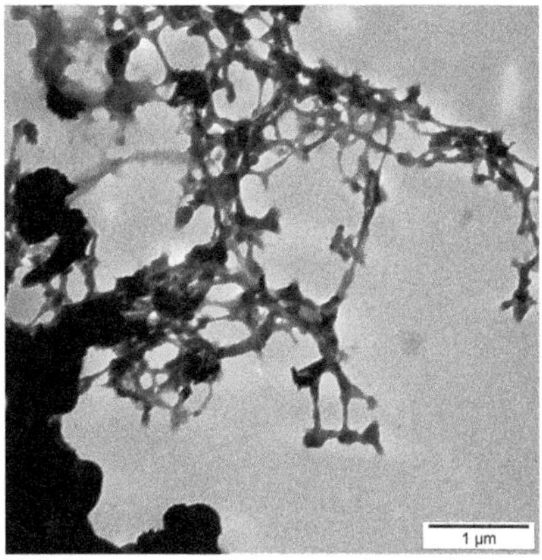

(b) TEM images of $M_{OF\text{-}DP}$ - charged pollutants

Fig. 13. TEM morphostructure investigation of $M_{OF\text{-}DP}$ charged with pollutants - case of sewage waters (images acquired with Philips S208 TEM microscope)

The GC-MS investigations revealed the fact that the filtered water (the water charged with pollutants) contains a wide range of organic pollutants, a significant fraction of them being partially decelable. Partially decelable pollutants have clear, distinct peaks, whose variation can be quantified depending on the quantity of filtered water (table 1). A great part of the chemical compounds present in the investigated water could not be sensitized in relation to the retention time and the specific mass fragments. The quantification of these types of pollutants was entirely determined, by quantifying the variation degree of the whole sensitization area. It has been noticed that the concentrations of decelable pollutants and of the partially decelable ones decrease rapidly with the increase of the filtered water quantity (fig.11). Each pollutant attains a certain value for which the membrane is saturated through specific reticulation. After attaining the saturation peak, the filtering layer loses its ability to separate the pollutants which saturated it, thus becoming permeable in relation to them.

The saturation processes are most likely due to the neutralization through reticulation of the functional groups with which the pollutants come into contact. This fact was pointed out indirectly by SEM (fig.10 a), TEM (fig.13) and EDX (fig.10b, fig.12) investigations. Thus, we observed the severe degeneration of the specific reticulation surface of $M_{OF\text{-}DP}$ (fig.13b). The degradation and reticulation processes have also been confirmed by EDX investigations, during which we noticed the fact that the microelement fractions (especially C and O) found in the depth and on the surface of $M_{OF\text{-}DP}$ increase significantly (fig.12). This fact can be explained by admitting the fact that organic pollutants reticulated on the surface or in the depth of $M_{OF\text{-}DP}$ form covering layers which lead to the masking of the detection of the structural elements of the composite.

The determinations in fig.12 have been carried out by supposing that the statistical quantity of silicon in the composite structure is approximately constant. Thus, to obtain the same quantity of silicon on the investigated microsurfaces, both before and after the depollution process, a normalization factor of 20.247 was added (20.247 x c%Si(after) = c%Si(before)), to minimize the errors of the statistical distribution of microelements. In order to obtain results as accurately as possible, the normalization microelement must have variations of the concentration distributions only on one of the investigated microsurfaces. These errors may be due to the different distribution probabilities of morphochemical patterns on distinct microsurfaces, as well as to the different periods of time of data acquisition.

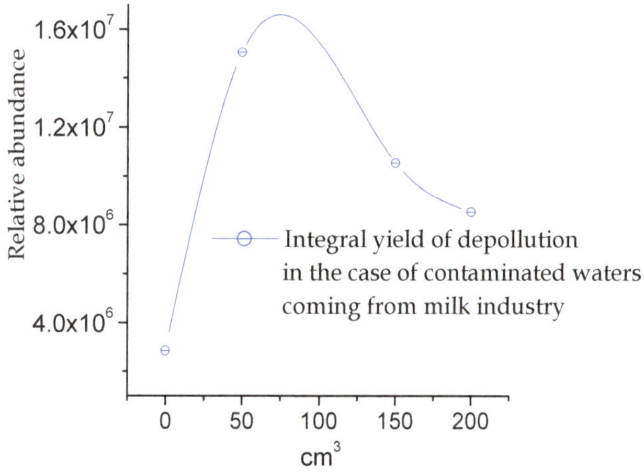

Fig. 14. Integral yield of sewage water depollution in the presence of 71.43 grams of M_{OF-DP}

The integral retention spectrum of the pollutants has been defined as the whole recorded sensitization area, minus the area of the background signal. This represents an indirect measure of the pollutant retention spectrum size, offering relative information regarding the variation of the quantity of organic compounds, which are generated or disappear from the collected depolluted water, before and after the depollution process. The integral retention curve shown in figure 14 presents the maximum variation at approximately the same value as that of the saturation point of MOF-D with decelable and partially decelable pollutants. This suggests that during the depollution process, the reticulated pollutants are partially or completely degraded, at the level of functionalized micro surfaces of the structural components of M_{OF-D}. Degradation processes can influence the dynamics of the depollution processes, as well as the achievement of specific saturation points. This explains the analytical structure of the depollution curves of pollutants 12.93 (specific retention time), 24 and PL2 (fig.11). The same tendency is also shown by the decontamination curve of the entire spectrum of pollutants (fig.14).

4.2 Depollution of waters coming from milk industry using M_{OF-DP}

As in the previous case, this type of investigations had as objective the determination of depollution parameters of polluted water from milk industry, as well as the morphostructural and morphochemical stability of M_{OF-DP}. The investigated samples came from a milk processing plant. Complex investigations were carried out by scanning electron microscopy (SEM), energy-dispersive X-ray spectroscopy (EDX).

Pollutants	**A_r-R_P**	**A_r-P1**	**A_r-P2**	**A_r-P3**	**A_r-P4**
Fenil carbamat	5786268	0	0	0	0
p cresol	28206310	0	3291097	2866582	4749335
5H-1 pyridine	9622808	0	0	0	0
NN dodecilamina	5710090	0	0	0	0
4 methylindolina	2179991	0	0	0	0
Heneicosane	589354	0	0	0	0
N[3[N-Azydiril]propilidena] 3-dymethylaminopropylamina	1378132	0	0	0	0
Octadecylacetate	6770653	0	0	0	0
Acidhexadecanoicisopropilester	5107004	0	0	0	0
Adogen	6786347	0	0	0	0
Acidoctadecanoictertbutilester	7061859	0	0	0	0
All-trans-squalene	20340307	0	0	0	0
Region with undecelable pollutants	198277663	0	0	0	0

Table 2. The relative abundance of decelable and partially decelable pollutants - case of waters coming from milk industry

Depolluted volume of water (cm^3)	Total relative abundance (A_r)
50	2494088578
100	1286801634
150	2164651861
200	5424524876

Table 3. Integral fraction of sensitized pollutants (58.6 grams of M_{OF-DP}) in the cases of P1, P2, P3 and P4 stages of depollution – case of waters coming from milk industry

To determine the relative degree of depollution, the total area of the contents of pollutants in the unfiltered water (R_{P0} = 6227542517) was measured. Also, the total area of the contents of pollutants contained in 50 cm^3 of distilled water (R_{PA} = 2371295106) passed through the filtering layer was measured, to correct and determine the level of contaminants released by M_{OF-DP}.

C	13.83	S	2.42	C	2.83	S	1.11
Sc	1.32	Cd	1.48	Sc	0.16	Cd	0.15
Zr	1.25	Os	1.32	Zr	0.34	Os	0.22
Fe	24.62	Ca	1.72	Fe	1.29	Ca	0.64
K	4.37	Si	23.69	K	0.56	Si	2.72
Al	6.8	Mg	1.09	Al	1.05	Mg	0.22
Na	0.78	O	0.99	Na	0.36	O	1.01
N	0.39	Ti	1.56	N	0.1	Ti	0.14
Co	2.03	Zn	0.70	Co	0.13	Zn	0.06
Rh	1.56			Rh	0.15		
(a) before depollution				(a) after depollution			

Table 4. Fractions of microelements (arbitrary u.m) measured on surface and in the depth of M_{OF-DP} in the cases of contaminated water depollution coming from milk industry

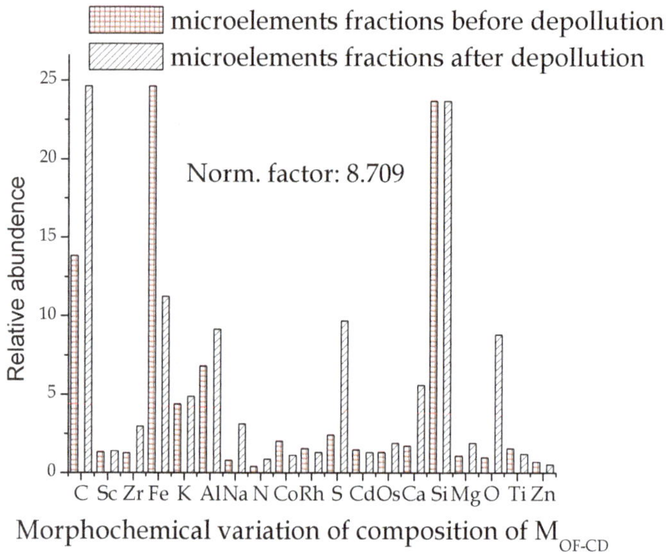

Fig. 15. Relative variation of chemical microelements fraction in the depth and on the surface of M_{OF-DP} (according to tab.4)

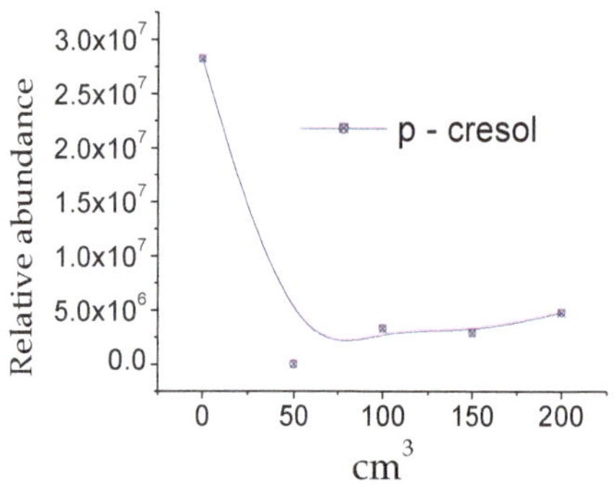

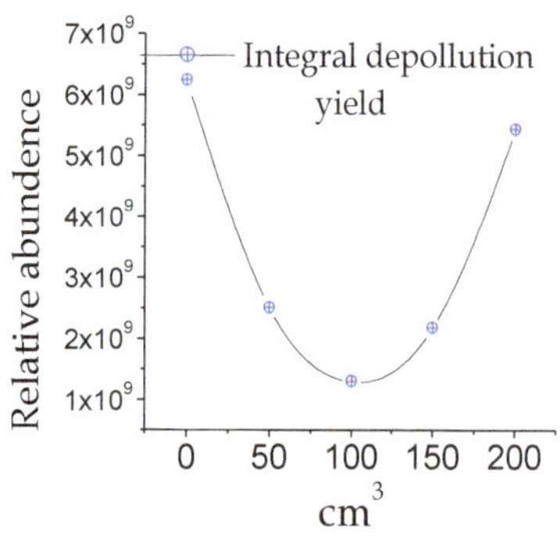

(a) Relative yield of p-cresol (b) Integral depollution yield

Fig. 16. Yield and efficiency of 58.6 grams of M_{OF-DP} – case of waters coming from milk industry

By determining R_{P0} şi R_{PA}, we had in mind checking the morphochemical stability of M_{OF-DP}, as well as determining the contribution of the pollutants coming from the filtering layer to the total pollution spectrum of filtered water (fig.19). Also, the aim of determining R_{P0} and

R_{PA} was to the determine the background signal level, found in the analytical structure of the aquired GC-MS spectra. The results of the investigations for determining the depollution yield and efficiency carried out on waters charged with pollutants coming from the milk industry have shown that $M_{OF\text{-}DP}$ has high depollution yields. In the case of decelable organic pollutants (table 2, fig.19), the individual retention yields of pollutants can reach the maximum value (~ 100%).

(a) SEM images of pollutant-charged $M_{OF\text{-}DP}$ (b) EDX images of pollutant-charged $M_{OF\text{-}DP}$

Fig. 17. SEM investigation of pollutant-charged $M_{OF\text{-}DP}$ - case of water depollution coming from milk industry (images acquired with VEGA II LMU SEM microscope)

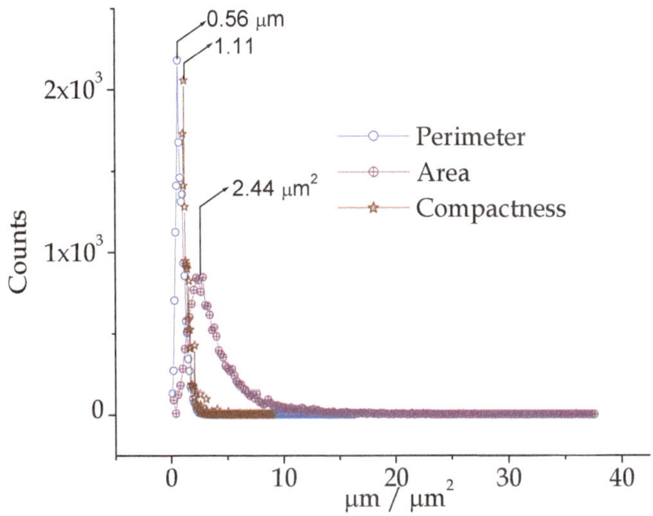

Fig. 18. Elementary morphostructural parameters of $M_{OF\text{-}DP}$ in the case of depollution of water coming from milk industry (following the depollution process)

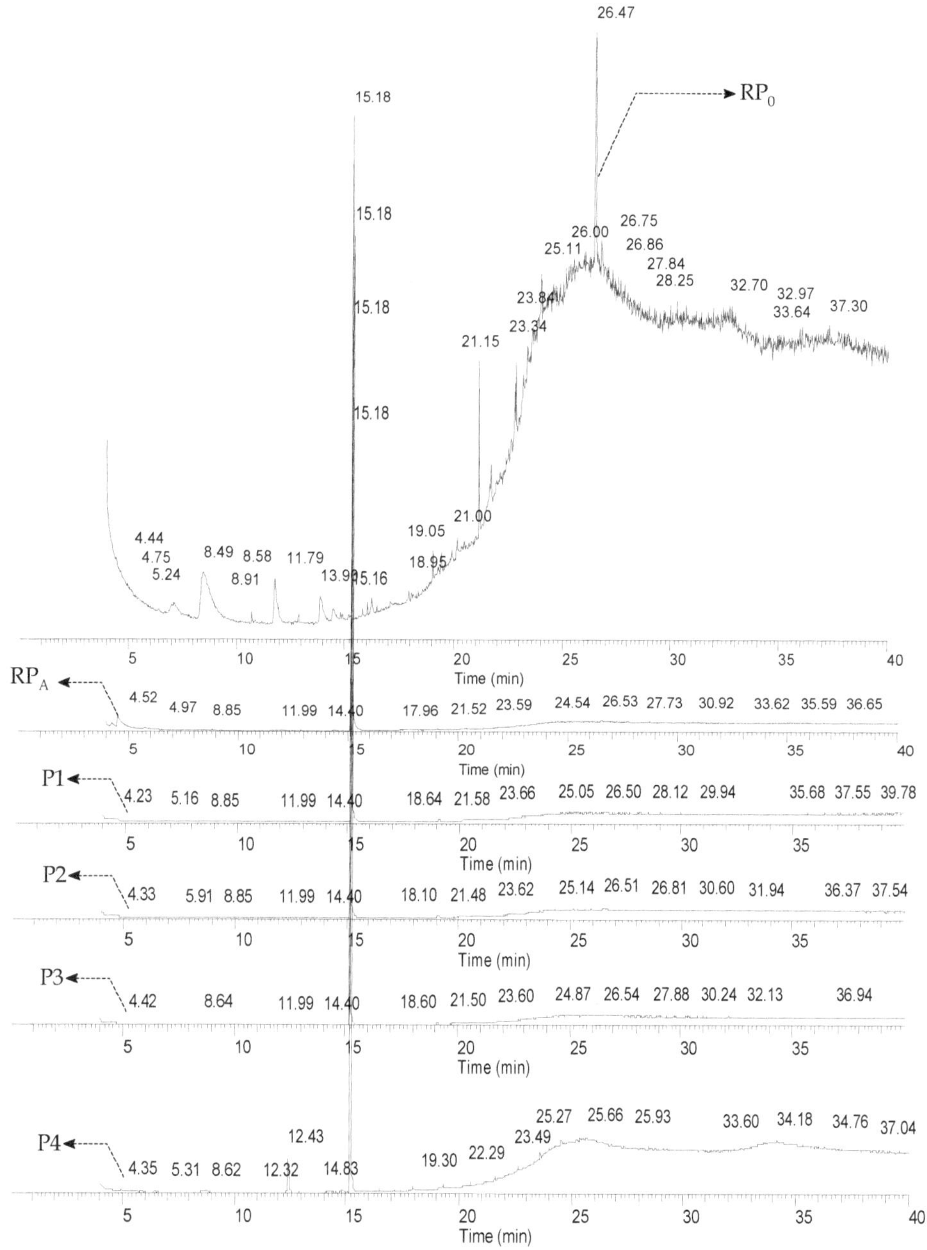

Fig. 19. GC-MS spectra of collected samples – evolution of integral depollution yield as a function of volume of filtered water (η = f(V(cm^3), S(p1, ... , pn); V – depolluted water volume; S – type of the contained p1, ... , pn pollutants)

In figure 16.a one can observe that p-cresol has a higher saturation capacity (fig.16), most likely, due to the high content of this pollutant, either due to the lack of chemical affinity of the specific functional groups of M_{OF-DP} in relation to p-creso. The obtained results demonstrate that the mechanisms of reticulation and separation processes of the pollutants are uniform in relation to the functionality type. These findings also remain valid in the case of the investigation of waters coming from milk industry.

In the case of the contaminated waters coming from milk industry, several features have emerged, due, most likely, to the type of pollutants and to their concentrations in contaminated water. It is noted that the integral depollution spectrum (fig.16b) presents a more predictable analytical behaviour, as compared to the integral depollution spectrum of sewage waters. Pollutants contained by water generate specific morphostructural reticulation and agglomeration effects (fig.17, fig.18). We have observed that the characteristic agglomeration domains have the following values: a. average surface: ~ 2.44 μm2 (distributed on an agglomeration domain that ranges from tens of nanometers to 35 μm^2); b. maximum perimeter: 13 μm; c. compactness value distributed on the 1 - 10 range (fig.18). The determined values for compactness show that M_{OF-DP} charged with pollutants has a compact structure, which has been severely degraded from a morphostructural point of view, as a result of the reticulation and internal morphochemical modification processes. The obtained numerical results in the case of polluted waters coming from milk industry demonstrate the high capacity of M_{OF-DP} to separate, degrade and encapsulate pollutants, on the surface and in the depth of its composite structure (tab.4, fig.15).

GC-MS investigations have demonstrated the ability of the obtained composite to reticulate a large number of pollutants (table 2), having a complex and varied functionality, as follows: saturated compounds with large linear chain (heneicosane), organic compounds with functionality and nitrogen content (adogen, 5H-1 pyridine, NN dodecilamina, N [3 [N-Azydiril] propilidena] dymethylaminopropylamina 3, 4 methylindolina), organic acids, esters, derivatives of benzene, carbamates, cyclic organic compounds, etc.

4.3 Depollution of waters coming from medicine industry using M_{OF-DP}

The investigated samples came from an antibiotics production company. The results of the investigations carried out by GC-MS are presented in table 5, demonstrating that they are heavily contaminated and degradated, containing a wide range of organic pollutants, varied from a morphochemical, functional and morphostructural viewpoint.

As it can be seen in the analytical representations in figures 20 and 22, the obtained material has a high potential of separation of organic pollutants, being able to reduce their concentrations below the limit of analytical technique detection (at ppm scale), given that they are found in appreciable fractions in the contaminated water.

It is to be noted that the investigations revealed the undifferentiated action of M_{OF-DP} on the the entire spectrum of partially decelable, decelable and undecelable pollutants found in the investigated polluted waters. This feature of the functionalized nanocomposite material is reflected by the total decontamination yield, which has a decreasing analytical variation, with the increasing of the water passed through the retention layer. It is also to be noted that the relative depollution yields, specific to each pollutant type, has different analytical behaviours, in accordance to the specific chemical affinity in relation to the functionalized filtering support (fig.20, fig.22).

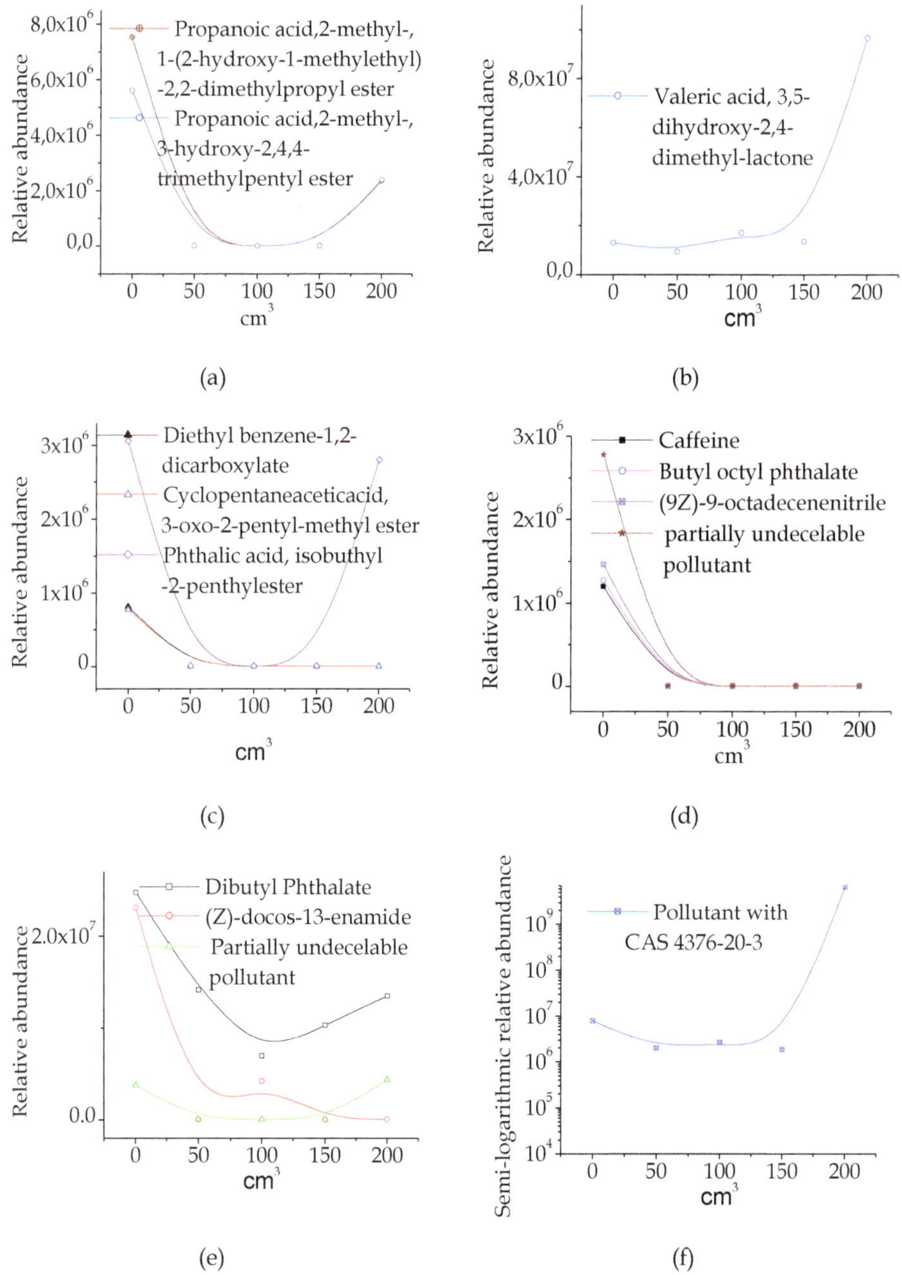

Fig. 20. Relative yield of decelable and partially decelable pollutants retained by 67.4 grams of M_{OF-DP} (according to tab.5)

Pollutants	A_r-R_P	A_r-P1	A_r-P2	A_r-P3	A_r-P4
Integral pollutants abundance	39093281641	12969701158	11684530024	9414486816	13743895083
Acetonyl dimtehylcarbinol	268513312	295489231	736052142	570281431	503499232
Propanoic acid,2-methyl-, 1-(2-hydroxy-1-methylethyl)-2,2-dimethylpropyl ester	7515830	0	0	0	2353024
Propanoic acid,2-methyl-, 3-hydroxy-2,4,4-trimethylpentyl ester	5585726	0	0	0	2379634
Valeric acid, 3,5-dihydroxy-2,4-dimethyl-lactone	12891275	9221762	16911174	13363342	96422370
Partially undecelable	3741563	0	0	0	4326558
Diethyl benzene-1,2-dicarboxylate	815869	0	0	0	0
Cyclopentaneaceticacid, 3-oxo-2-pentyl-methyl ester	772606	0	0	0	0
Phthalic acid, isobuthyl-2-penthylester	3049818	0	0	0	2799809
Caffeine	1202278	0	0	0	0
Butyl octyl phthalate	1270036	0	0	0	0
Dibutyl Phthalate	24757815	14138926	6956899	10335640	13505478
(9Z)-9-octadecenenitrile	1458619	0	0	0	0
Partially undecelable pollutant	2776681	0	0	0	0
CAS 4376-20-3	7729544	1965075	2632588	1815752	6402313990
(Z)-docos-13-enamide	22996078	0	4152090	0	0

Table 5. The relative abundance of the decelable and partially decelable pollutants - case of polluted waters coming from medicine industry

C	13.83	S	2.42	C	2.31	S	0.23
Sc	1.32	Cd	1.48	Sc	0.16	Cd	0.14
Zr	1.25	Os	1.32	Zr	0.19	Os	0.16
Fe	24.62	Ca	1.72	Fe	1.23	Ca	0.49
K	4.37	Si	23.69	K	0.4	Si	2.48
Al	6.8	Mg	1.09	Al	0.62	Mg	0.14
Na	0.78	O	0.99	Na	0.11	O	1.76
N	0.39	Ti	1.56	N	0.08	Ti	0.15
Co	2.03	Zn	0.70	Co	0.14	Zn	0.07
Rh	1.56			Rh	0.32		
(a) before depollution				(a) after depollution			

Table 6. Fractions of microelements (arbitrary measure units) measured on surface and in the depth of $M_{OF\text{-}DP}$ in the cases of depollution of contaminated waters coming from milk industry

As expected, due to its polyvalent functionality, $M_{OF\text{-}DP}$ saturates discretizedly, each pollutant separately saturating the composite, without masking or cancelling its functionality and the integral depollution potential. This implies the existence of specific differentiated reticular mechanisms, which take place in the composite nanostructures, leading to the selective separation of pollutants, depending on the specific chemical reactivity and on the achieved density of functional groups on $M_{OF\text{-}DP}$ microsurfaces.

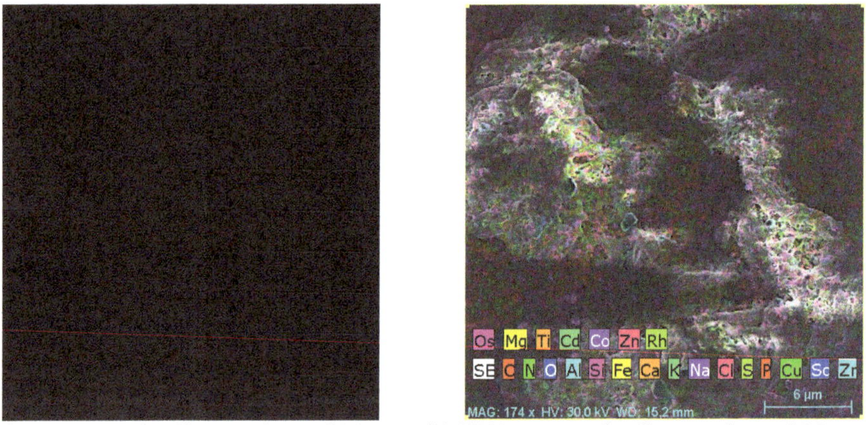

(a) SEM images of pollutant-charged $M_{OF\text{-}DP}$ (b) EDX images of pollutant-charged $M_{OF\text{-}DP}$

Fig. 21. SEM investigation of $M_{OF\text{-}DP}$ charged with pollutants - case of depollution of water coming from medicine industry (VEGA II LMU SEM microscope)

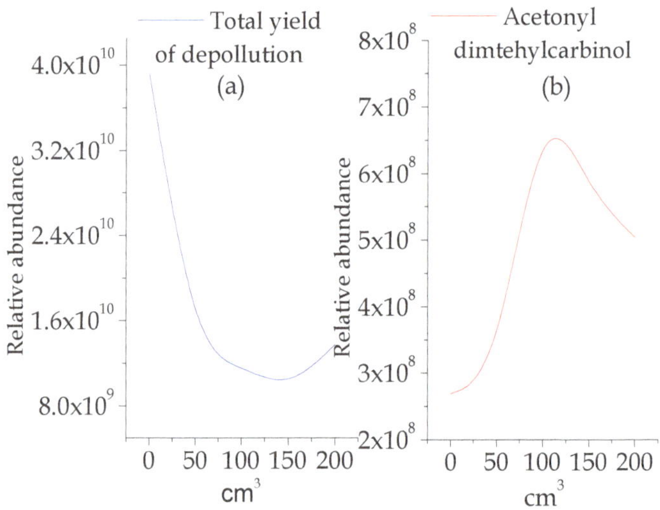

Fig. 22. Integral yield and efficiency of the depollution of water coming from medicine industry in the presence of 67.4 grams of $M_{OF\text{-}DP}$ ((a) - integral yield of depollution; (b) - relative yield of acetonil dimethylcarbinol)

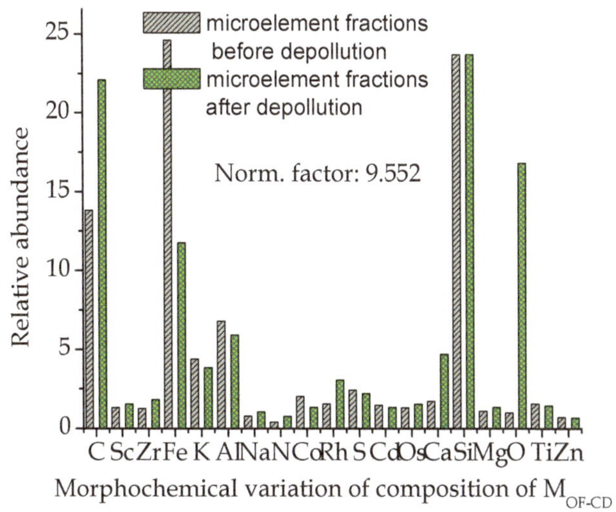

Fig. 23. Relative variation of microelements in the depth and on the surface of M_{OF-DP} (according to tab.6)

The separation investigations of the pollutants found in waters coming from medicine industry shows the manifestation of the dynamic degradation phenomenon of the pollutants. This is demonstrated by the analytical behaviour of the depollution curves of the valeric acid, 3,5-dihydroxy-2,4-dimethyl-lactones (fig.20b), the partially decelable pollutant CAS 4376-20-3 (fig.20f) and acetonyl dimethylcarbinol (fig.22b), which, after reaching the saturation point of the functionality of M_{OF-DP}, have growth rates that exceed by far the originally detected values. This indicates the presence of a pollutant in the contaminated water, which, following the reticulation processes, degrades and generates chemical structures that cannot be reticulated, or that are partially reticulated by the composite. All pollutants entering the structure of the integral spectrum, sensitized by pollution, of the investigated waters, have lower or higher separation rates. Thus, the lack of the reticulation potential, in relation to the degenerated pollutants listed above, is due, most likely, to: a. the saturation of M_{OF-DP} with the previously existing fractions of pollutants; b. large quantities of pollutants present in depolluted water, which quickly saturate the functionality of M_{OF-DP}; the degeneration of the separated pollutants in the depth of M_{OF-DP}. Moreover, the morphostructural (agglomeration, compactness) and morphochemical (functionality saturation processes, masking, chemical degradation processes of reticulated pollutants) degeneration processes were emphasized in all investigated depollution cases. The effects of pollutant charging are also showed in the case of polluted waters coming from medicine industry. In this sense, the processed data presented in figure 15 (tab.6) indicate a significant increase in the quantity of chemical microelements, specific to the structures and organic pollutants (C, N). This is indirectly confirmed by morphostructural investigations (fig.21, fig.21b), indicating the increase of the agglomeration and compactness degree of the composite, as well as its degeneration under the action of separated organic pollutants.

5. Conclusions

In this chapter we have presented and investigated the specific mechanisms underlying the reticulation, separation and encapsulation processes of organic pollutants. Specific depollution mechanisms, as well as the main depollution parameters (yield, efficiency), have been investigated by testing a polyfunctionalized nanocomposite material ($M_{OF\text{-}DP}$) on three types of water charged with organic pollutants. The polluted waters on which the depollution features of $M_{OF\text{-}DP}$ have been investigated came from various pollution sources: milk processing industry, medicine industry, sewage waters. The investigations' secondary aim was to determine the effects of organic pollutants on the morphological and morphochemical structure of the obtained $M_{OF\text{-}DP}$, as well as to determine how they influence the depollution parameters. It was intended, in particular, to check the range of pollutants that $M_{OF\text{-}DP}$ is able to separate, in order to determine the type of chemical functionality of pollutants capable to establish reticulation relations with the functional structure of the composite. The reticulated functionalities are a direct measure of the depollution efficiency of the obtained material, by defining the range of separated pollutants and the extent to which the depollution was carried out. The investigation processes included various analytical techniques, in order to confirm, through complementarity (SEM, TEM, EDX, FTIR, GC-MS), the depollution parameters and the morphostructural characteristics of $M_{OF\text{-}DP}$, both before the testing and after its completion. The investigation results confirm the reticulation, encapsulation and degradation mechanisms of the pollutants that the authors proposed in chapters „New methods and new types of functionalised nanocomposites intended for the ecological depollution of waters" and „Nanocomposite materials with oriented functionalized structure". The acquired experimental data have shown that the obtained $M_{OF\text{-}DP}$ presents high pollutant separation yield and efficiency. The concentrations of most decelated pollutants and of those partially decelated in depolluted water were below the detection limit, showing the applicative potential of this type of materials. The aim of the investigations of the morphological and morphochemical structure which we have carried out was to identify and quantify the critical parameters that influence the processes of encapsulation and degradation of pollutants. The acquired experimental data indicate that $M_{OF\text{-}DP}$ is able to reduce the level of organic pollution in surface waters by up to two orders of magnitude (the observed typical values have a separation factor of $20 \div 50$ x). The oxide materials with a functionalised mixed structure represent a viable solution for the removal of organic pollutants in surface waters, having the most favourable applicative potential in terms of economy, depollution yield and efficiency.

6. Acknowledgements

The authors are grateful for the financial support granted by the CNMP (National Research, Development and Innovation Plan II - Program 4) inside project 32-165/2008, as well as for the logistic support granted by the Scientifical Research Center for CBRN Defence and Ecology. Also, for their contribution to the obtaining of results, we are grateful to Dionesie Bojin, Filipescu Laurențiu, Toader Constantin and Mihaela Muraşan. For her contributions to text editing and translating, we are grateful to Loredana Iordache.

7. References

Lu, A. H.; Zhao, D. & Wan, Y. (2010). *Nanocasting - A Versatile Strategy for Creating Nanostructured Porous Materials*, RS•C, ISBN 978-0-85404-188-6

Cheremisinoff, N. P. (2002). *Handbook of water and wastewater treatment technologies*, Butterworth-Heinemann, ISBN 0-7506-7498-9, Boston

Tolgyessy, J. (2003). *Water, air and soil - Fundamental sources of the biosphere*, In: Chemistry and biology of water, air and soil - Environmental Aspects, J. Tolgyessy, (Ed.), 3-13, Elsevier, ISBN 0-444-98798-3, Amsterdam-London-New York-Tokyo

Evanghelou, V. P. (1998). *Environmental soil and water chemistry - Principles and applications*, John Wiley & Sons, ISBN 0-471-16515-8 (cloth: alk. paper), New York

Rosenfeld, P. E. & Feng, L. G. H. (2011). *Risks of Hazardous Wastes*, Elsevier, ISBN 978-1-4377-7842-7, Amsterdam

Bayliss, C. R. & Langley, K. F. (2003). *Nuclear Decommissioning, Waste Management, and Environmental Site Remediation*, Elsevier, ISBN 978-0-7506-7744-8, Amsterdam

Socolow, R. H.; Andrews, C.; Berkhout, F. & Thoma, V. (2003). *Industrial ecology and global change*, Cambridge University Press, ISBN 0-521-47197-4, New York

Lal, R. & Shukla, M. (2005). *Principles of soil physics*, Marcel Dekker, ISBN 0-8247-5324-0, New York, Basel

Haider, K. & Shäffer, M. (2009). *Soil Biochemistry*, Science Publishers, ISBN 978-1-57808-579-8, Enfield (NH), Jersey, Plymouth

Asano, T.; Burton, F. L.; Leverenz, H. L.; Tsuchihashi, R. & Tchobanoglous, G. (2007). *Water Reuse - Issues, Technologies, and Applications*, McGrow-Hill, ISBN 978-0-07-145927-3, Enfield (NH), New York

Cheremisinoff, N. P. (2003). *Handbook of solid waste management and waste minimization technologies*, Butterworth-Heinemann, ISBN 0-7506-7507-1 (alk. paper), Amsterdam, Boston, London

Cheremisinoff, N. P. & Cheremisinoff, P. N. (1995). *Hazardous materials and waste management*, Noyes Publications, ISBN 0-8155-1372-0, Park Ridge, New Jersey, U.S.A.

Hester, R. E. & Harrison, R. M. (2002). *Environmental and health impact of solid waste management activities*, RS•C, ISBN 0-85404-285-7, Manchester

Baud, I.; Post, J. & Furedy, C. (2004). *Solid waste management and recycling*, Kluwer Academic, ISBN 1-4020-1975-0, New York, Boston, Dordrecht, London, Moscow

Yu, T. R. & Wang, E. J. (1997). *Chemistry of variable charge soils, In: Chemistry of variable charge soils*, T. R. Yu, (Ed.), 309-338, Oxford University Press, ISBN 0-19-509745-9, New York, Oxford

Ojovan, M. I. & Lee, W. E. (2005). *An introduction to nuclear waste immobilisation*, Elsevier, ISBN 0-080-44462-8, Amsterdam, Boston, Heidelberg

Iordache, P. Z.; Lungu, R. M.; Epure, G.; Mureşan, M.; Petre, R.; Petrea, N.; Pretorian, A.; Dionezie, B.; Mutihac, L. & Ordeanu, V. (2011). The determination of the nanostructurated materials' morphology, by applying the statistics of the structural element maps, *Journal of Optoelectronics and Advanced Materials*, Vol. 13, No. 5, pp. 550-559, ISSN 1454-4164

Jain, A. K. (1989). *Fundamentals of digital image processing*, Prentice-Hall, ISBN 0-13-336165-9, New Jersey

Part 3

State-of-the-Art Polymers

8

Polymer/Montmorillonite/Silver Nanocomposite Micro- and Nanoparticles Prepared by *In-Situ* Polymerization and Electrospraying Technique

Jeong Hyun Yeum[1], Jae Hyeung Park[1], Jae Young Choi[1]
Jong Won Kim[2], Sung Kyou Han[3] and Weontae Oh[4]
[1]Department of Advanced Organic Materials Science & Engineering, Kyungpook National University, Daegu,
[2]R&D Division, Korea Dyeing Technology Center, Daegu,
[3]R&D Division, Taihan Textile Co. Ltd., Daegu,
[4]Department of Materials and Components Engineering, Dong-eui University, Busan, Korea

1. Introduction

Polymer nanocomposite materials have the capacity for producing a synergetic association with surpassing merit properties which cannot be obtained from the individual components. Such materials can be obtained by simply mixing required organic and inorganic components. The introduction of inorganic material into the polymer matrix has proved to be an effective and low-cost method to improve the performance of the existing polymer materials (Okamoto et al., 2000; Ramos et al., 2000; Wu et al., 2006; Zhu et al., 2000). It has various applications, such as new biological materials (biosensors and biochips), biocompatible thin coatings for medical applications, biodegradable scaffolds, drug delivery system and filter systems (Matsumoto et al., 2005; Okuda et al., 1996; Salata et al., 1997; Sinha et al., 2004).

The synthesis of polymer nanocomposites can be carried out by many different methods. For example, in-situ polymerization of monomers inside the galleries of the inorganic host has been one of the common methods for preparation of nanocomposites. Another common methods are based on melt intercalation, solventless melt intercalation, using microwaves, latex-colloid interaction, solvent evaporation, spray drying, shirasu porous glass membrane emulsification technique and electrospraying, etc. (Berkland et al., 2001; Ma et al., 1999; Messersmith et al., 1993; Mu et al., 2001; Oriakhi et al., 1995; Rosca et. al.,2004; Usuki et al., 1993). This article is concerned with fabrication of polymer nanocomposite particles by *in-situ* suspension polymerization and electrospraying which are represented as a traditional and current general method for synthesis of nanocomposite. The effectiveness of these nanocomposite particles are demonstrated with a field emission-type scanning electron microscope (FE-SEM), a transmission electron microscopy (TEM), an optical microscope, a reflection type X-ray diffraction (XRD), a

Fourier transform spectroscopy (FT-IR), a thermogravimetric analysis (TGA), a nuclear magnetic resonance Spectrometer (^{1}H-NMR), and the anti-bacterial performance was also discussed.

2. *In-Situ* suspension polymerization for polymer/silver and polymer/montmorillonite nanocomposite microparticles

For a long time, quite a range of polymer/inorganic nanocomposites have been investigated via *in-situ* suspension polymerization. Using suspension polymerization to prepare polymer/inorganic nanocomposite particle is interesting in terms of an easy manipulation, low cost, and controllable particle size. It is well known that a range of variables, such as the type and amount of initiator and suspending agent, the polymerization temperature, the monomer to water ratio, and the agitation speed, affect the molecular weight of the polymer synthesized (Gotoh et al., 2000; Giannetti et al.,1986; Hatchett et al., 1999; Huang et al., 1991; Lee et al., 2004).

In this section, we are reporting the *in-situ* synthesis of poly(vinyl acetate) (PVAc)/silver (Ag) and poly(vinyl alcohol) (PVA)/PVAc/montmorillonite (MMT) nanocomposite particles using a suspension polymerization technique (Jung et al., 2006; Jung et al., 2007; Yeum et al., 2005). The effects of MMT and Ag on the morphology, size, conversion rate, and polymerization time of nanocomposite particles were examined. According to the inorganic materials and polymerization process, diffrent type of particles such as spherical, rugged surface and golf ball shape microparticles are successfully prepared. The X-ray diffraction analysis illustrated that the clay silicate layer are intercalated in the polymer matrix.

2.1 Experimental

2.1.1 Materials

Vinyl acetate (VAc) purchased from Sigma-Aldrich was washed with an aqueous solution of $NaHSO_3$ and water, and then dried with anhydrous $CaCl_2$ followed by distillation under nitrogen atmosphere at a reduced pressure. The initiator, 2,2′-azobis (2,4-dimethylvaleronitrile) (ADMVN) (Wako Co.) was recrystallized twice in methanol before use. PVA with number average molecular weight of 127,000 and degree of saponification of 88% (Aldrich Co.) was used as a suspending agent. Aqueous Ag nanoparticles dispersion (AGS-WP001,10 000 ppm) with diameters ca. 15–30 nm were purchased from Miji Tech. Co., Ltd., Korean. Pristine Na-MMT, Kunipia-F from Kunimine, Japan, was used as received. It consists of exchangeable sodium ions with cationic exchange capacity of ca. 119 meq/100 g. The average particle size and surface area of Na-MMT were 100–1,000 nm and 750 m^2/g, respectively. Polyethylene glycol dipolyhydroxystearate (Arlacel P135), purchased from Uniquema, and cationic surfactant, such as cetyltrimethylammonium bromide (CTAB), obtained from Aldrich, are used as surfactants in order to disperse the Ag and MMT in polymer matrix, respectively. Deionized water was used for all of the experiments.

2.1.2 Preparation of PVAc/Ag nanocomposite microparticles

To prepare PVAc/Ag nanocomposite microparticles, a suspension polymerization approach in the presence of aqueous Ag nanoparticle was used. The polymerization was conducted in a 250 ml three-neck round bottom flask fitted with a condenser. In a typical run, the suspending agent was first dissolved in water under nitrogen atmosphere with constant stirring. Surfactant was dissolved in VAc monomer. After the surfactant was dissolved, Ag nanoparticle in water suspension and desired initiator was mixed with VAc monomer under ultrasonification for 5 min using Bandelin UW 3,100 equipment. The above mixture was poured into a flask and charged at once at a fixed polymerization temperature. After prefixed reaction time, the reaction liquor was cooled and kept for more than 24 h to allow complete sedimentation of PVAc/Ag nanocomposite microparticles. The PVAc/Ag nanocomposite microparticles were then filtered and washed with warm water to eliminate residual VAc and suspending agent, and dried under vacuum. Conversion was calculated by gravimetric method. In the case of calculating of conversion, weight of Ag was ignored because the total weight of used Ag nanoparticles is less than 0.5% of polymer weight. Conversions were averages of three measurements. The detailed polymerization conditions are listed in Table 1.

2.1.3 Preparation of PVA/PVAc/MMT nanocomposite microparticles

Suspending agent was dissolved in water under a nitrogen atmosphere with constant stirring in a 250 ml reactor fitted with a condenser. CTAB was mixed with the MMT in the monomer phase prior to suspension polymerization. The ADMVN was added at a fixed polymerization temperature. After predetermined times, the reaction mixture was cooled and kept for 24 h to allow the precipitation and separation of the PVAc/MMT nanocomposite microparticles. The collected PVAc/MMT nanocomposite microparticles were further washed with warm water. Conversion was calculated by measuring the weight of the PVAc/MMT using the average value from three determinations. The detailed polymerization conditions are listed in Table 1.

Condition	PVAc/Ag	PVA/PVAc/MMT
Type of initiator	ADMVN	ADMVN
Type of suspending agent	PVA	PVA
Initiator concentration	10^{-4}, $5x10^{-4}$, 10^{-3} mol/mol of VAc	10^{-4}, $5x10^{-4}$, 10^{-3} mol/mol of VAc
Suspending agent concentration	1.5, 5, 9 g/dl of water	1.5 g/dl of water
VAc/water	0.5 l/l	0.5 l/l
Rpm	500	500
Temperature	30, 40, 50 ^{o}C	30, 40, 50 ^{o}C
Inorganic material concentration	3 wt.% of VAc	1 wt.% of VAc
Surfactant concentration	0, 0.06 wt.% of VAc	0.5 wt.% of VAc

Table 1. Reaction conditions for the suspension polymerization

2.1.4 Heterogeneous saponification of PVA/PVAc/MMT nanocomposite microparticles

To prepare PVA/PVAc/MMT nanocomposite microparticles, heterogeneous saponification of PVAc/MMT nanocomposite microparticles was conducted in a flask equipped with a reflux condenser, a thermocouple, a dropping funnel, and a stirring device. The alkali solution used for saponification contains 10 g of sodium hydroxide, 10 g of sodium sulfate, 10 g of methanol and 100 g of water. The prepared PVAc/MMT nanocomposite microparticles were slowly added into the alkali solution at 50 °C with gentle stirring. The saponification was stopped at the required time and a PVA shell was formed on the surface of the PVAc/MMT microparticles. After the required reaction time, the mixture was poured into cold water and kept for 1 day to allow the precipitation of spherical core/shell PVA/PVAc/MMT nanocomposite microparticles. Finally, the solid saponification product was filtered and washed several times with water and dried in a vacuum at 40 °C for 1 day.

2.1.5 Characterizations

The morphology of nanocomposite microparticles were examined using a SEM(Hitachi S-570). To obtain the average size and size distribution, five SEM photographs and more than 200 particles were collected by computer, which linked with the SEM, followed by statistical analysis of data by computer. FT-IR spectrum of the sample was obtained using a Perkin-Elmer 1650 that cast on potassium bromide. XRD measurements were performed at room temperature on a Rigaku (D/Max IIIB) X-ray diffractometer using Ni-filtered CuKα (λ=1.54 Å) radiation. The core/shell structure of PVA/PVAc/MMT nanocomposite microparticles was examined using an optical microscope (Leica DC 100). The degree of saponification (DS) of PVA/PVAc/MMT nanocomposite microparticles was determined by the ratio of methyl and methylene proton peaks in the ^{1}H-NMR spectrometer (Varian, Sun Unity 300).

2.2 Results and discussion

2.2.1 PVAc/Ag nanocomposite microparticles

2.2.1.1 Suspension polymerization behaviour

Figure 1(a,b,c) presents the conversion-time relationship at different polymerization temperatures with initiator concentration at 10^{-4} mol/mol of VAc. Generally speaking, the reaction conversion increased with increasing polymerization temperature, regardless of the presence of Ag nanoparticles. Although a low initiator concentration (10^{-4} mol/mol of VAc)was used, the conversion increased steadily with the reaction time at a reaction temperature of 30 °C until ca. 95% of conversion was arrived. In contrary, such a high conversion is impossible in the bulk polymerization of PVAc. The high molecular weight and high conversion obtained suggest that the chain transfer and termination reactions were not significant at the conditions used in this study. As the slope of the curve on the conversion-time plot represents the polymerization rate, the bigger the slope (steeper), the higher the polymerization rate. As can be seen from Figure 1, the introduction of unmodified silver nanoparticles into the system slightly decreased the polymerization rate, in comparison with the absence of them. Otherwise, the rates of polymerization with modified silver nanoparticles are slightly higher than these without silver nanoparticles. The actual reason for the increase in the polymerization rate when the silver nanoparticles were dispersed in the water phase, but the decrease in the polymerization rate when the silver nanoparticles were dispersed in monomer phase is not clear and more work is needed(Lee et al., 2008; Yeum et al., 2005; Yeum et al., 2005). As shown in Figure 1(d.e.f), the polymerization rate, at a reaction temperature of 30 °C, increased with increasing initiator concentration, in accordance with theoretical prediction (Odian, 1981).

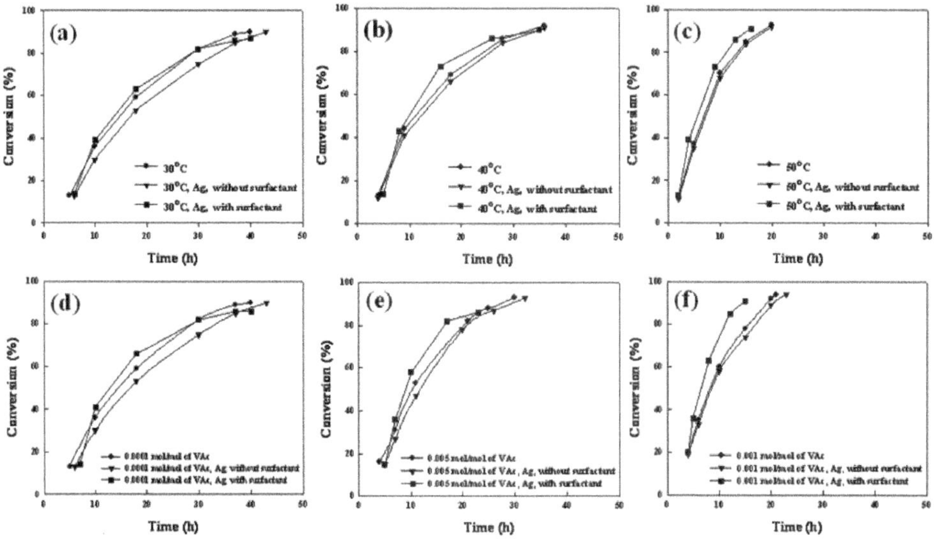

Fig. 1. Conversion vs. polymerization time (a,b,c) at different polymerization temperatures with [ADMVN]=10^{-4} mol/mol of VAc and (d,e,f) at different [ADMVN] with reaction temperature at 30 °C. (suspending agent concentration: 1.5 g/dl of water)

2.2.1.2 Morphology and structure property

SEM photographs of (a) pure PVAc microparticles, (b) PVAc/Ag nanocomposite microparticles without surfactant and (c) PVAc/Ag nanocomposite microparticles with surfactant suspension-polymerized with a suspending agent concentration of 5.0 g/dl of water are presented in Figure 2, respectively. It is surprising that three different appearances of microparticles, one with smooth surface, and another with rugged surface and the other with golf ball shape were observed. From the SEM observation, it is estimated that ca. 10–20% of particles when polymerized without surfactant are rugged surface. We believe this phenomenon must be related to the aggregation of Ag nanoparticles during the polymerization. It should be noted that the Ag nanoparticles used in this study are in an aqueous suspension form, so the Ag nanoparticles are relatively hydrophilic. For this reason, the Ag particles aggregate to small clusters when they are encapsulated in polymer phase. If these aggregated silver particles stay on the surface of the polymer microparticles, rugged microparticles may be obtained. On the other hand, PVAc/Ag nanocomposite microparticles which were polymerized with modified Ag nanoparticles have golf ball shape. Because of the low hydrophile-lyophile balance number of Arlacel P135, the surface of the Ag nanoparticles was converted from hydrophilic to hydrophobic by the adsorption of Arlacel P135, which resulted in a good dispersion of the Ag nanoparticles in monomerdroplets and polymer matrix. The detailed mechanism, such as why PVAc/Ag nanocomposite microparticles have different appearances according to surfactant, is shown in Figure 3. But the reason why only about 10-20% of the microspheres are in rugged surface, is not clear. Further work in this area definitely is needed.

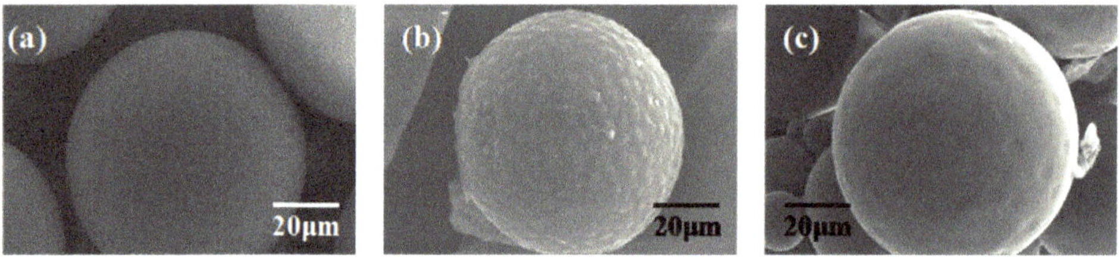

Fig. 2. SEM images of (a) pure PVAc microparticles, (b) PVAc/Ag nanocomposite microparticles without surfactant and (c) PVAc/Ag nanocomposite micropartices with surfactant (ADMVN concentration of 10^{-4} mol/mol of VAc, suspending agent concentration of 5.0 g/dl of water)

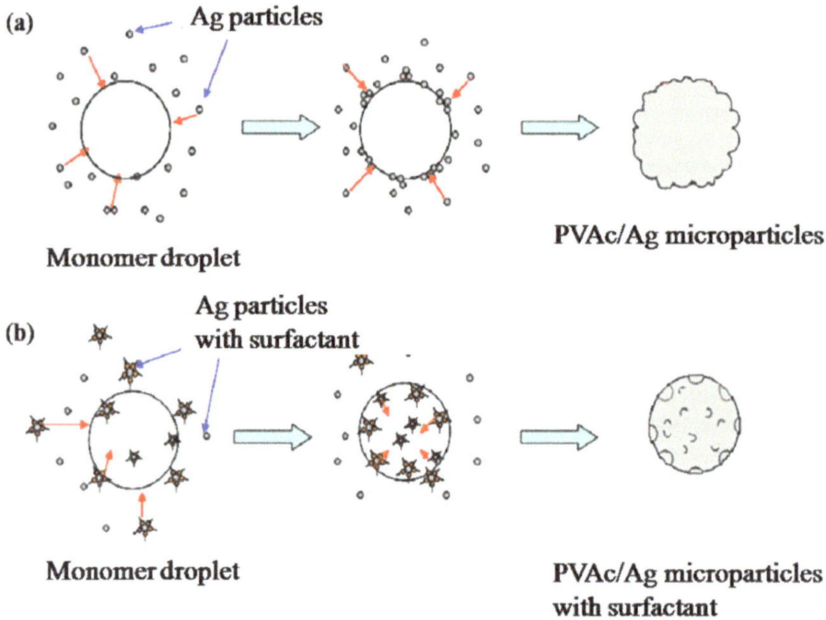

Fig. 3. Schematics of suspension polymerization of PVAc/Ag nanocomposite microparticles (a) without surfactant and (b) with surfactant

The PVAc microparticles and PVAc/Ag nanocomposite microparticles with modified silver nanoparticle contents of 3 wt% were characterized by XRD, and the results are shown in Figure 4. The XRD pattern of PVAc/Ag nanocomposite microparticles shows diffraction peaks at 2θ of ca. 15°, 22.7°, 38.2°, 44.6°, and 52°, respectively. Except the diffraction peaks of PVAc(15°, 22.7°), all the other peaks are corresponding to the silver phase. The XRD pattern clearly indicates that PVAc/Ag nanocomposite microparticles were successfully prepared with modified silver nanoparticles. These peaks are corresponding to the (111) and (200) planes of the silver nanocrystals with cubic symmetry (Dong et al., 2002). XRD results indicate that modified Ag nanoparticles are well dispersed in the PVAc matrix.

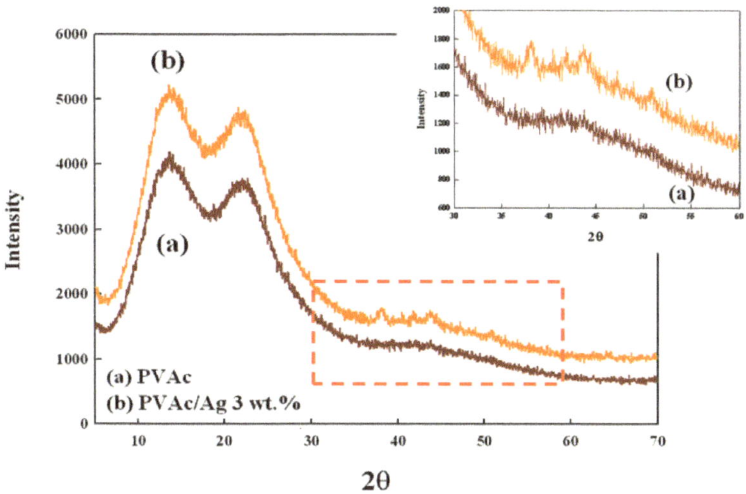

Fig. 4. XRD patterns of (a) pure PVAc and (b) PVAc/Ag nanocomposite microparticles with surfactant

2.2.2 PVA/PVAc/MMT nanocomposite microparticles

2.2.2.1 Suspension polymerization behavior

Figure 5(a) presents the conversion of the polymerization as a function of reaction time in the presence of MMT. The initiator concentration used in these reactions is 10^{-4} mol/mol of VAc. The results indicate that the rate of the polymerization decreased with increasing MMT addition. The reduction in the polymerization rate is probably due to the reduction of the diffusion rate of both monomer molecules and polymer chains in the intergalleries of nanoclay particles. For the polymerization without MMT, Figure 5(a) indicates that the conversion of the polymerization continually increased up to ~40 h, and reached a conversion of ~90% in spite of the low polymerization temperature (30 °C), which suggests that ADMVN is an effective initiator for low-temperature polymerization.

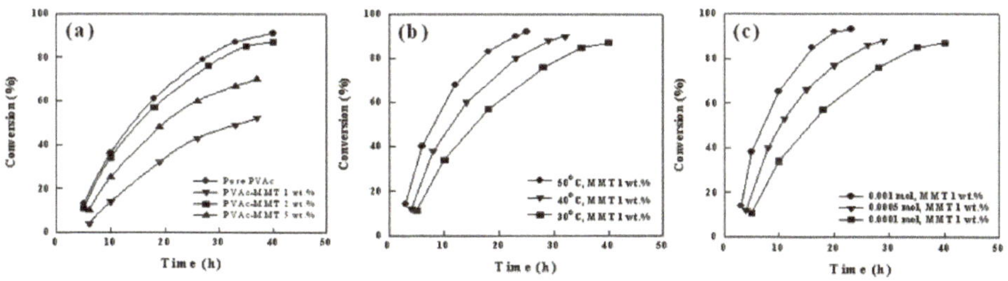

Fig. 5. Conversions of VAc into PVAc/MMT suspension polymerized using ADMVN concentration of 0.0001 mol/mol of VAc (a) with different MMT contents, (b) with different polymerization temperatures, and (c) suspension polymerized using 30 °C with different initiator concentrations with polymerization time, respectively

In a free radical polymerization, the rate of polymerization(R_p) could be expressed by Eq. (1)

$$R_p = K_p[M][I]^{0.5}(fk_d/k_t)^{0.5} \quad (1)$$

where f is the initiator efficiency; $[M]$ and $[I]$ are the concentrations of monomer and initiator; and k_d, k_p, and k_t are the rate constants of initiator decomposition, polymer propagation, and termination, respectively (Odian, 1981). This expression predicts that the rate of polymerization increases as the efficiency and concentration of the initiator are increased. The reduction in the polymerization rate in the presence of MMT suggests that either the efficiency of the initiator, f, or the propagation rate constant, k_p, were decreased when MMT was added into the suspension polymerization system. We believe the reduction of the diffusion rate in the nanoclay intergallerries may reduce both f and k_p in Eq. 1. It has been known that ADMVN is an effective low temperature initiator (the 10h half-life decomposition temperature of ADMVN is 51 °C) for preparing highmolecular-weight polymer with high yield. In this study, ADMVN was used for preparing PVAc/MMT nanocomposite microspheres at room temperature. Figure 5(b) presents the conversion-time relationship for the reactions carried at different temperatures with a constant initiator concentration. Although a low initiator concentration (10^{-4} mol/mol of VAc) and low reaction temperatures were employed, the conversion increased steadily up to 85-95%, depending on the MMT content in the system. The effect of initiator concentration on the polymerization rate is shown in Figure 5(c). It can be seen that the polymerization rate, at a reaction temperature of 30 °C, increased with the increasing initiator concentrations, in accordance with theoretical prediction (Odian, 1981).

2.2.2.2 Morphology and structure property

SEM photographs of pure PVAc and PVAc/MMT nanocomposite microspheres with 1 wt.% MMT contents are presented in Figure 6. As expected, the surface of pure PVAc microparticles shown in Figure 6(a) is smooth. Figure 6(b) shows that the roughness of the microsphere surface is increased by adding the MMT. To further study the distribution of MMT particles in the PVAc microspheres, the fracture surfaces of the PVAc/MMT nanocomposite microparticles were investigated. Although it is not shown here, MMT particles were embedded inside the composite microspheres, which indicate that PVAc/MMT microspheres could be prepared by *in-situ* suspension polymerization.

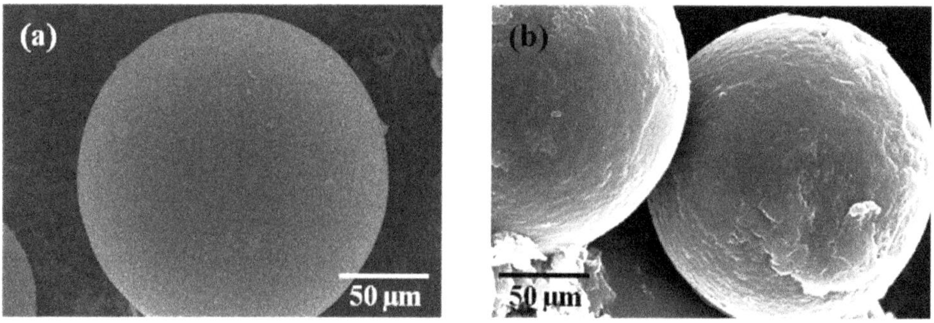

Fig. 6. SEM images of (a) pure PVAc microparticles and (b) PVAc/MMT nanocomposite microparticles with MMT concentration of 1 wt.%

The detail structure of the clay nanocomposites has been established using XRD analysis. Figure 7 shows the XRD patterns of the PVAc, MMT, and PVAc/MMT with MMT content of 1 wt.%. The MMT shows a diffraction pattern peak at 2θ =7.2°, which corresponds to the average basal spacing (d-spacing) of 12.3 Å. In the PVAc/MMT nanocomposite microparticles, the peak moved to a lower angle, i.e., 2θ =2.7°. The basal spacing increased from 12.3 to 32.6 Å. This spacing indicates that long alkyl (cetyltrimethyl) ammonium ions were inserted into the gallery of MMT; as a result, an intercalated structure formed. The inserted long alkyl chains caused the hydrophilic nature of the clay to decrease, and this effect improved the dispersion of silicates in the polymer matrix.

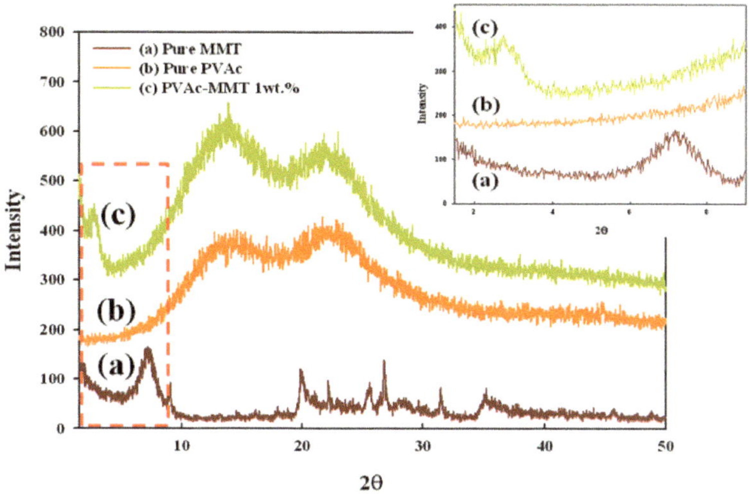

Fig. 7. XRD patterns of (a) pure PVAc microparticles, (b) pure MMT, and (c) PVAc/MMT nanocomposite microparticles with MMT concentration of 1 wt.%

2.2.2.3 Heterogeneous saponification of PVAc/MMT microparticles

The PVA obtained by the saponification of poly(vinyl ester) is a linear semicrystalline polymer and has been widely used as fibers for textile industries, films, membranes, and drug delivery systems. In this study, the PVA/MMT nanocomposite microspheres were prepared by a simple heterogeneous saponification method. To preserve the spherical shapes of PVAc/MMT nanocomposite particles, the saponificaion was carried out by disperse PVAc/MMT nanocomposite particles in alkali aqueous solution with very gentle agitation. The effect of MMT on the PVAc saponification rate was recorded by optical microscope observation. In this study, the heterogeneous saponifications of pure PVAc and PVAc/MMT microspheres were conducted under the same conditions. Degree of saponification (DS) is defined by the volume ratio of PVAc/PVA.

Figure 8 shows the optical micrographs of PVAc/PVA (Figure 8a,b) and PVA/PVAc/MMT (Fig. 8. c,d) nanocomposite microspheres prepared by heterogeneous saponification at different reaction times. As shown in Figure 8c,d, partially saponified nanocomposite microparticles (MMT content of 1 wt.%, degrees of saponification are 18.2 and 51.3%, respectively) with a PVAc core and PVA shell structure could be obtained by controlling the saponification degrees. It can be seen that the MMT presented in the PVAc microparticles significantly increase the DS of PVAc. It is well known that MMT is high swollable in alkaline solution, which results in a faster diffusion of base molecules into the PVAc particles to accelerate the saponification rate of the PVAc microparticles. Figure 9 shows the effect of MMT contents on the DS of PVAc microparticles, which indicates that the saponification rate increased remarkably with increasing MMT addition.

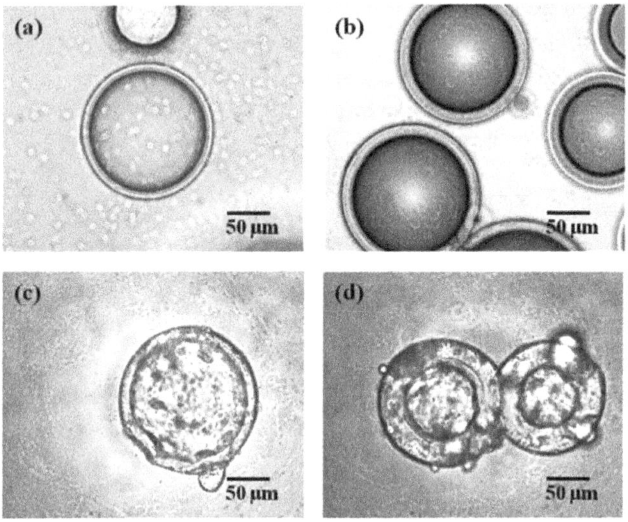

Fig. 8. Optical micrographs of (a,b) PVAc/PVA nanocomposite microparticles and (c, d) PVA/PVAc/MMT nanocomposite microparticles with MMT concentration of 1 wt.%. The saponification times and DS values were (a) 2 h and 14.7%, (b) 4 h and 17.5%, (c) 2 h and 18.2%, and (d) 4 h and 51.3%

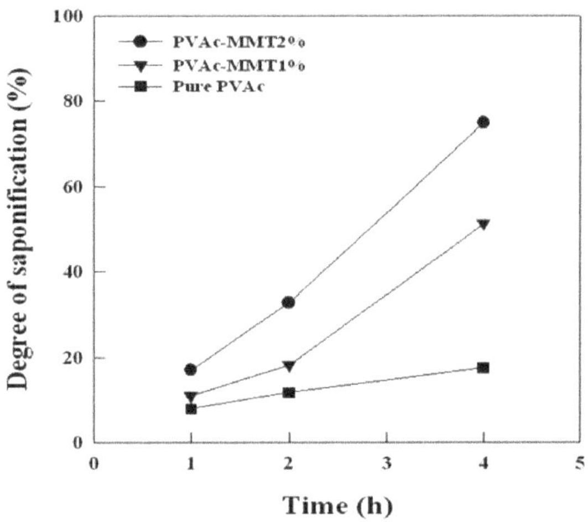

Fig. 9. Effect of MMT contents on the DS of PVAc/PVA microparticles with saponification time

The structures of the PVAc/MMT and PVA/PVAc/MMT nanocomposite microparticles were analyzed by using FT-IR spectroscopy. Figure 10 shows the FT-IR spectra for (A) pure MMT, pure PVAc microspheres, PVAc/MMT nanocomposite microparticles with MMT content of 1 wt.%, and (B) saponified PVA/PVAc/MMT nanocomposite microparticles. It is well known that pure MMT shows three strong peaks at 455, 520, and 1045 cm^{-1}. These peaks are associated with the bending mode of Si-O, the stretching vibration of Al-O, and the stretching vibration of Si-O, respectively. One can easily see peaks of both PVAc and MMT component in the spectrum of PVAc/MMT nanocomposite microparticles. It should be noted that MMT incorporated polymer particles could be prepared by the in situ suspension polymerization of VAc in the presence of organophilic MMT nanoparticles. From the generation of the -OH stretching vibration in the region of 3000-3600 cm^{-1} after the saponification process Figure 10(B), it is obvious that the surfaces of the PVAc/MMT nanocomposite microparticles were saponified with the hydroxyl groups.

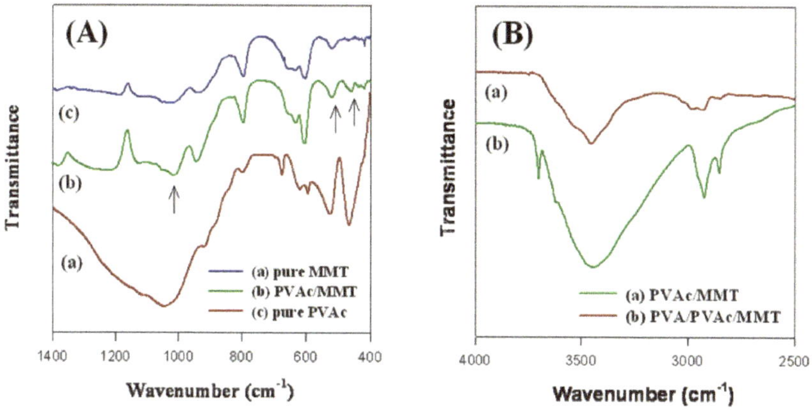

Fig. 10. FT-IR spectra of (A) pure MMT, pure PVAc microparticles, PVAc/MMT nanocomposite microparticles and (B) PVAc/MMT, PVA/PVAc/MMT nanocomposite microparticles (MMT concentration of 1 wt.%)

3. Fabrication and characterization of several types of polymer nanocomposite particles by one step electrospraying technique

The great advantage offered by electrospraying than other commonly used are simplicity and possibility for continuous one step process. The principle of electrospraying is same as electrospinning. High voltage is applied to a polymer fluid such that charges are induced within the fluid. When charges within the fluid reached a critical amount, a fluid jet will erupt from the droplet at the tip of the needle resulting in the formation of a Taylor cone. The electrospraying jet will travel towards the region of lower potential, which in most cases, is a grounded collector. The most important variable distinguishing electrospraying and electrospinning is solution parameter such as polymer molecular weight, concentration and viscosity, etc..

The major purpose of this section is to design an optimum solution parameter for electrospraying system and evaluate inorganic material effects such as antibacterial performance and thermal properties. PVA, PVA/MMT, PVA/Ag, and PVA/MMT/Ag nanocomposite nanoparticles which have an average diameter about 350 nm are successfully fabricated by one step electrospraying process in aqueous solution (Park et al., 2011; Park et al., 2011). In order to investigate the effect of MMT and Ag on the morphology, size, physical properties and anti-bacterial efficacy of composites, experiments were performed varying MMT and Ag concentrations. Transmission electron microscopy and reflection type X-ray diffraction analysis shows MMT and Ag were well dispersed in PVA nanoparticles. Those inorganic nanoparticles simultaneously enhanced thermal properties and anti-bacterial performance of the composite materials. The results obtained in this study may help to fabricate nanocomposite nanoparticle by one step process that can be utilized in biomedical application such as drug delivery system and functional nanoparticle for diagnosis and treatment.

3.1 Experimental

3.1.1 Materials

PVA with P_n = 500 (low-molecular-weight LMW) and 1700 (medium-molecular-weight, MMW) [fully hydrolyzed, degree of saponification = 99.9%] was obtained from DC Chemical Co., Seoul, Korea and MMT was purchased from Kunimine Industries Co., Japan. Aqueous silver nanoparticle dispersion (AGS-WP001, 10,000 ppm) with diameters ca.15-30 nm was got from Miji Tech., Korea. Doubly distilled water was used as a solvent to prepare all solutions.

3.1.2 Preparation of electrospraying solution

At first, 5 wt.% MMT powder was dispersed in distilled water under magnetic stirring for 1 h at room temperature. After making MMT dispersion, various concentrations of both LMW and MMW type of PVA were added in that solution. The solution was heated in a water bath at 80 °C under magnetic stirring for 2 h followed by cooling to room temperature. The PVA/MMT/Ag solution was made ready by mixing of PVA/MMT solution and Ag nanoparticles dispersion with different concentrations of Ag (1 and 3 wt.%) by stirring for more 2h at room temperature. The content of 5 wt.% of MMT in PVA nanocomposite nanofiber revealed best properties in our previous report (Ji et al., 2009; Lee et al., 2009; Park et al., 2009) which was predominant reason for using same concentration of MMT in this experiment.

Condition	Type of nanoparticles				
	LMW-PVA	MMW-PVA	PVA/MMT	PVA/Ag	PVA/MMT/Ag
Polymer concentration	2.5, 5, 7.5, 10 wt.%	2.5, 5, 7.5, 10 wt.%	2.5 wt.% of LMW PVA	2.5 wt.% of LMW PVA	2.5 wt.% of LMW PVA
MMT concentration*	-	-	5 wt.%	-	5 wt.%
Ag concentration*	-	-	-	1, 3 wt.%	1, 3 wt.%
Voltage	15 kV				
Tip collector distance	15 cm				

* based on polymer concentration

Table 2. Electrospraying condition of PVA/MMT/Ag nanocomposite nanoparticles

3.1.3 Electrospraying

During electrospraying, a high voltage power (CHUNGPA EMT Co., Ltd., Seoul, Korea; model CPS-60K02VIT) was applied to the electrospraying solution contained in a syringe via an alligator clip attached to the syringe needle. The applied voltage was adjusted at 15 kV. The solution was delivered to the blunt needle tip via syringe pump to control the solution flow rate. When a high electric field is applied to the solution in the syringe, droplets form at the nozzle tip in the form of a cone called Taylor cone. Eventually electrical forces overcome the surface tension of the solution, and a straight jet ejects from the apex of the cone. This jet breaks up into particles while traveling towards a grounded collector placed at 15 cm vertical distance to the needle tip due to inherent instabilities generated by the electric field. Particles were collected on an electrically grounded aluminum foil placed at 15 cm vertical distance to the needle tip.

3.1.4 Preservation test

The preservation efficacy of the nanocomposite nanoparticles was tested to evaluate the antimicrobial property. Samples were prepared by dispersing the nanoparticles in a viscous aqueous solution containing 0.01 wt.% of neutralized polyacrylic acid (Carbopol 941, Noveon Inc.). A mixed culture of microorganisms, *Staphylococcus aureus* (ATCC6538), and *Escherichia coli* (ATCC25922) were obtained on tryptone soya broth after 24h incubation at 32 °C. Then, 20 g of samples were inoculated with 0.2 g of the microorganism suspensions to adjust the initial concentration of bacteria to 107 cfu/g. Then, the inoculants were mixed homogeneously with the samples and stored at 32 °C. The microbial counts were carried out using the pour plate count method.

3.1.5 Characterizations

The morphology and structure of nanocomposite nanoparticles were observed with FE-SEM after gold coating and TEM analysis. XRD was used the Cu Kα radiation with wavelength of 0.154 nm. The scanning rate was 2°/min ranging 2 to 50° (2θ). The thermal behavior of nanocomposite nanoparticles was studied with TGA techniques (model Q-50) from TA instruments, USA at the rate of 10 °C/min from room temperature to 600 °C under the nitrogen gas atmosphere. The antibacterial performance was investigated to examine the biological function of PVA/MMT/Ag nanocomposite nanoparticles by KSM 0146 (Shake flask method) using *Staphylococcus aureus* and *Escherichia coli*.

3.2 Results and discussion

3.2.1 Effect of molecular weight and concentration

The nanoparticle formation conditions are strictly dependent on the polymer molecular weight and concentration. Nanoparticles could be obtained when either the molecular weight of the polymer and concentration of the polymer solution were too low and when it was used as the precursor in the electrospraying process. To the contrary, cone jet of polymer solution with sufficient resistance to shear stress will not break up to form them after solvent evaporation by a process called electrospinning (Li et al., 2004). Also, it must be considered that, when working in the electrospraying, the usual shape of the forming spherical particles is due to the physical properties of the system (Rayleigh et al., 1882).

To obtain the nanoparticles, a series of experiments with different solution concentrations of LMW-PVA (2.5 to 15 wt %) were conducted, and the results are presented in Figures 11 and 12. From Figures 11 and 12, it is clearly evident that nanoparticles are formed in case of

2.5 wt. % of LMW-PVA concentration. Polymer concentration above 2.5 wt. %, particles exhibit a spindle shape with one or two tail and fibrous shape. The polymers with higher molecular weight tended to attain a highly elongated shape like fibrous structure easily than low molecular weight polymer.

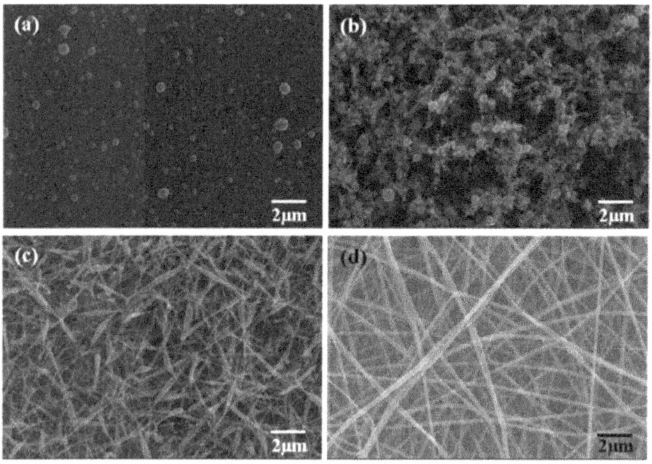

Fig. 11. FE-SEM images of LMW-PVA nanostructure syntheses by using different concentration of (a) 2.5 wt.%, (b) 5 wt.%, (c) 10 wt.% and (d) 15 wt.% (TCD= 15 cm and applied voltage= 15 kV; inset: high magnification morphologies of related images)

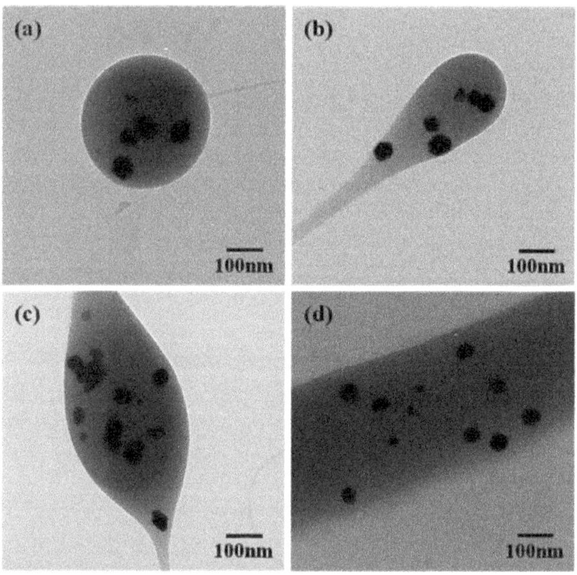

Fig. 12. TEM images of LMW-PVA/Ag nanostructure syntheses by using different concentration of PVA (a) 2.5 wt.%, (b) 5 wt.%, (c) 7.5 wt.% and (d) 15 wt.% (TCD= 15 cm and applied voltage= 15 kV and Ag concentration= 1 wt.%)

It is suggested that increasing entanglement of polymer chains caused highly by molecular weight and polymer concentration that contribute to the formation of the fibrous structure. Actually, MMW-PVA formed not only nanoparticle but also just bring forth spindle like shape at low concentration. On the contrary, in case of LMW-PVA, spherical nanoparticles were formed at 2.5 wt.% of polymer concentration. Another concentration of LMW-PVA showed more beaded morphology than MMW-PVA due to the low molecular weight and low viscosity of the spraying solution. The reason for different shape at each concentration is the Raleigh forces, which assist in particle formation, are able to overcome the viscous forces to enable the formation of spherical particles (Shin et al., 2001). Particles, which are obtained from PVA concentration above 2.5 wt.%, exhibit a spindle shape with one or two tail and fiber shape.

3.2.2 Morphology and structure

Successful formation of the PVA/MMT/Ag nanocomposite nanoparticles are illustrated by TEM images (Figure 13). The pure polymer nanoparticles were generally spherical with smooth surfaces at 2.5 wt.% concentration of LMW-PVA. The mean size of nanoparticles varied between 300 to 450 nm depending on the process parameter. When MMT clay was added, the nanoparticle morphology and shape are dramatically changed from sphere to rugged particle. TEM images support that the coexistence of Ag nanoparticles and PVA matrix for LMW-PVA/Ag nanocomposite nanoparticles. Small dark dots indicate the Ag nanoparticles which are seen from TEM images in nanocomposite nanoparticles. Due to the strong interactions between polymer and metal, Ag nanoparticles are well embedded in the PVA matrix. Also, it can be clearly observed that each silicate platelet forms a dark line in the

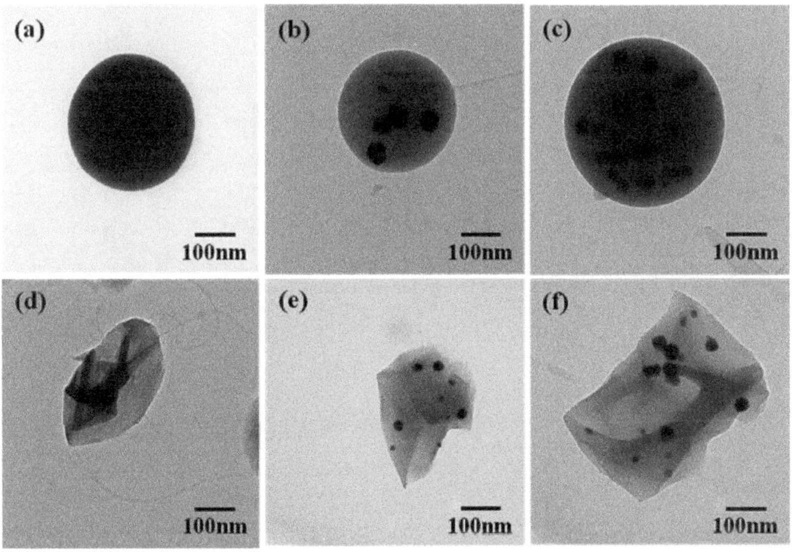

Fig. 13. TEM images of (a) PVA nanoparticles, (b) PVA/Ag 1wt.%, (c) PVA/Ag 3wt.%, (d) PVA/MMT (e) PVA/MMT/Ag 1wt.%, and (f) PVA/MMT/Ag 3wt.% nanocomposite nanoparticles (LMW-PVA solution concentration= 2.5 wt.%, MMT concentration= 5 wt.%, TCD= 15 cm and applied voltage= 15 kV)

nanoparticle compare with the pure PVA spherical nanoparticle. The size of the dark line is about 1-3 nm in width and 100–200 nm in length, and it indicated the good dispersion and exfoliation of MMT layers in the nanoparticle. Well-dispersed Ag and MMT were shown in Figure 13 from where we can see Ag and MMT were dispersed inside the polymer matrix without agglomeration. TEM results reveal that the MMT and Ag are distributed uniformly in the PVA/MMT/Ag nanocomposite nanoparticles.

The XRD patterns of nanocomposite nanoparticles show three diffraction peaks at 2θ of ca. 19.3°, 38.2° and 44.6° (Figure 14). The pure PVA shows a significant crystalline peak at about 19.3°, which is because of the occurrence of string inter- and intramolecular hydrogen bonding (Figure 14a). In case of clay-polymer composite, unexfoliated or intercalated MMT usually shows a peak in the range 3~9° (2θ). In exfoliated nanocomposites, generally single silicate layers (1-3 nm thick) are homogeneously dispersed in the polymer matrix, and XRD pattern with no distinct diffraction peak in the range of 3~9° (2θ) could be observed. The peaks of silver particles become gradually enlarged with increasing the content of Ag nanoparticles and crystallinity of PVA/MMT (2θ =19.3°) is lower in comparison with Figure 14(a) and (b). These peaks are corresponding to the (111) and (200) planes of the silver nanocrystals with cubic symmetry. XRD results indicate that MMT and Ag particles are well dispersed in the PVA matrix and MMT is predominantly exfoliated. Moreover, TEM images confirm MMT and Ag nanoparticles in the PVA/MMT/Ag hybrid nanoparticles (Figure 13).

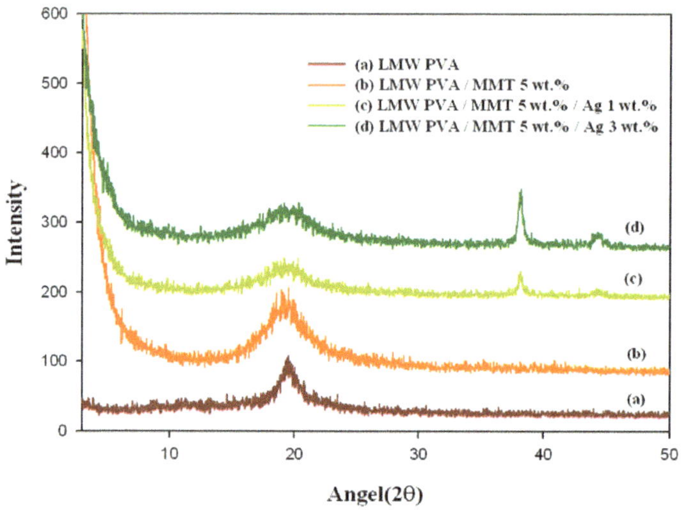

Fig. 14. XRD data of (a) PVA nanoparticles, (b) PVA/MMT nanocomposite nanoparticles, and PVA/MMT/Ag nanocomposite nanoparticles prepared with different Ag contents of (c) 1 wt.%, (d) 3 wt.% (LMW-PVA solution concentration= 2.5%, MMT concentration= 5 wt.%)

3.2.3 Thermal stability

Thermal stability of nanocomposite nanoparticles was measured using TGA in nitrogen atmosphere. The actual decomposition temperature depends on the structure, molecular weight and conformation of the polymer. Typical three weight loss peaks were observed in the TGA curve for bulk PVA. The first peak at 50-70 °C was due to its moisture vaporization, the second peak at 260-380 °C was due to the thermal degradation of PVA, and the third peak at 430-460 °C was due to the byproduct formation of PVA during the TGA thermal degradation process. Figure 15 shows TGA thermograms of different decomposition temperature with Ag contents of 0, 1 and 3 wt.%. Within up to 225 °C, there is an increased in thermal stability from the pure LMW-PVA nanoparticle to PVA/MMT/Ag nanocomposite nanoparticles. The higher thermal stability might be attributed due to its higher contents of Ag nanoparticles in the PVA/MMT/Ag nanocomposite nanoparticles.

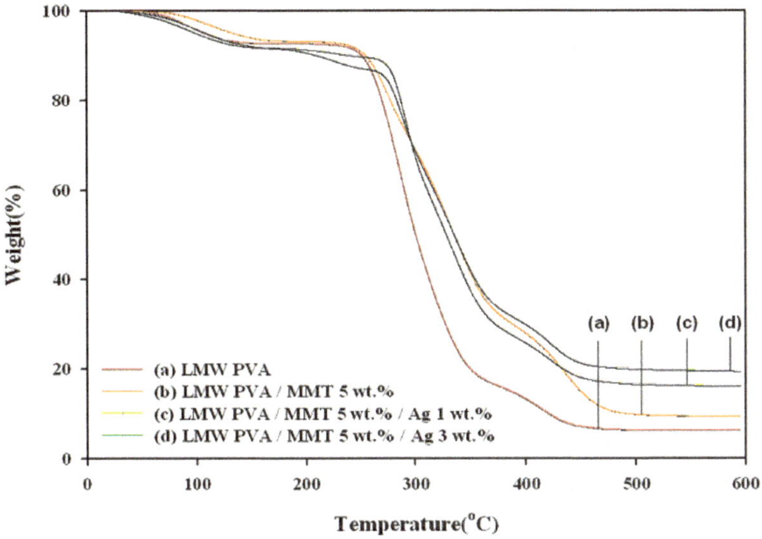

Fig. 15. TGA data of (a) LMW-PVA nanoparticle, (b) PVA/MMT nanocomposite nanoparticles, and PVA/MMT/Ag nanocomposite nanoparticles prepared with different Ag contents of (c) 1 wt.%, (d) 3 wt.% (LMW-PVA solution concentration= 2.5%, MMT concentration= 5 wt.%)

3.2.4 Anti-bacterial efficacy

Ag nanoparticles have been known to have strong inhibitory and antibacterial effects as well as a broad spectrum of antimicrobial activities (Dowling et al., 2001). In this study, in order to provide useful information for the biological function of PVA/MMT/Ag nanocomposite nanoparticles, the anti-bacterial performance of PVA/MMT/Ag nanocomposite nanoparticles were evaluated in viscous aqueous test samples and shown in Figure 16. The antibacterial activity of PVA/MMT/Ag nanocomposite nanoparticle is assessed by counting the number of bacteria in the sample with the storage time at 32 °C. As shown in Figure 16, in the absence of Ag nanoparticles, the number of bacteria remained constant. However, after adding Ag nanoparticles into the nanoparticles, all of bacteria decreased dramatically.

The increase in the concentration of the Ag accelerates diminishing in bacteria. With only a small amount of Ag, almost all the initially inoculated bacteria could be sterilized within a week.

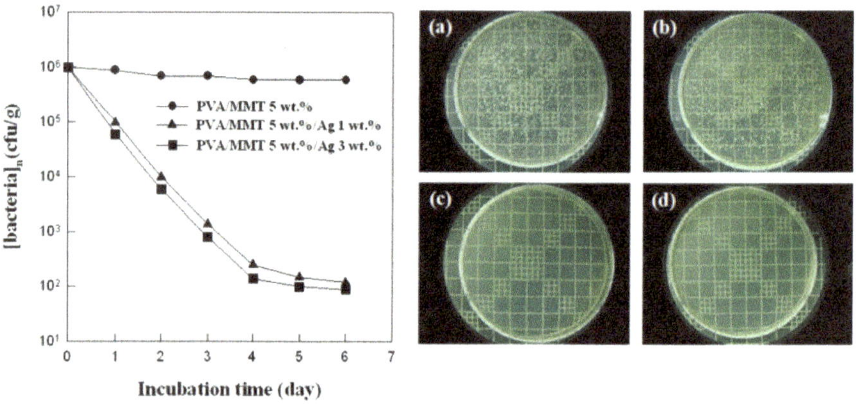

Fig. 16. Antibacterial ability test with Staphylococcus aureus (a) control and PVA/MMT/Ag hybrid nanoparticles containing with different Ag amounts of (b) 0 wt. %, (c) 1 wt.%, and (d) 3 wt.% (after 1week)

3.2.5 Application of polymer/natural extracts nanoparticles for children's products

Based on the above results, we prepared polymer/natural colorant nanoparticles for using children's products. Excellent antibacterial ability of the PVA/MMT/Ag nanocomposite nanoparticles has been proven, so polymer nanocomposite nanoparticles containing natural colorant, Cyanidin-3-O-glucoside, extracted from sorghum, has advantage of safety for using children's products over existing compound materials. Figure 17 is chemical structure of Cyanidin-3-O-glucoside and Figure 18 shows TEM image of prepared polymer nanocomposite nanoparticles containing Cyanidin-3-O-glucoside.

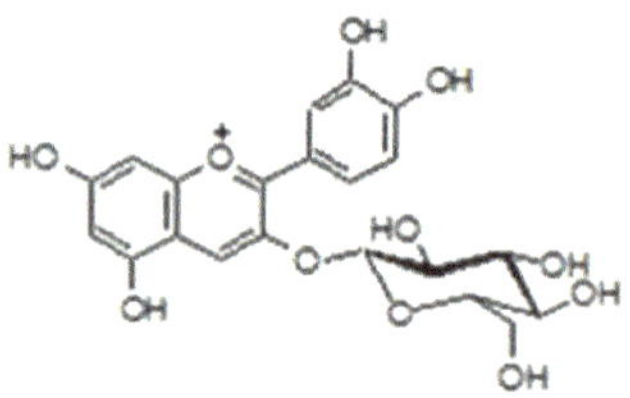

Fig. 17. Chemical structure of Cyanidin-3-O-glucoside

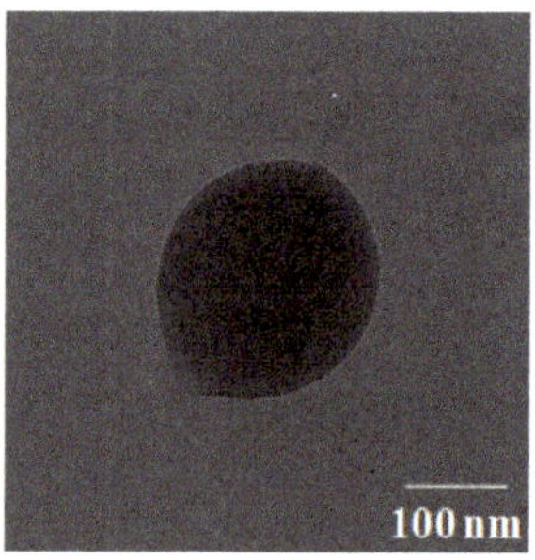

Fig. 18. TEM image of polymer/Cyanidin-3-O-glucoside nanoparticle

4. Conclusion

In this work, PVAc/Ag and PVA/PVAc/MMT nanocomposite microparticles were prepared by low temperature *in-situ* suspension polymerization of VAc in the presence of Ag and MMT and PVA/MMT, PVA/MMT/Ag nanocomposite nanoparticles are successfully fabricated from aqueous solutions by one step electrospraying technique. The morphology of PVAc/Ag nanocomposite microparticles was influenced by surfactant. By a heterogeneous saponification method, PVA/PVAc/MMT nanocomposite microparticles with core/shell structure could be prepared. Morphology of nanocomposite nanoparticles prepared by electrospraying technique can be affected by the molecular weight of polymer and concentration of polymer solution. The thermal properties of prepared nanocomposite micro- and nanoparticles enhanced with the addition of MMT and Ag. Moreover, it has good anti-bacterial performance. The results obtained in this study may help to fabricate various type of nanocomposite micro- and nanoparticle by *in-situ* polymerization and one step process that can be utilized in not only biomedical application such as drug delivery system and functional nanoparticle for diagnosis and treatment, but bioaffinity materials like children and food industries.

5. Acknowledgements

This research was financially supported in part by grants from the Agenda Program (PJ0073852010) of National Institute of Crop Science, Rural Development Administration (RDA), Korea. Also, this research was partially supported by a grant(70011324) from Daegu Technology Development Program of the Daegu Regional Government.

6. References

Berkland, C.; Kim, C. & Pack, D.W. (2001). Fabrication of PLG microspheres with precisely controlled and monodisperse size distributions. *J. Control. Release,* 73, 1, 59-74, 0168-3659

Dong, A.G.; Wang, Y.J. & Tang, Y. (2002). Fabrication of compact silver nanoshells on polystyrene spheres through electrostatic attraction. *Chem. Commun.,* 4, 350-351, 1359-7345

Dowling, D.P.; Donnelly, K.; McConnell, M.L.; Eloy, R. & Arnaud, M.N. (2001). Deposition of anti-bacterial silver coatings on polymeric substrates. *Thin Solid Films,* 398-399, 602-606, 0040-6090

Giannetti, E. & Mazzocchi, R. (1986). High conversion free-radical suspension polymerization: End groups in poly(methyl methacrylate) and their influence on the thermal stability. *J. Polym. Sci. Pol. Chem.,* 24, 10, 2517-2551, 0087-624X

Gotoh, Y.; Igarashi, R.; Ohkoshi, Y.; Nagura, M.; Akamatsu, K. & Deki, S. (2000). Preparation and structure of copper nanoparticle/poly(acrylic acid) composite films. *J. Mater.Chem.,* 10, 11, 2548-2552, 0959-9428

Hatchett, D.W.; Josowicz, M.; Janata, J. & Baer, D.R. (1999). Electrochemical formation of au clusters in polyaniline. *Chem. Mater.,* 11, 10, 2989-2994, 0897-4756

Huang, C.J.; Yen, C.C. & Chang, T.C. (1991). Studies on the preparation and properties of conductive polymers. III. Metallized polymer films by retroplating out. *J. Appl. Polym. Sci.,* 42, 8, 2237-2245, 0021-8995

Ji, H.M.; Lee, H.W.; Karim, M.R.; Cheong, I.W.; Bae, E.A.; Kim, T.H.; Islam, M.S.; Ji, B.C. & Yeum, J.H. (2009). Electrospinning and characterization of medium-molecular-weight poly(vinyl alcohol)/high-molecular-weight poly(vinyl alcohol)/montmorillonite nanofibers. *Colloid Polym. Sci.,* 287, 7, 751-758, 0303-402X

Jung, H.M.; Lee, E.M.; Ji, B.C.; Deng, Y.; Yun, J.D. & Yeum, J.H. (2007). Poly(vinyl acetate)/poly(vinyl alcohol)/montmorillonite nanocomposite microspheres prepared by suspension polymerization and saponification. *Colloid Polym. Sci.,* 285, 6, 705-710, 0303-402X

Jung, H.M.; Lee, E.M.; Ji, B.C.; Sohn, S.O.; Ghim, H.D.; Cho, H.; Han, Y.A.; Choi, J.H.; Yun, J.D. & Yeum, J.H. (2006). Preparation of poly(vinyl acetate)/clay and poly(vinyl acetate)/poly(vinyl alcohol)/clay microspheres. *Fiber. Polym.,* 7, 3, 229-234, 1229-9197

Lee, E.M.; Lee, H.W.; Park, J.H.; Han, Y.A.; Ji, B.C.; Oh, W.; Deng, Y. & Yeum, J.H. (2008). Multihollow structured poly(methyl methacrylate)/silver nanocomposite microspheres prepared by suspension polymerization in the presence of dual dispersion agents. *Colloid Polym. Sci.,* 286, 12, 1379-1385, 0303-402X

Lee, H.W.; Karim, M.R.; Ji, H.M.; Choi, J.H.; Ghim, H.D.; Park, J.H.; Oh, W. & Yeum, J.H. (2009). Electrospinning fabrication and characterization of poly(vinyl alcohol)/montmorillonite nanofiber mats. *J. Appl. Polym. Sci.,* 113, 3, 1860-1867, 0021-8995

Lee, J.E.; Kim, J.W.; Jun, J.B.; Ryu, J.H.; Kang, H.H.; Oh, S.G. & Suh, K.D. (2004). Polymer/Ag composite microspheres produced by water-in-oil-in-water emulsion polymerization and their application for a preservative. *Colloid Polym. Sci.,* 282, 3, 295-299, 0303-402X

Li, D. & Xia, Y.N. (2004). Electrospinning of nanofibers: Reinventing the wheel?. *Adv. Mater.,* 16, 14, 1151-1170, 0935-9648

Ma, G.; Nagai, M. & Omi, S. (1999). Preparation of uniform poly(lactide) microspheres by employing SPG emulsification technique. *Colloid Surface. A,* 153, 1-3, 383-394, 0927-7757

Matsumoto, H.; Mizukoshi, T.; Nitta, K.; Minagawa, M.; Tanoika, A. & Yamagata, Y. (2005). Organic/inorganic hybrid nano-microstructured coatings on insulated substrates by electrospray deposition. *J. Colloid Interf. Sci.,* 286, 1, 414-416, 0021-9797

Messersmith, P.B. & Giannelis, E.P. (1993). Polymer-layered silicate nanocomposites: In situ intercalative polymerization of ε-caprolactone in layered silicates. *Chem. Mater.,* 5, 8, 1064-1066, 0897-4756

Mu, L. & Feng, S.S. (2001). Fabrication, characterization and in vitro release of paclitaxel (Taxol®) loaded poly(lactic-co-glycolic acid) microspheres prepared by spray drying technique with lipid/cholesterol emulsifiers. *J. Control. Release,* 76, 3, 239-254, 0168-3659

Odian, G. (1981). Principles of polymerization. 2nd edn. *John Wiley & Sons, Inc.,*

Okamoto, M.; Morita, S.; Taguchi, H.; Kim, Y.H.; Kotaka, T. & Tateyama, H. (2000). Synthesis and structure of smectic clay/poly(methyl methacrylate) and clay/polystyrene nanocomposites via in situ intercalative polymerization. *Polymer,* 41, 10, 3887-3890, 0032-3861

Okuda, H. & Kelly, A.J. (1996). Electrostatic atomization–experiment, theory and industrial applications. *Phys. Plasmas.*, 3, 5, 2191-2196, 1070-664X

Oriakhi, C.O. & Lerner, M.M. (1995). Poly(pyrrole) and poly(thiophene)/clay nanocomposites via latex-colloid interaction. *Mater. Res. Bull.*, 30, 6, 723-729, 0025-5408

Park, J.H.; Kim, I.K.; Oh, W.; Deng, Y.; Kim, J.W. & Yeum, J.H. Electrospraying fabrication and characterization of low molecular weight poly(vinyl alcohol)/silver composite nanospheres for antibacterial applications. *Polym. Polym. Compos., in Press,* 0967-3911

Park, J.H.; Lee, H.W.; Chae, D.K.; Oh, W.; Yun, J.D.; Deng, W. & Yeum, J.H. (2009). Electrospinning and characterization of poly(vinyl alcohol)/chitosan oligosaccharide/clay nanocomposite nanofibers in aqueous solutions. *Colloid Polym. Sci.*, 287, 8, 943- 950, 0303-402

Park, S.M.; Park, J.H.; Islam, M.S.; Choi, J.H.; Yoon, N.S. & Yeum, J.H. (2011). Preparation of low molecular weight poly(vinyl alcohol)/montmorillonite composite nanoparticles using electrospraying technique. *Polym. Polym. Compos.*, 19, 1, 1-6, 0967-3911

Ramos, J.; Millan, A. & Palacio, F. (2000). Production of magnetic nanoparticles in a polyvinylpyridine matrix. *Polymer,* 41, 24, 8461- 8464, 0032-3861

Rayleigh, L. (1882). On the equilibrium of liquid conducting masses charged with electricity. *Philos. Mag.*, 14, 87, 184-186, 1478-6435

Rosca, I.D.; Watari, F. & Uo, M. (2004). Microparticle formation and its mechanism in single and double emulsion solvent evaporation. *J. Control. Release,* 99, 2, 271-280, 0168-3659

Salata, O.V.; Hull, P.J. & Dobson, P.J. (1997). Synthesis of nanometer-scale silver crystallites via a room-temperature electrostatic spraying process. *Adv. Mater.*, 9, 5, 413-417, 0935-9648

Shin, Y.M.; Hohman, M.M.; Brenner, M.P. & Rutledge, G.C. (2001). Electrospinning: A whipping fluid jet generates submicron polymer fibers. *Appl. Phys. Lett.*, 78, 8, 1149-1151, 0003-6951

Sinha, V.R.; Bansal, K.; Kaushik, R. & Trehan, A. (2004). Poly- ε -caprolactone microspheres and nanospheres: an overview. *Int. J. Pharmaceut.*, 278, 1, 1-23, 0378-5173

Usuki, A.; Kojima, Y.; Kawasumi, M.; Okada, A.; Fukushima, Y.; Kurauchi, T. & Kamigaito, O. (1993). Synthesis of nylon 6-clay hybrid. *J. Mater. Res.*, 8, 5, 1179-1184, 0884-2914

Wu, W.; He, T. & Chen, J.F. (2006). Study on in situ preparation of nano calcium carbonate/PMMA composite particles. Mater. Lett., 60, 19, 2410-2415, 0167-577X

Yeum, J.H. & Deng, Y. (2005). Synthesis of high molecular weight poly(methyl methacrylate) microspheres by suspension polymerization in the presence of silver nanoparticles. *Colloid Polym. Sci.*, 283, 11, 1172-1179, 0303-402X

Yeum, J.H.; Qunhui, S. & Deng, Y. (2005). Poly(vinyl acetate)/silver nanocomposite microspheres prepared by suspension polymerization at low temperature. *Macromol. Mater. Eng.*, 290, 1, 78-84, 1438-7492

Zhu, Z.K.; Yin, J.; Cao, F.; Shang, X.Y. & Lu, Q.H. (2000). Photosensitive polyimide/silica hybrids. *Adv. Mater.*, 12, 14, 1055-1057, 0935-9648

9

Designing Nanostructured Carbon Xerogels

Esther G. Calvo, J. Ángel Menéndez and Ana Arenillas
Instituto Nacional del Carbón (CSIC), Oviedo, Spain

1. Introduction

Until the discovery of carbon gels in 1890, inorganic gels had dominated sol-gel literature. Traditional inorganic gels based upon hydrolysis-condensation reactions of metal alkoxides are well known as a result of their high specific surface areas and their unique morphological and structural properties (De Sousa et al., 2001). Some of the precursors used for the preparation of such inorganic gels are aluminates, titanates and borates but the most frequently employed alkoxides are alkoxysilanes, leading to the extensively investigated silica gels (Mauritz, 1998; Zareba-Grodz et al., 2004). The reaction mechanism of silica gels is based on the hydrolysis of alkoxysilanes to yield silicic acid moities, $Si(OH)_4$, which spontaneously condensate to generate, after a sequence of specific stages, the final silica gel consisting of siloxane groups (Si-O-Si) within its framework and silanol groups (Si-OH) on its surface (Salazar-Hernández et al., 2009). Some of the attractive features of silica gels are: the tailored textural and structural properties, abundance and low cost, high sorption capacity, very high thermal shock resistance, insolubility in most solvents and lower index of refraction compared to other inorganic gels. It is for these reasons that they are used in a wide range of industrial applications including catalysis, chromatography, drug delivery and ion exchange (Qu et al., 2008; Teng et al., 2010).

It was the great interest aroused by inorganic gels, especially in the case of silica gels, in different fields of application together with the advantages associated with sol-gel methods, i.e. low temperature processing, the high homogeneity of final products and the possibility of controlling their surface properties (Houmard et al., 2009), that persuaded Pekala and co-workers to go an step further and apply this sol-gel methodology to the synthesis of organic gels (Pekala, 1989). The first organic gel was obtained by sol-gel polymerization of resorcinol and formaldehyde under alkaline conditions and supercritical drying. This produced a material called organic aerogel, consisting of interconnected colloidal particles approximately 10 nm in diameter. Properties such as low density, highly porous material and high versatility of the sol-gel process turned the carbon gel into a prominent member of the "carbon family".

Basically, an organic gel is a solid nanostructure comprised of nano-sized pores and interlinked primary particles obtained by means of polymerization reactions between hydroxylated benzenes and aldehydes, and then subjected to a drying process. The most commonly used monomers are resorcinol and formaldehyde (Al-Mutasheb & Ritter, 2003; Czzakel et al., 2005; Job et al., 2004; Tian et al., 2011a; Zhang et al., 2007; Zhu et al., 2007; Zubizarreta et al., 2008a) but, there are other potential combinations such as phenol/formaldehyde (Mukai et al., 2005a; Scherdel & Reichenauer, 2009; Teng & Wang,

2000), phenol/furfural (Dingcai & Ruowen, 2006; Long et al., 2008a; Pekala et al., 1995), phenol/melamine/formaldehyde (Long et al., 2008b), cresol/formaldehyde (Li et al., 2001; Zhu et al., 2006), etc. As shown in Figure 1, the formation of organic gels involves the following stages: (i) formation of a three-dimensional polymer in a solvent, known as gelation step, (ii) curing period where the crosslinking of previously formed polymer clusters (particles) takes place and, finally, (iii) drying step, that can be performed under subcritical, supercritical or freezing conditions, resulting in xerogels, aerogels and cryogels, respectively (Al-Mutasheb & Ritter, 2003; Czzakel et al., 2005; Job et al., 2004; Zubizarreta et al., 2008a). The last step needed to produce carbon gels is carbonization, with the purpose of removing any remaining oxygen and hydrogen groups, yielding a thermally stable nanostructure mainly composed of carbon. This carbonization is usually performed at high temperatures (approximately 600-1000 °C) under an inert atmosphere, such as N_2, He or Ar (Al-Mutasheb & Ritter, 2003; Calvo et al., 2011a; Lin & Ritter, 2000; Job et al., 2004). However, as will be discussed in Section 2.2 of this chapter, it is possible to use other reactive gases in order to modify the chemical composition of carbon gels (Kang et al., 2009).

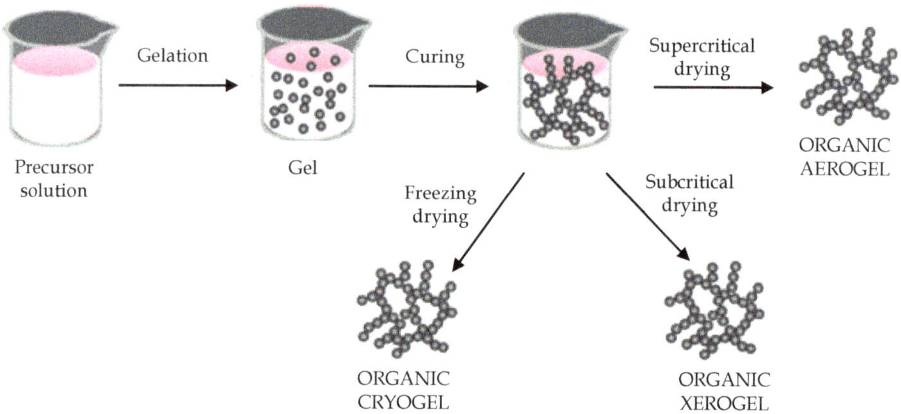

Fig. 1. Schematic representation of the steps involved in the synthesis of organic gels.

All the synthesis steps, referred to in the preceding paragraph, interfere in the development of the porosity of carbon gels. This gives an idea not only of the versatility of this type of carbonaceous materials but also of their complexity, since most of variables involved in the synthesis procedure are not independent, i.e. each one influences not only in certain properties of the gel but may also modify the effect of other variables (Job et al., 2006; Quin & Guo, 2001). The most important parameters that affect the properties of carbon gels are the pH of the precursor solution and the nature and concentration of the reactants, since variations in these parameters cause significant changes in the final porous properties of the carbon gel, making it possible to go from a totally non-porous material to a very highly porous carbon gel by only modifying one of these parameters. However, there are also other variables to consider such as temperature and time of gelation and curing stages, the nature of the solvent and addition of surfactants during the drying step, as these also have an important influence on the development of different properties of carbon gels (Al-Mutasheb & Ritter, 2003; Job et al., 2007a; Kraiwattanawong et al., 2011; Matos et al., 2006; Tian et al., 2011b). In so far as the carbonization step is concerned, there are several parameters that have a notable impact on the final characteristics of carbon gels, these include the carbonization temperature and time or the nature and flow rate of the carrier gas (Al-Mutasheb & Ritter, 2003; Lin & Ritter, 2000; Tamon et al., 1998).

The increasing popularity of carbon gels is largely due to their unique and controllable physicochemical properties such as their specific surface areas ranging from about 500 to 1200 m^2 g^{-1}, high pore volumes, low density, excellent electrical conductivity, high purity and the possibility of synthesizing them in the form of monoliths, powders, microspheres or thin films, with high packing densities (some of these shapes are shown in Figure 2) (Al-Mutasheb & Ritter, 2003; Juárez-Pérez et al., 2010; Mahata et al., 2008). The combination of these properties makes carbon gels the perfect candidates for diverse applications such as supercapacitors, fuel cells, desalination systems, catalyst supports, liquid and gas-phase adsorbents, etc. (Calvo et al., 2008; Frackowiak & Béguin, 2001; Moreno-Castilla et al., 2005; Zheivot et al., 2010; Zubizarreta et al., 2010). However, despite the large number of advantages associated with carbon gels, there are still some applications where it is preferable to use activated carbons as a result of their low production costs. The method of synthesis of carbon gels is the main hindrance to their implantation at industrial scale because with conventional methods, where gelation, curing and drying stages are performed in conventional furnaces, several days are required to produce the final materials. Of the three stages involved in the synthesis of organic gels, drying is the most expensive. This is due to, except in the case of subcritical drying, under supercritical and freezing conditions, it is necessary to perform solvent exchanges, which requires several days, and in the most extreme conditions of drying entails a substantial increase in production costs (Liu et al., 2006; Tamon et al., 2000; Zhang et al., 2007). Consequently, the research in this field is being addressed to the development of faster and cheaper methods of synthesizing carbon gels in order to make them more attractive and competitive than the activated carbons used until now (Calvo et al., 2008, 2011; Conceição et al., 2009; Tonamon et al., 2006; Zubizarreta et al., 2008b). Some of these works are based on the use of different types of electromagnetic radiation as a heating source for one or several stages of the synthesis process. As will be explained in more detail in Section 2.4, microwave and ultrasonic radiation are the most widely investigated of the new synthesis techniques, being the results very promising not only because they meet the target of lower production times and costs but also they produce carbon gels with properties similar to those obtained using more established methods (Calvo et al., 2008, 2011; Zubizarreta et al., 2008b).

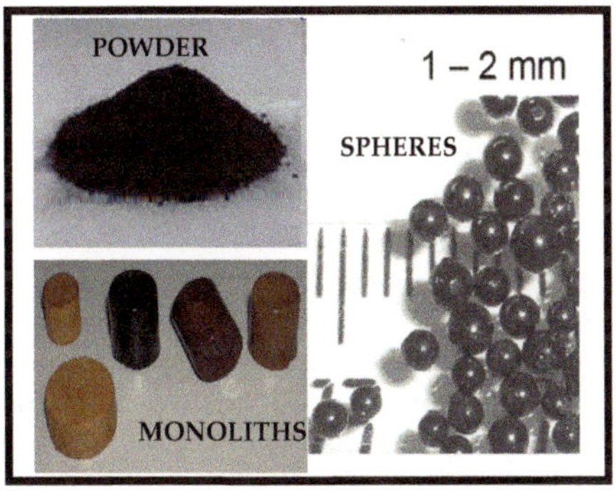

Fig. 2. Carbon xerogels presented in different shapes.

To increase the porosity of carbon gels or to enrich their chemical structure, carbon gels can be subjected to various activation, oxidation and doping processes. Activation processes can be performed during or after the carbonization step and the properties of the final carbonaceous material will be greatly influenced by the sequence used. The purpose of activation is to increase the surface area and pore volume created during the synthesis of organic gels and to promote pore widening, especially in the case of narrow pores (Contreras et al., 2010; Silva et al., 2009; Zubizarreta et al., 2008c). In Section 2.2, the different types of activation processes and the corresponding properties of carbonaceous materials produced are reviewed in more detail but, by way of introduction it may be said that there are two types of activation processes, chemical and physical activation, both with their respective advantages and disadvantages, and each of them generating carbon gels with specific textural and structural characteristics (Contreras et al., 2010). The surface areas of most carbon gels vary between 600-700 $m^2 g^{-1}$, but after an activation process, this value may increase to 2000-3000 $m^2 g^{-1}$. In other words, the porosity may be tripled which is a great advantage in applications that require highly porous materials such as supercapacitors, hydrogen storage or catalysis.

There are several published works that deal with ways to modify the chemical nature of carbon gels by means of doping or oxidizing processes in order to widen their range of applications (Gryzb et al., 2010, Job et al., 2007b; Lee et al., 2011; Sepheri et al., 2009; Silva et al., 2009; Zubizarreta et al., 2010). The porous texture of carbon gels is a crucial property in most of the fields of application. However, it is not the only one that determines the performance as the surface chemistry is also a key factor. Thus, several studies focus on the incorporation of oxygen functional groups by means of different oxidation processes (Mahata et al., 2008; Silva et al., 2009), the incorporation of nitrogen groups by using nitrogen-containing monomers or post-synthesis treatments (Gorgulho et al., 2009; Kang et al., 2009; Long et al., 2008b; Pérez-Cadenas, 2009) and the modification of carbon gels with the incorporation of metal species into the carbon framework (Bekyarova & Kaneko, 2000; Chandra et al., 2011; Cotet et al., 2006; Job et al., 2007b; Liu et al., 2006; Tian et al., 2010). In the synthesis process of carbon gels based on sol-gel methodology, these above-mentioned treatments can be performed using different reagents and conditions, making it necessary to optimize the operating conditions to meet the requirements of each individual case. In Section 2.3, different types of oxidative treatments and doping processes will be discussed in the light of the characteristics of the carbon gels produced. There are a lot of variables suitable to be adjusted in order to tailor the properties of the final carbon gel. All these tailored characteristics of the designed material (i.e. porous, chemical, mechanical characteristics) are described in Section 3 and directly connected to both the operating conditions of the synthesis of the materials and the suitability of the further application. Due to the great versatility of these kind of materials there area a wide range of applications in very different fields like adsorption (in gas and liquid media), catalysis, energy storage, etc. A review of these possible application fields are presented in the Section 4 of this chapter.

2. Synthesis of nanostructured carbon xerogels

Carbon gels are polymeric nanostructured carbon materials that can be synthesized by different procedures, all of them based on a hydrolysis-condensation reaction between hydroxybenzenes and aldehydes. There are several precursors that can be used to develop carbon gels including phenol, resorcinol or cresol in the case of hydroxybenzenes whereas as aldehyde it is possible to use formaldehyde, furfural, etc. (Czzakel et al., 2005; Dingcai &

Rouwen, 2006; Long et al., 2008a, 2008b; Pekala et al., 1995 ; Scherdel & Reichenauer, 2009). Amongst all the possible variations, probably the most commonly synthesized carbon gels are those based on resorcinol and formaldehyde, although in order to reduce the cost of the materials involved in the synthesis process, some less expensive precursors, such as phenol (Mukai et al., 2005a; Scherdel & Reichenauer, 2009; Teng & Wang, 2000) or cellulose (Gryzb et al., 2010), have attracted interest in recent years. Another important parameter for the preparation of carbon gels is the reaction media because there are several available solvents such as deionised water (Job et al., 2004; Pekala, 1989; Zhu et al., 2007), acetone (Berthon et al., 2001) or methanol (Zubizarreta et al., 2008a). It goes without saying that the least expensive reaction media is water, although other solvents are preferred in certain conditions in order to obtain specific final properties. Once the reagents involved in the sol-gel process have been selected, the recipe for producing carbon gels is the following. First, hydroxybenzene, aldehyde, solvent and catalyst are mixed in suitable molar ratios and then the solution is heated in order to obtain a stable crosslinked gel, which is saturated with solvent and it must next be dried. As will be seen throughout this section, there are several drying methods resulting in materials with different properties (Czzakel et al., 2005; Job et al., 2005). The last essential step for obtaining carbon gels is thermal stabilization, i.e. treatment at high temperature under inert atmosphere, yielding a thermally and chemically stable carbon gel.

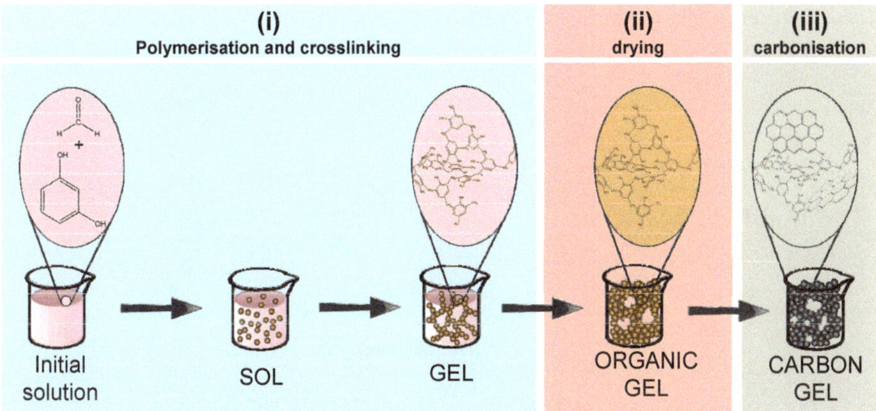

Fig. 3. Different stages involved in the synthesis process of carbon gels.

Figure 3 shows the main stages involved in the synthesis process of carbon gels by means of the polycondensation of a hydroxilated benzene and an aldehyde (resorcinol and formaldehyde in this particular case). These stages are as follows: (i) sol-gel reaction, i.e. the formation and crosslinking of polymeric particles, (ii) drying of the solvent-saturated gel and, finally, (iii) carbonization of the organic gel to yield the final carbon gel.

2.1 Synthesis steps for organic gels

As already mentioned at the beginning of Section 2, the synthesis process of organic xerogels is clearly divided into two main steps: (i) polymerization and crosslinking reactions between resorcinol-formaldehyde aggregates (gelation and curing stages) and, (ii) drying process. Each step plays an important role in determining the final properties of the xerogel, and therefore deserves a detailed description.

2.1.1 Polymerization and crossliking

The polymerisation and crossliking reactions, also referred as the gelation and curing processes, take place during the sol-gel reaction between resorcinol and formaldehyde. According to some published works (Lin & Ritter, 1997; Pekala & Alviso, 1992), the polymerisation mechanism includes two steps: (i) addition reaction to form hydroxymethyl derivatives of resorcinol and, (ii) condensation of hydroxymethyl derivatives to form methylene or methylene ether bridged compounds. After these reactions a polymer is formed and, as a consequence, the initial solution loses fluidity, producing a special material called gel, and the time which it takes for the gel formation is named *gelation time*. The curing step of the gel is an extension of the process whereby the crosslinking of polymeric aggregates previously formed in the gelation stage is favoured, so that a three-dimensional crosslinked polymer is obtained.

As can be seen in the reaction scheme represented in Figure 4, the formation of resorcinol anions via the abstraction of hydrogen is enhanced by the basic media. Resorcinol is a trifunctional fenolic compound that is able to add formaldehyde molecules in three different positions (2, 4 and 6), but these uncharged molecules are less reactive than the corresponding resorcinol anions. Therefore, when there are OH^- anions in the reaction media, the hydrogen of resorcinol molecules is abstracted promoting the formation of hydroxymethyl derivates. In the second stage, the condensation of these hydroxymethyl derivates proceeds via acidic media, generating colloidal particles that start to crosslink producing aggregates with a diameter of around 7-10 nm (Al-Mutasheb & Ritter, 2003).

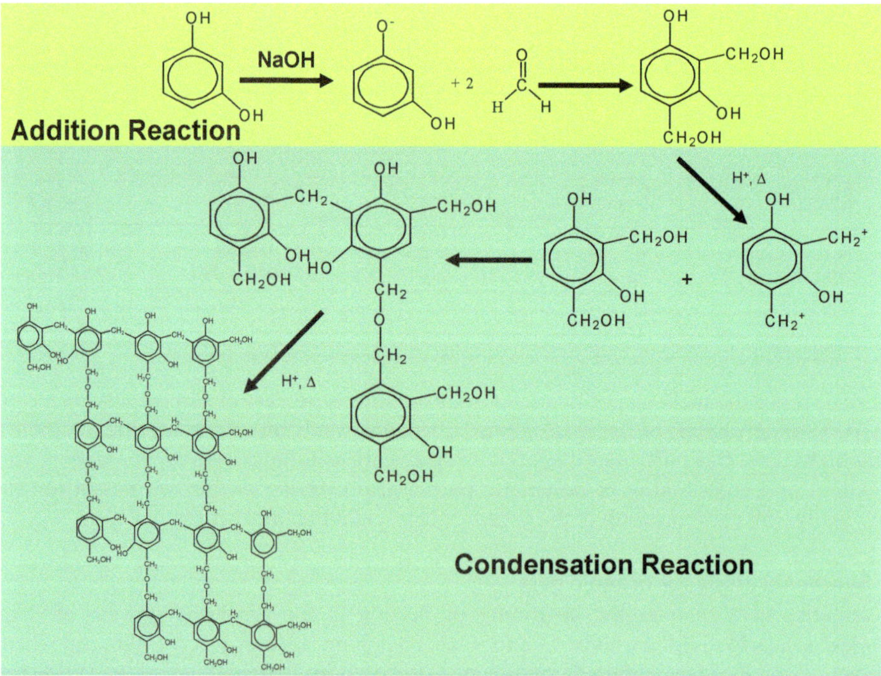

Fig. 4. Mechanism of resorcinol-formaldehyde polymerization
(based on reference Al-Mutasheb & Ritter, 2003).

According to this mechanism, the initial pH of the solution is a very important operating condition because it controls the polymerisation, the subsequent crooslinking and therefore, the final porous texture of carbons gels. It is for this reason that some compounds are usually added to the reactants in order to modify the pH of the media. Some authors claim that these compounds are simply pH regulators and therefore not the usually named catalysts (Job et al., 2005) while others postulate that the nature of the compound used (i.e., an ion) is going to have a direct bearing on the pH and therefore it is a real catalyst (Fairén-Jiménez et al., 2006; Tian et al., 2011b). In any case, independently on the nature of the compound used the R/C (resorcinol/catalyst molar ratio), is a parameter usually mentioned in the synthesis receipt and it is directly related with the pH of the reaction media. Therefore, at high catalyst concentrations, i.e. as the pH increases, the first addition reaction is favoured and therefore, very branched and unstable aggregates are formed, leading to smaller more interconnected polymer particles (see Figure 5b). The condensation reaction resulting from such small particles produces materials with smaller pores (Lin & Ritter, 1997; Job et al., 2004). In contrast, when the initial pH decreases, the formation of hydroxymethyl derivates is slow; this results in a smaller number of resorcinol anions. Naturally, polymeric particles have enough time to grow, producing large but weakly branched aggregates. The condensation of such less interconnected polymeric particles leads to higher pore sizes, illustrated in Figure 5c. Therefore, it is possible to affirm that the pH of the precursor solution plays a very important role in the sol-gel reaction because it determines the size of the polymeric particles formed during the gelation and curing stages and, accordingly, the size of the pores in the final carbonaceous material (Calvo et al., 2011a; Lin & Ritter, 1997; Job et al., 2004; Zubizarreta et al., 2008a).

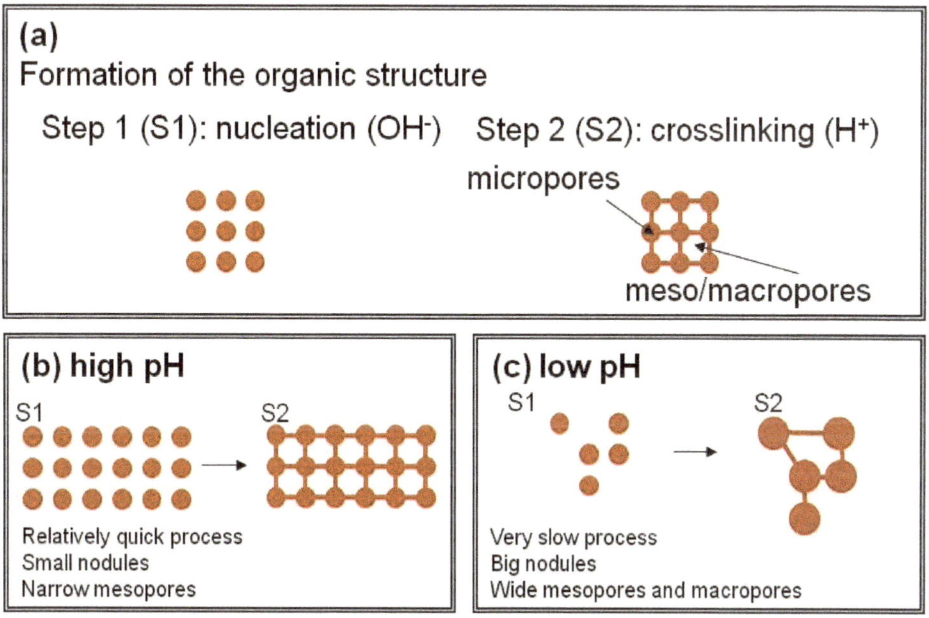

Fig. 5. Schematic representation of the influence of the pH of the precursor solution on the final nanostructure of the carbon gels obtained.

2.1.2 Drying

Once a stable three-dimensional polymer has been obtained, it is necessary to remove the solvent used as the reaction media. There are different types of drying methods, each of which produces materials with different properties. Therefore drying is another synthesis condition that needs to be taken into account when trying to control the final properties of the nanostructure organic gel (Czzakel et al., 2005; Job et al., 2005). The most widely drying methods used are: (i) subcritical drying, i.e. drying the gels by simple evaporation of the solvent (at ambient pressure and temperatures of around 100-150 °C); (ii) supercritical drying, which means eliminating the solvent in supercritical conditions (high pressures and temperatures), and (iii) freeze-drying, i.e. the solvent is frozen and then removed by sublimation. A scheme of the different ways of eliminating the solvent is presented in Figure 6, whilst a more detail description of each of them and their influence on the final properties of the carbon gels is described below.

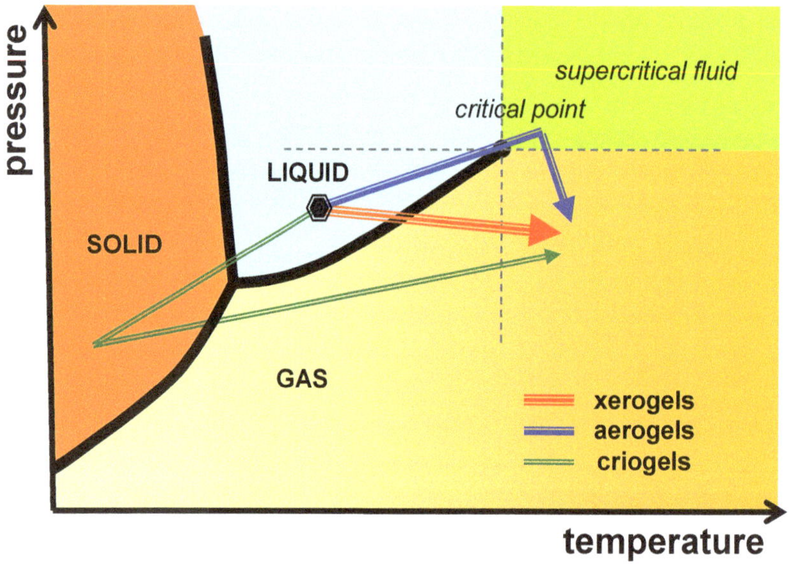

Fig. 6. Scheme of the different ways of removing the solvent used in the synthesis of organic gels.

The drying of organic gels under subcritical conditions is based on the evaporation of the solvent. Unlike other types of drying, by means of subcritical conditions, the formation of a liquid-vapour interface takes place. Therefore, when the solvent inside the pores of the material evaporates, the pores are subjected to high tension which causes the partial collapse of the structure. In order to reduce the capillary forces responsible for the partial destruction of porous texture, according to some pusblished works (Kraiwattanawong et al., 2011; Lee et al., 2010) before performing the evaporation of the water used as solvent, it can be exchanged with another solvent with a lower surface tension. Possible candidates to substitute for water could be, for example, acetone or cyclohexane (the surface tension of water at 20 °C is 72.25×10^3 Nm^{-1} while in the case of acetone and cyclohexane it is 23.7×10^3 and 25.5×10^3 Nm^{-1}, respectively). However, it must be said that in some cases the partial shrinkage of the structure when water is used as a solvent may not be of any significant if the porous texture desired for a specific application is micro or micro-mesoporosity and an easy and cheap process is required. When the solvent is eliminated by evaporation, the resulting organic gel is called xerogel. This kind of drying is the cheapest and easiest to perform, and usually the process is quite rapid (i.e. hours). Furthermore, according to some recent studies (Calvo et al., 2011a; Job et al., 2006; Leonard et al., 2005; Zubizarreta et al., 2008b) if the different evaporation conditions (i.e. the type of the solvent used, the temperature of evaporation, the carrier gas used during evaporation, use of vacuum, microwave heating, etc.) are correctly selected, it is possible to control the porous texture in order to obtain either microporous, micro-mesoporous or micro-macroporous materials.

Traditionally, although the most widely drying method used is supercritical drying, it is also the most expensive and complicated. Based on the removing of the solvent under high pressure and temperature, it is the best way to preserve the porous texture and structural properties created during the synthesis of the gels. The materials prepared by this method are called aerogels. In order to soften the conditions of supercritical drying, the solvent is usually replaced by CO_2 before its elimination under supercritical conditions. The temperature needed to remove CO_2 is lower than that of any organic solvent due to its lower surface tension, so most of the published studies on carbon aerogels make a preliminary exchange of solvent (Carrot et al., 2007; Liu et al., 2006; Tian et al., 2011a). The main problem is that the most commonly solvent used (i.e., water) needs to be exchange with an organic solvent, prior to be replacing by CO_2, complicating the procedure with several steps. Thus, the aqueous gels must to be placed in an organic solvent (i.e. acetone) for several days to completely remove the water and after that, the organic solvent is exchanged with liquid CO_2 which will finally be removed under supercritical conditions (Carrot et al., 2007). Another option is to remove the organic solvent directly under supercritical conditions, thereby by-passing the CO_2 exchange step (Liang et al., 2000; Wiener et al., 2004). However, although this second procedure avoids one of the steps in the drying process, the process usually requires more drastic operating conditions with respect to pressure and temperature apart from the fact that the aerogels obtained undergo a greater structure collapse and density than aerogels dried following the first recipe. Carbon aerogels are characterized by high pore volumes, which make these materials applicable in a huge number of application fields. However, the main disadvantage of supercritical drying, in addition to the difficulty and extremely high costs associated with the method, is the time needed to obtain dry gels as the solvent exchange steps require several days (Carrot et al., 2007; Job et al., 2005; Liu et al., 2006).

The third drying method, which results in materials called cryogels, is based on the freezing and subsequent removal of the solvent by sublimation. This method is an effective way of preparing gels with a controlled pore structure. The costs associated with freeze-drying are lower than those of supercritical drying, but it is still a more expensive method than drying by simple evaporation of the solvent (Yamamoto et al., 2001). In the case of aqueous gels, it is possible to freeze the water directly by placing the wet gel in a liquid nitrogen bath. Afterwards, the solvent is removed by sublimation under low pressures. This procedure can lead to dramatic changes in the density of the cryogels after freezing and also to the formation of megalopores or voids as a result of the creation of crystals inside the structure of the gels (Job et al., 2005; Kocklenberg et al., 1998). To prevent the formation of crystals which may deform the designed nanostructure of the polymer, the solvent is replaced before drying the gel. The most common solvent used for this purpose is t-butanol (Feaver

& Cao, 2006; Mukai et al., 2005b; Yoshimune et al., 2008), as it causes fewer changes in density than water and the vapour pressure is higher, which reduced the drying time. The advantage of this type of drying process is the possibility of obtaining high mesopore volumes (Yoshimune et al., 2008). As for the disadvantages, it should be highlighted that it is very difficult to prepare monoliths and in addition, in order to avoid the appearance of megalopores, it is necessary to perform a solvent exchange which entails an extra step and therefore and increase in time and costs.

To conclude this section of drying methods, it needs to be affirmed that the choice of drying conditions will determine the textural and structural properties of the final materials. Moreover, none of the drying methods are perfect, all three have their advantages and disadvantages (reported in Table 1), and the choice of method will depend on the requirements of the final applications.

DRYING METHOD	MATERIAL	ADVANTAGES	DISADVANTAGES
SUBCRITICAL	Xerogel	- Simple, rapid and cheap method - High surface areas and pore volumes can be achieved	- Presence of capillary forces that destroy part of the initial porosity
SUPERCRITICAL	Aerogel	- No shrinkage of pore texture	- Requires high temperatures and pressures - Extremely high cost - Long times required - Necessary to perform, at least, one solvent exchange with CO_2
FREEZE	Cryogel	- Low shrinkage - Materials with higher pore diameter	- High cost, long time and complicated method - With aqueous gels, it is essential to exchange the solvent

Table 1. Summary of the main characteristics of the drying methods used in the field of carbon gels.

2.2 Thermal treatments of organic gels

The thermal treatment of the organic gels, i.e. carbonization, allows the removal of non-crosslinked organic chains, labile oxygen and hydrogen surface groups, resulting in thermally stable nanostructured materials formed mainly by carbon. This type of nanostructure endows the materials with a series of electrical, thermal and mechanical properties that are very useful in applications related to catalysts, energy storage, etc. (Al-Mutasheb & Ritter, 2003). The main targets of the carbonization step are to obtain thermally stable carbonaceous materials and promote the development of microporosity in the gels. The nanostructure developed during the synthesis of the organic gel usually remains intact during the carbonization step, and therefore the meso and macroporosity obtained is preserved as it was previously designed. However, during the carbonization step the elimination of volatiles and labile matter leads to the formation of microporosity, mainly localized in the nodules of the polymer (see Figure 5a). Thus, the meso-macroporosity is controlled during the synthesis of organic gels while the microporosity is controlled independently during the carbonization step, which supposes a great advantage respect to design and control the porous texture of these materials. Usually, organic gels are carbonized by heating the samples in a furnace under inert atmosphere (i.e. N_2, Ar or He) for a specific period of time. Although most of the published works on carbon gels use N_2 or Ar as inert gas during the carbonization step, others use a reactive gas to modify the internal structure of the material. For example, K.Y. Kang et al. (Kang et al., 2008), showed that it was possible to perform the carbonization under an ammonia atmosphere (process known as ammonization). As a result, in addition to develop the microporosity of the samples, nitrogen functionalities are incorporated into the structure.

Like other steps involved in the synthesis process of organic gels, carbonization has also an important influence on the final properties of the material. The most important variables that have a significant influence on the characteristics of carbon gels are: temperature and time of carbonization, type of gas and flow rate used, heating device used, etc. Several authors have studied the influence of carbonization temperature on the porosity of carbon gels. Whereas some of these studies affirm that an increase in carbonization temperature leads to a loss of microporosity (Lin & Ritter, 2000). Others affirm the opposite, i.e. an improvement in microporosity when the temperature increases (Kang et al., 2008). C. Lin and J.A. Ritter (Lin & Ritter, 2008) evaluated the dependence of porous texture with carbonization temperature by performing the carbonization under N_2 flow at several temperatures (600, 750, 900, 1050 y 1200 °C). Their results reflected than the optimum temperature was 600 °C because as the temperature increased, smaller micropore volumes were reported. In addition, the present work shows that carbonization temperature does not influence the mesoporosity of the samples, which corroborates the affirmation that meso-macroporosity is developed during the synthesis of the organic gels while the appearance of micropores occurs during the subsequent carbonization stage (Al-Mutasheb & Ritter, 2003). Other work on the influence of carbonization temperature has also been carried out by K.Y. Kang et al. (Kang et al., 2008) but, in that case, the process was performed at 650, 850, and 950 °C under flows of nitrogen and ammonia. Results showed that higher temperatures promote the development of porosity. The bibliography, therefore, apparently contains contradictions regarding the real influence of the carbonization temperature on microporosity development of carbon gels. The reason for these divergences is probably that many other variables, such as the R/F molar ratio, the pH of the precursor solution, the type of catalyst used, etc., lead to different nanostructure materials that react in a different way with the increase of temperature. The nature of gas used during the carbonization may also have an effect on the porous and structural properties of carbon gels. Starting with the porous texture, most of the literature about carbon xerogels uses N_2 or Ar atmosphere as carrier during carbonization, resulting in materials with specific surfaces areas of approximately 600-700 $m^2 g^{-1}$ (Calvo et al., 2011a; Job et al., 2004; Matos et al., 2006; Zhu et al., 2007). However, it has been reported that the treatment of organic xerogels with ammonia produces carbon xerogels with specific surface areas above 1000 $m^2 g^{-1}$ (Kang et al., 2008). Moreover, differences in the chemical characteristics of the samples may also appear due to the nature of the atmosphere. For example, carbon xerogels prepared by means of ammonia-assisted carbonization contain approximately 6-7 wt. % of nitrogen while in the case of organic xerogels carbonized under nitrogen atmosphere, the amount of nitrogen does not exceed 1 wt. %. These examples show the huge influence of the carbonization process on the final properties of carbon gels and, as in the case of the other synthesis variables, a correct choice of carbonization conditions allows to obtaining a material with the appropriate properties (both porous and chemical) for a specific application.

Carbon gels can be also subjected to activation processes after or during the carbonization step. The aim of activation processes is to increase the surface area and volumes of the pores created during the synthesis, and also to promote their widening, especially in the case of the narrower pores. Therefore, it is very common to activate carbon xerogels, where the porosity is narrower and lower, and they are usually used in applications where besides narrow mesopores a high volume of micropores are needed. It is generally accepted that there are two types of activation methods: physical activation, by means of CO_2, steam, or a combination of both (Lin & Ritter, 2000) and chemical activation, where the activation agent may be KOH (Fang & Binder, 2006; Macia-Agullo et al., 2007; Zubizarreta et al., 2008c), H_3PO_4 (Conceição et al., 2009; Jagtoyen et al., 1993), $ZnCl_2$ (Olivares-Martín et al., 2006), etc. In all cases, it is essential to optimize a number of variables due to their notable influence on the final porosity. Some of these variables are: temperature and time of activation, activating agent and precursor used (i.e., organic or carbon gels as precursor), amount of activating agent, gas flow and heating rate, etc. (Fang & Binder, 2006; Lozano-Castelló, 2002). Chemical activation processes take place in two stages: (i) the precursor is mixed with the activating chemical agent and this can be done in two different ways, by physical mixture, i.e. the two solid products are directly mixed in a mortar, or by wet impregnation, when the sample is mixed with a concentrated solution of the selected chemical agent for a specific time at low temperature (< 100 °C); in the latter case the slurry formed must be subsequently dried before the process is continued; (ii) the mixture is subjected to thermal treatment under an inert atmosphere up to a selected temperature that may range from 300 to 900 °C, depending on the activated agent used. When this thermal treatment is completed, the sample must be washed with water several times in order to remove traces of chemical agent. Finally, the sample is dried. One of the main disadvantages of this type of activation, apart from the higher cost of the activating agents (KOH, H_3PO_4 vs. CO_2 and steam water, for instance), is the washing stage since, in addition to lengthen the process, sometimes it is extremely difficult to completely remove all the traces of the residual activating agent. Despite these drawbacks, chemical activation has several advantages compared to physical activation including the lower temperature and activation time, higher yield and higher development of porosity achieved (Lozano-Castelló, 2002; Molina-Sabio et al., 2004; Teng & Wang, 2000). Physical activation consists of (i) thermal treatment of the precursor in an inert atmosphere and the successively controlled gasification of the carbonaceous material or (ii), the direct activation of the raw material in the presence of the activating gas. The main characteristics of physical activation are: higher temperatures than chemical activation (between 800-1100 °C), a more heterogeneous micropore size and very simple method (Okada et al., 2003). The effect of the type of precursor on the porosity of final materials is well known in the case of chemical activation, whereas in physical activation processes, further studies are needed to determine the relevance of this variable. Published works on chemical activation of organic and carbon gels have shown that in the case of carbon gels, chemical activation produces an increase in micropores volume without modifying the mesoporous structure formed during the synthesis of the gels. However, when the chemical activation of organic gels is performed, mainly microporous materials are obtained since the mesoporosity created during the synthesis is severely damaged. This phenomenon again shows the versatility of these carbonaceous materials because, by means of chemical activation processes, only by varying the precursor used, it is possible to prepare mainly microporous materials with a small amount of mesopores or materials characterized by a high micropore volume but also by a significant amount of mesopores with a controlled size depending on the pH of the initial solution (Zubizarreta et al., 2008c). In chemical activation, several works in the literature evaluate the influence of the amount of chemical agent used on the final characteristics of the carbon gels. Usually, as the activating agent/precursor ratio increases, a further development of the microporosity takes place. However, it seems that there is a limit to this ratio, above which the specific surface area begins to decline. Thus, Zubizarreta et al. (Zubizarreta et al., 2008c) showed that in the chemical activation of different carbon xerogels with KOH, the activating agent/precursor mass ratio greatly influences (but not always in the same way) the porous texture of the activated carbon xerogels. The authors have used ratio values of 1, 2, 3 and 4 and observed that in some cases the maximum surface area was achieved with a ratio of 3, whilst with other samples the maximum was achieved using a ratio of 4. Different results were obtained depending on the pH of the initial solutions used to synthesize the materials. Highly microporous carbon xerogels can also be prepared by chemical activation with phosphoric acid (Conceição et al., 2009). In this work, several impregnation ratios were used and their influence on the final characteristics on the material was evaluated. It is noteworthy that for all the impregnation ratios employed, the mesoporosity of original samples was destroyed, which is consistent with the findings of other scientific studies (Zubizarreta et al., 2008c). In addition, the higher the impregnation ratio, the greater the development in porosity. However, it should be noted that differences between samples was not very significant, since with ratios of 1, a carbon xerogel with a specific surface area of 1525 $m^2 g^{-1}$ was obtained while if the impregnation ratio was tripled, an increase in specific surface area of barely 200 $m^2 g^{-1}$ was achieved.

From the examples showed in this section 2.2, both in carbonization and activation processes, is clear that the porous texture of carbon xerogels can be designed by modifying several synthesis conditions. This is of huge relevance as it reveals the carbon gels as a nanostructured material with a great potential as it is possible to tailor specific properties of this kind of materials to adequate them for an optimum behavior in a wide variety of scientific fields and applications.

2.3 Another post-synthesis treatments

As already mentioned, the most interesting characteristic of carbon gels is the possibility of tailoring the final properties in order to prepare them for a specific application (al-Mutasheb & Ritter, 2003; Czzakel et al., 2005; Job et al., 2004, 2005; Pekala, 1989; Zhang et al., 2007; Zhu et al., 2007; Zubizarreta et al., 2008a), control and design achieved by selecting the appropriate variables involved in the synthesis. There are many possible *bottoms to be pressed* to obtain the right results. Nevertheless, the porous texture is not the only factor that determines the optimum performance of the carbon gels in a specific application. Surface chemistry also plays a relevant role, due to the interactions between the fluids and the carbon surface (i.e. the possibility of redox reactions, charge transfer, different wettabilities depending on the surface chemistry, the blockage of reactive sites, etc. (Serdich et al., 2008)).

Therefore, there has been an increase in the number of published works focused on the tuning of the surface chemistry of carbon gels in recent years and these processes can be performed in several ways. For example, carbon gel functionalisation can be carried out with post-synthesis treatments, such as oxidation using different oxidizing agents, HNO_3, H_2O_2, air, etc. (Gryzb et al., 2010; Mahata et al., 2008), treatments with compounds such as ammonia, melamine or ammonia borane (Gorgulho et al., 2009; Pérez-Cadenas et al., 2009; Sepheri et al., 2009), with the aim of incorporating nitrogen and borane groups, respectively, into the structure of carbon gels. Alternatively, heteroatoms can be directly added during the synthesis using heteroatom-containing polymeric precursors, i.e. melamine, urea, cellulose acetate, etc. (Gryzb et al., 2010; Long et al., 2008b). Rather than in post-synthesis treatment, this is a modification of the synthesis receipt and therefore, the polymerization process would vary since the initial reactants are different.

Regarding to oxidative processes, oxygenated surface groups are incorporated using liquid-phase oxidants (e.g. nitric acid, hydrogen peroxide, ammonium persulphate) or gas-phase oxidants (i.e. air, steam, oxygen, etc.). The nature of the oxygenated functionalities incorporated depends not only on the type of oxidizing agent used, but also on the conditions in which the oxidative process is carried out. For example, N. Mahata and co-workers (Mahata et al., 2008), studied the type of oxygen surface groups created in carbon xerogel structures by means of three different oxidation treatments (oxygen plasma, nitric acid and diluted air). Of the three oxidative treatments, HNO_3 oxidation produces largest amount of carboxylic acid groups, but the amount of oxygen groups created is difficult to control and besides, high concentrations of HNO_3 are needed. Each of the three processes produces carbon xerogels with different surface chemistry. Consequently, the choice of oxidizing agent should be made according to the application requirements for which the carbon gels are intended. Another work focusing on the oxidation conditions of carbon xerogels with HNO_3 was published by Silva et al (Silva et al.., 2009). In this work, HNO_3-hydrothermal oxidation was carried out using several concentrations of nitric acid (from 0.01 to 0.30 mol l^{-1}) at different operating temperatures (between 120 and 200 °C) and the results show that both the concentration of HNO_3 and temperature notably influence on the level of oxygen functional groups created on the surface of carbon xerogels. There is a clear correlation between the degree of functionalization and the HNO_3 concentration used. The temperature of functionalisation may also affect the porosity of the final carbon xerogels. Unlike the partial blockage of the pores produced in oxidative processes with concentrated HNO_3, the HNO_3-hydrothermal method not only maintained the porous texture of carbon xerogels but also, when the operating temperature was fixed at 200 °C, there was even an increase in the specific surface area of carbon xerogels oxidized with a high concentration of HNO_3 solution. Another published study covering several types of carbon xerogels post-treatments, was performed recently by Grzib and co-workers (Gryzb et al., 2010). A series of nitrogen and oxygen functionalised carbon xerogels were synthesized by means of different oxidative processes with HNO_3 and H_2O_2, treatments with gaseous ammonia at high temperature and co-heating of carbon xerogels with melamine. All of these treatments give rise to a wide range of carbon xerogels, with different amounts and type of oxygen and nitrogen groups. This represents a breakthrough for applications that, besides a good porosity development, require materials with a rich surface chemistry. Out of all the post-synthesis treatments studied, it was found that ammonisation produces the most basic carbon xerogels not only because of the incorporation of basic N-groups, but also due to the reduction of acidic oxygen functionalities within the chemical structure of the carbon xerogels. Oxidation with nitric acid or hydrogen peroxide introduces almost amount of oxygen (about 5 wt. %), while the character of these functionalities is quite different. XPS data reported in this work show that H_2O_2 oxidation mainly produces carbon xerogels with oxygen groups like alcohols and ethers, whereas nitric acid treatment incorporates esters, lactones and carboxylic groups into the structure of the carbon xerogels, results which are consistent with other scientific works.

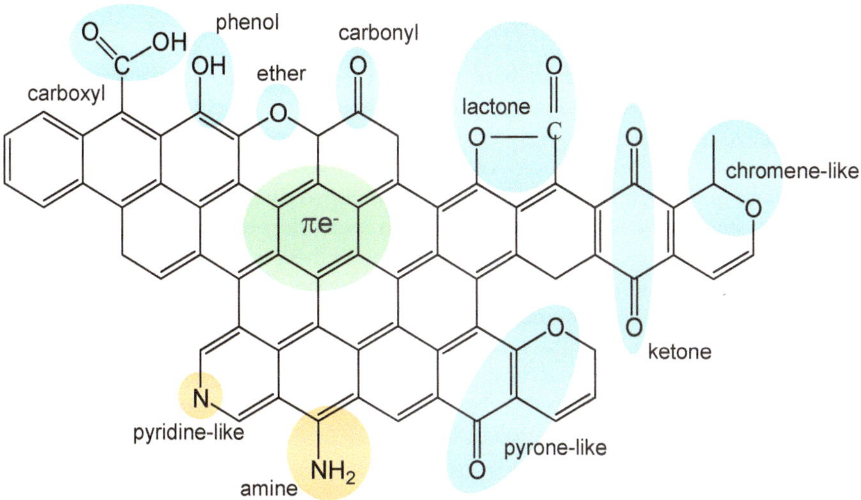

Fig. 7. Main functionality groups that can be incorporated into carbon gels by post-synthesis treatments.

The doping of carbon gels with different metal species (i.e. metals, metal oxides, metal carbides) is another common post-synthesis process in carbon gels, since by means of this technique it is possible to modify their chemical nature, enlarge their surface area and improve their conductivity, thereby increasing the already known potential of this kind of carbonaceous materials. Despite the great interest in carbon gels due to the possibility of tailoring their porous texture, the presence of metal particles acting as active sites is required for specific applications. Consequently, several studies on the doping of carbon gels with transition and non-transition metals have emerged in recent years (Chandra et al., 2011; Job et al., 2007b; Lee at al., 2011; Liu et al., 2006). For example, in the particular case of energy storage in supercapacitors, carbon gels are very promising materials due to their attractive features. Supercapacitors assembled with these carbonaceous materials stores energy mainly because of the formation of a double electric layer on electrodes surface. However, it is well known that specific capacitance can be increased as a result of redox reactions. Hence, in relation to this particular application, numerous studies have emerged on the synthesis of doped carbon gels. The species that are usually incorporated into carbon gel structures to improve their energy capacitance via pseudocapacitive processes are Ni, B, P, Cu, Ca, Co, Mn, etc. (Chandra et al., 2011; Cotet et al., 2006; Job et al., 2007b; Tian et al., 2010). Carbon gels have also received a great deal of attention as hydrogen storage media due to their good adsorption properties that are the result of their remarkable textural development. It has generally been accepted that the hydrogen storage capacity of carbon materials is directly related with their specific surface area and micropore volume (Tian et al., 2010; Zubizarreta et al., 2010). However, the storage capacity values at room temperature are too low, which is a very serious problem for their application in mobile devices. In order to overcome this limitation, researchers have proposed a doping process of carbon materials to improve the interaction between the hydrogen and the carbon surface and, consequently, to enhance their hydrogen storage capacities. Zubizarreta and co-workers (Zubizarreta et al., 2010), reported the preparation of Ni-doped carbon xerogels by two methods (both post-synthesis treatments): dry impregnation and strong electrostatic adsorption, SEA. Their results showed that both treatments produce Ni-doped carbon xerogels with a small particle size (2-9 nm) and a good nickel dispersion, but dry impregnation method produces carbon xerogels with the higher amount of nickel incorporated. In contrast, the SEA method produces materials with a high interaction between C-Ni and besides, Ni particles with a very homogeneous size (around 2 nm) are obtained. Some of the samples prepared showed very good hydrogen storage capacities, with higher values than those of undoped carbon gels, making Ni-doped carbon xerogels good candidates for hydrogen storage systems. In another study on metal-doped carbon gels performed by Tian et al. (Tian et al., 2010), carbon gels were doped with metallic cobalt particles by means of two methods: (i) the addition of cobalt acetate solution to the carbon gels followed by thermal treatment and a reduction process, and, (ii) ion-exchange method. It was found that the size of the Co particles incorporated into the structure of the materials was between 2 and 8 nm when cobalt acetate solution was used and, besides that, these samples had higher surface area and micropore volume than the Co-doped carbon gels obtained by the ion-exchange method. The hydrogen storage capacity of both types of doped-materials was compared with that of the original undoped material and the results show higher values of hydrogen storage in both of the metal-doped carbon gels, although the process that presented the better results was doping with a solution of cobalt acetate.

All the studies listed in the bibliography point out the huge versatility of carbon gels and the great potential of these materials not only because their tuneable nanostructure but also because their chemical nature can be easily altered by post-synthesis treatments.

2.4 Novel and underdevelopment alternative synthesis conditions

Over the past few decade, a number of works on alternative methods of manufacturing carbon gels by means of different types of electromagnetic radiation (i.e. infrared, ultrasonic or microwave) have been published (Calvo et al., 2008; Kang et al., 2008; Tonamon et al., 2005, 2006; Wu et al., 2004). In some cases, irradiation is applied in order to improve the porous texture of the final products, whereas in other cases, the aim is to shorten the synthesis process, in order to make carbon gels more cost-competitive materials.

Wu et al. (Wu et al., 2004) prepared carbon gels from resorcinol, furfural and hexamethylenetetramine (HMTA), by means of a drying process that combines drying at room temperature, infrared lamp and high-temperature drying. This particular drying method gives rise to low-density carbon gels with a well-developed porosity development. However, although no solvent exchange or high operating pressures are involved, it is still a long drying procedure (about 34 hours). There are many works about polymerization reactions and other types of chemical processes that are accelerated by ultrasonic radiation (Neppolian et al., 2008; Riera et al., 2010; Suslick et al., 1999; Tonamon et al., 2005). Several authors have employed ultrasonic radiation in one or several steps to synthesize carbon gel. Ultrasonic has been found to be very a helpful strategy to increase reaction rates, yields of products and, thereby shortening the reaction time required. One of the studies that applies sonication to carbon gel is that published by Tonamon and co-workers in 2005 (Tonamon et al., 2007). These authors synthesized resorcinol-formaldehyde carbon xerogels under different catalyst concentrations by means of ultrasonic irradiation using several intensities in order to evaluate the influence of sonication conditions on the porous texture of the materials. First, ultrasounds were applied to the resorcinol-formaldehyde mixture until a highly viscous sample was observed, and then the rest of the gelation and curing stages were completed by heating in a conventional furnace for 7 days. In all the samples prepared, the researchers found that ultrasonic irradiation improved the mesoporosity of the carbon gels, even in samples for which only a minor presence of mesopores could be expected (i.e. S_{BET} = 650 $m^2 g^{-1}$ and V_{meso} = 0.53 $cm^3 g^{-1}$ when the carbon xerogel was synthesized in the absence of sonication vs. 660 $m^2 g^{-1}$ and 0.93 $cm^3 g^{-1}$ if the intensity of the ultrasonic radiation was increased to 106 W cm^{-2}). In addition, it was observed that the higher the ultrasonic intensity, the shorter the gelation times and the higher the mesopore volume. In 2006, the same group of researches introduced a new variable in the synthesis process of carbon gels, i.e., drying by microwave technology (Tonamon et al., 2006). In some cases, the applied ultrasonic radiation to RF aqueous solutions and then they dried the samples by means of microwave radiation (after a water-exchange step using t-butanol in order to minimize the shrinkage of the structure). The results showed that the presence of mesopores was favored when ultrasonic and microwave radiations were combined (i.e. V_{meso} of 0.59 $cm^3 g^{-1}$ and 0.46 $cm^3 g^{-1}$ for resorcinol-formaldehyde carbon gels dried in a microwave oven with and without previous sonication process, respectively). Therefore, one of the conclusions of that work was that microwave drying was a new and efficient drying method for carbon gels because it resulted in time saving without destroying the meso-macroporosity of samples.

Microwave heating has been used to obtain different organic reactions for several years (Kappe, 2004; Menéndez et al., 2010). There are many processes involving carbon materials where microwave radiation is the main heating source used because of the good capacity of most carbon materials to absorb microwaves. The main advantage of microwave-assisted thermal processes is the saving of time, resulting in a reduction in the energy consumed, as a result of the different mechanism involved in the heating process (Menéndez et al., 2010).Thus, microwave radiation promises to be an effective technology in the field of carbon gels allowing their long synthesis time to be reduced and opening up a way to their application to industrial scale. As mentioned above, the first works that combine microwave heating and carbon gels are based on the application of this type of radiation in some stage of the synthesis process, i.e. during the drying stage (Zubizarreta et al., 2008b) or during the gelation and curing stages followed by drying in a vacuum oven (Kang et al., 2008). In all these cases, carbon gels with a good texture development were prepared in a considerably shorter time than by means of conventional heating but even so, the time required for their manufacture were too long. The next step would be the utilization of microwave heating for the entire synthesis process of carbon xerogels, i.e. during gelation, curing and drying stages. The first researchers that used microwave heating in the three stages involved in the synthesis process were Calvo et al. (Calvo et al., 2008), when in 2008 they have showed the possibility of obtaining RF carbon xerogels with similar characteristics to conventionally synthesized xerogels but with a considerable saving of time and energy (i.e. 3-4 hours under microwave radiation compared to several days by conventional heating). In addition, there are several other advantages of using microwave technology to fabricate carbon xerogels. According to a recent publication (Calvo et al., 2011a; Juárez-Pérez et al., 2010), by means of microwave heating is possible to determine the gelation point of the carbon xerogels during the sol-gel synthesis, which is essential for controlling the viscosity of the reaction media in order to obtain the carbon xerogel in an specific form (i.e., sphere, monolith, film, etc.). Another advantage of microwave radiation is the possibility of preparing mesoporous carbon xerogels over a wider pH range than when using conventional methods (Calvo et al., 2011a). It was pointed out that, in addition to time saving and the ability to accurately determine the gelation time, mesoporous materials with a tailored mesopore size can be synthesized in a wider range of pH than in the case of conventional heating. For example, with microwave heating, a pH between 4.5 and 6.5 can be used to produce micro-mesoporous carbon xerogels and micro-macroporous materials when the initial pH is fixed at 3.1, while in the case of conventional methods, only a pH range of 5.8-6.5 serves micro-mesoporous carbon xerogels.

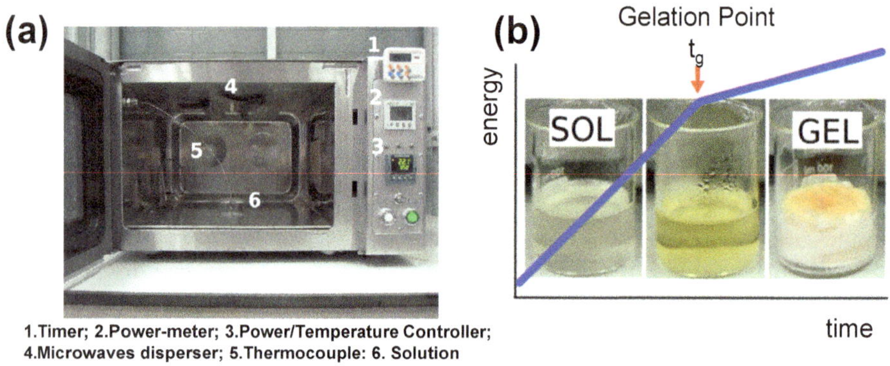

Fig. 8. The microwave device used in the synthesis of carbon xerogels (a) and a scheme showing the determination of the gelation point (b).

3. Properties of carbon gels

As already pointed out, the main advantage of carbon gels lies in the possibility of tailoring their porous texture to accomplish the requirements of the final application, and this can be achieved by selecting the appropriate synthesis conditions. While there is an abundant literature about the effect of the different stages of carbon gel synthesis (i.e. gelation, curing, drying, carbonization and activation) on the surface area, pore volume and pore size distribution (Czzakel et al., 2005, Job et al., 2004, 2005; Lin & Ritter, 2000; Matos et al., 2006; Zubizarreta et al., 2008a), the situation regarding the mechanical and chemical properties is quiet different. Although, in some cases, these parameters may be slightly influenced by the synthesis procedure followed, it is usually necessary to carry out additional processes to produce any significant changes in the mechanical or chemical properties of carbon gels (Gryzb et al., 2010; Gorgulho et al., 2009; Job et al., 2007b; Lee et al., 2011; Pérez-Cadenas et al., 2009; Silva et al., 2009).

In this section, the main factors that have any impact on the final properties of carbon gels are reported. Each of the stages involved in the synthesis process is highlighted and explained. As can be seen in Table 2, there are several variables to be considered in the synthesis of carbon gels, which explains the versatility of these materials since they can be obtained with very diverse characteristics depending on the application which they are intended. However, this also has its downside because there are a large number of variables to consider and control. Moreover, these variables are on many cases interrelated, which complicates their optimization. A different section will be assigned to each group of characteristics (textural, chemical and structural). Nevertheless, taking into account the information collected in the Table 2, it can be affirmed that: (i) all the variables involved in the different stages of synthesis have a remarkable effect on the porous texture of carbon gels but, in many cases, their influence on the chemistry and structure of these carbonaceous materials remains unknown; (ii) as regards the porosity of carbon gels, it might be said that each stage involved in the synthesis is very selective to a specific pore size. For example , it is possible to modify the macro/mesoporosity of samples by changing the pH of the precursor solution without affecting the development of microporosity, which is conditioned by the carbonization and activation steps (Calvo et al., 2008; Conceição et al., 2009; Job et al., 2004; Lin & Ritter, 2000; Zubizarreta et al., 2008c); (iii) in order to modify the chemistry of carbon gels, for example, by incorporating oxygenated functional groups, it is necessary to perform extra treatments (Gryzb et al., 2010; Silva et al., 2009).

3.1 Porous texture

In this section, the variables that influence the final porous texture of carbon gels are grouped according to the synthesis step and the corresponding operating conditions, in order to clarify the role that each stage plays.

3.1.1 Sol- gel process: pH, RF concentration or temperature and time of gelation and curing

Figure 9 shows a picture of four organic xerogels synthesized from resorcinol-formaldehyde solutions with a different initial pH (between 5.8 and 6.5). Intuitively it may be said that the pH of the precursor solution must be an important factor in the synthesis process of carbon gels because of the different colors of these four materials. Actually, numerous works (Al-Mutasheb & Ritter, 2003; Job et al., 2004; Zubizarreta et al., 2008a) have already established that the pH, or hydroxilated benzene/catalyst molar ratio, is the key factor that determines the meso/macroporous texture of carbon gels. As a general rule, as the initial pH increases, both the volume and the diameter of meso/macropores decrease, while there is no effect on the microporosity of the sample. That it is to say, by means of a slight increase in the initial pH it is possible to go from micro-macroporous to micro-mesoporous materials and from these to exclusively microporous carbon gels (Calvo et al., 2011a; Zubizarreta et al., 2008a).

Fig. 9. Resorcinol-formaldehyde organic xerogels prepared from precursor solutions with different pHs: from left to right 5.8. 6.0, 6.2 and 6.5.

The effect of pH on the porosity of carbon gels can be explained by the polymerization mechanism between hydroxilated benzene and aldehyde molecules. Polymerization reaction between these two species occurs in two different stages: (i) the formation of hydroxymethyl derivatives and (ii) the condensation of these hydroxymethyl derivatives and their clustering. At a high initial pH the first stage is favored, giving rise to the formation of hydroxymethyl derivatives and then to the formation of highly branched clusters. These cross-linked and unstable clusters lead to small and interconnected particles that, after condensation, give rise to the formation of pores, mainly mesopores. On the other hand, a low initial pH leads to the formation of fewer but bigger clusters that finally result in materials with bigger pores in the macropore range (Job et al., 2004; Lee et al., 2010). This is shown in Figure 5.

From the abundant literature that shows the dependence of the porosity of carbon gels on the initial pH (Calvo et al., 2008, 2011a; Job et al., 2004; Zubizarreta et al., 2008a), it would appear that the maximum pH that can be used to synthesize porous carbon gels is about 7.0. What is more, when the carbon gels synthesized in this study were compared, it was observed that variations of only 0.2 units in the initial pH promote significant changes in the size of mesopores. For example, the maximum mesopore diameter for a RF carbon xerogel can change from 26 nm to 14 nm when an initial pH of 5.8 and 6.0 were used. The same trend was observed for two series of carbon xerogels synthesized under microwave and conventional heating (Calvo et al., 2008). In both cases, conventional and microwave-assisted synthesis, carbon xerogels with a higher pH have smaller mesopore size. Moreover, a more recent study (Calvo et al., 2011a) shows that for microwave-assisted synthesis, the pH window used for producing micro-mesoporous materials of different mesopore size is about two units of pH (between 4.5 and 6.5), whereas synthesis in a conventional stove has a smaller pH window (between 5.8 and 6.5). The pore size distributions of these xerogels are collected in Figure 10 in order to see the porous texture in relation to the initial pH and the type of heating device used. This example confirms that there are interactions between the variables involved in the synthesis process. It can be seen that the same variations in initial pH do not produce similar carbon xerogels due to a new variable, i.e. the type of heating device used.

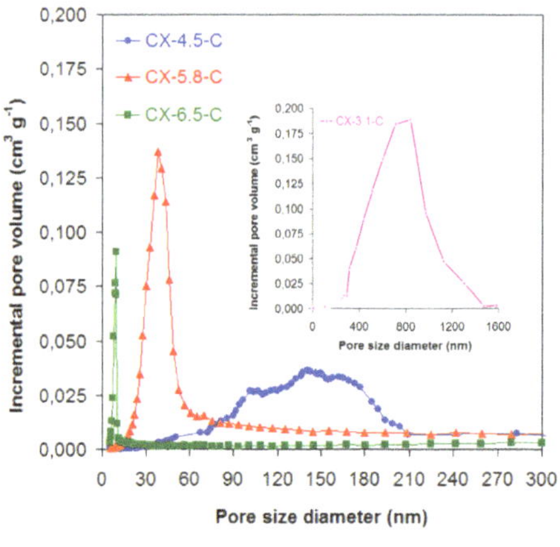

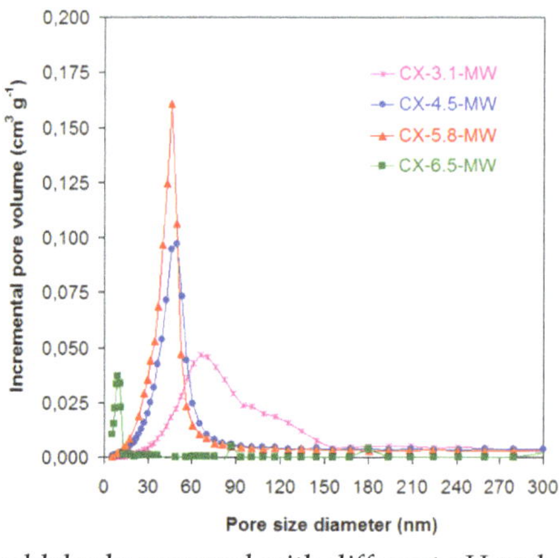

Fig. 10. Pore size distribution of resorcinol-formaldehyde prepared with different pH and heating mechanisms: conventional (a) and microwave radiation (b) (adapted from reference Calvo et al., 2011a).

Conversely, the effect of the concentration of reagents on the porosity of carbon gels has not been studied in depth. Most published works on carbon gels synthesized from resorcinol-formaldehyde mixtures use a molar ratio R/F = 0.5 (Al-Mutasheb & Ritter, 2003; Job et al., 2004; Zhang et al., 2007; Zhu et al., 2007; Pekala, 1989). This is because formaldehyde only has one aldehyde group able to interact with other groups while resorcinol molecules display two hydroxyl groups. Nevertheless, some studies use a lower molar ratio R/F, that it is to say, a more diluted RF solution, leading to carbon gels with smaller particle sizes and, therefore, smaller pore sizes (Al-Mutasheb & Ritter, 2003).

The synthesis procedure of carbon gels found in most of the published literature involves the heating of the monomers solution at temperatures ranging between 70 and 90 °C for 3-5 days in order to perform the gelation and curing stages (Pekala, 1989). The synthesis temperature and time affect the porosity of resulting materials. However, according to the results reported by Job et al. (Job et al., 2006), the effect on the porous structure is not the same in every carbon gel because the most important variable in this sol-gel synthesis, i.e. the initial pH, again comes into play. These authors studied the effect of several combined parameters, gelation and ageing temperature (50, 70 and 90 °C) and ageing time (between 0 and 72 hours), in three series of organic xerogels obtained from precursor solutions with different pH. In all the examples shown, the increase in temperature leads to lower gelation times, which is consistent with other works (Al-Mutasheb & Ritter, 2003; Job et al., 2007a; Kim et al., 2001). Moreover, as the synthesis temperature increases, the ageing time may be shortened (i.e. samples synthesized at 50 °C do not reach stability after 72 hours while in the case of 70 or 90 °C, the polymerization reactions are completed after 24 or 48 hours, depending on the initial pH). In general, it can be said that increasing the synthesis temperature yields narrower pores but this statement does not apply in all cases due to the influence of the pH. If the ageing time is increased, higher surface areas and pores volume are achieved, but once the polymerization reaches stability, the increase in ageing time scarcely has any influence on the porosity of materials. However, it should be noted that it is not easy to establish trends for each of the variables involved in the sol-gel process because a slight change in one variable, like the pH, implies a variation in the effect of other parameters, for example the synthesis temperature.

3.1.2 Drying: Heating device and drying temperature

It is particularly difficult to establish a direct relationship between the drying method and the textural properties of carbon gels because the effects produced by other variables such as the initial pH, synthesis temperature and time can conceal it. However, according to some works in which aero, cryo and xerogels were compared, it was reported that carbon gels with a higher pore size, large mesopores and macropores, are mainly achieved under supercritical and freezing conditions. In contrast, carbon xerogels are susceptible to a high degree of shrinkage of their porous nanostructure due to the surface tensions caused by the solvent upon the vapor-liquid interface (Czzakel et al., 2005; Job et al., 2005; Qin et al., 2001). As a result of the partial collapse of their structure, carbon xerogels possess a smaller pore size than their aero and cryogel counterparts. However, it should be noted that several works suggest that it is possible to prepare meso-macroporous carbon xerogels by employing low values of initial pH, although with lower pores volume than aerogels and cryogels (Calvo et al., 2011a). Subcritical drying is the cheapest, simplest and fastest procedure, since the other two methods require extra processes of solvent exchange and extreme operating conditions. However, in general, the choice of the drying method is conditioned by the requirements of porosity in the final application.

Czakkel and co-workers (Czzakel et al., 2005) published in 2005 a study on carbon gels dried under different conditions, keeping the rest of variables involved in the synthesis process constant. They obtained a S_{BET} of 2650, 1010 and 891 $m^2 g^{-1}$ and a V_p of 2.05, 0.79 and 0.44 $cm^3 g^{-1}$, for cryogels, aerogels and xerogels, respectively, and concluded that freeze-drying promotes the formation of materials with a larger textural development. The same trend was also observed in another work in 2005 (Job et al., 2005), where three types of drying methods were studied but in this case, precursor solutions with different pH were used. Carbon cryogels presented a higher specific surface area and pore volume than the other types of carbon gel in three out of the five series studied, those with lower pH. No clear trend was observed in the sizes of the meso and macropores of aerogels and cryogels synthesized. Nevertheless, in the pH range evaluated, carbon xerogels were the materials with the smallest meso and macropores.

In the case of carbon xerogels, some studies report the porous texture differences resulting from evaporative drying under different conditions (Kraiwattanawong et al., 2011; Zubizarreta et al., 2008a). For example, carbon xerogels with a good textural development and significantly reduced synthesis time (up to 98%) were obtained using microwave heating (Zubizarreta et al., 2008b). Furthermore, as illustrated in Figure 11, the microwave process allows the tailoring of the porous texture of the carbon xerogels depending on the initial pH, or R/C.

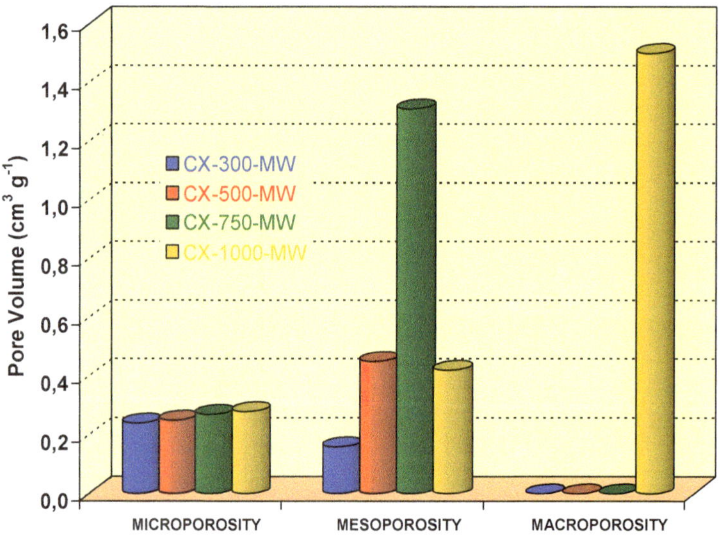

Fig. 11. Influence of R/C molar ratio on the pore size distribution of different carbon xerogels dried by microwave heating (adapted from reference Zubizarreta et al., 2008b).

Another factor to be taken into account is the drying temperature. A number of studies show that the removal of the solvent should be performed with a smooth gradual increase in temperature in order to avoid the collapse of the porosity. In most of the works, the drying stage was carried out at approximately 85-150 °C for the necessary time to remove the solvent inside the structure of the material, but using very slow temperature programs (Job et al., 2005).

3.1.3 Carbonization: Temperature, heating rate and atmosphere used

The aim of the carbonization stage is, on the one hand, to obtain thermally stable materials, mostly composed of carbon atoms and, on the other hand, to develop the microporosity of carbon gels (Al-Mutasheb & Ritter, 2003; Quin & Guo, 2001). Variables such as final temperature or heating rate have a significant influence on the textural properties of carbon gels while the type of atmosphere (inert or reactive), in addition to its impact on porosity, has a notable repercussion on their surface chemistry.

Most published works on the influence of carbonization temperatures on the textural characteristics of carbon gels agree that increasing the temperature leads to materials with a lower surface area and pore volume (Lin & Ritter, 2000). Thus, the microporosity of carbon gels is reduced when the carbonization temperature is increased. However, it is necessary to discriminate between ultramicropores, pores with diameter less than 0.7 nm and supermicropores, in the 0.7-2.0 nm range. Lin and Ritter (Lin & Ritter, 2000), evaluated the effect of carbonization temperature (600, 750, 900, 1050 and 1200 °C) on these two groups of micropores and concluded that the decrease in the surface area at higher temperatures is mainly due to the smaller number of micropores in 0.4-0.7 nm range. The carbon gels reported in this work are also composed of micropores with a diameter between 1-2 nm, whose volume decreases at low temperatures (600-900 °C), while when the carbonization was performed at 1050 and 1200 °C, their volume increased indicating that high temperatures destroy ultramicropores but create a greater amount of larger micropores.

The carbonization temperature is the most important parameter in the pyrolysis processes. However, other factors such as heating rate and carbonization atmosphere have also some influence on the textural properties of carbon gels (Kang et al., 2009). Overall, the heating rates used during pyrolysis are around 5-15 °C (Wu et al., 2004; Yoshimune et al., 2008; Zubizarreta et al., 2008a) since higher heating rates would mean a lower microporosity development, as reflected by several works that evaluate the different conditions in the carbonization processes of several carbon materials (Kuo et al., 2005; Liou, 2004).

Regarding the influence of the carbonization atmosphere, it seems that inert gases produce the materials with the lowest textural development. For example, carbon gels prepared under a N_2 atmosphere usually have specific surface areas about 600-700 $m^2\ g^{-1}$ (Calvo et al., 2008; Kraiwattanawong et al., 2011; Zubizarreta et al., 2008a), whereas this parameter can reach 1000 $m^2\ g^{-1}$ when the pyrolysis step is carried out under ammonia or CO_2 atmosphere (Kang et al., 2009).

3.1.4 Activation: Physical or chemical activating agent, activating agent/precursor ratio, type of precursor and activation time and temperature

Usually the temperature used for physical activation range between 700-1100 °C for different lengths of time, ranging from few hours (Guo et al., 2009; Lin & Ritter, 2000; Nabais et al., 2008; Zhao et al., 2007) to 24 hours (Contreras et al., 2010). The activation time and temperature are very important variables since they can notably modify the microporosity of the resulting materials, and even the narrower mesoporosity. Usually, increasing the temperature promotes higher development of carbon gel microporosity. As an example, Lin and Ritter (Lin & Ritter, 2000), performed the physical activation of resorcinol-formaldehyde carbon xerogels with CO_2 at 1050 °C for several activation times: 0.5, 1, 2 and 3 hours. The carbon xerogels reported in this study were found to have significant volume of micropores in the 0.6 nm range, micropores whose diameter is centered at approximately 1.3 nm and also mesopores ranging from 2-10 nm. The increase in the activation time produced a more developed pore structure since the total pore volume increased. The amount of ultramicropores increases until it reaches its maximum value after 2 hours of activation, while the volume of large micropores and smaller mesopores rises consistently with activation time. These results lead to the conclusion that in more prolonged activation processes, not only does the creation of micropores take place but also the destruction of the narrower micropores which are sometimes converted into large micropores and small mesopores. This phenomenon has also been found in another study performed by Contreras and co-workers (Contreras et al., 2010), where physical activation processes with CO_2 at 900 °C for 4, 8 and 16 hours resulted in carbon xerogels with specific surface areas of 1015, 1365 and 2180 $m^2\ g^{-1}$, respectively, and micropores of size 1.0, 1.1 and 1.4 nm, corroborating the development of porosity and the widening of micropores with the increase in activation time. In the same publication, the influence of another variable involved in the activation processes, i.e. temperature, upon the porosity of materials was evaluated. Two activation temperatures were studied, 800 and 900 °C, and it was found that the highest temperature produced a greater development in microporosity, as reflected by the increase in S_{BET}, V_{DUB-N2} and $V_{DUB-CO2}$. With these examples, it is possible to affirm that the development of microporosity increases with higher activation temperatures and times, i.e. as the burn-off increases, a phenomenon consistent with other works that deal with physical activation processes of different type of materials (Guo et al., 2009; Roman et al., 2008).

As expected, chemical activation processes are also conditioned by many variables such as the activation time and temperature, the amount of activating agent or type of precursor (Conceição et al., 2009; Contreras et al., 2010; Zubizarreta et al., 2007). With regard to the activating agent/precursor ratio, several studies have reported that the increase in the amount of chemical agent leads to materials with a higher textural development, although it should be pointed out that there is a maximum value above which surface area begins to decrease (Zubizarreta et al., 2008c). Conceição et al. (Conceição et al., 2009) published a work on the chemical activation with H_3PO_4 of resorcinol-formaldehyde aerogels using different impregnation ratio values (H_3PO_4/gel = 1, 2 and 3). The increase in the amount of H_3PO_4 leads to a higher total pore volume (i.e. 0.65, 0.94 and 1.41 $cm^3\ g^{-1}$ for samples obtained using impregnation ratios of 1, 2 and 3, respectively). The narrower micropores widen and, therefore, the volume of the ultramicropores decreases (i.e. 0.16, 0.15 and 0.13 $cm^3\ g^{-1}$), unlike the volume of supermicropores and narrower mesopores that increases with higher impregnation ratios, a phenomenon similar to that explained in the preceding paragraph regarding the effect of temperature on physical activation processes. This trend was also observed in chemical activation processes of RF xerogels by adding different amounts of KOH (Zubizarreta et al., 2008c). As the activating agent/precursor mass ratio increases, carbon xerogels evolve from exclusively microporous to micro-mesoporous materials when the A/P ratio is 4. As the KOH/precursor mass ratio increases, both the specific surface area and micropore volume increase but sometimes, when A/P > 3, both parameters decrease as a result of the widening of the micropores, which are then conerted to narrow mesopores.

Another noteworthy variable in chemical activation processes is the type of precursor used. According to different works (Calvo et al., 2011b; Zubizarreta et al., 2008c), the chemical activation of carbon gels develops the microporosity notably without modifying the pristine meso-macroporosity, as a result of the thermal stability of carbon gels. However, when carbonization and activation are simultaneous processes, the reactivity of the organic gels causes the destruction of most of the meso and macropores created during sol-gel synthesis, while the microporosity undergoes intense developed. When chemical activation is performed after the carbonization step, it is possible to synthesize materials with a specific surface area of about 1500 $m^2 g^{-1}$ and large mesopores volumen (S_{BET} = 1540 $m^2 g^{-1}$, V_{DUB-N2} = 0.69 $cm^3 g^{-1}$ and V_{meso} = 0.25 $cm^3 g^{-1}$ for a carbon xerogel activated with KOH, (Zubizarreta et al., 2008c)) whereas when organic gels are used as activating precursors, there is an intense development of microporosity, which is reflected by S_{BET} values ranging from 2000 to 3000 $m^2 g^{-1}$, although this is accompanied by a lower mesopores volume (i.e. S_{BET} = 2037 $m^2 g^{-1}$, V_{DUB-N2} = 0.82 $cm^3 g^{-1}$ and V_{meso} = 0 $cm^3 g^{-1}$ (Zubizarreta et al., 2008c)).

As mentioned above, during the activation processes narrow mesopores may be created depending on the activation conditions. Therefore, it is important to discriminate between this type of mesopores and the meso- macroporosity that appears during the sol-gel synthesis of organic gels. The latter type of porosity is unable to remain intact when chemical activation of organic gels is carried out. However, in a recent work it has been shown that, under certain conditions, the chemical activation with KOH when microwave radiation is used as a heating source of organic xerogels, leads to materials with the meso-macroporosity of the original samples, in addition achieving remarkable amount of micropores (Calvo et al., 2011b).

Usually, chemical activation processes are performed at lower temperatures than physical activations, i.e. 700-850 °C when metal alkoxides, such as KOH or NaOH, are used as activating agent (Macia-Agulloet al., 2007; Raymundo-Piñero et al., 2005; Zubizarreta et al., 2008c), and temperatures ranging from 450-650 °C in the case of chemical activations with H_3PO_4 (Conceição et al., 2009, Qin et al., 2001). The influence of temperature on the pore structure of carbon gels has been more extensively studied in the case of physical activation processes, although there are also some studies about its effect on chemical activations. As the activation temperature increases, materials with a more developed porosity (i.e. higher surface area, micropore and also mesopore volumes) are formed. However, it is noteworthy that, as with the effect of the amount of activating agent, there is a maximum temperature above which porosity begins to decrease (Niu & Wang, 2008; Okada et al., 2003).

To sum up, by choosing suitable synthesis and post-processing conditions, it is possible to tailor the porosity of carbon gels, discriminating between micropores, mainly produced during carbonization and activation stages and meso-macropores, created during the sol-gel synthesis. It is therefore possible to obtain exclusively microporous carbon gels (suitable for H_2 storage and adsorption applications (Cabria et al., 2007; Mahata et al., 2008)), micro-mesoporous materials (for use as electrodes in supercapacitors (Calvo et al., 2008; Escribano et al., 1998; Frackowiak & Béguin, 2001)) or even meso-macroporous samples (for water treatments (Sanchez-Polo et al., 2007)).

3.2 Chemical properties

Unlike the porous textural properties, which can be easily controlled with the synthesis and processing conditions, the chemical nature of carbon gels is not usually influenced by the synthesis protocol followed during the sol-gel synthesis. Apart from the obvious influence of the chemical nature of the monomers used, it is only possible to create different chemical characteristics when activation, oxidation or doping processes are employed (Contreras et al., 2010; Silva et al., 2009; Zubizarreta et al., 2008c).

Carbon gels are composed of approximately 92-98 wt. % carbon and the rest of the composition is divided between hydrogen and oxygen, regardless of the synthesis conditions such as initial pH, operating time and temperature, dilution, etc. (Calvo et al., 2011a). However, Zubizarreta et al. (Zubizarreta et al., 2008a) affirm that it is possible to synthesize carbon gels with a larger amount of oxygen by performing the drying step by means of microwave heating. The oxygen content of this type of carbon xerogels was about 6-8 wt. %, probably due to the fact that with microwave radiation, several secondary reactions take place, which would favour greater crosslinkage between organic gel and the more stable oxygenated groups (Caddick, 1995).

Activation processes, both physical and chemical, besides increasing considerably the microporosity, are able to increase the quantity of oxygen present in carbon gels of around 3-4 wt. % and, according to several published works, the higher the temperature and time of activation, the higher the amount of oxygenated groups created (Contreras et al., 2010). However, since the amount of oxygen inside the carbon gel structure is still low, the determination of the nature of these surface groups is no straightforward task (Contreras et al., 2010).

The pH_{PZC}, (i.e. the pH value at which the electrical charge density on the carbon surface is zero), of carbon gels synthesized from resorcinol-formaldehyde solutions generally ranges from 8-9.5, whatever the synthesis conditions used (Calvo et al., 2011a; Zubizarreta et al., 2008a; Lambert et al., 2009). This basic character is may be due to the presence of delocalized π electrons on the surface of the carbon gels, because of their aromatic character (Montes-Morán et al., 1998), and the presence of pyrone or chromene-like structures (Fuentes et al., 1998). When activation and/or oxidation processes are carried out, the point of zero charge is irreversibly modified, increasing or decreasing with respect to the value of the pristine material, depending on the nature of the oxygenated functionalities created (Mahata et al., 2008). Thus, Lambert et al. (Lambert et al., 2009) show that oxidation by nitric acid is able to modify the surface chemistry of resorcinol-formaldehyde carbon gels, by modifying their pH_{PZC} from ca 9.4 to 2.4.

The modification of surface chemistry of carbon gels by doping processes has been widely investigated because of the need for rich chemical nature and good conductivity of these materials in a wide range of areas (e.g. catalysis or energy applications) (Job et al., 2008; Lee et al., 2011; Moreno-Castilla et al., 1999, 2005; Zubizarreta et al., 2010). There are several methods for obtaining metal-doped carbon gels. Basically, they can be classified into three processes: (i) direct dissolution of the metal precursor in the resorcinol-formaldehyde mixture (Chandra et al., 2011; Maldonado-Hódar et al., 2003); (ii) use of a resorcinol derivative containing an ion exchange moiety that can be polymerized by sol-gel reactions (Baumann et al., 2002; Fu et al., 2005), and finally, (iii) deposition of the precursor metal either on organic or carbon gels (Mahata et al., 2008). In addition to the several types of doping processes, factors such as the nature and amount of metal precursor, operating temperature, etc. greatly determined the chemical properties of the resulting materials (Frackowiak & Béguin, 2001; Job et al., 2007b).

The effect of the amount of metal precursor on the final properties of carbon gels has been evaluated by Job et al (Job et al., 2007b), through the synthesis of RF carbon xerogels doped with Cu, Ni, Pd, and Pt, by the addition of complexing agents that react during the sol-gel process. The results of this work showed that the metal particle size varies with the concentration of the complexing agent, and these metal particles remain inserted into the structure of the materials after the drying and pyrolysis stages. For example, Pd-doped carbon xerogels had metal particle sizes of about 20 nm at low concentrations of metal precursor while the size decreased to 3-5 nm, as the concentration of complexing agent increased. In the case of Cu and Ni-loaded carbon gels, the metal particle size was unchanged with the concentration of complexing agent because they were not incorporated inside the polymeric network of the material and, therefore, they sintered during the pyrolysis stage.

The influence of the processing temperature on the surface chemistry of carbon gels was studied by Maldonado-Hódar (Maldonado-Hódar et al., 2003). In this work, they reported the preparation of tungsten-doped carbon aerogels by means of sol-gel reactions between resorcinol, formaldehyde and ammonium tungsten mixtures. Two carbonization temperatures were selected (500 and 1000 °C), and their influence on the resulting materials was evaluated. By means of SEM images, it was possible to determine the distributions of the metal species created, showing a more homogeneous distribution when the sample was carbonized at 1000 °C. Moreover, the temperature determined the type of tungsten particles inside the carbon structure, since carbonization at 500 °C produced materials composed of needle-like WO_3 particles with a radius of few nanometres, while when the temperature was set at 1000 °C, as well as the same type of particles, a denser species with a certain dendritic character were obtained.

The surface chemistry of metal-doped carbon gels are also conditioned by the pH of the precursor solution (Bekyarova et al., 2000) like the textural properties. Bekyarova and Kaneko (Bekyarova et al., 2000) prepared Ce,Zr-doped carbon aerogels by adding metal salts to the resorcinol-formaldehyde solutions. Two pH values were selected, 3 and 7, results in materials with different features. The TEM images collected in this work showed that in the case of pH 7, the carbon aerogel was composed of bound particles of about 20 nm and, there was a homogeneous distribution of the metal species. However, the sample synthesized from a solution with pH 3 gave rise to spherical particles with diameters of about 3 μm and no doping particles were detected.

3.3 Mechanical properties

Carbon gels are composed of interconnected quasi-spherical nodules, forming a three-dimensional matrix, as reflected by the SEM photograph presented in Figure 12. As already mentioned, the diameter of these nanospheres and, therefore, the pore size is influenced by several synthesis conditions (Czzakel et al., 2005; Silva et al., 2009; Zubizarreta et al., 2008a). As regards to the initial pH, lower microspheres size are obtained as the initial pH increases, corresponding to highly compact structures. Zubizarreta et al. (Zubizarreta et al., 2008a) determined the influence of the pH on the structure of the carbon xerogels, noting that precursor solutions with pH 9 and 7 produced carbon xerogels with a compact and uniform structure while the sample obtained using pH 6 exhibited a low-compact structure and besides, the microspheres have hardly intuit.

The influence of other synthesis variables, such as the drying procedure, on the structure of carbon gels has been determined by several research groups. Regardless of the drying conditions, the structure of carbon gels is composed of interconnected spherical particles, although it is necessary to emphasize that the size of such spheres changes depending on the method used (Czzakel et al., 2005). Sometimes it is difficult to discriminate between the structures of carbon aerogels and cryogels. However, by means of SEM or TEM photographs, it is possible to appreciate the smaller size of the microspheres of the carbon xerogels, as a result of the shrinkage of the structure (Czzakel et al., 2005).

One of the advantages associated with carbon gels is that they can be made with different morphologies, such as monoliths, although this monolithic shape is notably influenced by the synthesis conditions. There is a scientific work that has demonstrated the effect of the initial pH and the heating device used to evaporate the solvent, on the morphology of carbon xerogels (Zubizarreta et al., 2008a). With microwave heating, it was possible to preserve the monolithic shape of the carbon gel when a precursor solution with a high pH is used, whereas when the pH was low, the monolith broke into pieces as a result of its greater frailty. This confirms that the final carbon gel displays better mechanical properties as the initial pH increases.

Fig. 12. SEM image of resorcinol-formaldehyde carbon xerogel synthesised in a microwave oven using a precursor solution with a pH of 5.8.

Regarding the effect of the activation processes, both physical and chemical, on the mechanical properties of the resultant activated material, it has been observed that it is possible to obtain monoliths of physically activated carbon gels. However, when chemical activation is used, the process is more severe and it is not possible to retain the monolithic shape.

The porous, chemical and mechanical characteristics of carbon gels are related to the different synthesis steps and the operating variables involved in each stage. As a summary of this section this relationship is reported in Table 2. Although there are many works which deal with the designing of the carbon gel properties and the great potential of these materials, many variables require further study and adjustment with the aim of tailoring the properties of the final material for a specific application.

STEP	VARIABLES	POROUS TEXTURE	CHEMICAL PROPERTIES	STRUCTURAL PROPERTIES
SOLUTION, GELATION AND CURING	pH/catalyst	pH has no influence on microporosity and/or surface area (Calvo et al., 2011a; Job et al., 2004; Lin & Ritter,1997) Increasing pH yields a narrower mesoporosity (Calvo et al., 2011a; Job et al., 2004; Lin & Ritter,1997) Decreasing the pH increases the macroporosity (Zubizarreta et al., 2008a, Calvo et al., 2011a) As the pH increases, the gelation time (t_g) decreases (Calvo et al., 2011a; Juárez-Pérez et al., 2010)	Not influenced (Calvo et al., 2011a, Zubizarreta et al., 2008a)	Increasing the pH increases the mechanical resistance of the carbon xerogels (Zubizarreta et al., 2008a)
	RF concentration	A decrease in the RF concentration leads to a smaller pore size (Al-Mutasheb & Ritter, 2003; Petricevic et al., 2001)	Not influenced	Little studied
	Temperature	Very high synthesis temperatures cause a shrinkage of porosity	Not influenced	Little studied
SUBCRITICAL DRYING	Stove	Causes shrinkage of the pore structure (Al-Mutasheb & Ritter, 2003; Czzakel et al., 2005; Job et al., 2005)	Not influenced	Easy to obtain carbon xerogel monoliths
	Microwave	Slightly higher shrinkage of the pore structure (Zubizarreta et al., 2008b)	Higher oxygen content but few studies on this topic (Zubizarreta et al., 2008b)	It is difficult to obtain carbon xerogel monoliths directly, but the good control of gelation point leads to design the final shape (Calvo et al., 2011a; Mahata et al., 2008)
CARBONISATION	temperature	Increasing the temperature either increases (Husley et al., 1992; Lin & Ritter, 2000) or decreases the surface area and micropore volume (Kang et al., 2009) Little influence on meso or macroporosity	Increasing the temperature reduces the oxygen content (Al-Mutasheb & Ritter, 2003)	Little studied
	Heating rate	A high heating rate reduces the micropore volume (Kuo et al., 2005; Liou, 2004)	Little studied	Little studied
	Atmosphere	A carbonization atmosphere notably influences the porous texture of the carbon gels (Kang et al., 2009; Zubizarreta et al., 2008a)	Depending on the type of carbonisation atmosphere (Kang et al., 2009; Zubizarreta et al., 2008a)	Little studied

STEP	VARIABLES	POROUS TEXTURE	CHEMICAL PROPERTIES	STRUCTURAL PROPERTIES
ACTIVATION	Physical	Increasing activation the temperature and operating times leads to a more developed pore texture (Contreras et al., 2010; Lin & Ritter, 2000)	The oxygen content increases (Contreras et al., 2010)	Possible activation of the monoliths
	Chemical	Activation of the organic xerogels destroys the meso-macroporosity created during the synthesis (except in a MW oven), but increases the microporosity and surface area considerably (Calvo et al., 2011b; Zubizarreta et al., 2008c) Activation of the carbon xerogels increases the microporosity without altering the meso-macroporosity (Al-Mutasheb & Ritter, 2003; Molina-Sabio et al., 2004; Zubizarreta et al., 2008c) The activating agent/carbon ratio influences the micropore volume considerably (Conceição et al., 2009; Zubizarreta et al., 2008c)	The oxygen content increases	The monoliths are usually destroyed (Zubizarreta et al., 2008c)

Table 2. Effect of the synthesis conditions on the properties of resorcinol-formaldehyde carbon gels.

4. Applications of carbon xerogels

All the characteristics already mentioned along this chapter make carbon xerogels very promising candidates for a wide range of applications such as adsorption (Long et al., 2009; Maldonado-Hódar et al., 2007; Ying et al., 2005), water treatment (Girgis et al., 2011; Shánchez-Polo et al., 2007), gas separation (Yamamoto et al., 2004) or enzymes support (Chaijitrsakool et al., 2008). Besides the most common and referenced ones like catalysts support (Job et al., 2008; Lambert et al., 2010; Liu & Creager, 2010; Moreno-Castilla et al., 1999), electrode materials in electric double layer capacitors (Calvo et al., 2008; Frackowiak & Béguin, 2001; Sepheri et al., 2009; Zhang et al., 2007; Zhu et al., 2007) and hydrogen storage (Kang et al., 2009; Tian et al., 2010, 2011a, 2011b; Zubizarreta et al., 2010). In this section a brief review of all these applications of carbon xerogels are reviewed.

4.1 Catalysts support

Catalysis plays a decisive role in many reactions and technologies such as: energy supply by fuel cells, oxidation of organic compounds in liquid effluents, removal of SO_x and NO_x in order to reduce the pollution, synthesis of fine chemistry products, etc (Girgis et al., 2011; Machado et al., 2010; Moreno-Castilla et al., 2005; Pirard et al., 2011). Generally, two types of catalysis are distinguished: homogeneous catalysis, i.e. reactants and catalyst constitute just one phase and heterogeneous catalysis, it is to say, when catalyst is in a different phase respect to the reactants involved in the process. Both kind of catalysts display advantages and drawbacks. For example, heterogeneous catalysts are much important at industrial scale as a result of their simple preparation procedure, high stability and the possibility of separating them from the reaction media in an easy way. Nevertheless, the disadvantages of heterogeneous catalysis are the presence of two different phases and the control of the active sites nature. On the other hand, regarding to homogeneous catalysis, although reactions take place in a single phase and catalysts can be tailored, their carriage in industrial processes is lower due to their sometimes limited stability and their difficult recovery (Choplin & Quignard, 1998; Djakovitch et al., 2004; Fontecha-Cámara et al., 2011).

Most of published works on carbon gels used in catalysis field are focused on heterogeneous processes and the great interest of carbon gels in this field research is due to their attractive features such as: tailored pore texture, possibility of enriching their surface chemistry by different processes, high packing density, high durability under harsh conditions, etc. (Cotet et al., 2006; Job et al., 2005; Moreno-Castilla et al., 1999; Teng & Wang, 2000). The major challenge in catalysis, together with the reduction of costs is to increase their activity and selectivity. The selectivity is a feature very difficult to manage but catalytic activity has been extensively studied in many publications in order to achieve a relationship between this parameter and the properties of the carbon material. Both surface chemistry and porous texture of carbon gels play an important role on catalytic processes. However, catalytic activity depends on the nature, amount and accessibility of their active sites for the reactants, and it is not always straightforward to correlate it with intrinsic characteristics of carbonaceous materials.

One of the first works found about catalytic performance of metal-doped carbon gels were published in 1999 by Moreno-Castilla and co-workers (Moreno-Castilla et al., 1999). This work was focused on the study of catalytic activity of chromium, molybdenum and tungsten oxide-doped carbon aerogels in the isomerisation reaction of 1-butene and the results showed that the best catalyst for this specific reaction was carbon aerogel based on tungsten oxide because of the higher surface acidity. The isomerization reaction of 1-butene in other structures was a widely investigated process since a hydrocarbon with higher octanoic value can be obtained and therefore, several research groups have performed studies on the effect of using metal catalysts supported on carbon materials for accelerating this reaction (Álvarez-Merino et al., 2000).

Carbon gels have also been explored as metal catalyst support for fuel cell applications. The fact of using carbon supported catalysts in fuel cells allows production costs to be reduced and performance and durability of the electrochemical system to be improved (Arbizzani et al., 2007; Job et al., 2008; Liu & Creager, 2010). One the most recent works in this research field was carried out by Liu and Creager (Liu & Creager, 2010). Resorcinol-formaldehyde carbon xerogels was used as supports for Pt particles, by means of impregnation-reduction method using H_2PtCl_6 as platinum precursor and formaldehyde as the reducing agent. Electrochemical results of the membrane-electrodes fabricated from Pt-deposited carbon xerogels with Nafion as electrolyte, were compared with those of Pt catalyst supported on a commercial carbon black. Pt-carbon xerogel catalysts displayed good intrinsic catalyst activity due to the higher size of Pt particles in these catalysts and besides, the cell performance normalized by Pt loading was slightly higher when Pt particles were supported on resorcinol-formaldehyde carbon xerogels. Job et al. (Job et al., 2008) also prepared Pt-carbon xerogel catalysts for fuel cell applications. The catalysts were synthesized following two different recipes: (i) impregnation of carbon xerogel with Pt precursor solution followed by reduction process and (ii) strong electrostatic adsorption

(SEA method). This latter method produced higher dispersion of Pt particles than the impregnation and consequent reduction, improving the performance of fuel cells in terms of Pt mass activity. Moreover, there are research groups that used bimetallic catalysts for energy production in fuel cells. For example, Figueiredo and co-workers (Figueiredo et al., 2006) evaluated the electrochemical performance of direct-methanol fuel cells (DMFCs) assembled with Pt-Ru catalyst supported on oxidized carbon xerogels and the results reported in that study demonstrated that resorcinol-formaldehyde carbon xerogels were effective as supports for Pt/Ru particles. Pt-Ru bimetallic catalysts were also prepared by Arbizzani and et al (Arbizzani et al., 2007) for their use as anodes in DMFCs. In that case, the carbon supports were mesoporous cryo and xerogel carbons and Pt-Ru catalysts were obtained by impregnation of carbon materials with H_2PtCl_6 and $RuCl_3$ in ethylene glycol and later chemical reduction. The activity of the catalysts were related with structural and morphological properties of carbonaceous supports and electrochemical results of fuel cells were also compared to those obtained with a Pt-Ru supported on a commercial carbon support, showing again the good behaviour of carbon gels as support catalysts.
Other chemical processes and/or reactions catalyzed by carbon gels are for example, oxidation of several organic compounds with the aim of reducing pollution in liquids effluents (Girgis et al., 2011; Maldonado-Hódar et al., 2004), toluene combustion reactions (Gomes et al., 2008; Maldonado-Hódar et al., 2007), growth of carbon nanofilaments and nanotubes (Fu et al., 2003), conversion of D-glucose into D-gluconic acid (Pirard et al., 2011), and a long list since this type of catalysts has infinite number of applications in many different research fields.

4.2 Electric energy storage

A large number of publications can be found in the literature about carbon gels as electrode materials in supercapacitors as a result of being highly porous materials with a good electric conductivity (Kim et al., 2001).
In electrochemical double layer capacitors, EDLC´s, the main mechanism that governs charge storage processes is the formation of the electric double layer in the electrode/electrolyte interface (Frackowiak, 2007; Kötz & Carlen, 2000). Theoretically, as specific surface area increases, higher energy storage capacitance is achieved, but actually the situation changes because the whole surface area of the electrode material is not electrochemically accessible, and therefore useful, when electrodes are immersed in the electrolyte. Many studies about carbon gels and supercapacitors have concluded than charge storage is performed in micropores whereas mesopores with a specific size are needed for a fast diffusion of electrolyte ions (Frackowiak & Béguin, 2001; Salitra et al., 2000; Vix-Guterl et al., 2005). Electrochemical studies on carbon gels with a mesopores size between 3-13 nm have resulted in very high specific capacitance values and also, in a stable capacitive performance of the supercapacitor (Escribano et al., 1998). Therefore, it has been generally accepted that a balanced porosity between micro and mesopores is preferable to reach the optimum performance of the supercapacitor. Carbon gels are very promising materials for this application since besides their high microporosity, mesopores with a specific and tailored size can be obtained varying the synthesis conditions, absent feature in the case of activated carbons. Moreover, carbon gels display other advantages when they are used as electrodes in supercapacitors such as their high conductivity, enabling the removal of the usual additive to promote this property employed with active carbons, possibility of obtaining them in several morphologies directly without the need

of binders, i.e. it would be feasible to prepare carbon gel films in order to directly use as electrodes, and high cycling-life (Pandolfo & Hollenkamp, 2006). It has been found specific capacitance values of carbon gels in aqueous electrolyte between about 100 F g^{-1}, in the case of an untreated carbon gel and 300 F g^{-1}, data corresponding to carbon gels textural and chemically modified by means of several post-synthesis treatments. Some specific capacitance values found in the literature for untreated carbon gels are for example from 153 to 194 F g^{-1} in the case of several resorcinol-formaldehyde carbon xerogels with specific surface areas between 700-800 m^2 g^{-1} (Zhang et al., 2007); 150 F g^{-1} when the electrochemical performance were carried out using a basic aqueous media as electrolyte and RF carbon xerogel dried by microwave heating after a previous solvent exchange with acetone (Halama et al., 2010), or finally, 120 F g^{-1} for a RF carbon xerogel with a S_{BET} of 594 m^2 g^{-1} obtained by microwave-assisted synthesis (Calvo et al., 2011a). It should be noted that despite the lower energy storage capacitance shown in the case of this late carbon xerogel, it is a very promising material because it has been synthesized by microwave heating in few hours compared to several days needed in the other examples cited. Higher capacitance values are reported when activated carbon gels are used as electrode materials in supercapacitors. Zhu and co-workers (Zhu et al., 2007) and Wang et al. (Wang et al., 2008) performed the synthesis of activated carbon gels with KOH and studied their electrochemical performance as electrode materials in supercapacitors with basic aqueous electrolyte, and results reported by both research groups were 244 F g^{-1} (Wang et al., 2008) and 284 F g^{-1} (Zhu et al., 2007), comparable values to those shown by other carbonaceous materials commonly used as electrodes in supercapacitors (Kierzek et al., 2004; Lota et al., 2008; Shi, 1996).

The energy storage mechanism based on charges separation in the electrode/electrolyte interface is not the only mechanism that can carry out in supercapacitors. In fact, there is another type of energy storage, induced by faradaic reactions occurring in electrodes surface, which considerably enhances the capacitance of supercapacitors (Frackowiak & Béguin, 2001). These redox reactions promote so-called pseudocapacitance effects and they are caused due to the presence of heteroatoms in the surface of the carbon electrodes. Some of the heteroatoms which contribute to the energy storage by means of pseudocapacitive effects are, for example, O, N, P, B, some metals, etc. (Frackowiak, 2007; Kang et al., 2008; Tian et al., 2010). Moreover, together with the increase of energy capacitance, surface groups improve the wettability of electrodes in aqueous media, due to electrostatic interactions on the electrode surface with dipole moments of water molecules. There are many examples in the literature about the use of doped carbon gels as electrode materials for supercapacitors. In the case of nitrogen-doped carbon gels, the work presented by Kang and co-workers, mentioned in Section 2 (Kang et al., 2008), reported the preparation of nitrogen enriched carbon xerogels by means of ammonisation processes. In this study, three different carbon xerogels were compared both textural and electrochemically: RF carbon xerogel conventionally synthesized, RF carbon xerogel conventionally synthesized and carbonised under NH_3 atmosphere and, finally, a RF carbon xerogel prepared by microwave-assisted synthesis and carbonised with NH_3. The two samples subjected to NH_3-carbonization display similar nitrogen contents (between 2.6 and 3.2 wt. %), while the porous texture is noticeable different. Microwave assisted sample has around 1700 m^2 g^{-1} of specific surface area opposite to 1080 m^2 g^{-1} for its counterpart synthesized by conventional heating and also, the latter carbon xerogel has lower mesopore volume. This different porosity affects the electrochemical performance of samples since conventional sample display a specific capacitance of 148 F g^{-1} vs. 185 F g^{-1} in the case of sample with larger porosity development. When these two carbon xerogels were compared to the sample pyrolysed under N_2 atmosphere, it was clearly demonstrated the profitable effects of nitrogen doping since although this last sample has a specific surface area close to 800 m^2 g^{-1}, its energy storage capacitance is very poor as a result of the absence of nitrogen functionalities, i.e. the lack of reversible electrochemical reactions that increase the capacitance due to pseudo-faradaic processes. Other work that shows the improvement of energy storage due to the presence of heteroatoms was performer by Sepheri et al. (Sepheri et al., 2009). In that case, the functionalization was carried out during the synthesis process of organic gels since, once organic gels are synthesized but still wet, they were introduced in ammonia borane/trifluoroacetic acid solution in order to incorporate B and N atoms into the structure of resulting materials. This treatment promotes the presence of functional groups and besides, the increase of mesoporosity since the carbon gel mixed with AB solution possesses a mesopore volume of 1.57 cm^3 g^{-1} opposite to 1.17 cm^3 g^{-1} in the case of untreated carbon xerogel. Results about the chemical nature of these samples, show that the ammonia borane treatment allows the incorporation of borane and oxygen functionalities (2.2 and 11.4 wt. % of boron and oxygen for treated carbon xerogel vs. 0.0 and 3.8 wt. %, respectively, in the case of untreated sample) while nitrogen groups disappear during the carbonization step. The enrichment of surface chemistry together with the enhanced porous texture causes an increase of ca. 30 % in the specific capacitance values and also an improvement in the current density of supercapacitors.

Quin and other researchers (Qin et al., 2011), also reflect the improvement of capacitive performance of supercapacitors by means of nitrogen functionalization of electrodes. Unlike the other two mentioned examples, the electrochemical devices assembled were asymmetric supercapacitors, where $Ni(OH)_2/Co(OH)_2$ composite works as anode and an activated carbon gel/melamine resin composite as cathode material, strategy widely used in others research groups (Ganesh et al., 2006; Staiti & Lufrano, 2010). The followed recipe to performer the functionalization of materials was quite different respect to the other two published studies. On the one hand, resorcinol-formaldehyde carbon gel was synthesized and chemically activated with KOH and secondly, a melamine resin was prepared in order to use it as nitrogen source. Both samples were mixed in water with ultrasonic radiation and afterwards, the resulting material was pyrolysed and activated with KOH causing the activated nitrogen enriched carbon/carbon gel composite. Results show that activated organic gel has a specific surface area of 1670 m^2 g^{-1} opposite to 1848 m^2 g^{-1} for the nitrogen enriched composite, demonstrating that, as in previous cases (Kang et al., 2008; Sepheri et al., 2009), treated carbon gels display high porosity development. The composite also has an important amount of heteroatoms since XPS measurements show 2.1 and 13.3 % of nitrogen and oxygen content, for such sample. Electrochemical results reported in this work corroborate the higher energy storage as a result of redox reactions in the electrode surface since specific capacitance values were 103 F g^{-1} in the case of using the activated carbon gel as cathode material and 224 F g^{-1} when the supercapacitor was assembled with the nitrogen enriched composite. The difference in specific surface area between two samples was only 11 % while the electrochemical performance was enhanced about 50 %, indicating that the presence of pseudocapacitive interactions allow the global capacitance of supercapacitor to be increased.

As reflected in Section 2 of this chapter, there are lots of research groups working on doping processes of carbon gels with metal species in order to prepare higher porous and conductive materials. Therefore, one of the final applications of these metal-doped carbon gels is as energy storage systems. One of the more recent works dealing with doping processes of carbon gels for their use in supercapacitors was published by Lee and co-workers (Lee et al., 2011). The doping method was performed either by impregnation of manganese oxide on carbon gel power or by impregnation onto monolith material and the resulting carbon gels was electrochemically characterized. The best specific capacitance values were found in the case of impregnation of power form (i.e. 135 F g^{-1} for Mn-doped carbon aerogel obtained by impregnation onto power form opposite to 108 F g^{-1} in the best case when the impregnation was carried out with the monolithic material).

4.3 Hydrogen storage

As a result of the scarcity of fossil hydrocarbon resources, hydrogen is becoming a promising substitute for these fossil fuels in mobile applications. In addition to achieve the independence of fossil fuels, the use of hydrogen represents an environmentally friendly technology since allows the production of zero emission vehicles. However, the main requisite to a successful implementation is to store and transport the hydrogen in a safe and easy way (Dillon & Heben, 2001; Schimmel et al., 2004; Zubizarreta et al., 2009). Many researches have showed great attention to solve this hydrogen storage problem by means of several methodologies: high pressure, low temperature, metal or complex hydrides and porous materials, being the latter one of the most attractive solutions. High specific surface areas with narrow micropore size distributions are required for high-efficiency physical adsorption of hydrogen. As already mentioned, the possibility to tailor the micro-mesoporosity of carbon gels besides their surface chemistry makes them suitable materials for hydrogen storage devices (Kabbour et al., 2006; Tian et al., 2010, 2011, Zubizarreta et al., 2010).

There are a large number of published works about carbon gels and hydrogen storage that try to determine the interaction mechanism between H_2 and carbonaceous support and the relationship between adsorption capacities and textural and morphological properties of sorbent material. Regarding to the texture of nanoporous carbon materials, most of the studies conclude that there is a linear relationship between hydrogen storage capacity and surface area and micropore volume, but these are not the only influencing factors, since micropore diameter plays a key role in the final storage capacity. In other words, for hydrogen storage application is so important a high micropore volume as well as a narrow micropore size distribution (De la Casa-Lillo et al., 2002; Gadiou et al., 2005; Jordá-Beneyto et al., 2008). According to several published works (Gadiou et al., 2005; Jordá-Beneyto et al., 2008), micropores with a pore size of approximately 0.7-0.9 nm promotes higher hydrogen sorption capacities. In contrast, regarding to smaller mesopores, there is not a clear relationship between mesopore diameter and process of hydrogen storage. Zubizarreta et al. (Zubizarreta et al., 2009) have published a study based on the hydrogen sorption capacity of several carbon materials, including three resorcinol-formaldehyde carbon xerogels with different porous texture and morphology. In most samples, it has been found that in H_2 adsorption experiments performed at 77 K, the gravimetric storage capacity increases with the narrow micropore volume, analogous results to those reported in the literature about hydrogen storage on other kind of materials such as activated carbon (Akasaka et al., 2011; Cabria et al., 2007; Gadiou et al., 2007; Jordá-Beneyto et al., 2008; Xua et al., 2007) or metal organic frameworks, MOFs (Hirscher & Panella, 2007; Thomas, 2007), for example.

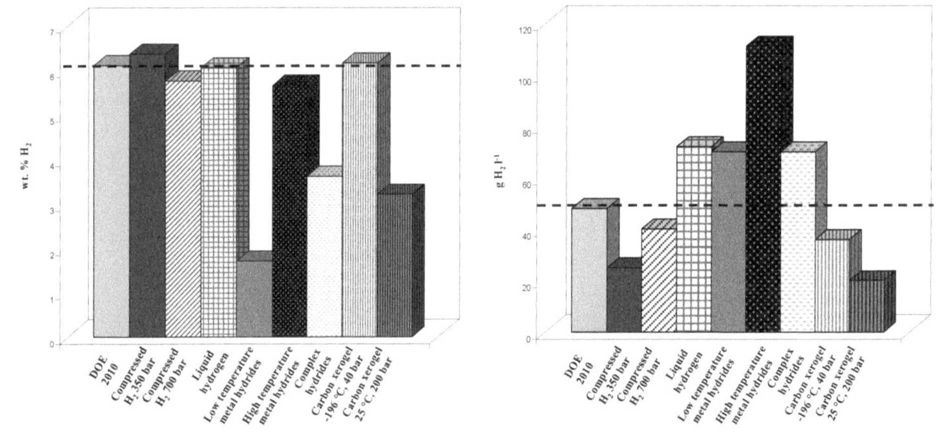

Fig. 13. Gravimetric and volumetric storage capacity of several hydrogen storage systems

For an optimum hydrogen storage capacity in solid materials, besides a well developed porous texture, there is another important feature to take into account, which is their surface chemistry. Experimental results about hydrogen storage on porous materials show that at low temperature, the dominant mechanism in the hydrogen storage process is based on microporous adsorption. However, as the temperature increases or the pressure decreases, chemical structure begins to have noticeable relevance in the mechanism of H_2 adsorption. Therefore, with the aim of changing the hydrogen/carbon interaction and therefore, enhancing the hydrogen storage capacity at room temperature or low pressure, several researches have used doped carbon materials. Some of the heteroatoms used to modify the surface properties and to achieve an enhanced hydrogen adsorption are N, B, Ni, Co, etc. (Kabbour et al., 2006; Tian et al., 2010; Zubizarreta et al., 2010). As already reflected in Section 2, carbon gels can undergo different doping processes, which is a great interest to solve the limitation of poor hydrogen uptake at room temperature. It can be found a remarkable number of studies related to hydrogen storage on doped carbon gels. However, it should be noted that, as a result of the large number of variables involved in the storage process (i.e. temperature and pressure of hydrogen storage, porous texture of carbon gels, amount and particle size of the heteroatom incorporated into the structure of carbon materials, etc.) it is very difficult to find several publications agreeing with the values of hydrogen storage capacity. For example, Kabbour et al (Kabbour et al., 2006) have published in 2006 a study about Co and Ni-doped carbon gels for hydrogen storage. The gravimetric hydrogen storage values reported were 2.1 and 2.3 wt.% for Co and Ni-doped carbon gels, respectively, when the hydrogen sorption experiments were performed at 77 K and low pressure (pressure between 0 and 2.5 bar). Other work about Co-doped carbon gels for H_2 storage shows a value of hydrogen storage capacity of 4.38 wt.% under lower temperature and high pressure conditions (Tian et al., 2010), but it should be said that the carbon gels used as adsorbent material in both works displays very different textural properties (i.e. S_{BET} = 1667 m^2 g^{-1} in 157 versus ca 1000 m^2 g^{-1} in 156), which could explain the notable difference in hydrogen storage capacities. Zubizarreta et al. (Zubizarreta et al., 2010) also investigated the performance of Ni-doped carbon gels in hydrogen storage systems and reported one of the higher values of gravimetric hydrogen storage capacity for this type of carbonaceous materials. The best Ni-doped carbon xerogel synthesized in that work, whose specific surface area is 1727 m^2 g^{-1} with a Ni content of 2.7 wt.%, exhibits a value of hydrogen storage capacity of 6 wt.% at 77 K and 40 bar. It is very complicate to compare all these results with those found with other doped carbonaceous materials used in hydrogen storage applications (i.e. activated carbons (Akasaka et al., 2011; De la Casa-Lillo et al., 2002; Takagi et al., 2004), nanotubes (Gao et al., 2010; Lamari et al., 2002; Schimmel et al., 2004; Surya et al., 2009) or nanofibers (De la Casa-Lillo et al., 2002; Kim et al., 2008)). The hydrogen storage capacity of Ni-doped carbon gels compared with other hydrogen storage systems is collected in Figure 13. Regarding to the gravimetric storage capacity, the Ni-doped xerogel shows analogous and even higher values to other materials reported in the figure when the experiments were carried out at -196 °C and 40 bar but, as reflected at the beginning of this paragraph, the increase of the temperature produces an important decrease of the hydrogen storage capacity, thus the Ni-doped carbon xerogel at 25 °C is far from that the DOE had proposed for the year 2010.

4.4 Other applications

As already reflected in this chapter, the sol-gel reaction allows textural and chemical properties to be tailored but, in addition, it is possible to control other important characteristic of these carbonaceous materials, such as their morphology. As pointed out in Section 1, carbon gels can be synthesized with several shapes: monolith, films, powder, spheres, etc. and morphology may be one of the key factors that could determine the good or bad performance of carbon gels in some specific applications. An example of the relevance of the morphology is the use of carbon xerogel spheres as columns filler for gas separation. A research group from INCAR-CSIC has recently prepared resorcinol-formaldehyde carbon spheres with different sphere size depending on the synthesis procedure followed and has studied their performance as material sieves for separation of N_2/CO_2 gas mixtures (some of their results reported in Figure 14). On the one hand, it can be seen pictures with carbon spheres of different size and on the other hand, N_2 and CO_2 monitoring curves that show that CO_2 can be separated as a result of the porosity and basicity of the carbon xerogel spheres.

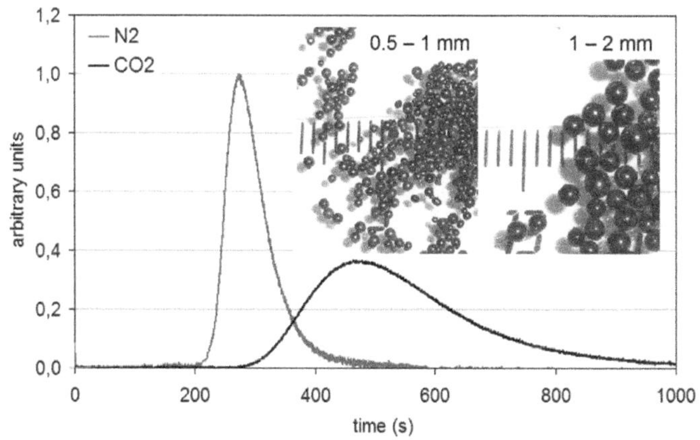

Fig. 14. Carbon xerogel spheres and their potential as fillers for gas separation systems.

In the literature, there is some example about carbon gels designed for gas separation applications. In fact, Zheivot et al. (Zheivot et al., 2010) published in 2010 a study about the production of several phenol-formaldehyde carbon xerogels and their use as adsorption materials in gas chromatography. Results reported showed that microporous carbon xerogels can be prepared varying the synthesis conditions and they can be successfully used as adsorbents to concentrate the impurities of light hydrocarbons and gaseous products of many catalytic reactions. Liquid phase adsorption can also be performed with carbon gels. In 2007, Sanchez-Polo and other researchers (Sánchez-Polo et al., 2007) analyzed the efficiency of Ag-doped carbon aerogels for the removal of several ions (Br^- and I^-) from drinking waters. The doping of resorcinol-formaldehyde aerogels was carried out by the addition of silver acetate as catalyst instead of the commonly used sodium carbonate. The study confirmed that the synthesis of mainly meso and macroporous carbon aerogels with an important amount of Ag particles could be carried out and the resulting materials displayed good adsorption capacities. The performance of carbon aerogels as adsorption materials was compared with the adsorption capacity of a commercial activated carbon that turned out to be lower. Other published work about carbon gels as liquid phase adsorbents was carried out by Girgis et al. (Girgis et al., 2011). In that study, the synthesis of carbon xerogels with difference porous texture was carried out and also the performance as adsorbent materials of two cationic dyes was evaluated. The main conclusion extracted from that publication is that carbon gels are good adsorbing materials in remediation processes since their adsorption capacities towards Methylene blue and Rhodamine B are comparable to conventional activated carbons. There is other recent work that includes the removal of several compounds from liquid effluents by using carbon gels as adsorbent materials (Figueiredo et al., 2011) but in this case, adsorption capacities collected are worse or similar than those shown by activated carbons, indicating that carbon xerogels do not seem to be an feasible option to remove colour from the effluents.

5. Conclusion

Carbon xerogels are nanostructured materials of a huge versatility since they can be synthesized with (i) different porous texture at micro-, meso- and macropore scale, (ii) different surface chemistry and (iii) different final shape. Moreover, all these properties can be tailored by adjusting the synthesis conditions, so it is possible, at least in theory, to obtain nanoporous carbon materials on-demand. However, the main problem to be solved in order to obtain a material with certain predetermined characteristics lies in the large number of variables that is necessary to control during their synthesis and to the fact that some of these variables are not independent. In consequence, one of the main drawbacks for the extensive use of these materials is to define the appropriate synthesis conditions and achieve them in a competitive way in terms of costs and operating time. Despite to the fact that great progress has been done in this sense -for example microwave induced synthesis has reduced the time of the carbon xerogels synthesis from days to hours with an important reduction in the costs of production- there is still need a step forward giving the opportunity to vast research field. Thus, it seems not risked to assure that as investigations on this matter reveals new information the great versatility of the carbon xerogels will make them important members of the family of the nanostructured materials with applications in numerous new, or existing, technologies ranging from electricity storage to drug delivery.

6. Acknowledgment

Authors would like to thanks the Spanish Ministerio de Ciencia e Innovación (Ref. MAT2008-00217/MAT) for the financial support. EGC also acknowledges FICYT for the research grand received.

7. References

Akasaka, H.; Takahata, T.; Toda, I.; Ono, H.; Ohshio, S.; Himeno, S.; Kokubu, T.; Saitoh, H. (2011). Hydrogen storaga ability of porous carbon materials fabricated from coffee bean wastes. *International Journal of Hydrogen Energy*, Vol. 36, (580-585).

Al-Mutasheb, S.A.; Ritter, J.A. (2003). Preparation and Properties of Resorcinol-Formaldehyde Organic and Carbon gels. *Advanced Materials*, Vol. 15 (101-114).

Álvarez-Merino, M.A.; Carrasco-Marín, F.; Moreno-Castilla, C. (2000). Tungsten catalysts supported on activated carbon: II. Skeletal isomerization of 1-butene. *Journal of Catalysis*, Vol. 192, (374-380).

Arbizzani, C.F.; Beninati, S.; Manferrari, E.; Soavi, F.; Mastragostino M. (2007). Cryo and xerogel carbon supported PtFu for DMFC anodes. *Journal of Power Sources*, Vol. 172, (578-586).

Baumann, T.F.; Fox, G.A.; Satcher, J.H.; Yoshizawa, R.; Fu, R.; Dresselhaus, M.S. (2002). Synthesis and characterization of copper-doped carbon aerogels. *Langmuir*, Vol. 18, (7073-7076).

Bekyarova, E.; Kaneko, K. (2000). Structure and Physical Properties of Tailor-Made Ce,Zr-Doped Carbon Aerogels. *Advanced Materials*, Vol. 12, (1625-1628).

Berthon, S.; Barbieri, O.; Ehrburger-Dolle, F.; Geissier, E.; Achard, P.; Bley, F.; Hecht, A.-M.; Livet, F.; Pajonk, G.M.; Pinto, N.; Rigacci, A.; Rochas, C. (2001). DLS and SAXS investigations of organic gels and aerogels. *Journal of Non-Crystalline Solids*, Vol. 285, (154-161).

Burdinova, T.; Ekinci, E.; Yardim, F.; Grimm, A.; Björnbom, E.; Minkova, V.; Goranova, M. (2006). Characterization and application of activated carbon produced by H_3PO_4 and water vapour activation. *Fuel Processing Technology*, Vol. 87, (899-905).

Cabria, I.; López, M.J.; Alonso, J.A. (2007). The optimun average nanopore size for hydrogen storage in carbon nanoporous materials. *Carbon*, Vol. 45, (2649-2658).

Caddick, S. (1995). Microwave Assisted Organic Reactions. *Tetrahedrom*, Vol. 51, (10403-10432).

Calvo, E.G.; Ania, C.O. ; Zubizarreta, L. ; Menéndez, J.A.; Arenillas, A. (2008). Exploring new routes in the synthesis of carbon xerogels for their applications in electric double-layer capacitors. *Energy & Fuels*, Vol. 24 (3334-3339).

Calvo, E.G.; Juárez-Pérez, E.J..; Menéndez, J.A.; Arenillas, A. (2011a). Fast microwave-assisted synthesis of tailored mesoporous carbon xerogels. *Journal of Colloid and Interface Science*, Vol. 357 (541-547).

Calvo, E.G.; Ferrera-Lorenzo, N.; Arenillas, A.; Menéndez, J.A. (2011b). Mesopores role in electrochemical performance of activated organic xerogels obtained by microwave radiation. *III Symposium on Hydrogen, Fuel Cells and Advanced Batteris, Hyceltec*, Abstract Book, p. 465, PO-15 (Zaragoza) 2011, ISBN: 978-84-938668-8-4.

Carrot, P.J.M.; Conceição, F.L.; Carrot, M.M.L.R. (2007). Use of n-nonane pre-adsorption for the determination of micropore volume of activated carbon aerogels. *Carbon*, Vol. 45, (1310-1313).

Chaijitrsakool, T.; Tonamon, N.; Tanthapanichakoon, W.; Tamon, H.; Prichanont, S. (2008). Effects of pore characters of mesoporous resorcinol-formaldehyde carbon gels on enzyme immobilization. *Journal of Molecular Catalysis B: Enzymatic*, Vol. 55, (137-141).

Chandra, S.; Bag, S.; Bhar, R.; Pramanik, P. (2011). Effect of transition and non-transition metals during the synthesis of carbon xerogels. *Microporous and Mesoporous Materials*, Vol. 138, (149-156).

Choplin, A.; Quignard, F. (1998). From supported homogeneous catalysts to heterogeneous molecular catalysts. *Coordination Chemistry Reviews*, Vol. 17, (1679-1702).

Conceição, F.L.; Carrot, P.J.M..; Ribeiro-Carrot,, M.M.L. (2009). New carbon materials with high porosity in the 1-7 nm range obtained by chemical activation with phosporic acid of resorcinol-formaldehyde aerogles. *Carbon*, Vol. 47, (1874-1877).

Contreras, M.S.; Paez, C.A.; Zubizarreta, L.; Leonard, A.; Blacher, S.; Olivera-Fuentes, C.G.; Arenillas, A.; Pirard, J.-P.; Job, N. (2010). A comparison of physical activation of carbon xerogels with carbon dioxide with chemical activation using hydroxides. *Carbon*, Vol. 48, (3157-3168).

Cotet, L.C.; Gich, M.; Roig, A.; Popescu, I.C.; Cosoveanu, V.; Molins, E.; Danciu, V. (2006). Synthesis and structural characteristics of carbon aerogels with a high content of Fe, Co, Ni, Cu and Pd. *Journal of Non-Crystalline Solids*, Vol. 352, (2772-2777).

Czzakel, O.; Marthi, K.; Gueissler, K.; Laszlo, K. (2005). Influence of drying on the morphology of resorcinol-formaldehyde based carbon gels. *Microporous and Mesoporous Materials*, Vol. 86, (124-133).

De la Casa-Lillo, M.A.; Lamari-Darkrim, F.; Cazorla-Amorós D.; Linares-Solano, A. (2002). Hydrogen storage in activated carbons and activated carbon fibers. *Journal of Physical Chemistry*, Vol. 106, (10930-10934).

De Sousa, E.M.B.; Guimares, A.P.; Mohallem, N.D.S.; Lago, R.M. (2001). The effect of thermal treatment on the properties of sol-gel copper silica catalysts. *Applied Surface Science*, Vol. 183, (216-222).

Dillon, A.C.; Heben, M.J. (2001). Hydrogen storage using carbon adsorbents: past, present and future. *Applied Physics A: Material Science Progress*, Vol. 72, (133-142).

Dingcai, W.; Ruowen, F. (2006). Synthesis of organic and carbon aerogels from phenol-furfural by two-step polymerization. *Microporous and Mesoporous Materials*, Vol. 96, (115-120).

Djakovitch, L.; Wagner, M.; Hartung, C.G.; Beller, M.; Koehler K. (2004). Pd-catalyzed Heck arylation of cycloalkenes: Studies on selectivity comparing homogeneous and heterogeneous catalysts. *Journal of Molecular Catalysis A: Chemical*, Vol. 219, (121-130).

Escribano, S.; Berthon, S.; Ginoux, J.L.; Achard, P. (1998). Characterization of carbon aerogels. *In: Extended Abstract, Eurocarbon*, Strasbourg, France (841-842).

Fairén-Jiménez, D.; Carrasco-Marín, F.; Moreno-Castilla, C. (2006). Porosity and surface area of monolithic carbon aerogels prepared using alkaline carbonates and organic acids as polymerization catalysts. *Carbon*, Vol. 44, (2301-2307).

Fang, B.; Binder, L. (2006). A modified activated carbon aerogel for high-energy storage in electric double layer capacitors. *Journal of Power Sources*, Vol. 163 (616-622).

Feaver, A.; Cao, G. (2006). Activated carbon cryogels for low pressure methane storage. *Carbon*, Vol. 44, (590-593).

Figueiredo, J.L.; Pereira, M.F.R.; Serp, P.; Kalck, P.; Samant, P.V.; Fernandes, J.B. (2006). Development of carbon nanotube and carbon xerogels supported catalysts for the electro-oxidation of methanol in fuel cells. *Carbon*, Vol. 44, (2516-2522).

Figueiredo, J.L.; Sousa, J.P.S.; Orge, C.A.; Pereira, M.F.R.; Orfão, J.J.M. (2011). Adsorption of dyes on carbon xerogels and templated carbons: influence of surface chemistry. *Adsorption*, Vol. 17, (431-441).

Fontecha-Cámara, M.A.; Álvarez-Reinoso, M.A.; Carrasco-Marín, F.; López-Ramon, M.V.; Moreno-Castilla, C. (2011). Heterogeneous and homogeneous Fenton processes using activated carbon for the removal of the herbicide amitrole from water. *Applied Catalysis B: Environmental*, Vol. 101, (425-430).

Frackowiak, E. (2007). Carbon materials for supercapacitor applications. *Physical Chemistry Chemical Physics*, Vol. 9 (1774-1785).

Frackowiak, E.; Béguin, F. (2001). Carbon materials for the electrochemical storage of energy capacitors. *Carbon*, Vol. 39, (937-950).

Fu, R.; Baumann, T.F.; Cronin, S.; Dresselhaus, G. (2005). Formation of Graphitic Structures in Cobalt- and Nickel-Doped Carbon Aerogels. *Langmuir*, Vol. 21, (2647-2651).

Fu, R.; Dresselhaus, M.S.; Dresselhaus, G.; Zheng, B.; Liu, J.; Satcher, J.H.; Theodore, F.B. (2003). The growth of carbon nanostructures on cobalt-doped carbon aerogels. *Journal of Non-Crystalline Solids*, Vol. 318, (223-232).

Fuentes, E.; Menéndez, J.A.; Suárez, D.; Montes-Morán, M.A. (1995). Basic surface oxides on carbon materials: A global view. *Langmuir*, Vol. 19, (3505-3511).

Gadiour, R.; Texier-Mandoki, N.; Piquero, T.; Saadallah, S.; Parmentier, J.; Patarin, J. (2005). The influence of microporosity on the hydrogen storage capacity of ordered mesoporous carbons. *Adsorption*, Vol. 11, (823-827).

Ganesh, V.; Pitchumani, S.; Lakshminarayanan, V. (2006). New symmetric and asymmetric supercapacitors based on high surface area porous nickel activated carbons. *Journal of Power Sources*, Vol. 158, (1523-1532).

Gao, L.; Yoo, E.; Nakamura, J.; Zhang, W.; Chua, H.T. (2010). Hydrogen storage in Pd-Ni doped defective carbon nanotubes through the formation of CH_x (x = 1, 2). *Carbon*, Vol. 48, (3250-3255).

Girgis, B.S.; Attia, A.A.; Fatthy, N.A. (2011). Potential of nano-carbon xerogels in the remediation of dye-contamined water discharges. *Desalination*, Vol. 265, (169-176).

Gomes, H.T.; Machado, B.F.; Ribeiro, A.; Moreira, I.; Rosario, M.; Silva, A.M.T.; Figueiredo, J.L.; Faris, J.L. (2008). Catalytic properties of carbo materials for wet oxidations of aniline. *Journal of Hazardous Materials*, Vol. 159, (420-426).

Gorgulho, H.F.; Gonçcalves, F.; Pereira, M.F.R.; Figueiredo, J.L. (2009). Synthesis and characterization of nitrogen-doped carbon xerogels. *Carbon*, Vol. 47, (2032-2039).

Gryzb, B.; Hildenbrand, C.; Berthon-Frabry, S.; Béguin, D.; Job, N.; Rigacci, A.; Achard, P. (2010). Functionalisation and chemical characterization of cellulose-derived carbon aerogels. *Carbon*, Vol. 48, (2297-2307).

Guo, S.; Peng, J.; Li, W.; Yang, K.; Zhang, L.; Zhang, S.; Xia, H. (2009). Effects of CO_2 activation on porous texture strucutres of coconut shell-based activated carbons. *Applied Surface Science*, Vol. 255, (8443-8449).

Halama, A.; Szubzda, B.; Pasciak, G. (2010). Carbon aerogels as electrode material for electric double layer supercapacitors: Synthesis and properties. *Electrochimica Acta*, Vol. 55, (7501-7505).

Hirscher, M.; Panella, B. (2007). Hydrogen storage in metal-organic frameworks. *Scripta Materalia*, Vol. 56, (809-812).

Houmard, M.; Vasconcelos, D.C.L.; Vasconcenlos, W.L.; Berthomé, G.; Joud, J.C.; Langlet, M. (2009). Water and oil wettability of hybrid organic-inorganic titanate-silicate thin films deposited via sol-gel route. *Surface Science*, Vol. 603, (2698-2707).

Husley, S.S.; Alviso, C.T.; Kong, F.M.; Pekala, R.W. (1992). The effect of Pyrolysis Temperature and Formulation on Pore Size Distribution and Surface Area of Carbon Aerogels. *In: Peprint of Lawrence Livermore National Laboratory*, 1992.

Jagtoyen, M.; Groppo, J.; Derbyshire, F. (1993). Activated carbon from bituminous coals by reaction with H_3PO_4: The influence of coal cleaning. *Fuel Proccesing Technology*, Vol. 34, (85-93).

Job, N; Pirard, R.; Marien, J.; Pirard, J.-P. (2004). Porous carbon xerogels with texture tailored by pH control during sol-gel process. *Carbon*, Vol. 42, (619-628).

Job, N.; Marie, J.; Lambert, S.; Berthon-Fabry, S.; Achard, P. (2008). Carbon xerogels as catalyst supports for PEM fuel cell cathode. *Energy Conversion and Management*, Vol. 49, (2461-2470).

Job, N.; Panariello, F.; Marien, J.; Crine, M.; Pirard, J.P.; Leonard, A. (2006). Synthesis optimization of organic xerogels produced from convective air-drying of resorcinol-formaldehyde gels. *Journal of Non-Crystalline Solids*, Vol. 352, (24-34).

Job, N.; Panariello, R.; Crine, M.; Pirard, J.-P.; Leonard, A. (2007a). Rheological determination of the sol-gel transition during the aqueous synthesis of resorcinol-formaldehyde resins. *Colloids and Surfaces A: Physicochemical and Engineering Aspects*, Vol. 293, (224-228).

Job, N.; Pirard, R.; Vertruyen, B.; Colomer, J.-F.; Marien, J.; Pirard, J.-P. (2007b). Synthesis of transition metal-doped carbon xerogels by cogelation. *Journal of Non-Crystalline Solids*, Vol. 353, (2333-2345).

Job, N.; Thery, A.; Pirard, R.; Marien, J.; Kocon, L.; Rouzaud, J.-N.; Beguin, F.; Pirard, J.-P. (2005). Carbon aerogels, cryogels and xerogels: Influence of the drying method on the textural properties of porous carbon materials. *Carbon*, Vol. 43, (2481-2494).

Jordá-Beneyto, M.; Lozano-Castelló, D.; Suárez-García, F.; Cazorla-Amorós, D.; Linares-Solano, A. (2008). Advanced activated carbon monoliths and activated carbons for hydrogen storage. *Microporous and Mesoporous Materials*, Vol. 112, (235-242).

Juárez-Pérez, E.J.; Calvo, E.G.; Arenillas, A.; Menéndez, J.A. (2010). Precise determination of the point of sol-gel transition in carbon gels synthesis using a microwave heating method. *Carbon*, Vol. 48, (3305-3308).

Kabbour, H.; Baumann, T.F.; Satcher, J.H.; Saulnier, J.A.; Ahn, C.C. (2006). Toward new candidates for hydrogen storage: High-Surface-Area Carbon Aerogels. *Chemistry of Materials*, Vol. 18, (6085-6087).

Kang, K.-Y.; Lee, B.I.; Lee, J.S. (2009). Hydrogen adsorption on nitrogen-doped carbon xerogels. *Carbon*, Vol. 47, (1171-1180).

Kappe, C.O. (2004). Microwave Heating in Modern Organic Synthesis. *Angewandte Chemie International Edition*, Vol. 43, (6250-6284).

Kierzek, K.; Frackowiak, E.; Lota, G.; Gryglewick, G.; Machnikowski, J. (2004). Electrochemical capacitors based on highly porous carbons prepared by KOH activation. *Electrochemica Acta*, Vol. 49, (515-523).

Kim, B.-J.; Lee, Y.-S.; Park, S.-J. (2008). A study on the hydrogen storage capacity of Ni-plated porous carbon nanofibers. *International Journal of Hydrogen Energy*, Vol. 33, (4112-4115).

Kim, S.Y.; Yeo, D.H.; Lim, J.W.; Yoo, K.; Lee, K.; Kim, H. (2001). Synthesis and Characterization of Resorcinol-Formaldehyde Organic Aergoel. *Journal of Chemical Engineering of Japan*, Vol. 34, (216-220).

Kocklenberg, R.; Mathieu, B.; Blacher, S.; Pirard, R.; Pirard, J.-P.; Sobry, R.; Van den Bossche, G. (1998). Texture control of freeze-dried resorcinol-formaldehyde gels. *Journal of Non-Crystalline Solids*, Vol. 225, (8-13).

Kötz, R.; Carlen, M. (2000). Principles and applications of electrochemical capacitors. *Electrochimica Acta*, Vol. 45, (2483-2498).

Kraiwattanawong, K.; Tamon, H.; Praserthdam, P. (2011). Influence of solvent species used in solvent exchange for preparation of mesoporous carbon xerogels from resorcinol-formaldehyde via subcritical drying. *Microporous and Mesoporous Materials*, Vol. 138, (8-16).

Kuo, H.H.; Lin, J.H.C.; Pu, C.P. (2005). Effect of carbonization rate on the properties of a PAN/phenolic-based carbon/carbon composite. *Carbon*, Vol. 43, (229-239).

Lamari, F.L.; Malbrunot, P.; Tartaglia, G.P. (2002). Review of hydrogen storage by adsorption in carbon nanotubes. *International Journal of Hydrogen Energy*, Vol. 27, (193-202).

Lambert, S.; Job, N.; D'Souza, L.; Pereira, M.F.R.; Pirard, R.; Heinrichs, B.; Figueiredo, J.L.; Pirard, J.-P.; Regalbuto, J.R. (2009). Synthesis of very highly dispersed platinum catalysis supported on carbon xerogels by the strong electrostatic adsorption method. *Journal of Catalysis*, Vol. 261, (23-33).

Lee, Y.J.; Chung, J.C.; Yi, J.; Baeck, S.-H.; Yoon, J.R.; Song, I.K. (2010). Preparation of carbon aerogels in ambient conditions for electric double-layer capacitor. *Current Applied Physics*, Vol. 10, (682-686).

Lee, Y.J.; Jung, J.C.; Park, S.; Seo, J.G.; Baeck, S.-H.; Yoon, J.R.; Yi, J.; Song, I.K. (2011). Effect of preparation method on electrochemical property of Mn-doped carbon aerogel for supercapacitor. *Current Applied Physics*, Vol. 11, (1-5).

Leonard, A.; Job, N.; Blacher, S.; Pirard, J.-P.; Crine, M.; Jomaa, W. (2005). Suitability of convective air drying for the production of porous resorcinol-formaldehyde organic and carbon xerogels. *Carbon*, Vol. 43, (1808-1811).

Li, W.-C.; Lu, A.-H.; Guo, S.-C. (2001). Characterization of the microstructures of organic and carbon aerogels based upon mixed cresol-formaldehyde. *Carbon*, Vol. 39, (1989-1994).

Liang, C.; Sha, G.; Guo, S. (2000). Resorcinol-formaldehyde aerogels prepared by supercritical acetone drying. *Journal of Non-Crystalline Solids*, Vol. 271, (161-170).

Lin, C.; Ritter, J.A. (1997). Effect of synthesis pH on the structure of carbon xerogels. *Carbon*, Vol. 35, (1271-1278).

Lin, C.; Ritter, J.A. (2000). Carbonization and activation of sol-gel derived carbon xerogels. *Carbon*, Vol. 38, (849-861).

Liou, T.-H. (2004). Evolution of chemistry and morphology during the carbonization and combustion of rice husk. *Carbon*, Vol. 42, (785-794).

Liu, B.; Creager, S. (2010). Carbon xerogels as Pt catalyst supports for polymer electrolyte membrane fuel-cell applications. *Journal of Power Sources*, Vol. 195, (1812-1820).

Liu, N.; Zhang, S.; Fu, R.; Dresselhaus, M.S.; Dresselhaus G. (2006). Carbon aerogel spheres prepared via alcohol supercritical drying . *Carbon*, Vol. 44, (2430-2436).

Liu, Z.; Wang, A.; Wang, X.; Zhang, T. (2006). Reduction of NO by Cu-carbon and Co-carbon xerogels. *Carbon*, Vol. 44, (2330-2356).

Long, D.; Qiao, W.; Zhang, L.; Liang, X. (2009). Bimolecular adsorption behaviour on spherical carbon aerogels with various mesopore size. *Journal of Colloid and Interface Science*, Vol. 331, (40-46).

Long, D.; Zhang, J.; Yang, J.; Hu, Z. ; Li, T. ; Chen, G. ; Zhang, R. ; Ling, L. (2008a). Preparation and microstructure control of carbon aerogels produced using m-cresol mediated sol-gel polymerization of phenol and furfural. *New Carbon Materials*, Vol. 23, (165-170).

Long, D.; Zhang, J.; Yang, J.; Hu, Z. ; Cheng, G. ; Liu, X.; Zhang, R.; Zhang, L.; Qiao, W.; Ling, L. (2008b). Chemical state of nitrogen carbon aerogels issued from phenol-melamine-formaldehyde gels. *Carbon*, Vol. 46, (1259-1262).

Lota, G.; Centeno, T.A.; Frackowiak, E.; Stoeckli, F. (2008). Improvement of the structural and chemical properties of a commercial activated carbon for its application in electrochemical capacitors. *Electrochimica Acta*, Vol. 53, (2210-2216).

Lozano-Castelló, D.; Cazorla-Amorós, D.; Linares-Solano, A. (2002). Can highly activated carbons be prepared with a homogeneous micropore size distribution?. *Fuel Proccesing Technology*, Vol. 77, (325-330).

Machado, B.F.; Gomes, H.T.; Serp, P.; Kalck, P.; Figueiredo, J.L.; Faria, J.L. (2010). Carbon xerogel supported noble metal catalysts for fine chemical applications. *Catalysis Today*, Vol. 149, (358-364).

Maciá-Agulló, J.A.; Moore, B.C.; Cazorla-Amorós, D.; Linares-Solano, A. (2007). Influence of carbon fibers crystallinities on their chemical activation by KOH and NaOH. *Microporous and Mesoporous Materials*, Vol. 101, (397-405).

Mahata, N.; Pereira, M.F.R.; Suárez-García, F.; Martínez-Alonso, A.; Tascón, J.M.D.; Figueiredo, J.L. (2008). Tuning of texture and surface chemistry of carbon xerogels. *Journal of Colloid and Interface Science*, Vol. 324, (150-155).

Maldonado-Hódar, F.J.; Moreno-Castilla, C.; Carrasco-Marín, F.; Pérez-Cadenas, A.F. (2007). Reversible toluene adsorption on monolithic carbon aerogels. *Journal of Hazardous Materials*, Vol. 148, (548-552).

Maldonado-Hódar, F.J.; Moreno-Castilla, C.; Pérez-Cadenas, A.F. (2004). Catalytic combustion of toluene on platinum-containing monolithic carbon aerogels. *Applied Catalysis B: Environmental*, Vol. 54, (217-224).

Maldonado-Hódar, F.J.; Pérez-Cadenas, A.F.; Moreno-Castilla, C. (2003). Morphology of heat-treated tungsten doped monolithic carbon aerogels. *Carbon*, Vol. 41, (1291-1299).

Matos, I.; Fernandes, S.; Guerreiro, L.; Barata, S.; Ramos, A.M.; Vital, J.; Fonseca, I.M. (2006). The effect of surfactants on the porosity of carbon xerogels. *Microporous and Mesoporous Materials*, Vol. 92, (38-46).

Mauritz, K.A. (1998). Organic-inorganic hybrid materials: perfluorinated ionomers as sol-gel polymerization templates for inorganic alkoxides. *Materials Science and Engineering: C*, Vol. 6, (121-133).

Menéndez, J.A.; Arenillas, A.; Fidalgo, B.; Fernández, Y.; Zubizarreta, L.; Calvo, E.G.; Bermúdez, J.M. (2010). Microwave Heating processes involving carbon materials. *Fuel Processing Technology*, Vol. 91, (1-8).

Molina-Sabio, M.; Rodríguez-Reinoso, F. (2004). Role of chemical activation in the development of carbon porosity. *Colloids and Surface A: Physicochemical Engineering Aspects*, Vol. 241, (15-25).

Montes-Morán, M.A.; Menéndez, J.A.; Fuentes, E.; Suárez, D. (1998). Contribution of the Basal Planes to Carbon Basicity: An Ab Initio Study of the H_3O^+–π Interaction in Cluster Models. *The Journal of Physical Chemistry B*, Vol. 102, (5595-5601).

Moreno-Castilla, C.; Maldonado-Hódar, F.J. (2005). Carbon aerogels for catálisis applications: An overview. *Carbon*, Vol. 43, (455-465).

Moreno-Castilla, C.; Rivera-Utrilla, F.J.; Rodríguez-Castellón, J. (1999). Group of 6 metal oxide-carbon aerogels. Their synthesis, characterization and catalytic activity in the skeletal isomerization of 1-butene. *Applied Catalysis*, Vol. 183, (345-356).

Mukai, S.R.; Tamitsuji, C.; Nishihara, H.; Tamon, H. (2005a). Preparation of mesoporous carbon gels from an inexpensive combination of phenol and formaldehyde. *Carbon*, Vol. 43, (2628-2630).

Mukai, S.R.; Nishimura, H.; Yoshida, T.; Taniguchi, K.; Tamon, H. (2005b). Morphology of resorcinol-formaldehyde gels obtained through ice-templating. *Carbon*, Vol. 43, (1563-1565).

Nabais, J.N.V.; Nunes, P.; Carrot, P.J.M.; Carrot, M.M.L.R..; García, A.M.; Díaz-Díez, M.A. (2008). Production of activated carbons from coffee endocarp by CO_2 and steam activation. *Fuel Processing Technology*, Vol. 89, (262-268).

Neppolian, B.; Wang, Q.; Jung, H.; Choi, H. (2008). Ultrasonic-assisted sol-gel method of preparation of TiO_2 nano-particles. Characterization, properties and 4-chlorophenol removal application. *Ultrasonics Sonochemistry*, Vol. 15, (649-658).

Niu, J.J.; Wang, J.N. (2008). Effect of temperature on chemical activation of carbon nanotubes. *Solid Stated Sciences*, Vol. 10, (1189-1193).

Okada, K.; Yamamoto, N.; Kameshima, Y.; Yasumori, A. (2003). Porous properties of activated carbons from waste newspaper prepared by chemical and physical activation. *Jounal of Colloid and Interface Science*, Vol. 262, (179-193).

Olivares-Martín, M.; Fernández-González, C.; Macías-García, A.; Gómez-Serrano, V. (2006). Preparation of activated carbon from cherry stones by chemical activation with $ZnCl_2$. *Applied Surface Science*, Vol. 252, (5967-5971).

Pandolfo, A.G.; Hollenkamp, A.F. (2006). Carbon properties and their role in supercapacitors. *Journal of Power Sources*, Vol. 157, (11-27).

Pekala, R.W. (1989). Organic aerogels from the polycondensation of resorcinol with formaldehyde. *Journal of Materials Science*, Vol. 24, (3221-3227).

Pekala, R.W.; Alviso, C.T. (1992). Carbon aerogels and xerogels. *Materials Research Society Symposium Proceedings*, Vol. 270, (3-14).

Pekala, R.W.; Alviso, C.T.; Lu, X.; Gross, J.; Fricke, J. (1995). New organic aerogels based upon a phenolic-furfural reaction. *Journal of Non-Crystalline Solids*, Vol. 188, (34-40).

Pérez-Cadenas, M.; Moreno-Castilla, C.; Carrasco-Marin, F; Pérez-Cadenas, A.F. (2009). Surface chemistry, porous texture and morphology of N-doped carbon xerogels. *Langmuir*, Vol. 25, (466-470).

Petricevic, R.; Glora, M.; Fricke, J. (2001). Planar fiber reinforced carbon aerogels for application in PEM fuel cells. *Carbon*, Vol. 39, (857-867).

Pirard, S.L.; Diverchy, C.; Hermans, S.; Devillers, M.; Pirard, J.-P.; Job, N. (2011). Kinetics and diffusional limitations in nanostructured heterogeneous catalyst with controlled pore texture. *Catalysis Communications*, Vol. 12, (441-445).

Qin, C.; Lu, X.; Yin, G.; Jin, Z.; Tan, Q.; Bai, X. (2011). Study of activated nitrogen-enriched carbon and nitrogen-enriched carbon/carbon aerogel composites as cathode materials for supercapacitors. *Materials Chemistry and Physics*, Vol. 126, (453-458).

Qin, G.; Wei, W.; Guo, S. (2001). Semi-continuous drying of RF with supercritical acetone. *Carbon,* Vol. 41, (851-853).

Qu, R; Wang, M.; Sun, C.; Zhang, Y.; Ji, C.; Chen, H.; Meng, Y.; Yin, P. (2008). Chemical modification of silica-gel with hydroxyl- or amino-terminated polyamine for adsorption of Au (III). *Applied Surface Science,* Vol. 255, (3361-3370).

Quin, G.; Guo, S. (2001). Preparation of RF organic aerogels and carbon aerogels by alcoholic sol-gel process. *Carbon,* Vol. 39, (1935-1937).

Raymundo-Piñero, E.; Azäis, P.; Cacciaguerra, D.; Cazorla-Amorós, D.; Linares-Solano, A.; Béguin, F. (2005). KOH and NaOH activation mechanisms of multiwalled carbon nanotubes with different structural organisation. *Carbon,* Vol. 43, (786-795).

Riera, E.; Blanco, A.; García, J.; Benedito, J.; Mulet, A.; Gallego-Juárez, J.A.; Blasco, M. (2010).High-power ultrasonic system for the enhancement of mass transfer in supercritical CO_2 extraction processes. *Ultrasonics,* Vol. 50, (306-309).

Roman, S.; González, J.F.; González-García, C.M.; Zamora, F. (2008). Control of pore develpment during CO_2 and steam activation of olive stones. *Fuel Processing Technology,* Vol. 89, (715-720).

Salazar-Hernandez, M.M.; Leyve-Ramirez, M.A.; Gutierrez, J. A. (2009). Neutral alkoxysilanes from silica gel and N-phenyldiethanolamine. *Polyhedrom,* Vol. 28, (4044-4050).

Salitra, G.; Soffer, A.; Eliad, L.; Cohen, Y.; Aurbach, D. (2000). Carbon electrodes for double-layer capacitors. Relation between ion and pore dimensions. *Journal of Electrochemistry Society,* Vol. 147, (2486-2493).

Sanchez-Polo, M.; Rivera-Utrilla, J.; Salhi, E.; von Gunten, U. (2007). Ag-doped carbon aerogels for removing halide ions in water treatment. *Water research,* Vol. 41, (1031-1037).

Scherdel, C.; Reichenauer, G. (2009). Carbon xerogels synthesized via phenol-formaldehyde gels. *Microporous and Mesoporous Materials,* Vol. 126, (133-144).

Schimmel, H.G.; Nijkamp, G.; Kearley, G.J.; Rivera, A.; de Jong, K.P.; Mulder, F.M. (2004). Hydrogen adsorption in carbon nanostructures compared. *Materials Science and Engineering B,* Vol. 108, (124-129).

Sepheri, S.; Garcia, B.B; Zhang, Q.; Gao, G. (2009). Enhanced electrochemical and structural properties of carbon cryogels by surface chemistry alteration with boron and nitrogen. *Carbon,* Vol. 47, (1436-1443).

Serdich, M.; Jurcakova, D.H.; Lu, G.Q.; Bandosz, T.J. (2008). Surface functional groups of carbons and the effects on their chemical character, density and accesibility to ions and electrochemical performance. *Carbon,* Vol. 46, (1475-1488).

Shi, H. (1996). Activated carbons and double layer capacitance. *Electrochimica Acta,* Vol. 41, (1633-1639).

Silva, A.M.T.; Machado, B.F; Figueiredo, J.L.; Faira, J.L. (2009). Controlling the surface chemistry of carbon xerogels using HNO_3-hydrothermal oxidation. *Carbon,* Vol. 47, (1670-1679).

Staiti, P.; Lufrano, F. (2010). Investigation of polymer electrolyte hybrid supercapacitor based on manganese oxide-carbon electrodes. *Electrochimica Acta,* Vol. 55, (7436-7442).

Surya, V.J.; Iyakutti, K.; Rajarajeswari, M.; Kawazoe, Y. (2009). Functionalisation of single-walled carbon nanotube with borane for hydrogen storage. *Physica E: Low-dimensional Systems and Nanostructures,* Vol. 41, (1340-1346).

Suslick, K.S.; Price, G.J. (1999). Applications of ultrasound to materials chemistry. *Annual Review of Materials Science*, Vol. 29, (295-326).

Takagi, H.; Hatori, H.; Yamada, Y.; Matsuo, S.; Shiraishi, M. (2004). Hydrogen adsorption properties of activated carbons from modified surfaces. *Journal of Alloys and Compounds*, Vol. 385, (257-263).

Tamon, H.; Ishizaka, H.; Araki, T.; Okazaki, M. (1998). Control of mesoporous structure of organic and carbon aerogels. *Carbon*, Vol. 36, (1257-1262).

Tamon, H.; Ishizaka, H.; Yamamoto, T.; Suzuki, T. (2000). Influence of freeze-drying conditions on the mesoporosity of organic gels as carbon precursors. *Carbon*, Vol. 38, (1099-1105).

Teng, H.; Wang, S.-C. (2000). Preparation of porous carbons from phenol-formaldehyde resins with chemical and physical activation. *Carbon*, Vol. 38, (817-824).

Teng, Z.; Han, Y.; Li, J.; Yan, F.; Yang, W. (2010). Preparation of hollow mesoporous silica spheres by a sol/gel emulsion approach. *Microporous and Mesoporous Materials*, Vol. 127, (67-72).

Thomas, K.M. (2007). Hydrogen adsorption and storage on porous materials. *Catalysis Today*, Vol. 120, (389-398).

Tian, H.Y.; Buckley, C.E.; Paskevicius, M.; Wang, S.B. (2011b). Carbon aerogels from acetic acid catalysed resorcinol-furfural using supercritical drying for hydrogen storage. *The Journal of Supercritical Fluids*, Vol. 55 (1115-1117).

Tian, H.Y.; Buckley, C.E.; Paskevicius, M.; Sheppard, D.A. (2011a). Acetic acid catalysed carbon xerogels derived from resorcinol-furfural for hydrogen storage. *International Journal of Hydrogen Energy*, Vol. 36, (671-679).

Tian, H.Y.; Buckley, C.E.; Sheppard, D.A.; Paskevicius, M.; Hanna, N. (2010). A synthesis method for cobalt doped carbon aerogels with high surface area and their hydrogen storage properties. *International Journal of Hydrogen Energy*, Vol. 35, (13242-13246).

Tonamon, N.; Siyasukh, A.; Tanthapanichakoon, W.; Nishihara, H.; Mukai, S.R.; Tamon, H. (2005). Improvement of mesoporosity of carbon cryogels by ultrasonic irradiation. *Carbon*, Vol. 43, (525-531).

Tonamon, N.; Wareenin, Y.; Siyasukh, A.; Tanthapanichakoon, W.; Nishihara, H.; Mukai, S.R.; Tamon, H. (2006). Preparation of resorcinol-formaldehyde (RF) carbon gels: Use of ultrasonic irradiation followed by microwave drying. *Journal of Non-Crystalline Solids*, Vol. 352, (5683-5686).

Vix-Guterl, C.; Frackowiak, R.; Jurewick, K.; Friebe, M.; Parmentier, J.; Béguin, F. (2005). Electrochemical energy storage in ordered porous carbon materials. *Carbon*, Vol. 43, (1293-1302).

Whang, J.; Yang, X.; Wu, D.; Fu, R.; Dresselhaus, M.S.; Dresselhaus, G. (2008). The porous structures of activated carbon aerogels and their effects on electrochemical performance. *Journal of Power Sources*, Vol. 185, (589-594).

Wiener, M.; Reichenauer, G.; Scherb, T.; Frichke, J. (2004). Accelerating the synthesis of carbon aerogel precursors. *Journal of Non-Crystalline Solids*, Vol. 350, (126-130).

Wu, D.; Fu, R.; Zhang, S.; Dresselhaus, M.; Dresselhaus, G. (2004). Preparation of low-density carbon aerogels by ambient pressure drying. *Carbon*, Vol. 42, (2033-2039).

Xua, W.-C.; Takahashi, K.Y.; Matsuo, Y.; Hattoria, M. (2007). Investigation of hydrogen storage capacity of various carbon materials. *International Journal of Hydrogen Energy*, Vol. 32, (2504-2512).

Yamamoto, T.; Endo, A.; Ohmori, T.; Nakaiwa, N. (2004). Porous properties of carbon gel microspheres for gas separation. *Carbon*, Vol. 42, (1671-1676).

Yamamoto, T.; Nishimura, T.; Suzuki, T.; Tamon, H. (2001). Effect of drying conditions on mesoporosity of carbon precursors prepared by sol-gel polycondensation and freeze-drying. *Carbon*, Vol. 39, (2374-2376).

Yang, K.Y.; Hong, S.J.; Lee, B.I.; Lee, J.S. (2008). Enhanced electrochemical capacitance of nitrogen-doped carbon gels synthesized by microwave-assisted polymerization of resorcinol and formaldehyde. *Electrochemistry Communications*, Vol. 10, (1105-1108).

Ying, D.J.; Haji, Z.S.; Erkey, C. (2005). Dynamics of removal of organosulfur compounds from diesel by adsorption on carbon aerogels for fuel cell applications. *International Journal of Hydrogen Energy*, Vol. 30, (1287-1293).

Yoshimune, M.; Yamamoto, T.; Nakaiwa, M.; Haraya, K. (2008). Preparation of highly mesoporous carbon membranas via sol-gel process using resorcinol and formaldehyde . *Carbon*, Vol. 46, (1031-1036).

Zareba-Grodz, I.; Mista, W.O.; Strek, W.; Bukowska, E.; Hermanowick, K.; Maruszewski, K. (2004). Synthesis and properties of an inorganic-organic hybrid prepared by the sol-gel method. *Optical Materials*, Vol. 26, (207-211).

Zhang, L; Liu, H.; Wang, M.; Chen, L. (2007). Structure and electrochemical properties of resorcinol-formaldehyde polymer-based carbon for electric double-layer capacitors. *Carbon*, Vol. 45, (1439-1445).

Zhao, N.; He, C.; Jiang, Z.; Lin, J.; Li, Y. (2007). Physical activation and characterization of multi-walled carbon nanotubes catalytically synthesized from methane. *Materials Letters*, Vol. 61, (681-685).

Zheivot, V.I.; Molchanov, V.V.; Zaikovskii, V.I.; Krivoruchko, V.N.; Zaitseva, N.A.; Shchuchkin, M.S. (2010). Carbon xerogels: Nano and adsorption textures, chemical nature of the surface and gas chromatography properties. *Microporous and Mesoporous Materials*, Vol. 130, (7-13).

Zhu, Y.; Hu, H.; Li, W.; Zhang, X. (2007). Resorcinol-formaldehyde based porous carbon as electrode material for supercapacitors. *Carbon*, Vol. 45, (160-165).

Zhu, Y.; Hu, H.; Li, W.; Zhao, H. (2006). Preparation of cresol-formaldehyde carbon aerogels via drying aquagel at ambient pressure. *Journal of Non-Crystalline Solids*, Vol. 352, (3358-3362).

Zubizarreta, L.; Arenillas, A.; Dominguez, A.; Menéndez, J.A.; Pis, J.J. (2008a). Development of microporous carbon xerogels by controlling synthesis conditions. *Journal of Non-Crystalline Solids*, Vol. 354, (817-825).

Zubizarreta, L.; Arenillas, A.; Menéndez, J.A.; Pis, J.J.; Pirard, J.-P.; Job, N. (2008b). Microwave drying as an effective method to obtain carbon xerogels. *Journal of Non-Crystalline Solids*, Vol. 354, (4024-4026).

Zubizarreta, L.; Arenillas, A.; Pirard, J.-P.; Pis, J.J.; Job, N. (2008c). Tailoring the textural properties of activated carbon xerogels by chemical activation with KOH. *Microporous and Mesoporous Materials*, Vol. 115, (480-490).

Zubizarreta, L.; Arenillas, A.; Pis, J.J. (2009). Carbon materials for H_2 storage. *International Journal of Hydrogen Energy*, Vol. 34, (4575-4581).

Zubizarreta, L.; Menéndez, J.A.; Marco-Lozar, J.P.; Pirard, J.P.; Pis, J.J.; Linares-Solano, A.; Cazorla-Amorós, A.; Arenillas, A. (2010). Ni-doped carbon xerogels for H_2 storage. *Carbon*, Vol. 48, (2722-2733).

Part 4

State-of-the-Art Nano-Material Applications

10

Bioinspired Metal Nanoparticle: Synthesis, Properties and Application

Sujoy K. Das and Enrico Marsili
School of Biotechnology, Dublin City University, Dublin, Ireland

1. Introduction

Nanomaterials with a characteristic dimension in the range of 1-100 nanometers (nm) are at the leading edge of nanoscience and nanotechnology. In recent years nanomaterials, and specifically metal nanoparticles, have received particular interest in diverse field ranging from material science to biotechnology (Guo et al. 2005, Daniel and Astruc 2004, Huang et al. 2007). Although widespread interest in nanomaterials is recent, the concept was introduced over 40 years ago. Nanomaterials have actually been produced and used by humans for hundreds of years: for example, the beautiful ruby red colour of some glass is due to gold nanoparticles (AuNP) trapped in the glass matrix. In the decorative glaze known as luster, found on some medieval pottery, the special optical properties of the glaze arose from metallic spherical nanoparticles which were dispersed in the glaze in a random fashion. Michael Faraday in 1857 on his pioneering work "Experimental relations of gold (and other metals) to light" (Faraday, 1857) explain the properties of this glaze. Now with advances of science and technology, the morphology of this material, which contains metallic nanoparticles, has been understood. Because of extremely small size and high surface volume ratio of nanoparticles, the physicochemical properties of nanoparticles-containing materials are quite different to those of the bulk materials (El-Sayed 2001). Thus, nanomaterials have potential applications in electronics and photonics, catalysis, information storage, chemical sensing and imaging, environmental remediation, drug delivery and biological labelling (Guo et al. 2005, Daniel and Astruc 2004, Huang et al. 2007). It is well known that the optical, electronic, and catalytic properties of metal nanoparticles are greatly influenced by their size, shape, and crystal structure. For example, silver (Ag) and gold (Au) nanocrystals of different shapes possess unique optical scattering responses (Daniel and Astruc 2004, Roduner 2006). Whereas highly symmetric spherical particles exhibit a single scattering peak, anisotropic shapes such as rods, triangular prisms, and cubes exhibit multiple scattering peaks in the visible wavelengths due to highly localized charge polarizations at corners and edges (Mie 1908). Thus, synthesis of metal nanoparticles with defined morphology gained much interest. A variety of strategies have been developed for the synthesis of metal nanoparticles (MNPs) and nanomaterials. Optimizing the nanomaterial synthesis has now become a prolific area of investigation. In the first part of this chapter, we will review the synthetic methods for MNPs production, with particular regard to biosynthesis in viable organisms and protein extracts. In the second part, we will discuss the most recent finding on the biosynthetic mechanism, the properties of the nanobioconjugates, and recent applications of MNPs.

2. Synthesis of metal nanoparticles

Synthesis of MNPs is carried out by several physical and chemical methods that include laser ablation (Mafuné et al. 2001), ion sputtering (Raffi et al. 2007), solvothermal synthesis (Rosemary, T. Pradeep 2003), chemical reduction (Chaki 2002), and sol-gel (Shukla and Seal 1999) method. Basically, there are two approaches for nanoparticle synthesis, the top-down and bottom-up. Top-down approaches seek to create nanoscale objects by using larger, externally-controlled microscopic devices to direct their assembly, while bottom-up approaches adopt molecular components that are built up into more complex assemblies. The top-down approach often uses microfabrication techniques where externally controlled tools are used to cut, mill, and shape materials into the desired shape and size. Micropatterning techniques, such as photolithography and inkjet printing are well known examples of top-down approach. On the other hand, bottom-up approaches use the self-assembled properties of single molecules into some useful conformation. Different commonly used physical and chemical methods are described in the following paragraphs.

2.1 Laser ablation

Laser ablation (Mafuné et al. 2001) enables to obtain colloidal nanoparticles solutions in a variety of solvents. Nanoparticles are formed during the condensation of a plasma plume produced by the laser ablation of a bulk metal plate dipped in a liquid solution. This technique is considered as a 'green technique' alternative to the chemical reduction method for obtaining noble MNPs. However, the main drawbacks of this methodology are the high energy required per unit of MNPs produced and the little control over the growth rate of the MNPs.

2.2 Inert gas condensation

Inert gas condensation (IGC) is the most widely used methods for MNPs synthesis at laboratory-scale. Gleiter (1984) introduced the IGC technique in nanotechnology by synthesizing iron nanoparticles. In IGC, metals are evaporated in ultra high vacuum chamber filled with helium or argon gas at typical pressure of few hundreds pascals. The evaporated metal atoms lose their kinetic energy by collisions with the gas, and condense into small particles. These particles then grow by Brownian coagulation and coalescence and finally form nano-crystals. Recent application of this technique includes size-controlled synthesis of Au/Pd NPs (Pérez-Tijerina et al., 2008) and hetero-sized Au nanoclusters for non-volatile memory cell applications (Kang et al., 2011).

2.3 Sol-gel method

The sol-gel process is a wet-chemical technique developed recently in nanomaterial synthesis. The inorganic nanostructures are formed by the sol-gel process through formation of colloidal suspension (sol) and gelation of the sol to integrated network in continuous liquid phase (gel). Size and stability control quantum-confined semiconductor, metal, and metal oxide nanoparticles has been achieved by inverted micelles (Gacoin 1997), polymer blends (Yuan et al. 1992), block copolymers (Sankaran et al. 1993), porous glasses (Justus et al. 1992), and ex-situ particle-capping techniques (Olshavsky and Allcock 1997). However, the fundamental problem of aqueous sol-gel chemistry is the complexity of process and the fact that the as-synthesized precipitates are generally amorphous. In non-aqueous sol-gel chemistry the transformation of the precursor takes place in an organic solvent. The non-aqueous (or non-hydrolytic) processes are able to overcome some of the major limitations of aqueous systems, and thus represent a powerful and versatile alternative. The advantages are a direct consequence of the manifold role of the organic components in the reaction system (e.g., solvent, organic ligand of the precursor molecule, surfactants, or *in situ* formed organic condensation products). Nowadays, the family of metal oxide nanoparticles are synthesized by non-aqueous processes and ranges from simple binary metal oxides to more complex ternary, multi-metal and doped systems.

2.4 Hydrothermal and solvothermal synthesis

The hydrothermal and solvothermal synthesis of inorganic materials is an important methodology in nanomaterial synthesis. In hydrothermal method, the synthetic process occurs in aqueous solution above the boiling point of water, whereas in solvothermal method the reaction is carried out in organic solvents at temperatures (200-300°C) higher than their boiling points. Though development of hydrothermal and solvothermal synthesis has a history of 100 years, recently this technique has been applied in material synthesis process. Normally, hydrothermal and solvothermal reactions are conducted in a specially sealed container or high pressure autoclave under subcritical or supercritical solvent conditions. Under such conditions, the solubility of reactants increases significantly, enabling reaction to take place at lower temperature. Among numerous examples, TiO_2 photocatalysts were synthesized through hydrothermal process (Ren et al., 2007). Because of low cost and energy consumptiom, hydrothermal process can be scale-up for industrial production. Solvothermal process enables to choose among numerous solvents or mixture thereof, thus increasing the versatility of the synthesis. For example, well-faceted nanocrystals of TiO_2 with high reactivity were synthesized in a mixture of the solvents Hydrogen fluoride (HF) and 2-propanol (Yang et al., 2008)

2.5 Colloidal methods

The crystallographic control over the nucleation and growth of noble-metal nanoparticles has most widely been achieved using colloidal methods (Tao et al. 2008, Turkevich et al. 1951, Frens 1972, Brust et al. 1994). In general, metal nanoparticles are synthesized by reducing metal salt with chemical reducing agents like borodride, hydrazine, citrate, etc., followed by surface modification with suitable capping ligands to prevent aggregation and confer additional surface properties. Occasional use of organic solvents in this synthetic process often raises environmental questions. At the same time, these approaches produce multi-shaped nanoparticles requiring purification by differential centrifugation and consequently have low yield. Thus, the development of reliable experimental protocols for the synthesis of nanomaterials over a range of chemical compositions, sizes, and high monodispersity is one of the challenging issues in current nanotechnology. In this context, current drive focused on the development of green and biosynthetic technologies in for production of nanocrystals with desired size and shape.

2.6 Bio-inspired nanomaterial synthesis

In keeping with global efforts to reduce generation of hazardous waste and to develop energy-effective production routes, 'green' chemistry and biochemical processes are progressively integrating with modern developments in science and technology. Hence, any synthetic route or chemical process should address the fundamental principles of 'green chemistry' by using environmentally benign solvents and nontoxic chemicals (Anstas and Warner 1998). The green synthesis of MNPs should involve three main steps based on green chemistry perspectives, namely (1) the selection of a biocompatible and nontoxic solvent medium, (2) the selection of environmentally benign reducing agents, and (3) the selection of nontoxic substances for stabilization of the nanoparticles. Employing these principles in nanoscience will facilitate the production and processing of inherently safer nanomaterials and nanostructured devices. Green nanotechnology (Dahl et al. 2007) thus aims to the application of green chemistry principles in designing nanoscale products, and the development of nanomaterial production methods with reduced hazardous waste generation and safer applications. Further, biochemical process can occur at low temperatures, because of the high specificity of the biocatalysts. Hence, a synthetic route that include one or more biological steps will result in consistent energy saving and lower environmental impact with respect to conventional methods. To optimize safer nanoparticle production, it would be desirable to employ bio-based methods, which could minimize the hazardous conditions of materials fabrication and use. Inspiration from nature, where living organisms produce inorganic materials through biological guided process known as biomineralization, is adopted as a superior approach to nanomaterials assembly (Mann 1993). The biomineralization processes exploit biomolecular templates that interact with the inorganic material at nanoscale, resulting in extremely efficient and highly controlled syntheses. Typical examples of biomineralized products include siliceous materials synthesized by diatoms and sponges, calcite or aragonite (calcium carbonates) in invertebrates, and apatite (calcium phosphates and carbonates) in vertebrates. These biominerals are the phosphate and carbonate salts of calcium that form structural entities such as sea shells and the bone in mammals and birds in conjunction with organic polymers. The structures of these biocomposite materials are highly controlled both at nano- and macroscale level, resulting in complex architectures that provide multifunctional properties. Simpler organisms, such as bacteria, algae, and fungi, have also developed during hundreds of millions of years of evolution highly specialized strategies for biominerals synthesis. The role of the templating molecule in biomineralization is to provide a synthetic microenvironment in which the inorganic phase morphology is tightly controlled by a range of low-range interactions. Typical example includes silicatein filament, protease-like proteins termed silicateins, from a marine sponge *Thetya aurantia* direct the polymerisation of silica and silicones from tetraethoxysilane in vitro. Figure 1A-F demonstrates that silicatein filaments serve as scaffolds to organize the deposition of the resulting silica and silsesquioxanes.

Diatoms are unicellular algae that have the extraordinary capability to produce an enormous variety of biosilicates structures. Each diatom species is characterized by a specific biosilicate cell wall that contains regularly arranged slits or pores in the size range between 10 and 1,000 nm (nanopatterned biosilica). An important component of biomineralization is the protein or peptide template that controls the shape and crystal structure of biominerals as well as the assembly procedure. Biosilica morphogenesis takes place inside the diatom cell within a specialized membrane-bound compartment termed the silica deposition vesicle. Silica formation from a silicic acid solution is catalyzed by biosilica-associated peptides (silaffins) and long-chain polyamines in diatoms (Fig. 1G-J). Other examples of inorganic, nanostructured biosynthetic minerals include Fe_3O_4, $CaSO_4$, $BaSO_4$, etc., which are synthesized by a variety of microorganisms. These minerals often have special uses such as magnetic sensors in magnetotactic bacteria (Fe_3O_4), gravity sensing devices ($CaCO_3$, $CaSO_4$, $BaSO_4$) and iron storage and mobilization ($Fe_2O_3.H_2O$ in the protein ferritin). Magnetotactic bacteria (*Magnetospirillum magnetotacticum, Magnetobacterium bavaricum, Magnetospirillum gryphiswaldense*) have special organelles called magnetosomes that contain magnetic crystals, which enable the cells to orient themselves along the magnetic field lines of the Earth. Magnetotactic bacteria synthesize magnetic particles comprising iron oxide (Fe_3O_4), iron sulphides (Fe_3S_4), or both in their intracellular compartment. The magnetic particles are aligned in chains within the bacterium, and function as biological compass needles. This enables the bacterium to migrate along oxygen gradients in aquatic environments under the influence of the Earth's geomagnetic field. Fig. 2 depicts the formation of magenite in *M. magnetotacticum* and *M. Gryphiswaldens* and also orientation of isolated magnetosome particles under the magnetic field.

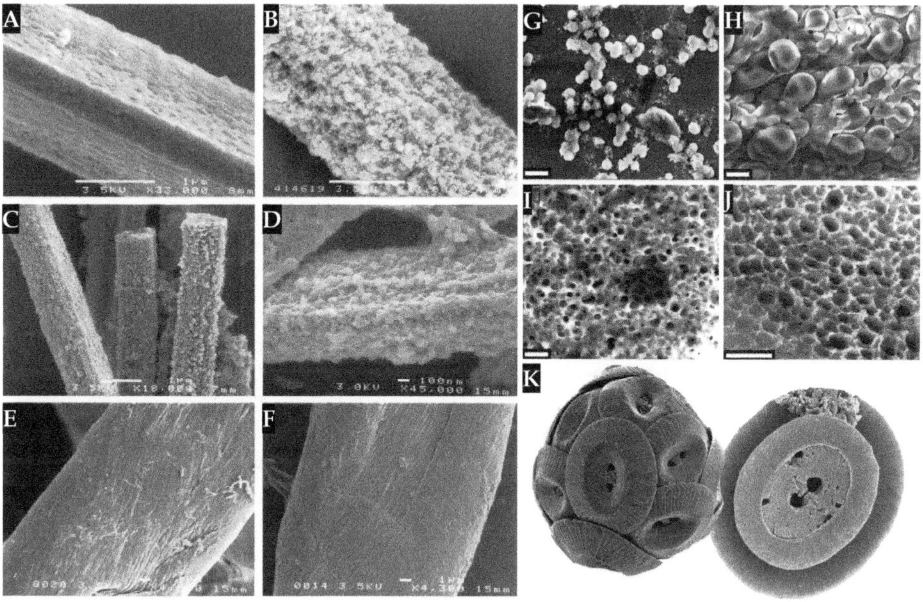

Fig. 1. SEM images of the products of the reaction between silicon alkoxides and silicatein or cellulose filaments. (*A*) Silicatein filaments before the reaction. (*B*) Silicatein filaments after a 12-h reaction with TEOS (1.0 ml; 4.5 mmol) plus Trisz-HCl buffer. (*C*) Air-dried silicatein filaments incubated with TEOS as in *B*, but with no additional water. (*D*) Silicatein filaments after an 8-h reaction with phenyltriethoxysilane (1.0 ml; 4.1 mmol) plus TriszHCl buffer. (*E*) Cellulose fiber. (*F*) Cellulose fiber after a 12-h reaction with TEOS as in *B*. Proc. Natl. Acad. Sci. USA. 1999, 96, 361–365. SEM analysis of silica precipitates formed by mixtures of natSil-1A and natSil-2. (G) natSil-2 at 0.5 units/µl, natSil-1A at 0.3 mM. (H) natSil-2 at 5.0 units/µl, natSil-1A at 0.3 mM. (I) natSil-2 at 2.0 units/µl, natSil-1A at 0.3 mM. (J) natSil-2 at 1.6 units/µl, natSil-1A at 0.2 mM. Proc. Natl. Acad. Sci. USA. 2003, 100, 12075–12080. SEM micrographs of coccosphere (K) showing their morphology and the composite nature.

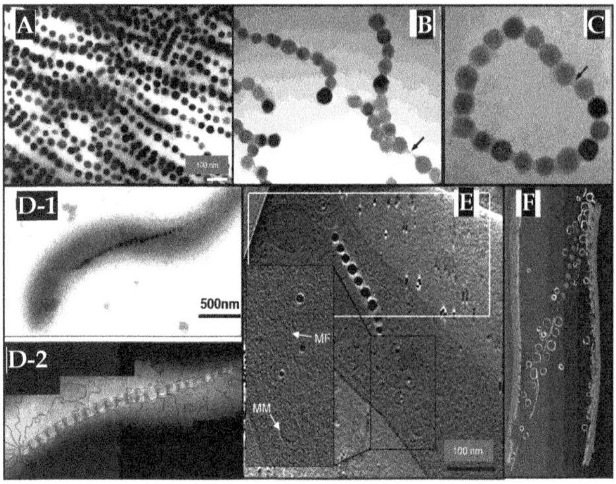

Fig. 2. Isolated magnetosome (A–C) particles form straight chains in weak ambient magnetic fields (A) but form bent chains (B) or flux-closure rings in zero fields (C). Magnetic microstructure (D) of a magnetosome chain. TEM bright-field image (D-1) of a single bacterial cell of *Magnetospirillum magnetotacticum*. Magnetic induction map (D-2) recorded using off-axis electron holography from the magnetosome chain of the same cell as shown in part D-1. Cryoelectron tomography image (E) shows organization of the magnetosome chain and the cytoskeletal magnetosome filament (MF) in *Magnetospirillum gryphiswaldense*, and Tomographic reconstruction (F) of a magnetic cell. Chem. Rev. 2008, 108, 4875-4898.

Desulfovibrio magneticus, a dissimilatory sulphate-reducing bacterium, produces unique irregular bullet-shaped magnetite. Formation of extracellular magnetic iron sulphide (Sakaguchi *et al.* 1993) and intracellular hematite (Posfai *et al.* 2006) by this bacterium has also been reported. The molecular mechanism of bacterial magnetite biomineralization is very complex and occurs in multistep process. Recent genome, proteome and transcriptome studies revealed that proteins located on the magnetosomial membrane are the key components in biomineralization process.

The natural occurrence of inorganic nanomaterials in microorganisms encourages the use of microorganisms as possible eco-friendly nanofactories for nanoparticle synthesis routes alternative to chemical methods. The metal reduction potential of several bacteria, yeast, and fungi has been known for at least 20 years. Microorganisms such as dissimilatory metal reducers and sulphate reducers play a key role in metal biogeochemistry and have been used with a certain success in metal bioremediation of contaminated subsurface environments. The microbial based biosynthetic process for nanoparticle production is a relatively new and largely unexplored area of research. In recent past, prokaryotic microorganisms such as bacteria and actinomycetes, as well as algae, yeast, and fungi (eukaryotes) have been employed in the biosynthesis of metal nanoparticles. In the next section we will review in details MNPs biosynthesis in microorganisms.

2.6.1 Bacterial nanoparticle synthesis

Bacteria play a crucial role in metal biogeochemical cycling and mineral formation in surface and subsurface environments (Lowenstam 1981, Southam and Saunders 2005). In presence of high concentration and even toxic metal ions, bacteria harbour numerous detoxification mechanisms like dissimilatory oxidation or reduction of metal ions, complexation and/or precipitation, impaired transport system, efflux system, etc (Mergeay et al. 2003). Thus microorganisms have been used with a certain success in metal bioremediation of contaminated subsurface environments. However, their role in nanoparticle synthesis has recently been observed. Among the microorganisms, prokaryotic bacteria have received the most attention in the area of metal nanoparticle biosynthesis. The formation of extracellular and intracellular metal nanoparticles by bacteria like *Escherichia coli, Pseudomonas stutzeri, Pseudomonas aeruginosa, Plectonema boryanum, Salmonells typlus, Staphylococcus currens, Vibrio cholerae*, etc., have been reported (Klaus et al. 1999, Konishi 2004, Beveridge and Murray 1980, Southam and Beveridge 1994). Marine Fe^{+3} reducing bacterium *Shewanella algae* reduces Au^{+3} ions in anaerobic environments. In the presence of *S. algae* and hydrogen gas, Au ions are completely reduced and 10-20 nm AuNPs are formed (Konishi et al. 2004). The formation of AuNP was indicated by the change in reaction mixture, which turned to light yellow after 1h. The AuNPs were synthesized adopting similar procedure using two *Pseudomonas aeruginosa* isolates (Husseiny et al. 2007). The synthesis of stable gold nanocubes by the reduction of aqueous $AgCl_4^-$ solution by *Bacillus licheniformis* was reported by Kalishwaralal et al. (2009). Biofilm formation of gram-negative β-proteobacterium *Cupriavidus metallidurans* is very common on Au grains. The isolated *C. metallidurans* from soils and sediments from temperate and tropical Australian sites interacted with Au^{+3} ions and form AuNPs distributed homogenously throughout cell wall (Reith 2009). The reduction of Au^{+3} complexes by *C. metallidurans* to AuNPs occurred through fast accumulation leading to the formation of intermediate Au^{+1}-S complexes followed by a slow biochemically-driven reduction and intra- and extracellular deposition of AuNPs (Fig. 3).

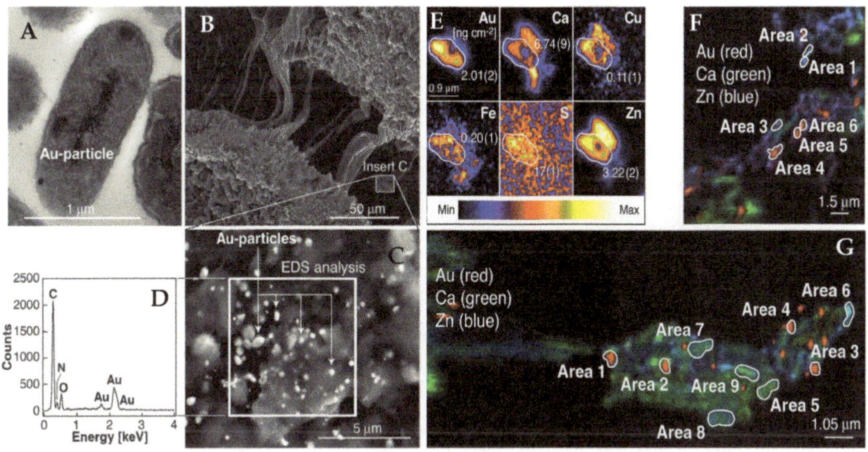

Fig. 3. TEM image of ultra-thin section *C. metallidurans* cell demonstrating formation of AuNPs in the periplasmic space (*A*). SEM image (*B* and *C*) with energy dispersive X-ray analysis (SEM-EDXA) (*D*) of a biofilm containing particulate Au (*C*) growing on Au grain from the Prophet Gold Mine, Australia. µXRF maps (E) showing the distribution of Au, Ca, Cu, Fe, S, and Zn in an *C. metallidurans* cell following 1 min exposure to Au^{+3} at pH 7.0. Overlay false colour quantitative µXRF maps (F-G) of the distribution of Au (red), Zn (blue), and Ca (green) in cell clusters after 72 h (F) and 144 h (G) of incubation at pH 7.0. Proc. Natl. Acad. Sci USA. 2009, 160, 17757-17762.

In brief, following adsorption of Au^{+3} complexes on cell surfaces, they reduced to Au^{+1} because of the high redox potential of Au^{+3} complexes [e.g., E_0 = 1.002 V for the reaction $AuCl_4^- + 3e^- = Au(s) + 4\ Cl^-$]. This results in the induction of oxidative stress due to transfer of electrons from suitable electron donors in the cell to Au^{+3} ions and ultimately results in the upregulation of oxidative stress response genes. The resulting Au^{+1}-complexes readily associated with nonpolar soft bases like S, present in membrane and periplasmic proteins and formed Au^{+1}-S species. The resulting Au^{+1}-S species reduced to Au^0 in a slower reductive pathway. The AuNPs were also synthesized on the surface of *Rhodopseudomans capsulate* by interaction bacterial cells with $HAuCl_4$ solution (He et al. 2007). The aqueous chloroaurate ions were reduced after 48 h of incubation and transformed to AuNPs. The pH value of the solution controlled the shape of AuNPs. The aqueous chloroaurate ions bound on the bacterial cell surface by interacting with different positively charged cell surface function groups such as amino, sulfhydryl and carboxylic. The electrons were then transferred from NADH by NADH-dependent reductase and reduced Au^{+3} to Au^0.

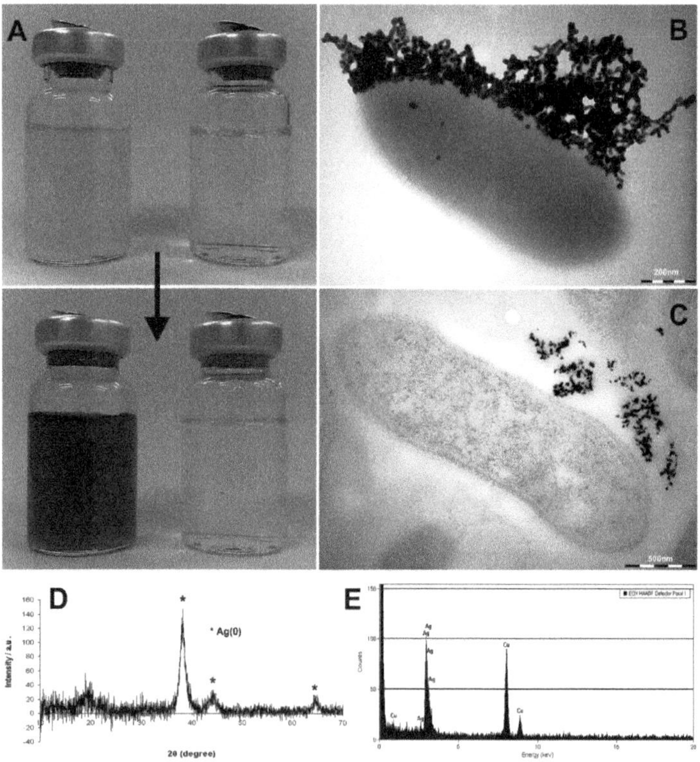

Fig. 4. Reduction of 200 μM Ag^{+1} by late-exponential-phase cultures of *G. sulfurreducens*. (A) Cultures (top left) and spent supernatant (top right) before addition of Ag^{+1} and 2-min incubation (bottom) in the dark after addition of Ag^{+1}. TEM of whole-mount (B) and ultrathin (C) section. (D) XRD profile of the cells and black precipitate, confirming the formation of Ag^0. (E) EDAX spectrum of extracellular precipitates from unstained sectioned samples. Appl. Environ. Microbiol. 2008, 74, 7090–7093.

Several bacterial strains cells are also known to synthesize silver nanoparticles (AgNPs) in a similar manner to that of AuNPs biosynthesis. *P. stutzeri* AG259 isolated from a silver mine reduced Ag^{+} ions and formed AgNPs within the periplasmic space. The obtained AgNPs have well defined size and morphology, ranging from few nm to 200 nm or more. The supernatant of gram positive, thermophilic bacterium *Bacillus licheniformis* synthesized AgNPs in the range of 50 nm (Kalishwaralal et al. 2008). Formation of nanoscale elemental silver sarticles through enzymatic reduction was reported in *Geobacter sulfurreducens* (Law et al. 2008). The AgNPs formation is indicated by the change in colour of the silver nitrate reaction mixture, which gradually changed from yellow to brown. The extracellular precipitation of the nanoscale Ag^{0} was observed in *G. sulfurreducens* planktonic cells (Fig.4). C-type cytochromes are involved in the transferring the electrons to metal ions, thus reducing Ag^{+1} ions to Ag^{0}. X-ray powder diffraction (XRD) and EDAX analysis confirmed that the obtained black precipitate was composed of AgNPs.
Kalimuthu and coworkers (2008) studied AgNPs synthesis using bacteria *B. licheniformis*, isolated from sewage collected from municipal wastes, and ultrasonically lysed bacterial cell. The synthesized AgNPs had average particle size of around 50 nm. Recently, a rapid method for synthesizing small (1-7 nm) monodiperse AgNPs has been described by electrochemically active biofilm (EAB) using sodium acetate as an electron donor (Kalathil et al. 2011). Electrochemically active biofilm was formed by dipping a carbon paper into a mineral salt medium containing sodium acetate. The electrode was then mounted in a conventional microbial fuel cell. Anaerobic sludge was added under strict anaerobic conditions. Following successive medium changes and washing, EAB was formed on carbon paper after a week. Under strict anaerobic conditions, interaction of EAB with $AgNO_3$ as precursor and sodium acetate as electron donor resulted in the synthesis of AgNPs. The XRD pattern demonstrated four characteristic peaks at 2θ values of 38°, 46.33°, 64.58° and 77.24°, which correspond to the 111, 200, 220 and 311 planes of face-centered cubic (fcc) silver, respectively.

2.6.2 Fungi in nanoparticle synthesis

The fungal mediated MNP synthesis is a relatively recent research area. Compared to bacteria, fungi hold good promises for large scale nanoparticles production because they secrete large amounts of the enzymes involved in the MNP biosynthesis and are simpler to grow both in the laboratory and at industrial scale. Different fungal and actinomycete species, i.e. *Fusarium oxysporum*, *Verticillium* sp., *Thermomonospora* sp., *Rhodococcus* sp., and *Rhizopus oryzae* have been reported to synthesize nanoparticles intra- or extracellularly (Ahmad et al. 2003, Mandal et al. 2006). The biomass of *Verticillium* turned to purple color following exposure to 10^{-4} M $HAuCl_4$ solution. This indicates the formation of AuNP (Mukherjee et al. 2001a) as confirmed by UV-visible absorption spectroscopy. The spectrum recorded from the gold-loaded biomass exhibited a SPR band at ~550 nm. However, this resonance band was not observed in the control biomass not exposed to gold ions. TEM analysis of the thin sections of the post treated cells clearly showed the intracellular formation of AuNPs. High resolution TEM image demonstrated that AuNPs with 20± 8 nm average size are formed both on the cell wall and in the cytoplasmic membrane of the fungal cell. Further, the powder diffraction pattern indicated the crystalline nature of AuNPs. The Bragg reflection was characteristic of face centered cubic (fcc) gold structure.

The exposure of *Verticillium* sp. to silver ions resulted in a similar intracellular growth of silver nanoparticles (AgNPs) (Mukherjee et al. 2001b). The intracellular formation mechanism of AuNPs and AgNPs has not been understood. However, it has been postulated that the gold and silver ions initially bind on the fungal cell surface through electrostatic interaction. The adsorbed metal ions are then reduced by enzymes present in the cell wall, leading to the formation of the metal nuclei, which subsequently grow through further reduction of metal ions. Absar and coworkers (2005) reported the extra- and intracellular biosynthesis of AuNPs by fungus *Trichothecium* sp. They observed that *Trichothecium* sp. reacted with gold ions during stationary phase and forms extracellular AuNPs of various morphologies, such as spherical, rod-like and triangular. However, under shaking conditions, the same fungal biomass forms intracellular AuNPs under shaking conditions. It was postulated that under shaking condition fungi secretes enzymes and proteins into the medium, however in shaking conditions these enzymes and proteins are not being released, thus resulting in the formation of extracellular or intracellular AuNP, respectively. Fungal templates has been used for noble-MNP synthesis (Bigall et al. 2008). Fungal cells were grown in presence of AuNP. Growth of a variety of fungi, such as *Penicillium citreonigrum, Trametes versicolor, Fusarium sp., Phanaerochaete crysosporium, Trichoderma viride, Neurospora crassa, Nematolona frowardii*, and *Bjerkandera adusta* was tested in citrate-stabilized colloidal medium containing different noble-metal nanoparticles. All these fungi grow well in the as-prepared gold, silver, platinum, and palladium nanoparticle solutions. Decoration of the fungal surface with MNPs (Fig. 5) takes place without any further functionalization. The metal nanoparticles bind on the surface of the fungi and forms nano-hybride system. Supercritical drying of these structures conserves the three-dimensional tubular shape.

The extracellular synthesis of AgNPs by a marine fungus *Penicillium fellutanum,* isolated from costal mangrove sediment, has been described by Kathiresan and co-workers (2009). They adopted a similar biosynthetic procedure to that described earlier. The synthesis of AgNPs in the reaction mixture was confirmed by the absorption peak at 430 nm. The particles had spherical shape with size ranging from 5 to 25 nm. The extracellular synthesis of stable AgNPs using the fungus *Penicillium brevicompactum* WA 2315 was demonstrated by Shaligram et al. (2009). The nanoparticles was obtained by incubating silver ions with the fungus supernatant for 72 h. TEM analysis showed that the average size of nanoparticles was 58.35 ± 17.88 nm. The FTIR spectrum of the freeze-dried powder AgNPs sample formed showed band at 3356 cm^{-1} and 2922 cm^{-1}, which were assigned to the stretching vibration of primary and secondary amines, respectively. The bands at 1622 cm^{-1} and 1527 cm^{-1} correspond to the stretching vibration of primary and secondary amines. The two bands at 1412 cm^{-1} and 1029 cm^{-1} were assigned to the C-N stretching vibrations of aromatic and aliphatic amines. This FTIR spectrum supports the involvement of proteins in the biosynthesis mechanism of AgNPs. Extracellular aromatic compounds such as naphthoquinones and anthraquinones act as electron shuttle in silver ions reduction as they have excellent redox properties.

The fungi morphology was different following binding with MNPs. The mycelia and spore in gold binding fungi (E07AS) was visible in both secondary electron and backscattering image, however, the spore and mycelia is almost invisible in the gold sensitive backscattering mode in the fungus O07AS (Fig. 5C-F). Though all the fungi tested grew in presence of metal nanoparticles, they had different affinity for different MNPs. Most of the fungi grew best in gold, followed by platinum and palladium NP solutions. Growth of fungi in AgNPs solution only occurred in the case of *B. Adusta* due to toxic effect of silver.

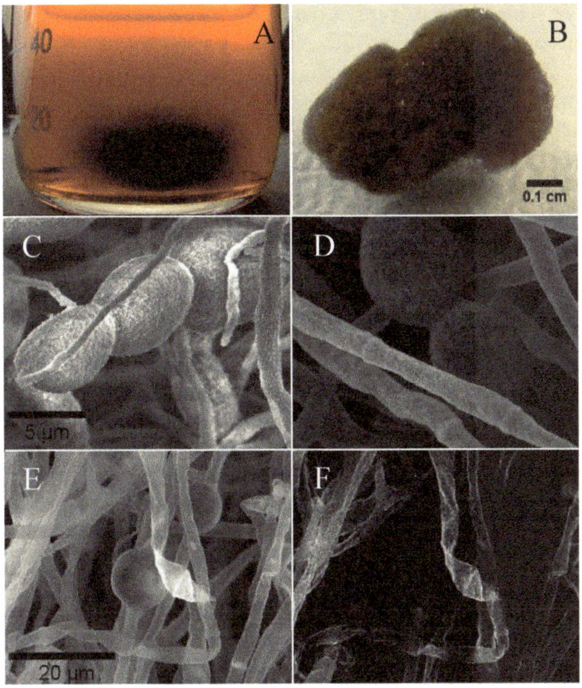

Fig. 5. Color photographs of a gold-*Penicillium citreonigrum* hybrid structure in a gold nanoparticle solution (A), and after critical point drying (B). Scanning electron micrographs (C-F) of similar regions within a hybrid structure (left: secondary electron mode, right: backscattering mode). The spores of the fungus E07AS (C-D) show significant assembly of AuNP, whereas the spores of the fungus O07AS (E-F) do not. Angew. Chem. Int. Ed. 2008, 47, 7876–7879.

It was also reported (Duran et al. 2005) that the enzyme hydrogenase is present in a filtrate broth obtained from *Fusarium oxysporum* growth. Thus, the silver nanoparticles production capacity is likely to be a result of the interaction between reductases and electron shuttle.
A single pot green chemical synthesis of AuNP by fungal strain *Rhizopus oryzae* has been reported by Das et al. (2009). *R. oryzae* synthesize AuNP on the mycelial surface, which exhibited SPR band at 540 nm. HRTEM image demonstrated formation of monodisperse gold nanoparticle with average size of 10 nm on the surface of the mycelia. FTIR study demonstrated that gold ions initially bound through electrostatic interaction with functional groups of proteins present on the surface of the mycelia, then got reduced to AuNPs. The surface bound protein molecules act both as reducing and as stabilizing agents. The extremophilic actinomycete, *Thermomonos* Sp. reduced the gold ions extracellularly, yielding ANPs (Sastry et al. 2003). Even the edible mushroom *Volvariella volvacea* can produce Au and Ag NPs through metal reducing compounds. The mushroom was boiled initially in water and then filtered. The filtrate was cooled to room temperature and used as reducing agent for AuNPs synthesis. Following reduction purple colored AuNPs was formed. The mushroom biomass also prevents NPs aggregation after their formation (Philip 2009).

2.6.3 Mechanism of nanoparticle synthesis

Synthesis of metal nanoparticles in different types of microbial species are reported, however, the exact mechanism of nanoparticle biosynthesis has not been established. The complex interaction process along with metabolic complexity of viable microorganisms complicates the analysis and identification of active species in the nucleation and growth of metal NPs. The elucidation of the biochemical pathways leading to gold biomineralization is necessary to develop a rational approach to NP biosynthesis. A number of issues need to be addressed from the nanotechnology and microbiological points of view before such biosynthetic procedures can compete with the conventional protocols. An important component of biomineralization is the protein or peptide template that controls the shape and crystal structure of biominerals as well as the assembly behavior. Thus the underlying mechanism can be applied in the synthesis of technologically important materials beyond those few existing biominerals in nature. Recent works by Xie et al. (2007) demonstrated that proteins are the principal biomolecules involved in the algal synthesis of gold NPs. Enzymatic reduction (Ahmad et al. 2003; He et al. 2007) of metal ions and subsequent growth of metal nanoparticles have postulated for fungal mediated synthetic mechanism. Ahmad et al. (2003) postulated that in *Fusarium oxysporum*, a NADH-dependent reductase is involved in AgNPs synthesis. Daniel et al. (2008) demonstrated that the FAD-dependent enzyme glutathione reductase (GR) catalyzes the NADPH-dependent reduction of $AuCl_4^-$. The synthesized AuNPs at the active site tightly bound through the catalytic cysteines group. Involvement of nitrate reductase in *B. licheniformis* in the fabrication of nanoparticles has also been reported. It has been postulated that NADH dependent nitrate reductase is an important factor in the biosynthesis of metal nanoparticles. However, the biochemical mechanism of metal ion reduction and subsequent NP formation remains unexplored.

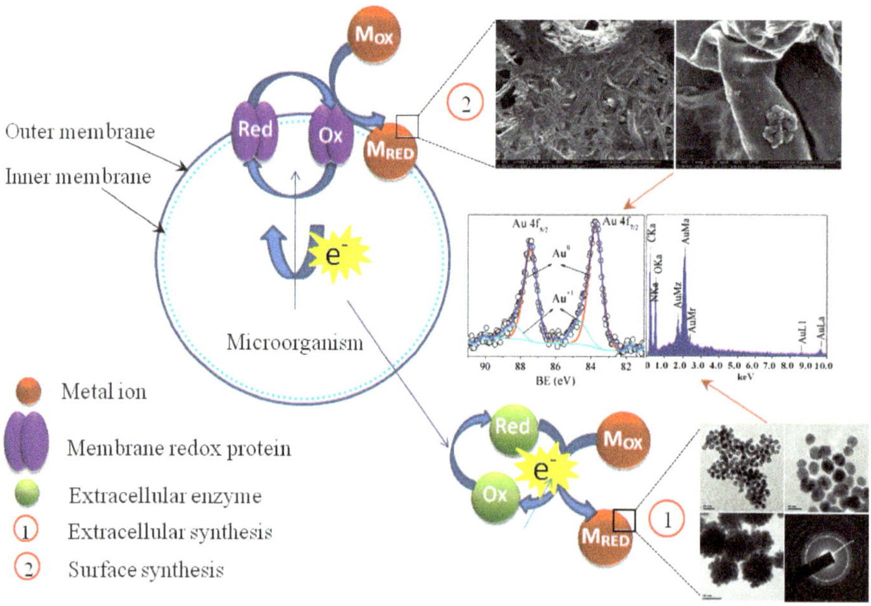

Fig. 6. Proposed mechanism of AuNP synthesis in *Rhizopus oryzae* (Rev. Environ. Sci. Biotechnol. 2010, 9, 199-204).

In recent research investigation on biosynthesis of AuNPs, we observed through TEM analysis that AuNPs are synthesized both in cytosolic portion as well as cell wall of *R. oryzae* (Unpublished data). Different surface protein(s) as well as cytosolic protein(s) have been extracted from whole mycelia and used for AuNPs synthesis. Based on protein induction and also through SDS-Gel electrophoresis experiment we hypothesized that both cytosolic and surface protein(s) of *R. oryzae* are involved in AuNPs synthesis (Fig. 6). X-ray photoelectron spectroscopy (XPS) and FTIR study demonstrated that protein(s) following reduction of gold ions bonded with the AuNPs and formed bioconjugate system. The identification of specific protein and or peptides molecules responsible for nanoparticle synthesis will broaden the possible application of biomineralization in nanoparticle process scale up.

3. Properties of metal nanoparticles

The physicochemical properties of the metal nanoparticle are completely different to that of the bulk material due to their very small size and high surface to volume ratio. While the properties of bulk materials are solely composition-dependent, as the size of the particles decreases to few nm, the the electronic structure is altered from the continuous band to discrete electronic levels Thus, the properties of the nanomaterial become size-dependent (Burda et al. 2005). In a small nanoparticles, a large fraction of the atoms will be on the surface. The atoms on the surface are chemically more active compared to the bulk atoms, because they usually have feweradjacent coordinate atoms and unsaturated sites. A striking properties deriving from the increasing surface energy of nanomaterials is the decrease of melting point. Further, the overlapping of different grain sizes affects the physical strength of the material. Also, when the crystallites of a material are reduced to the nanometer scale, there is an increase in the role of interfacial defects: grain boundaries, triple junctions, and elastically distorted layers. The long established Hall-Petch model (2005), that shows the inverse relationship between grain size and material yield strength, has been proven to hold with nanoparticles as well. A study of nanocrystalline Fe powder revealed that when grain size was decreased from 33nm to 8nm, hardness increased and fracture stress and elongation to failure decreased.

Hall Petch Model:

$$\sigma_y = \sigma_o + \frac{K}{\sqrt{d}}$$

(yield strength determined by a constant K and the mean grain size, d)

These results can be connected to the reduction of defects in the material such as micropores and other flaws (Andrievski and Glezer 2000).

Similarly, the yellow colored noble metal gold is known as a shiny, has a face centred cubic structure, is non-magnetic and melts at 1336 K. However, 10 nm gold particles is quite different, absorb green light and thus appear red. The melting temperature decreases dramatically as the size decreases. It is interesting to note that scientists have been using nanoparticles in tires for a very long time but without knowing the reason, besides that it turned the rubber black. For years, engineers have added carbon black nanoparticles in the rubber of tires to improve the quality. The addition of carbon black increases the strength and tensile properties, tear and abrasion resistance, and increased hardness because of the integration of carbon grains (Maiti et al. 2005). It has been found that with increased amounts of carbon black in the rubber compounds, the absolute strength initially increases because of reinforcement from the carbon grains but then decreases due to the dilution effect. The improved reinforcement results by the carbon black particles can be explained by exfoliation, intercalation, dispersion, and occlusion which are essentially the interactions of the grains with the material.

3.1 Optical properties

The Localized Surface Plasmon Resonance (LSPR) is referred as free to oscillation of electrons in the conduction band of metals upon excitation with incident radiation (El-Sayed 2001, Murphy et al. 2005 and reference therein). However, for nanoscale metals, the oscillation distance is restricted by the size of nanoparticle. For gold and silver nanoparticles, LSPR corresponds to photon energies in the visible wavelength regime, giving rise to significant interest in their optical properties. These optical characteristics include strong plasmon absorption, resonant Rayleigh scattering, and localized electromagnetic fields at the nanoparticle surface. The size, shape, and dielectric constant of the surrounding medium of nanoparticle strongly influence the Plasmon absorption. For instance, spherical gold and silver nanoparticles have single plasmon absorption bands at ~540 and ~400 nm, respectively. However, anisotropic gold and silver nanorods exhibit two principle plasmon absorption bands (Kelly et al. 2003); one at shorter wavelength corresponding to absorption and scattering of light along the short axis of the nanorod (transverse plasmon band), and the other band at longer wavelength corresponding to absorption and scattering of light along the long axis of the nanorod (longitudinal plasmon band). The longitudinal plasmon absorption bands are tunable with nanorod aspect ratio from the visible to the near-IR.

The excellent light-scattering properties of gold nanorods are sensitive to their orientation and Murphy and Orendorff (2005) used this property to determine the nanorod orientation in polymer films. The plasmonic AuNP have been used for sensing, imaging and Surface Enhanced Raman Scattering (SERS) detection. The light scattering properties of nanoparticle has been used in dark field microscopy to detect the uptake nanoparticles by cells (Maier 2007). Sometimes the combined dark field microscopy with the shifts associated with local dielectric constant changes used for analyte detection.

Inelastic visible light scattering from metal nanoparticles is also a useful means to gain chemical information about the nanoparticle's environment. SERS is a powerful analytical tool for determining chemical information for molecules on metallic substrates on the 10-200 nm size scale. Though the Raman vibrations of molecules are very weak, in presence nanoparticle the molecular Raman vibrations are enhanced by orders of magnitude. As SERS substrates colloidal nanoparticles are of special interest not only because of their strong light scatterering property but also their tunable optical properties, which depend on nanoparticle size, shape, and aggregation state. The largest surface enhancements, that seemingly lead to single molecule SERS, are observed for molecules adsorbed in the fractal space at junctions between nanoparticles, referred to as 'SERS hot spots'. This is a result of localized surface plasmon (LSP) coupling between nanoparticles and the enhanced electromagnetic field intensity localized at nanoparticle junctions. Anisotropic metallic nanoparticles have all of the characteristics that make them excellent candidates as SERS substrates. Plasmon absorption bands can be tuned with nanocrystal aspect ratio to be in resonance with common visible laser sources.

3.2 Properties of nanoparticle–biomolecule hybrid systems

The development of hybrid nanomaterials that incorporate the highly selective catalytic and recognition properties of biomaterials, such as enzymes, with the unique properties of nanoparticles, result in flexible building blocks for biosensors and drug delivery. In fact, these construct are the forefront of the developing bionanotechnology research. However, the understanding of these complex assemblies is still incomplete, and the modelling still inadequate to describe their behaviour (Katz and Willner 2004 and reference therein). The functionalization of NPs with biomolecules results in changes in the properties of the NPs and their interactions with the environment. Upon adsorption of vitamin C on TiO_2 NPs, the optical properties of the particles were red-shifted by 1.6 eV as a result of charge transfer that originates from the specific binding of the electron-donating modifier to corner defects on the surface of the nanoparticles. The solubility of NPs in water can be greatly improved by the functionalization of their surfaces with highly hydrophilic biomolecules.

Furthermore, tiopronin or coenzyme A modified Au NPs demonstrate excellent solubility in water (Templeton 1999). Subsequent association with metal nanoparticles the intrinsic properties of biomolecules can be changed. Recently it has been shown through Surface-Enhanced Resonance Raman (SERR) spectroscopic studies that vibrations of the biologically active prosthetic heme groups of myoglobin or hemoglobin are selectively enhanced relative to the vibrational modes characteristic of the protein backbone following their adsorption on metal NPs (Bizzarri and Cannistraro 2002).

4. Applications of metal nanoparticles

In the ever expanding field of nanomaterial research, metal nanoparticle received particular attention due to their wide application in catalysis, electronics, sensing, photonics, environmental cleanup, imaging, and drug delivery (Guo et al. 2005, Daniel and Astruc 2004, Huang et al. 2007).

4.1 Catalysis

The application of nanoparticles as catalysts is a rapidly growing field in nanoscience and technology. The properties of noble metal nanoparticles make them ideal materials for nanocatalysis, where reaction yield and selectivity are dependent on the nature of the catalyst surface. Compared to bulk materials, nanoparticles have high surface-area-to volume ratio and thus found to exhibit higher turnover frequencies. The catalytic activity of Au, Ag and Pt in the decomposition of H_2O_2 to oxygen is well known. Additionally they also catalyze luminal-H_2O_2 systems. It was observed that the chemiluminoscence emission from the luminal-H_2O_2 system was greatly enhanced by addition of Ag colloid (Guo, et al. 2008). The catalytic application of Ag in oxidation of ethylene to ethylene oxide and methanol to formaldehyde is also the most popular.

More interestingly, the nanoparticles shows shaped controlled catalytic activity. The shape-controlled catalytic properties have recently been observed in benzene hydrogenation by Pt catalyst (Brown 1997). Among different Pt nanocrystals, (cubes, tetrahedra, and spheres) tetrahedral nanocrystals, completely bound by {111} crystal facets, exhibited highest catalytic activity whereas cubic nanocrystals exhibit the lowest activity (Fig. 7). However, it is not clearly understood whether this observation is truly a shape-dependent effect, as in solution the surface reconstruction and shape changes of NPs is evident. To study the dynamics of adsorbates on solution phase nanocrystalline structures is required for direct surface measurements analogous to those commonly used for single-crystalline studies.

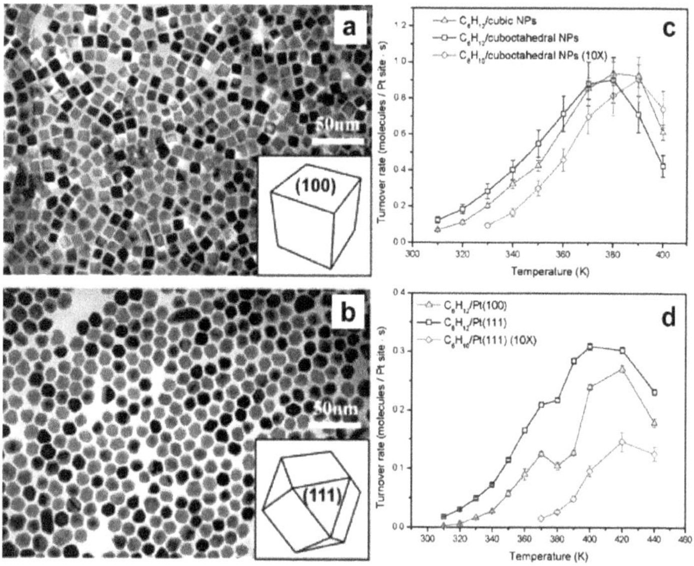

Fig. 7. Shape-controlled Pt nanocrystals show catalytic selectivity for benzene hydrogenation. The turnover rate for films of surfactant-stabilized Pt A) cubes and B) cuboactahedra, shown in these TEM images, were measured. C) Cubic nanocrystals are selective for cyclohexane, while cuboctahedra exhibit activity for both cyclohexane and cyclohexene. These results are consistent with the selectivity exhibited by single-crystalline Pt surfaces, shown in (D). Nat. Biotech. 1997, 15, 269.

Among other metal catalysts, Au has potentially more advantages is due to lower cost and greater stability. Au is being substantially cheaper and considerable more plentiful then Pt. AuNP less than 5 nm supported on base metal oxide or carbon demonstrated very high activity (Hvolbeck et al., 2007). High activity of AuNP for the oxidation many compounds, particularly CO and trimethylamine are also observed. AuNP based gas sensors have recently been developed for detecting a number of gases, including CO and NOx (Thompson 2007). Very recently Zeng et al. (2010) also demonstrated shape controlled catalytic activity of AuNPs for well known *p*-nitrophenol reduction in presence of sodium borohydride. Gold nanoboxes among other nanostructure (nanocages, and solid nanoparticles) have highest catalytic activity. The good intrinsic electrical connection across the entire surface of an Au nanocage makes it a much better catalyst than small Au solid nanoparticles for the redox reaction.

4.2 Biological application of metal nanoparticles

The application of metal nanoparticles in biological science showed very rapid progess the past decades is in the area of labelling, delivery, heating, and sensing. The SPER optical properties of colloidal AuNPs directed towards recent biomedical applications with an emphasis on cancer diagnostics and therapeutics.

4.2.1 Labelling

For labelling, electron absorbing properties of the metal nanoparticles are exploited to generate contrast. The AuNPs strongly absorb electrons, thus make them suitable as a as a contrasting agent in TEM. Besides, nanoparticles have the same size domain as proteins that make nanomaterials suitable for bio tagging or labelling (Sperling et. 2008). Due to their small size and functionalising properties, i.e. with antibodies (immunostaining), AuNPs provide extremely high spatial resolution and applied in many labelling applications (Salata 2004). Additionally optical detection techniques are wide spread in biological research because of change of their optical or fluoresce properties. The formations of nano-biomaterials construct are schematically presented below (Fig. 8).

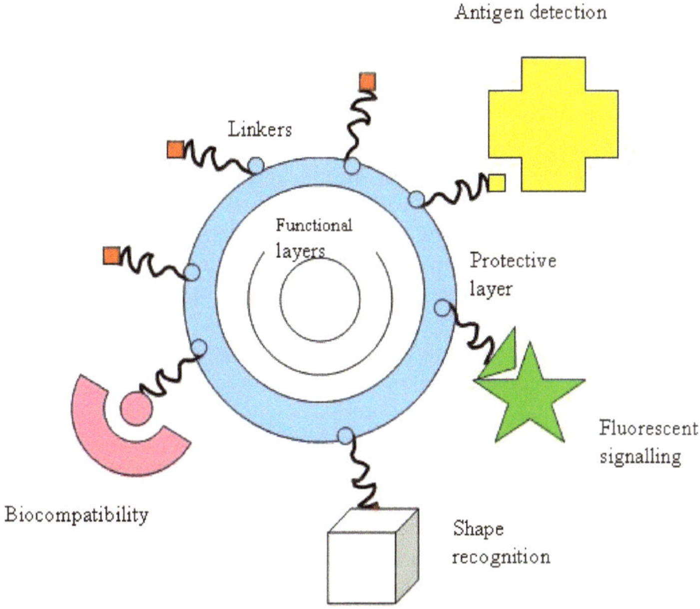

Fig. 8. Schematic presentation of application of nanoparticles in medical or biological problems. *J. Nanobiotechnol.* 2004, 2, 3.

Similarly, the particles' optical properties- strong absorption, scattering and especially plasmon resonance- make them of value for a large variety of light-based techniques including combined schemes such as photothermal or photo-acoustic imaging. In addition, AuNP can be radioactively-labelled by neutron activation, which allows for very sensitive detection, and used as an x-ray contrast agent.

4.2.2 Sensors

Metal nanoparticles can also be used as sensors. The optical and electronic sensing of biomaterials on surfaces is a common practice in analytical biochemistry. Thus, the immobilization of biomolecule-NP conjugates on surfaces provides a general route for the development of optical or electronic biosensors. Metal NPs such as Au or Ag NPs exhibit plasmon absorbance bands in the visible spectral region that are controlled by the size of the respective particles. Their optical properties can change upon binding to certain molecules, allowing the detection and quantification of analytes (Huo 2007). The absorption spectra of AuNP change drastically when several particles come close to each other. Numerous studies on the labelling of bioassays and the staining of biological tissues by metal particles as a means to image and visualize biological processes have been reported. The spectral shifts which originate from adjacent or aggregated metal nanoparticles, such as Au NPs, are of increasing interest in the development of optical biosensors based on biomolecule–NP hybrid systems. As an example, NPs that were functionalized with two kinds of nucleic acid, which were complementary to two segments of an analyzed DNA, were hybridized with the analyzed DNA. This led to the aggregation of the NPs and to the detection of a red shifted interparticle plasmon absorbance of the nanoparticle aggregate.

4.2.3 Drug delivery

Because of nontoxicity and nonimmunogenicity AuNPs is ideal for preparation of drug-delivery scaffold. Functionalization property of AuNP also makes it an excellent potential vehicle for the drug delivery. Functionalized AuNP represent highly attractive and promising candidates in the applications of drug delivery. Hamad-Schifferli (2008) groups recently developed druge delivery system AuNPs and infrared light. This delivery system released multiple drugs in a controlled fashion. They demonstrated that nanoparticles of different shapes respond to different infrared wavelengths. For example, nanobones and nanocapsules melt at light wavelengths of 1,100 and 800 nm, respectively. Thus excitation at one wavelength could selectively melt one type of Au nanorods and selectively release one type of DNA strand.

Brown et al. (2010) also reported AuNPs for the improved anticancer drug delivery of the active component of oxaliplatin. Naked AuNPs were functionalized with a thiolated poly(ethylene glycol) (PEG) monolayer capped with a carboxylate group. [Pt(1*R*,2*R*-diaminocyclohexane)$(H_2O)_2]_2NO_3$ was added to the PEG surface and yielding a supramolecular complex with drug molecules. The cytotoxicity, drug uptake, and localization in the A549 lung epithelial cancer cell line and the colon cancer cell lines HCT116, HCT15, HT29, and RKO were examined for platinum-tethered nanoparticles. The platinum-tethered nanoparticles showed significant improvement in cytotoxicity than oxaliplatin alone in all of the cell lines and an unusual ability to penetrate the nucleus in the lung cancer cells.

4.2.4 Cancer therapy

Nanotechnology is one of the most popular research areas, especially with regard to biomedical applications. Nanoparticles have very good opportunity in the form of targeted drug therapies (Ghosh at al. 2008). Nanoparticles also carry the potential for targeted and time-release drugs. A potent dose of drugs could be delivered to a specific area but engineered to release over a planned period to ensure maximum effectiveness and the patient's safety. The strong light absorbing properties of AuNPs makes it suitable as heat-mediating objects; the absorbed light energy is dissipated into the surroundings of the particles', generating an elevated temperature in their vicinity. This effect can be used to open polymer microcapsules, for example, for drug delivery purposes and even destroy the cancerous cells. The nanoparticles are functionalised with antibody specific to the cancerous cells. The functionalized nanoparticles specifically bind with the targeting cells, which was then killed by hyperthermal therapy through heating the particle-loaded tissue. However, for such in vivo applications, the potential cytotoxicity of the nanoparticles might become an issue and should be investigated with care. Due to biocompaticbility, hyperthermal activity

activity AuNPs find wide application now a days in killing of malignant cancerous cells (Dickerson et al. 2008). Recently, Melancon et al. (2008) demonstrated destruction of cancerous cell by photothermal effect of AuNPs (Fig. 9). The hollow gold nanoshells (HAuNS; average diameter, ~30 nm) was covalently attached to monoclonal antibody directed to the epidermal growth factor receptor (EGFR). The resulting anti-EGFR-HAuNS exhibited excellent colloidal stability and efficient photothermal effect in the near-infrared region. Anti-EGFR-HAuNS then bound in EGFR-positive A431 tumor cells. Irradiation of A431 cells treated with anti-EGFR-HAuNS with near-infrared laser resulted in selective destruction of these cells.

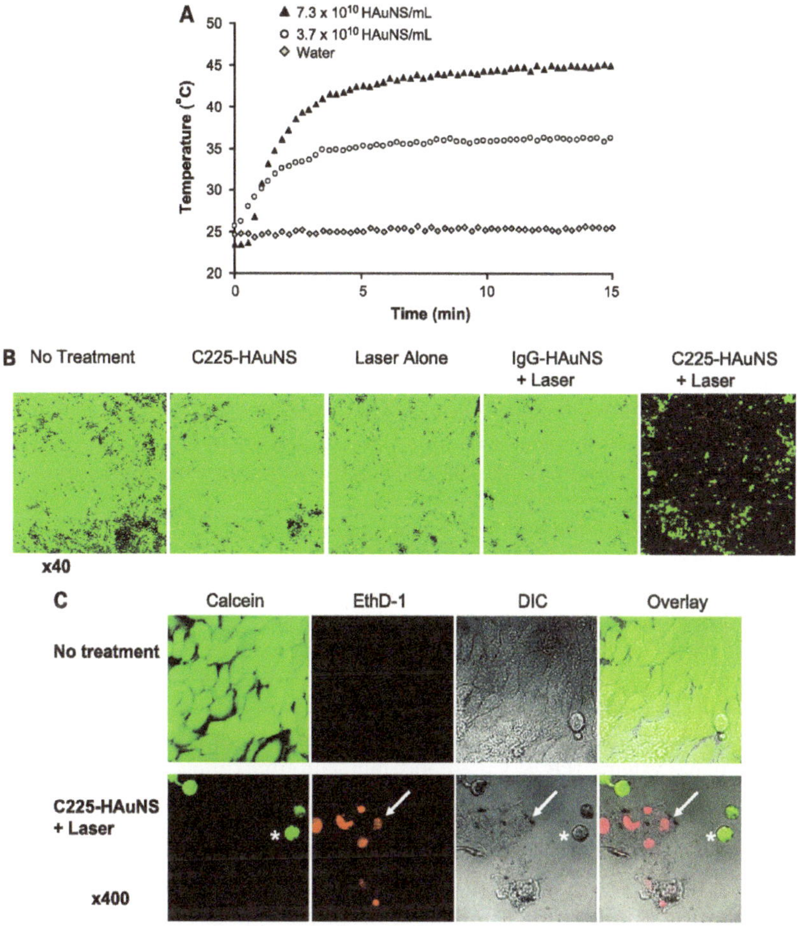

Fig. 9. **A,** heating of aqueous C225-HAuNS solutions exposed to NIR light centered at 808 nm at 8 W/cm². **B,** cell viability after various treatments. Cells retained normal morphology with no apparent death observed (stained green with calcein CM) when cells were not treated or treated with C225-HAuNS alone, NIR laser alone, or nontargeted IgG-HAuNS plus NIR laser. In contrast, most cells were dead after treatment with C225-HAuNS plus NIR laser. Dead cells were labeled red with EthD-1. *Mol. Cancer. Ther.* 2008, 7, 1730-1739.

AuNPs has also been applied to amplify the biorecognition of the anticancer drug (Shen et al. 2008). Dacarbazine [5-(3, 3-dimethy-1-triazenyl) imidazole-4-carboxamide; DTIC] is a commonly used anticancer drug. AuNPs were stabilized by PPh3 with negative charge. The oxidized DTIC is positive charged. Thus, DTIC could be easily assembled onto the surface of AuNPs. The specific interactions between anticancer drug DTIC and DNA or DNA bases were facilitated by AuNPs.

4.3 Environmental cleanup

Although MNPs are increasingly being employed in different emergent areas, their use in environmental biotechnology is still limited. One of the key environmental challenges is the contamination of water bodies by different chemicals due to diverse anthropogenic and industrial activities. The most interesting application of MNPS is purification of drinking water contaminated with heavy metals and pesticides. Current limitations in removal of heavy metals have been tried to overcome through adsorption process on MNPs due to alloy formation. Au and mercury exist in several phases such as Au_3Hg, AuHg, and $AuHg_3$. The interaction of AgNPs with Hg^{+2} ions was investigated because enhanced ability of Ag to form alloy in different phases. It was found that the surface plasmon of AgNPs blue shifted along with a decrease in the intensity, immediately after the addition of Hg^{+2} ions (Bootharaju and Pradeep 2010). Partial oxidation of AgNPs to silver ions is responsible for the decrease in intensity. The shift is attributed to the incorporation of mercury into the AgNPs. The mercury nanoparticle solutions exhibited plasmon absorption band below 300 nm. The Hg–Ag alloy nanoparticles, prepared by simultaneous reduction with sodium borohydride, exhibited a plasmon in the region of 300–400 nm. The potential of AgNPs to reduce a number of heavy metals can also be looked at as a method to prepare alloy nanoparticles; e.g., Ag–Hg bimetallic nanoparticles. A demonstration of the galvanic etching for the preparation of alloy nanoparticles is illustrated through the preparation of Pd–Ag and Pt–Ag nanoparticles.

Recently colorimetric detections of heavy metals like arsenic, mercury, lead, etc., have also been tried by using MNPs. One of the important properties exhibited by functionalized MNPs surfaces is the detection of heavy metals. In one such method, heavy metal specific biomolecule functionalized AuNP can be utilized. An example of this approach is the interaction of metal ions with nucleotides: Hg^{+2} promoted formation of thymine–thymine base pairs (Ono and Togashi 2004). In a similar approach, ligands functionalized MNPs have been used for specific detection of metal ions. This ligand-metal ion complexation leads to observable optical changes at concentrations in the ppm level. Examples of such ligands are gallic acid (Pb^{+2}), cysteine (Hg^{+2}, Cu^{+2}), and mercaptoundecanoic acid (Pb^{+2}, Cd^{+2}, Hg^{+2}). Carboxylate group modified surface of AuNP can be induced to aggregate in the presence of Hg^{+2} and pyridinedicarboxylic acid, which is manifested in the form of colorimetric response, fluorescence quenching and enhancement of hyper-Rayleigh scattering intensity (Huang et al. 2007, Darbha et al. 2008).

The removal of pesticides by MNPs is a new addition to this field. Among other contaminants, presence of pesticide residue in potable water above permissible limit is of great concern to public health. This happens due to indiscriminate use of pesticide, specially belonging to organophosphorus groups, in agricultural practices. It is essential to reduce the concentration of pesticide in potable water but difficult to achieve by conventional chemical methods due to wide variation of their chemical structures. To meet these environmental challenges, very recently researchers are focusing on the development of methods based on nanotechnology. Very recently, Das et al (2009). demonstrated adsorption of different organophosphorous pesticides AuNPs surface. AuNPs was synthesized on the surface of the *R. oryzae* mycelia in a single set. The AuNPs adsorbed on mycelia were then used for adsorption of different organophosphorous pesticides. Following adsorption of these pesticides the surface morphology conspicuously changed compared the unadsorbed nanomaterial (Fig. 10) as depicted from atomic force microscopic images.

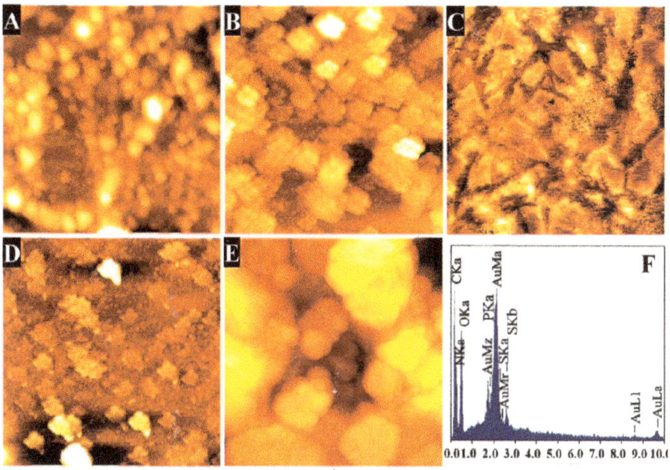

Fig. 10. AFM images of (A) the pristine nanogold bioconjugate (NGBC) and the (B) malathione, (C) parathione, (D) chlorpyrifos, and (E) dimethoate-adsorbed NGBC material. (F) EDXA spectrum of NGBC material after adsorption of organophosphorous pesticide. Langmuir 2009, 25(14), 8192–8199.

5. Conclusions

In this chapter, we provided an account of the chemical and biological methods for MNPs synthesis, as well as a short review of MNPs properties and of their most promising applications in biomedical devices, environmental processes, and chemical catalysis. From the considerations as outlined in this review, it emerges as biosynthesis represents a promising route for MNPs production. In fact, biosynthesis result in low energy use and environmental impact, with respect to conventional chemical synthesis methods. Further, the high specificity of biomolecules involved in the biosynthesis process may enable an efficient control of MNPs size and shape, whose tight control is critical to optimize MNP-based devices and applications. Biosynthetic MNPs have been observed in numerous fungal and bacterial species, and the molecular machinery needed for MNPs biosynthesis overlap significantly with that already developed and optimized for bioreduction and detoxification of soluble metals. However, most published studies deal with MNPs biosynthesis in viable microorganism, and the complexity of the system makes it difficult to identify the exact nature of the multiple biological agents responsible for the biosynthetic process. Further research with cell-free extract and biological fractions may lead to this identification and to a thorough understanding of the complex regulation underlying the expression of metal-reducing agents.

Future studies should also investigate the role of biological capping agents involved in MNPs biosynthesis on their cytotoxicity, catalytic and optoelectronics properties, ultimately to determine if biosynthetic MNPs are a viable commercial alternative to those synthesized through conventional chemical process.

6. References

Absar, A.; Satyajyoti, S.; Khan, M.I.; Rajiv, K.; Sastry, M. (2005). Extra-/intracellular biosynthesis of gold nanoparticles by an alkalotolerant fungus, *Trichothecium* sp., *J. Biomedical Nanotechnnol.* 1, 47-53

Ahmad, A.; Mukherjee, P.; Senapati, S.; Mandal, D.; Khan, M.I.; Kumar, R.; Sastry, M. (2003) Extracellular biosynthesis of silver nanoparticles using the fungus *Fusarium oxysporum*. *Colloids Surf B Biointerfaces* 28, 313–318

Andrievski, R.A.; Glezer. A.M. (2000). Size effects in Properties of Nanomaterials. Elsevier Science Ltd. 44, 1621-1623.

Anstas, P.T.; Warner, J.C. (1998). Green Chemistry: Theory and Practice, Oxford University Press, New York.

Beveridge, T.J.; Murray, R.G.E. (1980). Site of metal deposition in the cell wall of Bacillus subtilis. *J. Bacteriol.* 141, 876–887

Bigall, N.C.; Reitzig, M.; Naumann, W.; Simon, P.; van Pée, K.H.; Eychmüller. A. (2008). Fungal Templates for Noble-Metal Nanoparticles and Their Application in Catalysis. *Angew. Chem. Int. Ed.*, 47 7876–7879

Birringer, R.; Gleiter, H.; Klein, H.P.; Marquardt P. (1984). Nanocrystalline materials, an approach to a novel solid structure with gas like disorder. *Phys. Lett. A* 102, 365

Bizzarri, A.R.; Cannistraro, S. (2002). *Appl. Spectrosc.* 56, 1531–1537

Bootharaju, M.S.; Pradeep, T. (2010). Uptake of toxic metal ions from water by naked and monolayer protected silver nanoparticles: An X-ray photoelectron spectroscopic investigation. J. Phys. Chem. C 114, 8328–8336

Brown, S. (1997). *Nat. Biotech.* 15, 269

Brown, S.D.; Nativo, P.; Smith, J.-A.; Stirling, D.; Edwards, P.R.; Venugopal, B.; Flint, D.J.; Plumb, J.A.; Graham, D.; Wheate. N.J. (2010). Gold Nanoparticles for the Improved Anticancer Drug Delivery of the Active Component of Oxaliplatin. *J. Am. Chem. Soc. 132*, 4678–4684

Brust. M.; Walker, M.; Bethell, D.; Schiffrin ,D. J.; Whyman, R. (1994). Synthesis of Thiol-derivatised Gold Nanoparticles in a Two-phase Liquid-Liquid System. *Chem. Commun.* 7, 801.

Burda, C.; Chen, X.; Narayanan, R.; El-Sayed M.A. (2005). Chemistry and Properties of Nanocrystals of Different Shapes. *Chemical Rev.* 105, 1025-1102

Cha, J.N.; Shimizu, K.; Zhou, Y.; Christiansen, S.C.; Chmelka, B.F.; Stucky, G.D.; Morse, D.E. (1999). Silicatein filaments and subunits from a marine sponge direct the polymerization of silica and silicones *in vitro. Proc. Natl. Acad. Sci USA. 96, 361–365*

Chaki, N.K.; Sundrik, S.G.; Sonawane, H.R.; Vijayamohanan, K. (2002). *J. Chem. Soc. Chem. Commun.* 76.

Dahl, J.A.; Maddux, B.L.S.; Hutchison. J.E. (2007). Toward Greener Nanosynthesis. *Chem. Rev.* 107, 2228-2269

Daniel, M.-C.; Astruc, D. (2004). Gold Nanoparticles: Assembly, Supramolecular Chemistry, Quantum-Size-Related Properties, and Applications Toward Biology, Catalysis, and Nanotechnology. *Chem. Rev. 104*, 293–346

Darbha, G.K.; Singh, A.K.; Rai, U.S.; Yu, E.; Yu, H.; Ray, P.C. (2008). *J. Am. Chem. Soc.*130, 8038

Das, S.K.; Das, A.R.; Guha. A.K. (2009). Gold Nanoparticles: Microbial Synthesis and Application in Water Hygiene Management. *Langmuir* 25, 8192–8199

Das, S.K.; Marsili, E. (2010). A green chemical approach for the synthesis of gold nanoparticles: characterization and mechanistic aspect. *Rev. Environ. Sci. Biotechnol.* 9, 199-204

Dickerson, B.E.; Dreaden, C.E.; Huang, X., El—Sayed, H. I.; Chu, H.; Pushpanketh, S.; McDonald, H.I., El-Sayed, A.M. (2008). Gold nanorod assisted near-infrared plasmonic photothermal therapy (PPPT) of aquamous cell carcinoma in mice. *Cancer Lett.* 269, 57-66

Duran, N.; Marcato, D.P.; Alves, L.O.; De Souza, G.; Esposito, E. (2005). Mechanical aspect of biosynthesis of silver nanoparticles by several *Fusarium oxysporum* strains. *J. Nanobiotechnol.* 3, 8-15

El-Sayed, M.A. (2001). Some Interesting Properties of Metals Confined in Time and Nanometer Space of Different Shapes. *Acc. Chem. Res., 34*, 257–264

Faivre, D.; Schüler, D. (2008). Magnetotactic Bacteria and Magnetosomes. *Chem. Rev. 108*, 4875–4898

Faraday, M. (1857). Experimental relations of gold (and other metals) to light. *Phil. Trans. Roy. Soc. London* 147, 145–181

Frens, G. (1972). Particle size and sol stability in metal colloids. *Colloid Polym. Sci.* 250, 736-741

Gacoin, T.; Malier, L.; Boilot, J.-P. (1997). Sol–gel transition in CdS colloids *J. Mater. Chem.* 7, 859-860

Ghosh, P.; Han, G.; De, M.; Kim, K.C.; Rotello, M.V. (2008). Gold nanoparticles in delivery applications. *Advanced Drag Delivery Rev.* 60, 1307-1315

Guo, J-Z.; Cui, H.; Zhou, W.; Wang, W. (2008). Ag nanoparticle-catalyzed chemiluminescent reaction between luminal and hydrogen peroxide. *J. Photochem. Photobiol., A Chem.* 193, 89-96

Guo, R.; Song, Y.; Wang, G.; Murray, R.W. (2005). Does Core Size Matter in The Kinetics of Ligand Exchanges of Monolayer-Protected Au Clusters? *J. Am. Chem. Soc. 127*, 2752-2757

"Hall Petch Relationship, The." 25 Apr 2005. http://www.matsci.ucdavis.edu/MatSciLT/EMS-174L/Files/HallPetch.pdf

He, S.; Guo, Z.; Zhang, Y.; Zhang, S.; Wang, J.; Gu, N.; (2007). Biosynthesis of gold nanoparticles using the bacteria *Rhodopseudomonas capsulate. Materials Latter* 61, 3984-3987

Henglein, A. (1998). *Chem. Mater.* 10, 444

Huang, C.C.; Yang, Z.; Lee, K. H.; Chang, H. T. (2007). Synthesis of Highly Fluorescent Gold Nanoparticles for Sensing Mercury(II) *Angew. Chem., Int. Ed. 46*, 6824–6828

Huang, C.-C.; Yang, Z.; Lee, K.-H.; Chang, H.-T. (2007). *Angew. Chem. Int. Ed.* 46, 6824

Huo, Q. (2007). A perspective on bioconjugated nanoparticles and quantum dots. *Colloids Surfaces B: Biointerfaces* 59, 1-10.

Husseiny, I.M.; El-Aziz, A.M.; Badr, Y.; Mahmoud, A.M. (2007). Biosynthesis of gold nanoparticles using *Pseudomonas aeruginosa, Spectrochimica Acta Part A* 67, 1003-1006

Hvolbek, B.; Janssens, W.V.T.; Clausen, S.B.; Falsig, H.; Christensen, H.C.; Norskov, K.J. (2007). Catalytic activity of Au nanoparticles. *Nano Today* 2, 14-18

Justus, B.L.; Tonucci, R.J.; Berry, A.D. (1992). Discovery and development of antibodies. *Appl. Phys. Lett.* 61, 3151.

Kalathil, S.; Lee, J.; Cho, M.H. (2011). Electrochemically active biofilm-mediated synthesis of silver nanoparticles in water. *Green Chem. DOI: 10.1039/c1gc15309a*

Kalimuthu, K.; Babu, S.R.; Venkataraman, D.; Bilal, M.; Gurunathan, S. (2008). Biosynthesis of silver nanoparticles by *Bacillus licheniformis, Colloids Surfaces B: Biointerfaces* 65, 150-153

Kalishwaralal, K.; Deepak, V.; Ram Kumar Pandian, S.; Gurunathan, S. (2009). Biosynthesis of gold nanocubes from *Bacillus lichemiformis. Bioresour. Technol.* 100, 5356-5358

Kalishwaralal, K.; Deepak, V.; Ramkumarpndian, S.; Nellaiah, H.; Sangiliyandi, G. (2008). Extracellular biosynthesis of silver nanoparticles by the culture supernatant of *Bacillus licheniformis. Materials Letters* 62, 4411-4413

Kang, Il-S.; Kang, M.-H.; Lee, E.; Seo H.-S.; Ahn. C.W. (2011). Facile, hetero-sized nanocluster array fabrication for investigating the nanostructure-dependence of nonvolatile memory characteristics. *Nanotechnol.* 22, 254018 doi: 10.1088/0957-4484/22/25/254018

Kathiresan, K.; Manivannan, S.; Nabeel, A.M.; Dhivya, B. (2009). Studies on silver nanoparticles synthesized by a marine fungus *Penicillum fellutanum* isolated from coastal mangrove sediment, *Colloids Surfaces B: Biointerfaces* 71, 133-137

Katz, E.; Willner, I. (2004). Integrated Nanoparticle–Biomolecule Hybrid Systems: Synthesis, Properties, and Applications. *Angew. Chem. Int. Ed.* 43, 6042–6108

Kelly, K.L.; Coronado, E.; Zhao, L.L.; Schatz, G.C. (2003). *J. Phys. Chem. B 107*, 668-677.

Klaus, T.; Joerger, R.; Olsson, E.; Granqvist, C.-G. (1999). Silverbased crystalline nanoparticles, microbially fabricated. *Proc. Natl. Acad. Sci. USA* 96, 13611–13614

Konishi, Y.; Nomura, T.; Tsukiyama, T.; Saitoh, N. (2004). Microbial preparation of gold nanoparticles by anaerobic bacterium. *Trans. Mater. Res. Soc. Jpn.* 29, 2341–2343

Law, N.; Ansari, S.; Livens, F.R.; Renshaw, J.C.; Lloyd. J.R. (2008). Formation of Nanoscale Elemental Silver Particles via Enzymatic Reduction by *Geobacter sulfurreducens. Appl. Environ. Microbiol. 74, 7090–7093*

Lowenstam, H.A. (1981). Minerals formed by organisms. *Nature* 211, 1126–1131

Mafuné, F.; Kohno, J.; Takeda, Y.; Kondow. T. (2001). Formation of Gold Nanoparticles by Laser Ablation in Aqueous Solution of Surfactant. *J. Phys. Chem. B 105* 22, 5114–5120

Maier, S.A. (2007). Plasmonics: Fundamentals and Applications. Springer Science+Business Media LLC.

Maiti, M.; et al. (2005). Effect of Carbon Black on Properties of Rubber Nanocomposites. Wiley InterScience.

Mandal, D.; Bolander, M.E.; Mukhopadhyay, D.; Sarkar, G.; Mukherjee, P. (2006). The use of microorganisms for the formation of metal nanoparticles and their application. *Appl. Microbiol. Biotechnol.* 69, 485–492

Mann, S. (1993) Molecular tectonics in biomineralization and biomimetic materials chemistry. *Nature* 365, 499–505

Melancon, M.P.; Lu, W.; Yang, Z.; Zhang, R.; Cheng, Z.; Elliot, A.M.; Stafford, J.; Olson, T.; Zhang, J.Z.; Li, C. (2008). In vitro and in vivo targeting of hollow gold nanoshells directed at epidermal growth factor receptor for photothermal ablation therapy. *Mol. Cancer. Ther.* 7, 1730-1739

Mergeay, M.; et al. (2003) *Ralstonia metallidurans*, a bacterium specifically adapted to toxic metals: Towards a catalogue of metal-responsive genes. *FEMS Microbiol. Rev.* 27, 385–410

Mie, G. (1908). *Ann. Phys.* 25, 377–445

Mukherjee, P.; Ahmad, A.; Mandal, D.; Senapati, S.; Sainkar, S.R.; Khan, M.I.; Ramani, R.; Parischa, R.; Ajayakumar, P.V.; Alam, M.; Sastry, M.; Kumar, R. (2001a). Bioreduction of AuCl4− ions by the fungus, *Verticillium sp.* and surface trapping of the gold nanoparticles formed. *Angew. Chem. Int. Ed.* 40, 3585–3588

Mukherjee, P.; Ahmad, A.; Mandal, D.; Senapati, S.; Sainkar, S.R.; Khan, M.I.; Parischa, R.; Ajayakumar, P.V.; Alam, M.; Kumar, R.; Sastry, M. (2001b). Fungus mediated synthesis of silver nanoparticles and their immobilization in the mycelia matrix: a novel biological approach to nanoparticle synthesis. *Nano Lett.* 1, 515–519

Murphy, C.J.; Orendorff. C.J. (2005). Alignment of gold nanorods in polymer composites and on polymer surfaces. *Adv. Mater.* 17, 2173-2177

Murphy, C.J.; Sau, T.K.; Gole,A.M.; Orendorff, C.J.; Gao, J.; Gou, L.; Hunyadi, S.E.; Li, T. (2005). Anisotropic Metal Nanoparticles: Synthesis, Assembly, and Optical Applications. *J. Phys. Chem. B 109,* 13857-13870

Olshavsky, M.A.; Allcock, H.R. (1997). Small scale system for *in-vivo* drug delivery. *Chem. Mater,* 9, 1367.

Ono, A.; Togashi, H. (2004). *Angew. Chem. Int. Ed.* 43, 4300

Pérez-Tijerina, E.; Gracia-Pinilla, M.A.; Mejía-Rosales. S.; Ortiz-Méndez, U.; Torres .A.; José-Yacamán, M. (2008). Highly size-controlled synthesis of Au/Pd nanoparticles by inert-gas condensation. *Faraday Discuss.*138, 353-362

Philip, D. (2009). Biosynthesis of Au, Ag and Au-Ag nanoparticles using edible mushroom extract, *Spectrochemia Acta Part A* 73, 374-381

Posfai, M.; Moskowitz, B. M.; Arato, B.; Schuler, D.; Flies, C.; Bazylinski, D. A.; Frankel, R.B. (2006). Properties of intracellular magnetite crystals produced by Desulfovibrio magneticus strain RS-1. *Earth Planet. Sci. Lett.* 249, 444–455

Poulsen, N.; Sumper, M.; Kröger, N. (2003). Biosilica formation in diatoms: Characterization of native silaffin-2 and its role in silica morphogenesis. *Proc. Natl. Acad. Sci USA.* 100, 12075–12080

Raffi, M.; Rumaiz, A.K.; Hasan, M.M.; Shah. S.I. (2007). Studies of the growth parameters for silver nanoparticle synthesis by inert gas condensation. *J. Mater. Res.* 22, 3378- 3384

Reith, F.; Etschmann, B.; Gross, C.; Moors, H.; Benotmane, M.A.; Monsieurs, P.; Grass, G.; Doonan, C.; Vogt, S.; Lai, B.; Martinez-Criado, G.; George, G.N.; Nies, D.H.; Mergeay, M.; Pring, A.; Southam, G.; Brugger, J. (2009). Mechanisms of gold biomineralization in the bacterium *Cupriavidus metallidurans*. *Proc. Natl. Acad. Sci USA.* 160, 17757–17762

Ren, W., Ai, Z.;Jia, F.; Zhang, L.; Fan, X.; Zou, Z. (2007). *Low temperature preparation and visible light photocatalytic activity of mesoporous carbon-doped crystalline* TiO_2. *Appl. Catal. B* 69, 138–144

Roduner, E. (2006). Size matters: Why Nanomaterials are Different. *Chem. Soc. Rev. 35*, 583-592

Rosemary, M.J.; Pradeep. T. (2003). Solvothermal synthesis of silver nanoparticles from thiolates. *J. Colloid Interface Sci*. 268, 81-84

Sakaguchi, T.; Burgess, J.G.; Matsunaga, T. (1993). Magnetite formation by a sulphate-reducing bacterium. *Nature* 365, 47-49

Salata, O.V. (2004). Applications of nanoparticles in biology and medicine. *J. Nanobiotechnol.* 2, 3

Sankaran, V.; Yue ,J.; Cahen, R.E.; Schrock, R.R. Silbey, R.J. (1993). Advanced drug delivery advices. *Chem. Mater,* 5: 1133.

Sastry, M.; Ahmad, A.; Khan, I.M.; Kumar, R. (2003). Biosynthesis of metal nanoparticles using fungi and actinomycete, *Current Sci.* 85, 2, 162-170

Scott, D.; Toney, M.; Muzikár, M. (2008). Harnessing the Mechanism of Glutathione Reductase for Synthesis of Active Site Bound Metallic Nanoparticles and Electrical Connection to Electrodes. *J. Am. Chem. Soc. 130,* 865-874

Shaligram, S.N.; Bule, M.; Bhambure, R.; Singhal, S.R.; Singh, K.S.; Szakacs, G.; Pandey, A.; (2009). Biosynthesis of silver nanoparticles using aqueous extract from the compactin producing fungal, *Process Biochem.* 44, 939-943

Shen, Q.; Wang, X.; Fu, D. (2008). The amplification effect of functionalized gold nanoparticles on the binding of cancer drug decarbazine to DNA and DNA bases. *Applied Surface Sci.* 255, 577-580

Shukla, S.; Seal, S. (1999). Cluster size effect observed for gold nanoparticles synthesized by sol-gel technique as studied by X-ray photoelectron spectroscopy. *NanoStruct. Mater.* 11, 1181-1193,

Southam, G.; Beveridge, T.J. (1994). The in vitro formation of placer gold by bacteria. *Geochim. Cosmochim. Acta* 58, 4527-4530

Southam, G.; Saunders, J.A. (2005). The geomicrobiology of ore deposits. *Econ Geol* 100, 1067-1084

Sperling, A.R.; Gil, R.P.; Zhang, F.; Zanella, M.; Parak, J.W. (2008). Biological application of gold nanoparticles. *Chem. Soc. Rev. 37, 1896-1908*

Tao, A.R.; Habas, S.; Yang, P. (2008). Shape Control of Colloidal Metal Nanocrystals. *Small* 4, 310-325

Templeton, A.C.; Chen, S.; Gross, S.M.; Murray, R.W. (1999). *Langmuir* 15, 66-76

Thompson, T.D. (2007). Using gold nanoparticles for catalysis. *Nano Today*, 2, 40-43

Turkevich, J.; Stevenson, P.C.; Hillier, J. (1951). A study of the nucleation and growth processes in the synthesis of colloidal gold. *Discuss. Faraday. Soc.* 11, 55-75

Wijaya, A.; Schaffer, S.B.; Pallares, I.G.; Hamad-Schifferli, K. (2008). *ACS Nano* 3, 80

Xie, .;, Lee, J.Y.; Wang, D.I.C.; Ting, Y.P. (2007). Identification of Active Biomolecules in the High-Yield Synthesis of Single-Crystalline Gold Nanoplates in Algal Solutions. *Small* 3, 672 - 682

Yang, H.G.; Sun, C.H.; Qiao, S.H.; Zou, J.; Liu,G.; Smith, S.C.; Cheng, H.M.; Lu, G.Q. (2008). Anatase TiO_2 single crystals with a large percentage of reactive facets. *Nature* 453, 638-641

Yuan, Y.; Fendler, J.; Cabasso, I. (1992). Therapeutics antibiotics. *Chem. Mater.* 4, 312.

Zeng, J.; Zhang, Q.; Chen, J.; Xia, Y. (2010). A Comparison Study of the Catalytic Properties of Au-Based Nanocages, Nanoboxes, and Nanoparticles. *Nano Lett. 10*, 30-35

11

Carbon and Silicon Fluorescent Nanomaterials

Joaquim G. G. Esteves da Silva
Centro de Investigação em Química da Universidade do Porto (CIQ-UP), Department of Chemistry and Biochemistry, Faculty of Sciences, University of Porto, Porto, Portugal

1. Introduction

The most known fluorescence nanoparticles are the semiconductor nanocrystals usually called quantum dots (QD) (Smith & Nie, 2010). These nanoparticles, with typical sizes between 1 to 12 nm, are being used in many advanced technological applications, for example in bioimaging (Gerion et al., 2001; Murcia et al., 2008; Williams et al., 2009). However, some of the materials that compose the QD are rare in the earth crust and highly toxic (Lovric et al., 2005).

Recently, the elements carbon and silicon, which are among the most abundant elements in the earth crust and are intrinsically non-toxic, become quite important elements in nanochemistry because they originate fluorescent nanostructures with relatively high quantum yield. Bulk carbon and silicon materials are not fluorescent but the corresponding nanomaterials are strongly fluorescent, allow emission colour tuning and are non-blinking nanoparticles with high scientific and technological potential.

Carbon nanomaterials are already well known, like for example the fullerenes, carbon nanotubes (CNT), either single-wall (SWNT) or multiple-wall (MWNT), carbon nanofibers and graphene (Liu et al., 2010). Highly fluorescent carbon nanomaterials, here called carbon dots (CD), were only accidently discovered in 2004 during the electrophoretic purification of SWNT derived from arc-discharge soot (Xu et al., 2004). However, in the years 2000 and 2007, studies observed strongly fluorescence shortened MWNT and shortened SWNT (Luo et al., 2007; Riggs et al., 2000). Indeed, CD are carbon based nanomaterials that possesses similar size and surface functionality they constitute different families of nanomaterials and are constituted mainly by carbon with sp^2 hybridization characteristic of monocristaline graphite with relatively high oxygen contents (Baker & Baker, 2010; Esteves da Silva & Gonçalves, 2011; Fan & Chu, 2010; Xu et al., 2004). CD are different from nanodiamonds because these last nanoparticles are constituted by about 98% carbon with a sp^3 hybridization with small amounts of graphitic carbon on the surface that are synthesized from milling microdiamonds, chemical vapour deposition, shockwave or detonation processes (Baker & Baker, 2010). New simplified synthetic pathways are being proposed and the number of potential technological applications of CDs is increasing in the last years (Baker & Baker, 2010; Esteves da Silva & Gonçalves, 2011; Fan & Chu, 2010).

Another class of potentially non-toxic and biocompatible fluorescent nanoparticles are those of silicon (silicon dots - SD). One fluorescent silicon nanoparticle (silicon porous nanoparticles) were first prepared in the 1950s by electrochemical etching in hydrofluoric acid and ultraviolet irradiation (Parkhutik & Timashev, 2000) but its red luminescence was only discovered in 1990 (Canham, 1990). However, porous silicon is fragile and highly reactive limiting its straightforward applications. SD with typical sizes in the order of 1 to 10 nm are mechanically more rough and show high quantum yields. Also, SD has one common property to QD, i.e. the emitted fluorescence is red shift with increasing size particles (Canham, 1999). These properties confer SD a quite interesting role for optoelectronic devices.

The research and modulation of SD is more advanced than CD and the application of computational theoretical methods tom SD is a quit interesting research field (Pudzer et al., 2003; Belomoin et al., 2002; Trabi & Barone, 2011). Indeed, SD have been studied by density functional theory (DFT) and quantum Monte Carlo (QMC) methods to understand and simulate their structural and optical properties. This theoretical approach together with experimental confirmation will allow further developments in the understanding of the fluorescent properties tuning.

This chapter will focus on the description of the fluorescent nanoparticles carbon dots (CD) and silicon dots (SD) with a brief description of the synthesis methodologies, the current state of the art about the comprehension of the fluorescence mechanisms, their multiphoton excitation properties and its relevance to new bioimaging methodologies, and their most recent scientific and technological applications.

2. Synthesis of carbon and silicon nanomaterials

The synthesis of stable carbon and silicon based high fluorescent nanoparticles usually requires a two step methodology: the synthesis of the raw nanoparticle; and, reaction of the surface atoms with other substances to passivation and/or functionalization.

Raw nanomaterials can usually be obtained by top-down or bottom-up approaches. Top-down approaches usually include physical techniques, like for example laser ablation, or chemical reactions, like for example chemical catalyzed etching. Bottom-up approaches are usually chemical synthesis methods that use chemical precursors to synthesise the nanoparticles. Figures 1 and 2 show the main synthetic routes of CD and SD, respectively. (Baker & Baker, 2010; Esteves da Silva & Gonçalves, 2011; Fan & Chu, 2010; Gonçalves and Esteves da Silva, 2011; Kang et al., 2011)

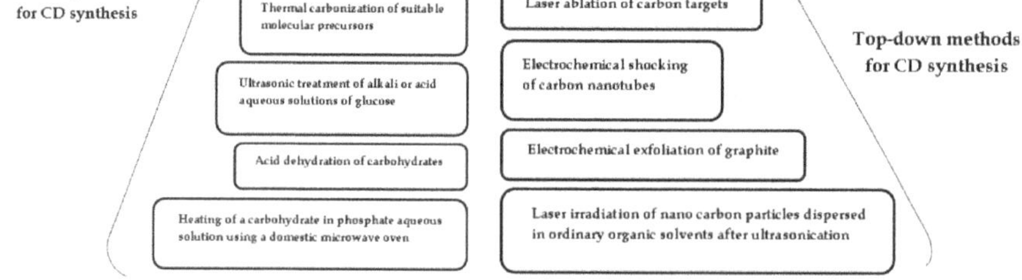

Fig. 1. Synthetic methodologies for CD.

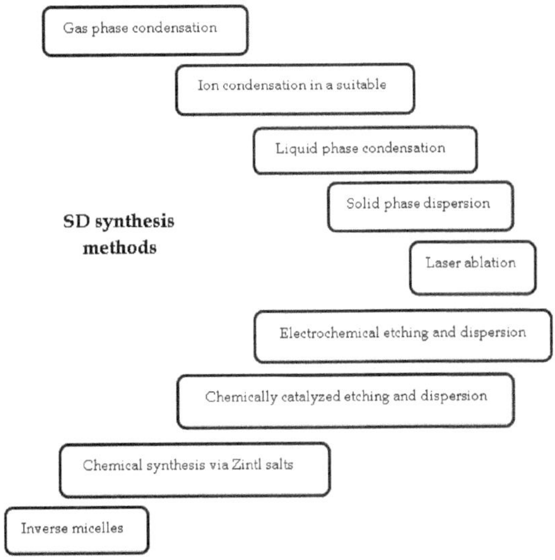

Fig. 2. Synthetic methodologies for SD.

3. Fluorescence mechanism of silicon and carbon nanomaterials

QD are usually constituted by elements from groups 12 to 16, for example III-V materials (GaN, InP, etc.), II-VI materials (ZnO, CdS, CdTe, etc.) and IV-VI materials (PbS, PbTe, etc.), (Smith & Nie, 2010). Two of the most spectacular properties of QD, which is not observed in the bulk material that is not fluorescent, are their quite strong luminescence and the dependence of their fluorescence properties with the size of the nanocrystal.

The dependence of the luminescence properties of QDs with their size is well established. QD show an electronic energy states distribution between that of a discrete molecule and of the bulk semiconductor, characterized by a bandgap energy (Eg) corresponding to the energy difference between the valence and conduction band (Fig. 3). The absorption of energy higher than Eg by an electron in the valence band provokes its excitation into the conduction band originating a negative charge and a positive hole in the valence band (the pair negative charge and positive hole is called the exciton). In a nanoparticle the annihilation of the exciton corresponds to the emission of fluorescence. However, the exciton has a finite nanometric size defined by the Bohr exciton diameter that, for relatively small nanocrystals, may be greater than the size of the nanoparticle (Esteves da Silva and Gonçalves, 2011). When this condition is observed the nanoparticle is under a quantum confinement regime and their electronic properties are size dependent - the nanocrystal growth provokes a redshift in the emission of the fluorescence (Fig. 3) (Smith & Nie, 2010).

Another important characteristic of nanoparticles is that, due to its small size, a relatively high fraction of the atoms that constitutes the nanoparticle are localized at the surface. The atoms at the surface shows reduced coordination number and there is corrosion and imperfections which may affect the electronic properties of a nanoparticle - the smaller the nanocrystal the more important become the surface defects (Esteves da Silva and Gonçalves,

2011). Consequently, the electronic properties of QD may have contribution of surface defects besides quantum confinement effects.

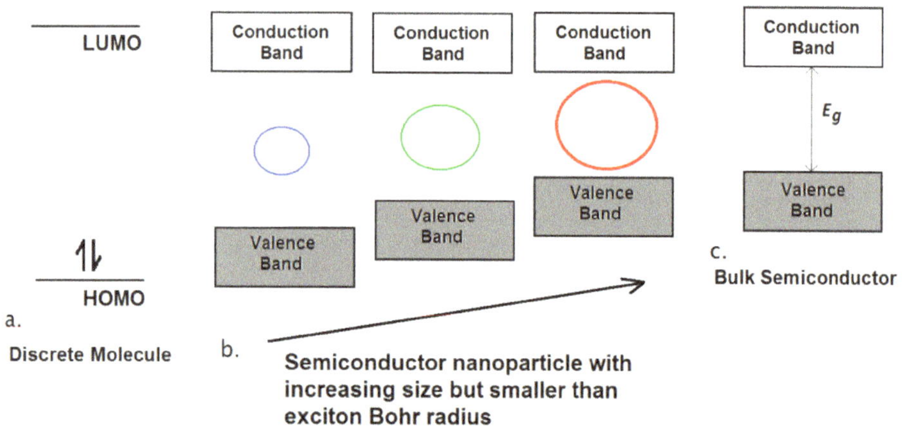

Fig. 3. Schematic representation of the electronic energy states in a semiconductor where quantum confinement is observed in the nanoparticles.

Silicon is a semiconductor with a bandgap energy of 1.14 eV (Shirahata et al., 2010; Wilcoxon et al., 1999). However, Silicon has no significant optical performance due to its indirect bandgap character (Fig. 4). The exciton Bohr radius of silicon is 4 nm and, under strong quantum confinement regime, the probability of radiative recombination through the direct bandgap transitions is increased and through phonon-assisted indirect bandgap transitions is reduced (Fig. 4) (Warner et al., 2005). Nanoparticles of silicon are highly luminescent and show quantum confinement size dependent properties resulting in the widening of the bandgap from 1.14 eV of the bulk to about 3.26 eV (380 nm) (Holmes et al., 2001; Shirahata et al., 2010).

Quantum confinement size dependent properties of SD have been observed for Si nanoparticles of size between 3 and 8 nm (Ledoux et al., 2002). Indeed, the energy of the emitted photons shifts to higher values when the size of the nanoparticles is reduced. However, for smaller SD, the surface composition becomes more important than the quantum confinement effect and the tuning of the emitted light becomes more difficult (Kang et al., 2011). SD of size between 1 and 4 nm showed quantum confinement effects, namely the maximum emission wavelength of 1 nm nanoparticles particles was 450 nm, 2 nm 520 nm, 3 nm 640 nm and 4 nm 740 nm (Kang et al., 2007; Kang et al., 2011). The emission of these SD could be tuned by chemically modification of the surface of the nanoparticles showing controllable surface effects.

The fluorescence of SD is still not complete understood but both quantum confinement effects and/or surface defects are involved in the fluorescence emission mechanism.

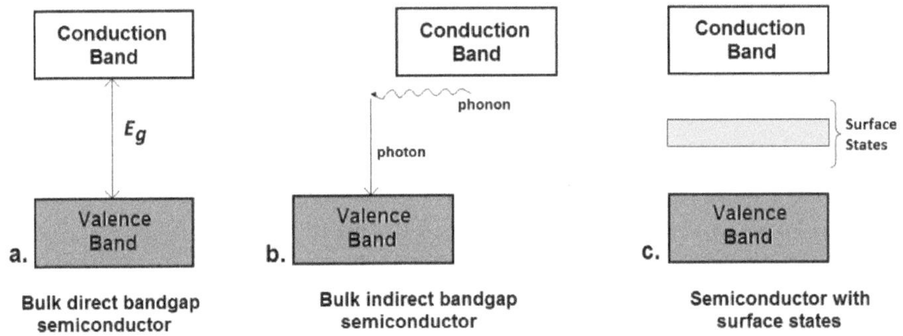

Fig. 4. Schematic representation of the electronic energy states in a direct (a.), indirect (b.) and with surface states semiconductor.

Similarly to QD and SD, CD are highly fluorescent nanoparticles. However, the fluorescence properties of CD do not show size dependency suggesting different fluorescence mechanism. Nevertheless, CD also shows the particular property of multicolour emission when different excitation wavelengths are used (Li et al., 2010; Liu et al., 2007; Liu et al., 2009; Mao et al., 2010; Peng & Travas-Sejdic, 2009; Zhu et al., 2009). Indeed, carbon is an insulator and quantum confinement effects are not expected for CD. Although the fluorescent mechanism of CD is still not clearly understood the fluorescence properties of these nanomaterials should be due to surface defects that originates energy levels that justifies the light emission (Fig. 4.c). These electronic levels distribution allow radiative recombination of excitons (Peng & Travas-Sejdic, 2009; Zhu et al., 2009). Also, and supporting the main effect of surface defects on the fluorescence properties, CD become fluorescent and the quantum yield increases when the nanoparticles are subjected polymer passivation and nitric acid treatment (Li et al., 2010; Liu et al., 2007; Liu et al., 2009; Mao et al., 2010; Peng & Travas-Sejdic, 2009; Zhu et al., 2009). Different starting materials and different fractions of similar size carbon nanoparticles also show different fluorescence emission properties.

4. Multiphoton excitation phenomenon

When two low-energy photons arrive 'simultaneously' at a fluorophore and, through interaction with it, provokes the excitation of an electron that normally is excited by one higher energy photon, a two-photon excitation phenomenon (TP) is observed (Fig. 5) (Kim & Cho, 2009; Xu et al., 1996; Williams et al., 2001). TP based fluorescence microscopy using lower energy radiations as excitation sources (red or NIR photons) is becoming a quite popular technique for in vivo analysis (bioimaging) because of the following: (i) deeper penetration depth (>500 μm); (ii) lower tissue auto-fluorescence and self-absorption; (iii) reduced photodamage and photobleaching; and, (iv) localized excitation. One characteristic of the TP probes is the excitation cross sections (δTP) usually expressed in GM (Göppert-Mayer) units (1 GM = 10-50 cm^4 s/photon). Larger δTP the better because the TP probe is more fluorescent and higher sensitivities are achieved. Apparently, the progress in this area is limited for the lack of appropriate probes (Kim & Cho, 2009).

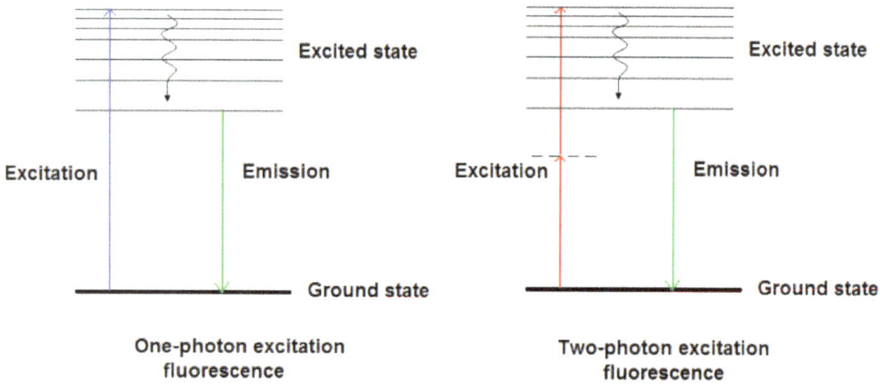

Fig. 5. Schematic representation of the electronic energy states of a fluorescent substance when one and two-photon excitation is observed.

Carbon and silicon nanoparticles can be excited by single (UV or near UV) or TP (red or NIR). TP and the low cytotoxicity and biocompatibility of carbon and silicon nanomaterials make them excellent sensors for bioimaging applications with a simple and straightforward application methodology (Fig. 6). CD and SD show luminescence with TP excitation in the NIR (800 nm fentosecond pulsed Ti:sapphire laser) (Cao et al., 2007) or long-wavelength light (from 500 to 1000 nm) (Akcakir et al., 2000; Li et al., 2010b). The maximum of the emission of fluorescence is in the 400 to 500 nm wavelength range as usually observed when one-photon excitation is used. The TP excitation is confirmed because a quadratic relationship between excitation laser power and the measured luminescence intensity is observed (Cao et al., 2007; Li et al., 2010b).

5. Applications of carbon and silicon nanomaterials

In this section some selected applications of fluorescent CD and SD are described. The selection criteria were their recent publication and potential for the future development of scientific and technological real applications. Indeed, taking into consideration the unlimited resources of both carbon and silicon in the earth crust, and their no toxicity and biocompatibility potential, major nanotechnological advances and future applications will have a strong contribution from CD and SD. These fluorescent nanoparticles will have an active role in the near future in bioimaging in biomedicine and bioanalytical applications, as nanocatalysts in technology and biochemistry, in sustainable energy devices and advanced analytical chemistry applications in general.

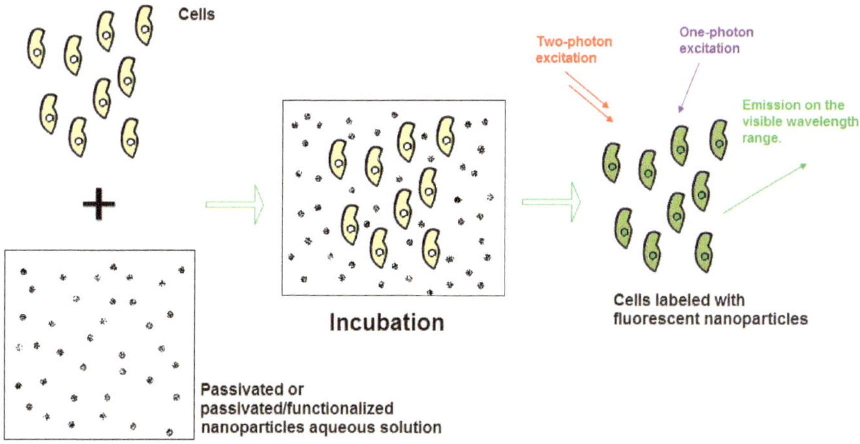

Fig. 6. Steps necessary to label cells with fluorescent nanoparticles before imaging with (either one-photon or two-photon excitation).

5.1 Carbon and silicon dots in bioimaging

Bioimaging sensors for in vivo diagnostics must be non-toxic and biocompatible. QDs based bioimaging methodologies have appeared in the last years together with toxicity concerns because they are constituted by intrinsically toxic elements like cadmium (Hardman, 2006; Lovric et al., 2005). CD and SD are constituted by intrinsically non-toxic elements which make them a particularly useful and promising bioanalytical tools as their performance approaches that of QDs. CD and SD have been demonstrated to have a very low cytotoxicity and were shown to internalize cells, probably by endocytosis mechanism, which, taking into consideration the MPE properties discussed above, makes these nanoparticles suitable for bioimaging purposes (Akcakir, 2000; Erogbogbo, 2011; Li et al., 2010; Li & Ruckenstein, 2004; Manhat, 2011; Ray et al., 2009; Sun et al., 2006; Veinot, 2006; Yang et al., 2009a; Yang et al., 2009b).

The development of bioimaging agents that are selectively uptake by cancer cells is a particularly active research field. Usually CD or SD are functionalized with an aminoacid (lysine or glutaric acid), folic acid, antimesothelin or transferring to be selectively uptake by cancer cells (Erogbogbo, 2011; Li et al., 2010a; Manhat, 2011).

5.2 Carbon dots as peroxidase mimetics

Surprisingly CDs have shown to possess enzymatic properties because they have a peroxidase like activity (Shi et al., 2011). Indeed, CDs catalyses the oxidation of peroxidase substrates by hydrogen peroxide. For example, 3,3',5,5'-tetramethylbenzidine (TMB) is oxidised by hydrogen peroxide in the presence of CDs and detected by the production of a blue coloured compound (oxidised TMB). This new function of CDs opens new perspectives in their bioanalytical potential as a nanosensor for hydrogen peroxide and glucose (Shi et al., 2011).

The comparison of CDs with horseradish peroxidase (HRP) showed that CDs required a hydrogen peroxide concentration about two orders of magnitude higher than HRP to reach a maximum level of peroxidase activity (Shi et al., 2011). This result suggested that CDs are more stable than HRP at high hydrogen peroxide concentration. Also, CDs are much more stable than HRP.

Because the colour development (absorbance at 652 nm) was found proportional to the hydrogen peroxide concentration when TMB was used as substrate a quantitative method was obtained with a limit of detection of 0.2 μM (linear plot in the hydrogen peroxide concentration between 0.0010-0.10 mM) (Shi et al., 2011).

The coupling of CDs with other enzymes can generate new bioanalytical methodologies, like for example for the quantification of glucose. Indeed, by combining CDs and TMB with glucose oxidase a straightforward colorimetric method for glucose is obtained with a limit of detection of 0.4 μM (linear plot in the glucose concentration between 0.0010-0.50 mM) (Shi et al., 2011). This method is sensible enough for serum glucose quantification because the concentration in healthy and diabetic individuals range from 3 to 8 mM and 9 to 40 mM.

5.3 Silicon dots as nanocatalyst and photodynamic therapy

SD shows great potential as nanocatalyst with strong and tunable chemical activity, specificity and selectivity (Kang et al., 2011). These properties of SD are attributed to their tunable band gap energy and photoconductivity properties.

Due to the larger exciton (electron/hole pair) energy of SD with 1 to 2 nm size, these nanoparticles induces the photochemical reduction of carbon dioxide to carbonate anion and degradation of methyl red. Also, SD with 3 to 4 nm size photocatalyses the hydroxylation of aromatic hydrocarbons - benzene is transformed to phenol with a 100% yield with high selectivity. (Kang et al., 2011)

These catalyst properties of SD are probably due to their efficient photosensitizers of singlet oxygen (Timoshenko et al., 2006). This property has been used to suppress the division of cancer cells as consequence of the oxidation of cell material by singlet oxygen (Steller, 1995). Also, the formation of superoxide ions was observed when SD are present (Fujii et al., 2005). The photochemical synthesis of reactive oxygen species appears promising for use in photodynamic therapy of cancer.

5.4 Metallic plasmons and carbon dots luminescence interaction

An important recent scientific result was the successful coupling of plasmonic metal nanoparticles with luminescent CD (Zhang et al., 2011; Li et al., 2011). Indeed, the coupling of CD with plasmon metals enhances the CD brightness (metal enhanced fluorescence - MEF) and photostability resulting into better detectability in bioanalytical applications. Two different coupling strategies have been successfully tested that show MEF: CDs immobilized in glass substrates containing silver islands films (SiFs) (Zhang et al., 2011); and, deposition of CDs on the surface of silver nanoparticles mediated with a silica layer (Li et al., 2011).

MEF results from the interaction, within the wavelength of light (near-field conditions), of the luminescent material with the metallic surface plasmons. This interaction provokes an enhancement of the fluorescence of the nanoparticle and reduces the corresponding excited decay times leading to enhanced photostability (Zhang et al., 2011). When the fluorescent nanoparticle is more than one wavelength of light way (far-field conditions) the corresponding quantum yield (Q_0) is given by (Zhang et al., 2011):

$$Q_o = \Gamma / (\Gamma + K_{nr})$$

where Γ is the fluorophores' radiative decay rate and K_{nr} are the nonradiative decay rates for excited state relaxation. Under near-field conditions the quantum yield (Q_m) is given by:

$$Q_m = (\Gamma + \Gamma_m) / (\Gamma + \Gamma_m + K_{nr})$$

where Γ_m is the system modified radiative rate. The corresponding far (τ) and near-field (τ_m) lifetimes are given by (Zhang et al., 2011):

$$\tau = 1 / (\Gamma + K_{nr})$$
$$\tau_m = 1 / (\Gamma + \Gamma_m + K_{nr})$$

These equations show that when Γ_m increases, the quantum yield increases (MEF) and the lifetime is reduced (enhanced photostability).

5.5 Silicon dots coupled with multi-polar plasmonic hot spots

The metal enhanced fluorescence (MEF) has been investigated for SD (Nychyporuk et al., 2011). Indeed, as discussed above for CD, the photo-stimulated emission of fluorescence of fluorophores suffers a significant enhancement when they are in the vicinity of metal nanoparticles. Stronger fluorescence enhancement is achieved when SD are localized in regions where the photo-induced electric fields from several metal nanoparticles are superimposed (hot spots) (Nychyporuk et al., 2011).

SD were embedded in a silicon nitride matrix (SiN_x) and deposited on a quartz substrate using a chemical vapour deposition technique followed by the fabrication of a monolayer of silver nanoislands on the surface of the SiN_x dielectric film and covered by a thin silicon nitride film (SiN_y) containing SD dispersed inside it (Nychyporuk et al., 2011). This procedure resulted in the preparation of 208 nm films (SiN_x - 170 nm; SiN_y - 38 nm) containing SD with mean diameters between 2 and 4 nm, silver nanoclusters with a mean diameter of 18±3 nm and two adjacent silver islands are separated by about 20 nm distance. Strong MEF was observed in the mean distance between two adjacent silver islands because these places correspond to the photoexcited plasmon hot spots (Nychyporuk et al., 2011). The local plasmon resonance of the silver islands could be tuned in the visible spectral range (400 to 600 nm) by adjusting the SiN_y dielectric constant allowing an optimization of the MEF for the SD.

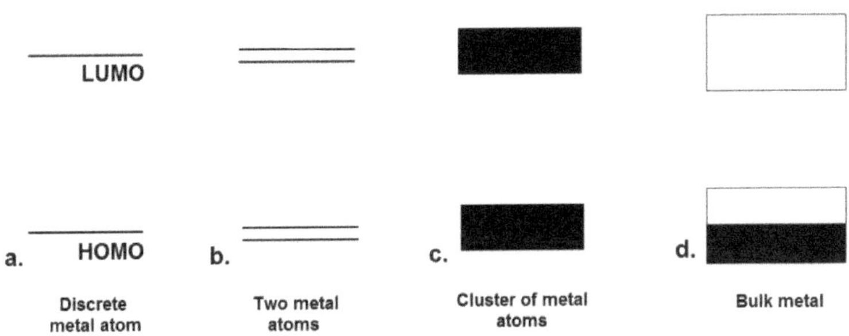

Fig. 7. Schematic representation of the electronic energy level diagram of discrete metal atoms (a.), two metal atoms (b.), clusters of metal atoms (c.) and bulk metallic element (d.).

5.6 Carbon dots immobilized in fiber optics

The use of optical sensors in chemical and biological analytical measurements is an expanding area of research with growing importance, especially in biomedical and environmental applications. Indeed, optical sensors are intrinsically immune to electromagnetic interferences and, if properly designed, can show an increased sensitivity, fast response times and suitability to remote monitoring when used with optical fibers (Jorge et al., 2005). The combination of optical fiber technologies with fluorescence spectroscopy has greatly contributed to the progress of optical chemical sensors (Gonçalves et al., 2010; Jorge et al., 2005; Maule et al., 2010). However, the design of sensing heads based on fiber optics requires high quantum yield fluorescent sensors and fluorescent nanomaterials are strong candidates to be coupled to fiber optics (Gonçalves et al., 2010; Jorge et al., 2005; Maule et al., 2010).

CD synthesised by laser ablation and and functionalized with N-acetyl-l-cysteine were successful immobilized (thin of about 750 nm) in the tip of a fiber optic using a sol-gel technology (Gonçalves et al., 2010). Due to the CD functionalization the sensing head selectively and reversibly responds to mercury(II) ions by a quenching mechanism. The response of the nanosensor is fast (less than10s) and stable. Also, submicronmolar concentrations of mercury(II) could be detected and quantified - Stern-Volmer constant (pH=6.8) of 5.3×10^5 M^{-1}.

5.7 Carbon dots for photocatalyst design

Photoexcited CD have been shown to have redox active properties namely as electron-donor capabilities in reduction reactions (Wang et al., 2009). This property was observed when silver ion was reduced to silver metal upon irradiating with a Xe arc lamp (450 and 600 nm) an aqueous solution of CD silver ions (Wang et al., 2009).

The upconverted photoluminescent and redox properties of CD opened interesting perspectives in its application as energy-transfer agents in photocatalyst design (Li et al., 2010). CD were dispersed on the surface of TiO_2 and SiO_2 and their potential in the photocatalytic degradation of methyl blue was assessed using a 50 mg/L solution - after 25 minutes the reduction of methyl blue was almost completed (Li et al., 2010). When undoped TiO_2 and SiO_2 or CD alone were used the reduction of methyl blue was nearly inexistent. The observed difference in the reduction yield is due to the relatively large intrinsic band gap of TiO_2 which has as consequence of only less than 5% of the sunlight is used. When TiO_2 is doped with CD the full spectrum of sunlight can be used because the nanoparticles absorb visible light and emit at shorter wavelength (325 to 425 nm) which can be used by TiO_2 and SiO_2 originating the exciton. This excitation species is responsible for the formation of reactive oxygen species that provokes the degradation of the dyes (Li et al., 2010). Moreover, the presence of CD attached to the oxides surfaces permits the transfer of electrons allowing charge separation, stabilization and hindered recombination with an overall higher activity of the photocatalyst (Li et al., 2010).

5.8 Carbon dots for white LED

White light-emitting devices (WLED) are currently an alternative as low energy consumption light sources and liquid-crystal displays among many other technological applications with significant economical advantages (Jang et al., 2010; Wang et al., 2011). CD have been successfully used in the fabrication of an electroluminescent (EL) device that emits a spectrum that approximates natural sun light (Wang et al., 2011).

WLED based CD were assembled using a tri-layer nanocrystal EL device (Wang et al., 2011): a 40 nm thick poly(3,4-ethylenedioxythiophene):poly(stryrenesulfonate) (PEDOT:PSS) layer on the anode; CD with 5 nm diameter with a quantum yield of more than 60%, dispersed in toluene, were spun-casted onto the PEDOT:PSS forming a 20 nm thick film; a 40 nm thick 1,3,4-tris(N-phenylbenzimidazol-2-yl) benzene (TPBI) was used as the electron transporting layer; finally, a 1 nm thick LiF and 120 nm thick Al electrodes were deposited by thermal evaporation. The EL spectrum of this device covers the entire visible zone with a turn-on voltage of about 6 V and a maximum brightness output of 35 cd m^{-2}.

5.9 Silicon dots for photovoltaic applications

The use of solar cells for the sustainable production of electric energy is a well known strategy that is characterized by its relatively high cost due to the cost of the cell material and relatively low efficiency. SD may contribute significantly for efficiency improving and cost reducing of solar cells (Luo et al., 2011).
One of the most important properties of SD useful for solar cells is their tunability because different bandgap layers, with chemical and thermal compatibility (the same material), can be assembled in multi-junction solar cells. Indeed, a large number of different bandgap layers that match the solar spectrum and raising the efficiency can be assembled using the same growth process because the same SD are being used. Another property of SD, that is common to QDs, is the multi-exciton generation (MEG) and consists in the generation of multiple electron-hole pairs from the absorption of a single photon. MEG may increases considerably the power conversion of solar cells. Also, the possibility of MPE of SD constitutes another factor for increasing the efficiency of solar cells.

6. Perspectives

The progresses that are being observed in the synthesis of high quantum yield fluorescent carbon and silicon nanoparticles are opening new technological applications for these intrinsically no toxic and highly abundant materials. Indeed, in the very near future CD and SD will have an important role in energy conversion systems, new illumination apparatus and catalytic processes.
In the human health sector, namely in medical diagnosis and imaging, CD and SD are becoming an important area of scientific and technological development. Indeed, these nanomaterials with a suitable functionalization can be used in bioimaging methodologies replacing the already established protocols including classical organic dyes and the potentially toxic cadmium based QD. Moreover, CD and SD show multiphoton excitation properties which markedly increase their bioimaging potential.
Consequently, fluorescent carbon and silicon nanomaterials will play an important role in a sustainable development of the earth.

7. Acknowledgments

Financial support from Fundação para a Ciência e a Tecnologia (FCT, Lisbon) (Programa Operacional Temático Factores de Competitividade (COMPETE) e comparticipado pelo Fundo Comunitário Europeu FEDER) (Project PTDC/QUI/71001/2006) is acknowledged.

8. References

Akcakir, O.; Therrien, J.; Belomoin, G.; Barry, N.; Muller, J.D.; Gratton, E. & Nayfeh, M. (2000) Detection of luminescent single ultrasmall silicon nanoparticles using fluctuation correlation spectroscopy. *Applied Physics Letters*, 76, 1857-1859, ISSN: 1077-3118.

Baker, S.N. & Baker, G.A. (2010) Luminescent Carbon Nanodots: Emergent Nanolights. *Angewandte Chemie International Edition*, 49, 6726 - 6744, ISSN: 1521-3773.

Belomoin, G.; Rogozhina, E.; Therrien, J.; Braun, P.V.; Abuhassan, L. & Nayfeh, M.H. (2002) Effects of surface termination on the band gap of ultrabright Si_{29} nanoparticles: Experiments and computational models, *Physical Review B*, 65, 193406, ISSN: 1550-235X.

Canham, L.T. (1990) Silicon quantum wire array fabrication by electrochemical and chemical dissolution of wafers, *Applied Physics Letters*, 57, 1046-1048, ISSN: 1077-3118.

Canham, L.T. (1999) Lewis Acid Mediated Hydrosilylation on Porous Silicon Surfaces. *Journal of the American Chemical Society*, 121, 11491-11492, ISSN: 1520-5126.

Cao, L.; Wang, X.; Meziani, M.J.; Lu, F.; Wang, H.; Luo, P.G.; Lin, Y.; Harruff, B.A.; Veca, L.M.; Murray, D.; Xie, S.Y. & Sun, Y.P. (2007) Carbon dots for multiphoton bioimaging. *Journal of the American Chemical Society*, 129, 11318-11319, ISSN: 1520-5126.

Erogbogbo, F.; Tien, C.-A.; Chang, C.-W.; Yong, K.-T.; Law, W.-C.; Ding, H.; Roy, I.; Swihart, M.T. & Prasad, P.N. (2011) Bioconjugation of luminescent silicon quantum dots for selective uptake by cancer cells. *Bioconjugate Chemistry*, 22, 1081-1088, ISSN: 1043-1802.

Esteves da Silva, J.C.G. & Gonçalves, H. (2011) Analytical and bioanalytical applications of carbon dots. *Trends in Analytical Chemistry*, doi:10.1016/j.trac.2011.04.009, ISSN: 0167-2940.

Fan, J & Chu, P.K. (2010) Group IV Nanoparticles: Synthesis, Properties, and Biological Applications. *Small*, 6, 2080–2098, ISSN: 1613-6829.

Fujii, M.; Kovalev, D.; Goller, B.; Minobe, S.; Hayashi, S. & Timoshenko, V.Y. (2005) Time-resolved photoluminescence studies of the energy transfer from excitons confined in Si nanocrystals to oxygen molecules. *Physical Review B*, 72, 165321-165329, ISSN: 1550-235X.

Gerion, D.; Pinaud, F.; Williams, S.C.; Parak, W.J.; Zanchet, D.; Weiss, S. & Alivisatos, A.P. (2001) Synthesis and Properties of Biocompatible Water-Soluble Silica-Coated CdSe/ZnS Semiconductor Quantum Dots. *Journal of Physical Chemistry B*, 105, 8861-8871, ISSN: 1520-5207.

Gonçalves, H. & Esteves da Silva, J.C.G. (2011) A New Insight on Silicon dots. *Current Analytical Chemistry*, In revision, ISSN: 1573-4110.

Gonçalves, H.; Duarte, A. & Esteves da Silva, J.C.G. (2010) Carbon dots based optical fiber sensor for Hg(II). *Biosensors and Bioelectronics*, 26, 1302-1306, ISSN: 0956-5663.

Hardman, R. (2006) A Toxicological Review of Quantum Dots: Toxicity Depends on Physicochemical and Environmental Factors. *Environmental Heath Perspectives*, 114, 165-172, ISSN: 0091-6765.

Holmes, J.D.; Ziegler, K.J.; Doty, R.C.; Pell, L.E.; Johnston K.P. & Korgel, B.A. (2001) Highly Luminescent Silicon Nanocrystals with Discrete Optical Transitions. *Journal of the American Chemical Society*, 123, 3743-3748, ISSN: 1520-5126.

Jang, E.; Jun, S.; Jang, H.; Lim, J.; Kim, B. & Kim, Y. (2010) White-Light-Emitting Diodes with Quantum Dot Color Converters for Display Backlights. *Advanced Materials*, 22, 3076–3080, ISSN: 1521-4095.

Jorge, P.A.S.; Caldas, P.; Esteves da Silva, J.C.G.; Rosa, C.C.; Oliva, A.G.; Farahi, F. & Santos, J.L. (2005) Luminescence-based optical fiber chemical sensors. *Fiber and Integrated Optics*, 24, 201-225, ISSN: 1096-4681.

Kang, Z.; Liu, Y. & Lee, S.T. (2011) Small-sized silicon nanoparticles: new nanolights and nanocatalysts. *Nanoscale*, 3, 777-791, ISSN: 0306-0012.

Kang, Z.; Tsang,C.; Wong, N.; Zhang, Z. & Lee, S. (2007) Silicon Quantum Dots: A General Photocatalyst for Reduction, Decomposition, and Selective Oxidation Reactions. *Journal of the American Chemical Society*, 129, 12090-12091, ISSN: 1520-5126.

Kim, H.M. & Cho, A.B.R. (2009). Two-photon probes for intracellular free metal ions, acidic vesicles, and lipid rafts in live tissues. *Accounts of Chemical Research*, 42, 863-872, ISSN: 1520-4898.

Ledoux, G.; Gong, J.; Huisken, F.; Guillois, O. & Reynaud, C. (2002) Photoluminescence of size-separated silicon nanocrystals: Confirmation of quantum confinement. *Applied Physics Letters*, 80, 4834-4836, ISSN: 1077-3118.

Li, Z.F. & Ruckenstein, E. (2004) Water-Soluble Poly(acrylic acid) Grafted Luminescent Silicon Nanoparticles and Their Use as Fluorescent Biological Staining Labels. *Nano Letters*, 4, 1463-1467, ISSN: 1530-6984.

Li, Q.; Ohulchanskyy, T.Y.; Liu, R.; Koynov, K.; Wu, D.; Best, A.; Kumar, R.; Bonoiu, A. & Prasad, P.N. (2010a) Photoluminescent carbon dots as biocompatible nanoprobes for targeting cancer cells in vitro. *Journal of the Physical Chemistry C*, 114, 12062-12068, ISSN: 1932-7447.

Li, H.; He, X.; Kang, Z.; Huang, H.; Liu, Y.; Liu, J.; Lian, S.; Tsang, C.H.A.; Xiaobao, X. & Lee, S.T. (2010b) Water-soluble fluorescent carbon quantum dots and photocatalyst design. *Angewandte Chemie International Edition*, 49, 4430-4434, ISSN: 1521-3773.

Li, J.; Zhang, B.; Wang, F. & Liu, C. (2011) Silver/carbon-quantum-dot plasmonic luminescence nanoparticles. *New Journal of Chemistry*, 35, 554-557, ISSN: 1369-9261.

Liu, H.; Ye, T. & Mao, C. (2007) Fluorescent Carbon Nanoparticles Derived from Candle Soot. *Angewandte Chemie International Edition*, 46, 6473-6475, ISSN: 1521-3773.

Liu, R.; Wu, D.; Liu, S.; Koynov, K.; Knoll, W. & Li, Q. (2009) An Aqueous Route to Multicolor Photoluminescent Carbon Dots Using Silica Spheres as Carriers. *Angewandte Chemie International Edition*, 48, 4598-4601, ISSN: 1521-3773.

Liu, Z.; Zhou X. & Qian, Y. (2010) Synthetic Methodologies for Carbon Nanomaterials. *Advanced Materials*, 22, 1963-1966, ISSN: 1521-4095.

Lovric, J.; Bazzi, H.S.; Cuie, Y.; Fortin, G.R.A.; Winnik, F.M. & Maysinger, D. (2005) Differences in subcellular distribution and toxicity of green and red emitting CdTe quantum dots. *Journal of Molecular Medicine*, 83, 377-385, ISSN: 1432-1440.

Luo, Y.; X. Xia, X.; Liang, Y.; Zhang, Y.; Ren, Q.; Li, J.; Jia, Z. & Tang, Y. (2007) Highly visible-light luminescence properties of the carboxyl-functionalized short and ultrashort MWNTs. *Journal of Solid State Chemistry*, 180, 1928-1933, ISSN: 0022-4596.

Luo, J.-W.; Stradins, P. & Zunger, A. (2011) Matrix-embedded silicon dots for photovoltaic applications: a theorectical study of critical factors. *Energy & Environmental Science*, doi:10.1039/CIEE01026C, ISSN: 1754-5706.

Manhat, B.A.; Brown, A.L.; Black, L.A.; Ross, J.B.A.; Fichter, K.; Vu, T.; Richman, E. & Goforth, A. (2011) One-step melt synthesis of water-soluble, photoluminescent, surface-oxidized nanoparticles for cellular imaging applications. *Chemistry of Materials*, 23, 2407-2418, ISSN: 1520-5002.

Mao, X.J.; Zheng, H.Z.; Long, Y.J.; Du, J.; Hao, J.Y.; Wang, L.L. & Zhou, D.B. (2010) Study on the fluorescence characteristics of carbon dots. *Spectrochimica Acta Part A*, 75, 553–557, ISSN: 1386-1425.

Maule, C.D.; Gonçalves, H.M.R.; Mendonça, C.; Sampaio, P.; Jorge, P.A.S. & Esteves da Silva, J.C.G. (2010) Wavelength encoded analytical imaging and fiber optic sensing with pH sensitive CdTe quantum dots. *Talanta*, 80 1932–1938, ISSN: 0039-9140.

Murcia, M.J.; Shaw, D.L.; Long, E.C. & Naumann, C.A. (2008) Fluorescence correlation spectroscopy of CdSe/ZnS quantum dot optical bioimaging probes with ultra-thin biocompatible coatings. *Optics Communications*, 28, 1771-1780, ISSN: 0030-4018.

Nychyporuk, T.; Zakharko, Y.; Serdiuk, T.; Marty, O.; Lemiti, M. & Lysenko, V. (2011) Strong photoluminescence enhancement of silicon quantum dots by their rear-resonant coupling with multi-polar hot spots. *Nanoscale*, 3, 2472-2475, ISSN: 0306-0012.

Parkhutik, V. & Timashev, S. (2000) Kinetics of porous silicon growth studied using flicker-noise spectroscopy. *Journal of Applied Physics*, 87, 7558-7560, ISSN: 1089-7550.

Peng, H. & Travas-Sejdic, J. (2009) Simple Aqueous Solution Route to Luminescent Carbogenic Dots from Carbohydrates. *Chemistry of Materials*, 21, 5563–5565, ISSN: 1520-5002.

Pudzer, A.; Williamson, A.J.; Grossman, J.C. & Galli, G. (2003) Computational studies of the optical emission of silicon nanocrystals. *Journal of the American Chemical Society*, 125, 2786-2791, ISSN: 1520-5126.

Ray, S.C.; Saha, A.; Jana, N.R. & Sarkar, R. (2009) Fluorescent carbon nanoparticles: synthesis, characterization, and bioimaging application. *Journal of the Physical Chemistry C*, 113, 18546-18551, ISSN: 1932-7447.

Riggs, J.E.; Guo, Z.; Carroll, D.L. & Sun, Y.P. (2000) Strong Luminescence of Solubilized Carbon Nanotubes. *Journal of the American Chemical Society*, 122, 5879-5880, ISSN: 1520-5126.

Shi, W.; Wang, Q.; Long, Y.; Cheng, Z.; Chen, S.; Zheng, H. & Huang, Y. (2011) Carbon nanodots as peroxidase mimetics and their applications to glucose detection. *Chemical Communications*, 47, 6695-6697, ISSN: 1364-548X.

Shirahata, N.; Hasegawa, T.; Sakka, Y. & Tsuruoka, T. (2010) Size-Tunable UV-Luminescent Silicon Nanocrystals. *Small*, 6, 915-921, ISSN: 1613-6829.

Smith, A.M. & Nie, S. (2010) Semiconductor nanocrystals: structure, properties, and band gap engineering. *Accounts of Chemical Research*, 43, 190-200, ISSN: 1520-4898.

Steller, H. (1995) Mechanisms and genes of cellular suicide. *Science*, 267, 1445-1449, ISSN: 0036-8075.

Sun, Y.P.; Zhou, B.; Lin, Y.; Wang, W.; Fernando, K.A.S.; Pathak, P.; Meziani, M.J.; Harruff, B.A.; Wang, X.; Wang, H.F.; Luo, P.J.G.; Yang, H.; Kose, M.E.; Chen, B.L.; Veca, L.M. & Xie, S.Y. (2006) Quantum-sized carbon dots for bright and colourful photoluminescence. *Journal of the American Chemical Society*, 128, 7756-7757, ISSN: 1520-5126.

Timoshenko, V.Y.; Kudryavtsev, A. A.; Osminkina, L. A.; Vorontsov, A. S.; Ryabchikov, Y.V.; Belogorokhov, I.A.; Kovalev, D. & Kashkarov, P.K. (2006) Silicon nanocrystals as photosensitizers of active oxygen for biomedical applications. *JETP Letters*, 83, 423-426, ISSN: 0370-274X.

Trani, F. & Barone, V. (2011) Silicon Nanocrystal Functionalization: Analytic Fitting of DFTB Parameters, *Journal of Chemical Theory and Computation*, DOI: dx.doi.org/ 10.1021/ct1006086, ISSN: 1549-9626.

Veinot, J.G.C. (2006) Synthesis, surface functionalization, and properties of freestanding silicon nanocrystals. *Chemical Communications*, 4160-4168, ISSN: 1364-548X.

Wang, X.; Cao, L.; Lu, F.S.; Meziani, M.J.; Li, H.; Qi, G.; Zhou, B.; Harruff, B.A.; Kermarrec, F. & Sun, Y.P. (2009) Photoinduced electron transfers with carbon dots *Chemical Communications*, 3774-3776, ISSN: 1364-548X.

Wang, F.; Chen, Y.H.; Liu, C.Y. & Ma, D.G. (2011) White light-emitting devices based on carbon dots' electroluminescence. *Chemical Communications*, 3502-3504, ISSN: 1364-548X.

Warner, J.H.; Dunlop, H.R. & Tilley, R.D. (2005) Highly Luminescent Silicon Nanocrystals with Discrete Optical Transitions. *Journal of the Physical Chemistry B*, 109, 19064-19067, ISSN: 1520-5207.

Wilcoxon, J.P.; Samara, G.A. & Provencio, P.N. (1999) Optical and electronic properties of Si nanoclusters synthesized in inverse micelles. *Physical Reviews B*, 60, 2704-2714, ISSN: 1550-235X.

Williams, Y.; Sukhanova, A.; Nowostawska, M.; Davies, A.M.; Mitchell, S.; Oleinikov, V.; Gun'ko, Y.; Nabiev, I.; Kelleher, D. & Volkov, Y. (2009) Probing cell-type-specific intracellular nanoscale barriers using size-tuned quantum dots. *Small*, 5, 2581-2588, ISSN: 1613-6829.

Williams, R.; Zipfel, W.R & Webb, W.W. (2001). Multiphoton microscopy in biological research. *Current Opinion in Chemical Biology*, 5, 603-608, ISSN: 1367-5931.

Xu, X.Y.; Ray, R.; Gu, Y.L.; Ploehn, H.J.; Gearheart, L.; Raker, K. & Scrivens, W.A. (2004) Electrophoretic Analysis and Purification of Fluorescent Single-Walled Carbon Nanotube Fragments. *Journal of the American Chemical Society*, 126, 12736-12737, ISSN: 1520-5126.

Xu, C.; Zipfel, W.; Shear, J.B.; Williams, R.M. & Webb, W.W. (1996). Multiphoton fluorescence excitation: new spectral windows for biological nonlinear microscopy. *Proceedings of the National Academy of Sciences*, 93, 10763-10768, ISSN: 1091-6490.

Yang, S.T.; Wang, X.; Wang, H.; Lu, F.; Luo, P.G.; Cao, L.; Meziani, M.J.; Liu, J.H.; Liu, Y.; Chen, M.; Huang, Y. & Sun, Y.P. (2009a) Carbon dots as nontoxic and high-performance fluorescence imaging agents. *Journal of the Physical Chemistry C*, 113, 18110-18114, ISSN: 1932-7447.

Yang, S.T.; Cao, L.; Luo, P.G.; Lu, F.S.; Wang, X.; Wang, H.F.; Meziani, M.J.; Liu, Y.F.; Qi, G. & Sun, Y.P. (2009b) Carbon dots for optical imaging in vivo. *Journal of the American Chemical Society*, 131, 11308-11309, ISSN: 1520-5126.

Zhang, Y.; Gonçalves, H.; Esteves da Silva, J.C.G. & Geddes, C.D. (2011) Metal-enhanced photoluminescence from carbon nanodots. *Chemical Communications*, 47, 5313-5315, ISSN: 1364-548X.

Zhu, H.; Wang, X.; Li, Y.; Wang, Z.; Yanga, F. & Yang, X. (2009) Microwave synthesis of fluorescent carbon nanoparticles with electrochemiluminescence properties. *Chemical Communications*, 5118–5120, ISSN: 1364-548X.

12

Biorecycling of Precious Metals and Rare Earth Elements

Kevin Deplanche, Angela Murray,
Claire Mennan, Scott Taylor and Lynne Macaskie
The University of Birmingham,
United Kingdom

1. Introduction

The six platinum group elements (PGEs), also known as the platinum group metals (PGMs) are platinum, palladium, rhodium, ruthenium, iridium and osmium. These, together with gold and silver, are considered to be "precious" metals due to their high demand coupled with relatively low abundance. Complex processing is required for their extraction and refining from primary ores (Bernardis *et al.*, 2005). Relative to the other precious metals PGMs have high technological importance. Valuable for their resistance to corrosion and oxidation, high melting points, electrical conductivity and catalytic activity, these elements have wide industrial applications (Xiao & Laplante, 2004). The major uses are found in the chemical, electrical, electronic, glass and automotive industries. For example their high catalytic activity for a wide range of substrates has resulted in their use in many industrial synthetic processes, reforming reactions in the petroleum refining industry, hydrogenation and dehydrogenation reactions in the pharmaceutical industry, and both organic and inorganic oxidation reactions (Bernardis *et al.*, 2005) to name but a few. The catalytic properties of PGM are also having a positive impact on the environment through the implementation of automotive emission control catalysts (Whiteley & Murray, 2003, Ek *et al.*, 2004, Zereini & Wiseman, 2009).

South Africa is currently the largest producer of PGMs, followed by Russia and North America. In 2008 South Africa supplied 76% of the worlds platinum, 33% of its palladium and 82% of its rhodium. Russia supplied more palladium than South Africa (51%) but its contribution of the other two metals was much smaller. With the exception of 2006 demand for PGMs has constantly exceeded supply, resulting in large price increases within the industry. The disparity in 2006 is thought to have been caused by significantly lower platinum usage in jewellery compared with previous years (in fact the lowest usage since 1993).

The last 75 years have seen the overall consumption and uses of platinum expand dramatically. Demand (and hence also its price) and uses are impossible to predict far into the future, but the resources and potential supply of platinum and palladium can be calculated with some degree of confidence. Cawthorn's calculations (Cawthorn, 1999) indicate about 204 and 116 million ounces of proven and probable reserves of platinum and palladium, respectively, and 939 and 711 million ounces of inferred resources, down to a depth of 2 km. These figures represent about 75 and 50% of the world's platinum and palladium resources, respectively. These figures for proven and probable reserves in the Bushveld Complex are sufficient for the next 40 years at current rate of production.

However, it has been estimated that if all 500 million vehicles in use today consuming fossil fuels, were substituted with hydrogen driven fuel cells, operating losses would mean that all the world's sources of platinum would be exhausted within 15 years (Gordon *et al.*, 2006, Anon., 2007). As we are moving away from a fossil fuel based economy into a hydrogen economy, it will be vital to conserve stocks of these metals by increasing recycling technologies and preferably by methodologies that remanufacture new materials within the same process.

The environmental impact of mining, extraction and primary production is uncertain but is likely to be large. The Resource Efficiency KTN (Anon., 2008) estimates that worldwide mining activities (of all minerals) are responsible for around 5% of global carbon dioxide emissions. A 2004 report by Earthworks (Anon. 2004) calculated a larger figure, stating that the metals mining industry is responsible for 7-10% of global energy consumption. Either figure demonstrates that the overall environmental impacts of the extractive industries are highly significant.

According to research commissioned by the UK Department for Transport (2006) 12.7 tonnes of ore is extracted per troy ounce of platinum produced. The Resource Efficiency KTN (2008) puts this figure at approximately 10 tonnes per ounce, depending on ore concentration and mining depth. This leads to large spoil heaps and high energy consumption. In fact 65 - 75% of the total cost of producing pure PGMs is accrued at the mining stage due to the large energy demand (Pincock, 2008). In the long-term carbon intensities will grow, as more energy is required to process lower grades of ore, unless technological change can offset the impact.

In contrast to PGMs and despite their name, Rare Earth Elements (REE) are relatively abundant in the earth's crust. Consisting of the 15 lanthanide elements of the periodic table (with the addition of yttrium and scandium), these metals have similar geochemical properties and, consequently, are often found together in mineral deposits. Although their average concentration in the earth's crust is similar to other commonly used industrial metals (e.g. chromium, nickel, zinc), these metals are often difficult to extract economically because of their low local concentrations.

The use of REE ranges from their incorporation in alloys for the metallurgy sector to "high-tech" applications such as laser-guided weapons and hybrid-car batteries. Their widespread incorporation in electronic devices (e.g. computers, mobile phones, digital cameras) and increasingly popular consumer goods (e.g DVDs) has caused demand to increase significantly in the past 5 years. As demand is expected to increase further with the development of electric cars and hybrid vehicles, there is growing concern that the world may soon face a shortage of the REE.

These concerns have recently intensified in response to the recent decrease in REE exports from China, who currently produces over 97% of the world's REE supply and is estimated to possess about half of the 110,000 metric tons of REE currently available worldwide. In the coming years, China is expected to build strategic reserves in REE to meet the rise in their domestic demand and potentially to influence global market prices. Alternative sources of REE are being sought, notably in Australia and California, but mining operations are still limited to date.

Significant amounts of REE can be found in the waste electronic equipment (WEE) network and, as such, these could form a partial answer to REE sourcing problems. In Japan alone, there is a very substantial amount of REE stored in electronic wastes. At present, REE are not often recycled, because of the small quantities present in many commercial products both on a total and a per-unit basis. In addition, their low market price per-kilogram relative to precious metals means that many common recycling technologies are not cost effective.

One alternative to traditional physicochemical reclaiming processes for PGM and REE is to use the ability of specialised bacteria to reduce or mineralise metals. Many studies have demonstrated that challenging bacteria with oxidised metal species resulted in the precipitation of solid phases of the metal. In the case of precious metals and gold, several studies showed that sulfate-reducing bacteria (SRB) and Fe(III) reducing bacteria are able to reduce palladium (Pd), platinum (Pt) and gold (Au) to the zerovalent form of the metal (Lloyd *et al.*, 1998; Kashefi *et al.*, 2001; Yong *et al.*, 2002a).

Bioreduction is a system involving enzymatically-assisted metal precipitation from a high valence to a zerovalent state. The identification of the enzymatic step responsible for metal deposition may permit its deposition in a growth-decoupled mode (Yong *et al.*, 2002b). Enzymatic systems may promote metal reduction under favourable conditions, independently of cell metabolism. In some cases, metal reduction to lower valence states is directly involved *via* electron transport reactions giving crystals of metal oxide or base metal coated on the cell surface. Bioreduction of metals is a relatively recent addition to the portfolio of metal biorecovery processes. Investigations into new recovery technologies were prompted by the increased market value of many precious metals and more recent studies showed a potential for bioreductive recovery of Pd (Lloyd *et al.*, 1998), Pt (Yong *et al.*, 2002a, b), Ag (Fu *et al.*, 2000) and Au (Kashefi *et al.*, 2001; Konishi *et al.*, 2006). The microbial reduction of Pd(II) to insoluble Pd(0) has attracted much interest as this process was not only successfully applied to the recovery of Pd from industrial automotive catalyst leachates (Yong *et al.*, 2003) but could also be coupled to synthesize nanoscale bioinorganic catalysts of considerable commercial potential (Mabbett *et al.*, 2006, Macaskie *et al.*, 2010). Studies on Au recovery (Lovley *et al.*, 1993) suggested that *c*-type cytochromes of the Fe(III) reducer *Geobacter metallireducens* transferred electrons to soluble Au(III); other work (Deplanche & Macaskie, 2008) has suggested the involvement of hydrogenase activity, in conjunction with another unidentified mechanism, in Au(III) reduction and the formation of Au(0) nanoparticles on the biomass. Other investigations showed the potential of a wide range of dissimilatory Fe(III) reducing mesophiles (*Geobacter ferrireducens, Shewanella alga*), thermophiles (*Thermotoga maritime*) and some Archea (*Pyrobacter* sp.) to reduce Au(III) to Au(0) (Kashefi *et al.*, 2001). As most strains precipitated Au(0) on the cell surface, bacterial Au recovery has the potential to become an easy, cheap and attractive process, especially since readily-available jewellery waste rich in Au provided an adequate source of the metal for biorecovery and conversion (Deplanche *et al.*, 2007)., while some urban road dusts were surprisingly rich in Au (H. Prichard and L.E. Macaskie, unpublished) adding to the concept of "urban biomining" (see later).

Research on bacterial metal reduction has been driven forward by the range of biotechnological applications that could derive from harnessing this unique microbial feature. Such applications include the remediation of radioactive contaminants (Lovley, 1991), immobilisation of problematic wastewater pollutants (Chen and Hao, 1998), degradation of xenobiotics (Larsson *et al.*, 1988), the development of new technologies in metal mining (Rawlings and Johnson, 2007) and metal recovery (Lloyd *et al.*, 1998; Kashefi *et al.*, 2001; Konishi *et al.*, 2006). Indirect applications could also include the development of biomagnets (J. R. Lloyd, personal communication) and manufacture of novel biocatalysts (see later), in addition to generation of electricity from sediments (Lloyd, 2003) and indeed from the bionanoparticles (Yong *et al.*, 2010) as well as the creation of metal detecting biosensors (Eccles *et al.*, 1997).

In metal and radionuclide remediation, dissimilatory reduction of Se, Tc, V, U and also possibly other metals (like plutonium (Pu) and neptunium (Np)) would convert soluble metal species to insoluble forms that can readily be removed from contaminated water and waste streams. So far, flow-through bioreactors containing immobilised cells of dissimilatory iron- or sulfate reducing bacteria have been shown to reduce and precipitate Cr(VI), Mo(VI), Tc(VII) and U(VI) (Lovley and Phillips, 1992; Lloyd *et al.*, 1997; Lloyd *et al.*, 1999; Mabbett *et al.*, 2002). Simple electron donors (formate, lactate or H_2) are fed to immobilised cells, which support continuous metal reduction and precipitation over many days. Lovley and coworkers demonstrated that *G. metallireducens* and *S. oneidensis* could couple U(VI) reduction to energy generation and thus could be ideal candidates for large scale remediation process (Lovley *et al.*, 1991). However, U(IV) is adsorbed by Fe(III) oxides forming large volumes of uranium-containing solid waste. Sulfate-reducing bacteria (SRB) were found to possess the same U(VI) reducing activity as Fe(III) reducers and can be cultured more quickly and cheaply, making them potential candidates for U(VI) remediation strategies although the production of H_2S may be problematic. Uranium exists mainly as U(VI) and U(IV) although some actinide contaminants can be stable and predominate as the intermediate (V) oxidation state. A novel approach to the remediation of Np(V) and its capture as biomineral phosphate was shown by Lloyd and Macaskie (2000). Although multiple electron transfer steps are involved in metal reduction, the exact mechanism remains unclear. A study by Renshaw *et al.*, (2005) showed that reduction of U(VI) proceeded *via* a U(V) intermediate showing single electron transfer.

The direct role of periplasmic hydrogenases of *D. desulfuricans* in Tc(VII) reduction was also demonstrated by Lloyd *et al.*, (1999a). In this study, cells treated with Cu(II) (a selective inhibitor of periplasmic -but not cytoplasmic- hydrogenases) lost their ability to reduce Tc(VII). Comparison of Tc(VII) reduction rates by *E. coli* and *D. desulfuricans* demonstrated the greater potential of the latter, and this was attributed to the high activity of the soluble, more accessible, periplasmic hydrogenases in the SRB. A subsequent study (Lloyd *et al.*, 1999b) identified *D. desulfuricans* as the ideal candidate for Tc(VII) reduction. H_2-supplied cells carried out continuous Tc(VII) reduction at far higher rates than shown in similar studies using *E. coli* (the rate of Tc(VII) reduction was 64-fold greater), *S. putrefaciens* (28-fold improvement) or *G. metallireducens* (36-fold improvement). Similarly, selenium has been shown to serve as terminal electron acceptor to support growth of some organisms with concomitant reduction of Se(VI) to Se(0) (Roux *et al.*, 2001; Yee *et al.*, 2007, Macaskie *et al.*, 2010). An economic evaluation of the use of selenate and selenite reducing microorganisms to remove selenium from waste waters concluded that the microbial process could potentially be cheaper than chemical reduction if a nutrient cheaper than peptone could be used in the culturing of the metal-reducing bacterial consortium (Lovley, 1994).

A second alternative to traditional physicochemical recycling techniques is to use bacterial enzymes to immobilise metallic species as metal precipitates. These can be as (e.g.) sulfides, carbonates and phosphates. Such metal 'biomineralisation' is a well-established technology for metal removal from wastewaters but until now the low value of the recovered metal has made such processes largely uneconomic. With the value of REEs (and also uranium) rising rapidly these biotechnologies are becoming attractive. With a view to metal recovery from metal-bearing wastewaters in a form that can be readily processed for subsequent re-use and avoiding energy-expensive commercial refining, an example of this technology has been demonstrated using a *Serratia* sp. originally isolated from metal polluted soil (Macaskie & Dean, 1982) for the recovery of heavy metals such as uranyl ion, lead, copper, cadmium, lanthanum, strontium, manganese, thorium, americium and plutonium (Macaskie & Dean, 1984; Macaskie & Dean, 1985; Tolley *et al.*, 1991; Macaskie, 1992; Macaskie *et al.*, 1994; Yong *et al.*, 1998; Forster & Wase, 2003; Paterson-Beedle *et al.*, 2004). In the case of uranium the deposited material can be used as nanocrystalline ion exchanger for the removal of radionuclides from nuclear wastes (Paterson-Beedle *et al.*, 2006), while a similar function can be achieved via biogenic hydroxyapatite (calcium phosphate) (Handley-Sidhu *et al.*, 2011) which has a crystallite size much smaller than that of commercial HA, and hence a much higher surface area for interaction with the 'target' metallic species.

The metal-accumulating *Serratia* strain (NCIMB 40259) atypically overproduces a PhoN acid type phosphatase. In the presence of heavy metals the phosphatase enzyme liberates inorganic phosphate ligand (HPO_4^{2-}) from a phosphate donor e.g. glycerol-2-phosphate (G2P) and deposits metal (M) as polycrystalline cell bound $MHPO_4$ (Macaskie *et al.*, 1992, Macaskie *et al.*, 2000).

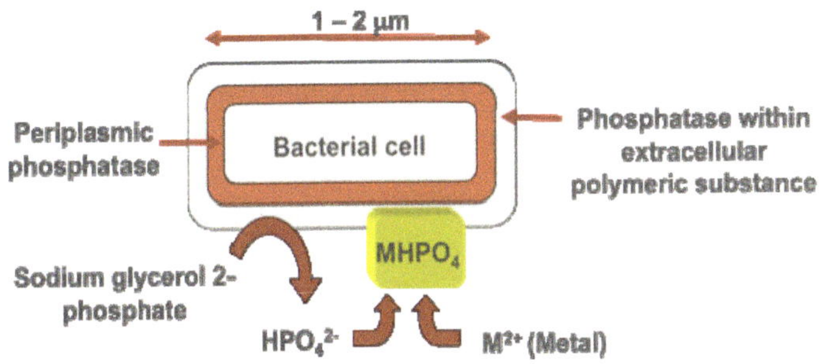

Fig. 1. Mechanism of metal phosphate accumulation by the action of acid phosphatase.

In order to achieve metal removal from solution, *Serratia* cells are immobilised on a support such as polyurethane foam and incorporated into a flow through column. Metal precipitation and crystallisation is *via* the production of locally high concentrations of phosphate ligand, which exceeds the solubility product of the metal phosphate in the vicinity of nucleation sites on the cell surface (Bonthrone *et al.*, 1996; Macaskie *et al.*, 1994). Studies using ^{31}P NMR attributed the nucleation sites to phosphate groups in the lipid A component of cell-surface lipopolysaccharide (Bonthrone *et al.*, 2000; Macaskie *et al.*, 2000). All heavy metals having insoluble phosphates should be amenable to bioremediation in this way.

As an alternative approach to the economic provision of phosphate ligand into the crystalline metal phosphate an interesting study exploited features of the 'enhanced biological phosphate removal' waste water treatment process, in which some bacteria (e.g. *Acinetobacter* spp) grown aerobically deposit intracellular reserves of polyphosphate (polyP). Upon transfer to anaerobic conditions (which can be spatial or temporal) the polyP is mobilised and inorganic phosphate is released from the cells. This was harnessed to the deposition of uranium (Dick *et al.*, 1995) and lanthanum (Boswell *et al.*, 2001) phosphates in a similar way to those biomanufactured using the phosphatase route (above), in a continuous process (Boswell *et al.*, 2001). As a third approach, the use of phytic acid (inositol phosphate, a component of plant wastes and a by product from biodiesel production) as the phosphate donor molecule has been shown to have good economic potential for metal biorecovery (Paterson-Beedle *et al.*, 2010)

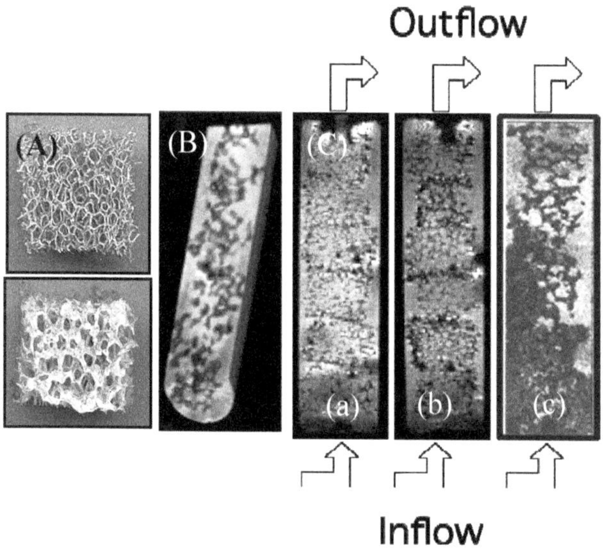

Fig. 2. (A) Polyurethane reticulated foam cubes before and after coating with biofilm of *Serratia* NCIMB 40259. (B) 3-D image of a flow- through a *Serratia* metal recovery column viewed using magnetic resonance imaging (Paterson-Beedle *et al.*, 2001). The reticulated foam matrix is clearly visible. (C) shows the centre of the column imaged in a 2-D slice: (a) before metal addition (glycerol 2-phosphate alone); (b) with metal and no substrate and (c) with metal and substrate. Metal phosphate deposition is shown as dark areas (Macaskie *et al.*, 2005).

Focusing on precious metal biorecovery for subsequent applications ('biorefining'), we now demonstrate in separate case studies the two potential pathways of biorecovery of PGMs *via* bioreduction (case study 1) and REE biorecovery *via* metal phosphates (case study 2).

2. Case study 1: Biorecovery of PGMs from spent car catalyst and biofabrication of new catalysts

Since the 1970s and the implementation of vehicle emissions regulations, worldwide consumption and market price of PGM have increased in parallel, driven by the widespread incorporation of Pd, Pt and rhodium (Rh) in automotive catalytic converters. The finite nature of PGM resources has led to the development of various recovery techniques which commonly include chemical processing of wastewaters but these treatments are often made difficult by unusual solution chemistry as these metals are usually present as complexes in solution (Demopoulos, 1989). Direct precipitation is often not readily applicable while electrochemical recovery requires electrodes with very large surface area for satisfactory metal deposition. Solvent extraction is often costly and non-selective; chelating ion exchange resins have been developed to increase separation of PGM from other metals (Streat and Naden, 1983) but, once eluted, individual metals still require separation using hydrometallurgical techniques. Until recently, there was no "clean" simple, selective and/or efficient system for precious metals recovery (Yong *et al.*, 2002b).

This goal remains elusive. It was shown (Dalrymple *et al.*, 2005) that while bacteria can selectively recover Pd(II) over Pt(IV), the period of selectivity before Pt(IV) is catalytically reduced by the Pd(0) seeds formed on the biomass is too short to form the basis of an industrial process. A selective process for metal recovery from electronic scraps was illustrated by Creamer *et al.*, (2006) but is probably too complex for industrial adoption. However, mixed metal deposits supported on bacterial cells have been reported to be catalytically active in reductive processes (Mabbett *et al.*, 2006) and as fuel cell electrocatalysts (Yong *et al.*, 2010). It is now accepted that biofabrication of new materials from wastes dictates the unavoidability of metal mixtures given the nature of the feedstocks. Since the ground-breaking work of Baxter-Plant *et al.*, (2003) and Mabbett *et al.*, (2006), the literature abounds with reports of using PGM-coated bacteria as catalysts in reactions of increasing complexity and industrial interest. Table 1 summarises the applications of bio-PGMs catalysts found in the literature biomanufactured from pure metal salts.

Metal	Microorganism support	Catalytic test	Comments	References
Au	*Desulfovibrio desulfuricans*	Selective oxidation of glycerol		Deplanche *et al.* (2007)
	Escherichia coli	Selective oxidation of glycerol Selective oxidation of alcohols		Deplanche *et al.* (2007) Deplanche *et al.* (2010b, 2011)
Au/Pd	*Desulfovibrio desulfuricans*	Selective oxidation of alcohols	Au/Pd core/shell NPs	Deplanche *et al.* (2010, 2011)
	Escherichia coli	Selective oxidation of alcohols	Au/Pd core/shell NPs	Deplanche *et al.* (2010, 2011)
Pd	*Arthrobacter oxydans*	Cr(VI) reduction Mizoroki-Heck coupling Hydrogenations		Deplanche *et al.* (2008a), Wood *et al.* (2010)
	Bacillus sphaericus	Hydrogenations		Creamer *et al.* (2007)
	Clostridium pasteurianum	Cr(VI) reduction	Process coupled to microbial H_2 production	Chidambaram *et al.* (2010)
	Cupriavidus necator	H_2 evolution from hypophosphite Suzuki-Miyaura and Mizoroki-Heck couplings Hydrogenations Reductive dechlorination		Søbjerg *et al.* (2009), Bunget *et al.* (2010), Søbjerg *et al.* (2011)
	Cupriavidus metallidurans	Cr(VI) reduction Mizoroki-Heck coupling Fuel cell electrocatalyst		Deplanche *et al.* (2008a), Yong *et al.* (2010)

Metal	Microorganism support	Catalytic test	Comments	References
Pd	*Desulfovibrio desulfuricans*	H_2 evolution from hypophosphite Cr(VI) reduction Reductive dechlorination Reductive debromination Suzuki–Miyaura and Mizoroki–Heck couplings Hydrogenations Fuel cell electrocatalyst	Free cells, flow through reactors, immobilised cells, e- donor provided by fermentation of sugar wastes	Yong *et al.* (2002b, 2007, 2010), Mabbett *et al.* (2002, 2004, 2006), Baxter-Plant *et al.* (2003), Macaskie *et al.* (2005), Humphries et *et al.* (2006), Deplanche *et al.* (2007, 2008, 2009), Harrad *et al.* (2007), Redwood *et al.* (2008). Bennett *et al.* (2010), Harrad *et al.* (2007)
	Desulfovibrio vulgaris	Reductive dechlorination Cr(VI) reduction	Free cells, flow through reactors, immobilised cells	Baxter-Plant *et al.* (2003, 2004), Humphries *et al.* (2006)
	Desulfovibrio sp. Oz-7	Reductive dechlorination		Baxter-Plant *et al.* (2003, 2004)
	Escherichia coli	Reductive dechlorination Cr(VI) reduction Mizoroki–Heck coupling Fuel cell electrocatalyst		Deplanche (2008), Yong *et al.* (2010)
	Micrococcus luteus	Cr(VI) reduction Mizoroki–Heck coupling		Deplanche (2008)
	Paracoccus denitrificans	H_2 evolution from hypophosphite Suzuki–Miyaura and Mizoroki–Heck couplings		Søbjerg *et al.* (2009), Bunge *et al.* (2010)
	Pseudomonas putida	H_2 evolution from hypophosphite Suzuki–Miyaura and Mizoroki–Heck couplings		Søbjerg *et al.* (2009), Bunge *et al.* (2010)
	Rhodobacter capsulatus	Hydrogenations		Wood *et al.* (2010)

Metal	Microorganism support	Catalytic test	Comments	References
Pd	*Rhodobacter sphaeroides*	Reductive dechlorination Fuel cell electrocatalyst		Redwood *et al.* (2008), Yong *et al.* (2010)
	Serratia sp.	Cr(VI) reduction Mizoroki–Heck coupling		Deplanche (2008)
	Shewanella oneidensis	Cr(VI) reduction Mizoroki–Heck coupling		DeWindt *et al.* (2005, 2006), Merten *et al.* (2007), Deplanche (2008), Hennebel *et al.* (2011).
		Reductive dechlorination	e- donor supplied by microbial fuel cell	
	Staphylococcus sciuri	Suzuki–Miyaura and Mizoroki–Heck couplings Hydrogenations Reductive dechlorination		Søbjerg *et al.* (2011)
Pt	*Desulfovibrio desulfuricans*	Fuel cell electrocatalyst		Yong *et al.* (2010)

Table 1. Applications of precious metal catalysts supported on bacteria and fabricated from metal salt precursors

Biosorption of precious metals onto biomass has been considered as an attractive alternative to traditional recovery techniques. Microorganisms used for biosorption processes include bacteria, fungi, yeasts and algae. Differences in metal uptake are due to the properties of each microorganism such as cell wall structure, nature of functional groups and surface area (Palmieri *et al.,* 2000). Beveridge (1989) showed that bacteria make excellent biosorbents because of their high surface-to-volume ratios and a high content of potentially active chemisorption sites such as teichoic acid present in the cell walls of Gram-positive bacteria. Various microbial species have been shown to be relatively efficient in the biosorption of metal ions from polluted effluents. The biosorption literature is very large over the last 30 years and the reader is referred to key reviews by Volesky (2007) and Gadd (2009). The application of biosorption to precious metals is less well developed due to their complex solution chemistry.

Volesky (1999) reported the effective utilisation of *Bacillus* biomass (the main active ingredient of the commercial biosorbent AMT-Bioclaim) in the biosorption of Au(III) from cyanide solution. In an early study, Yong *et al.,* (2002b) defined the optimum Pd(II) sorption pH to be between 2 to 4 and achieved Pd recovery using a combination of both biosorption and bioreduction. *D. desulfuricans* was shown to possess a high capacity towards Pd(II) biosorption. Cells challenged with acidic Pd(II) solution (2 mM) showed rapid uptake of metal ions (85% uptake after 10 mins) and up to 15% of the biomass dry weight. Saturation of the biosorption capacity was observed within 35 minutes. The effect of the nature of the Pd(II) complex on the bacterial biosorption capacity was also demonstrated. Optimum uptake from chloride salt ($PdCl_4^{2-}$) was at pH 4 and up to 50% of uptake from a 2 mM solution was observed at pH 2 but only 10% Pd was removed at neutral pH. Uptake from the amine salt ($Pd(NH_3)_4^{2+}$) was pH-independent but was lower; this was attributed to the strong complexing ability of ammonium ions for Pd(II). Furtermore, it was shown that cells of *D. desulfuricans* had greater affinity for Pd(II) than Pt(IV) and Rh(III) and it was suggested to exploit this hierarchy in a new "biological" separation technique (Yong *et al.*, 2002a; Yong *et al.*, 2002b). DeVargas *et al.*, (2004), investigated Pd(II) and Pt(IV) biosorption kinetics by three separate species of the genus *Desulfovibrio* and identified *D. desulfuricans* as the best biosorbent. A low pH is required for the protonation of cell surface functional groups to attract negatively charged species such as $PdCl_4^{2-}$ and the mechanism is probably an ion exchange of $PdCl_4^{2-}$ for biomass Cl^- groups (DeVargas *et al.*, 2005).

Other kinds of high metal-sorbing biomass such as yeast, fungi and algae have also been considered for precious metal recovery. Some chemical compounds of yeast cells can act as ion exchangers with rapid, reversible binding of cations whereas for metal removal and recovery from aqueous solutions, dead fungal biomass seems to offer several advantages; such biomass is often a by-product of industrial processes and may be obtained inexpensively, biosorption kinetics are not affected by metal toxicity since there is no metabolic activity, the biomass does not require nutrient supply and is not affected by the recovery of surface-bound metals (Volesky and Holan, 1995; Vieira and Volesky, 2000). Some studies (Volesky *et al.*, 1994) indicate that the most common yeast biomass (*Saccharomyces cerevisiae*) is an effective biosorbent material and it has been used for Pt(IV) recovery into a proton-exchange membrane fuel cell (PEMFC) to generate electricity following reduction of sorbed Pt(IV) (Dimitriadis *et al.*, 2007). Another industrially important organism, the fungus *Penicillium chrysogenum*, has been shown to extract Au(III) from cyanide solution through anionic gold cyanide species adsorption onto N^-, P^- or O^- containing functional groups on the biomass. However, the biosorption capacity was not encouraging (Niu and Volesky, 1999). Chitosan, which can be extracted from the cell wall of Zygomycetes (a group of phytopathogenic fungi) has a markedly cationic character and a positive charge at most pHs due to its significant content of free amino groups. Thus, chitosan has been found to be an excellent sorbent for the removal of numerous trace metals from wastewaters including Pd(II) and Pt(IV) in addition to base metal ions (see Schmuhl *et al.*, 2001). Catalytically active precious metal nanoparticles supported on chitosan have been reported previously (Guibal *et al.*, 2009 and references therein).

Despite the apparent promise of this type of technology, industry has been slow to endorse biosorption strategies. In addition to economic competitiveness, disadvantages include the perceived variation between batches of the biological product, a lack of specificity and sensitivity to changes in pH (Gavrilescu, 2004). Following discussions at both 2005 and 2007 International Biohydrometallurgy Symposia, the scientific community recognised the lack of commercial potential of biosorption despite extensive attempts and advised for a move towards alternative technologies. This position may, however, change in the case of highly valuable metals and in view of the increasing scarcity of resources.

The work of Lloyd *et al.*, (1998) first showed the potential of a Pd(II) bioreduction process using resting cells of *D. desulfuricans*. When contacted with Pd(II) solution, this organism was shown to precipitate soluble Pd(II) ions to elemental cell-bound Pd(0) using pyruvate, formate or H_2 as the electron donor. Under electron microscopy, opaque deposits of approximatively 50 nm were observed on the surface of challenged cells and these were confirmed to be elemental Pd(0) by energy-dispersive X-ray microanalysis. The authors gave an insight into the enzymatic nature of the Pd(II) reduction mechanism as cells pre-treated with Cu(II), a known inhibitor of periplasmic hydrogenases, lacked Pd(0) crystals. The role of periplasmic hydrogenases toward Pd(II) reduction was further investigated by Mikheenko *et al.*, (2008) and Deplanche *et al.*, (2010). In studies using several mutants of *D. fructosovorans* and *E. coli* with deleted periplasmic and/or cytoplasmic membrane-bound hydrogenases, mutants lacking periplasmic soluble hydrogenases reduced Pd(II) only at the site of the remaining cytoplasmic membrane-bound hydrogenase, confirming the role of the latter in the nucleation and growth of Pd(0) crystals.

Bioreduction of Pd(II) is affected not only by the same parameters affecting biosorption (e.g. pH, metal salt, competitive ions) but also by other factors such as the choice and concentration of the electron donor, the cell concentration and the presence/absence of O_2. In a study investigating the effect of both the electron donor and the cell concentration on the bioreduction of Pd(II) to cell bound Pd(0) by *D. desulfuricans*, Yong *et al.*, (2002a) found that reduction at the expense of H_2 was slower than when using formate. This was attributed to mass transfer limitations between gas in headspace and liquid. They also showed that the effect of the electron donor and cell concentrations synergistically affected reduction rates. Bioreduction was negligible when both cell and formate concentrations were low up to a threshold value (0.13 mg ml^{-1} of cells, 20 mM formate) where bioreduction dramatically increased. This was explained by the ability of Pd(0) to trap H_2 and hold it as stable, highly reactive atomic H•. Deposited Pd(0) on cells has a very high affinity for H_2 liberated from formate and itself acts as a catalyst in the chemical degradation of more formate (Yong *et al.*, 2002b). Thus, the initial biodeposition reaction is hydrogenase(s)-mediated until autocatalytic reduction takes over. This has been exploited in the recovery of Pd(0) from highly aggressive waste using cells lightly pre-palladised in an intial step under more permissive conditions (Table 2).

The rate of Pd(II) bioreduction was shown to be pH-dependent in at least two studies (Yong *et al.*, 2002a; Yong *et al.*, 2002b). However, the pattern differed according to the electron donor used. Bioreduction of Pd(II) at the surface of *Desulfovibrio* sp. using formate decreased to 50% when the pH was decreased from 7 to 6 and no reduction occurred at pH 2. H_2 allowed reduction at lower pH (optimum 4-7) with 50% reduction still retained at pH 2. The same pattern was observed with the influence of anions on bioreduction. Formate-supported reduction was inhibited up to 50% in the presence of low nitrate/chloride concentrations while H_2-supported reduction remained unaffected by 1 M nitrate at pH 2 and 75% activity was retained in 1 M chloride. Bioreduction was also shown to be O_2 sensitive with formate but not with H_2.

The early studies helped to define conclusively the role of microbial cells and enzymatic mechanisms in Pd reduction and inspired a large number of subsequent studies in this area (Table 1). Three key roles could be attributed. Cells were thought to: 1) act as an enzyme catalyst (formate dehydrogenase and hydrogenase(s) providing an initial pool of electrons), ;2) provide foci of metal deposition for subsequent growth of Pd(0) crystals and 3) act as a scaffold for Pd(0), the primary crystals ("seeds") of which autocatalysed further reaction, leading to stable Pd(0)-nanocluster growth.

Other early studies (Mikheenko *et al.*, 2003; Humphries *et al.*, 2004) suggest that the choice of an organism possessing a high number of active hydrogenases is a requisite to improve Pd recovery yields. Hence most studies on Pd biorecovery have used SRB and notably *Desulfovibrio* spp. as organisms of choice for Pd(II) precipitation. As mentioned earlier, *Desulfovibrio* spp. are known to have a high metal-reduction capacity *via* hydrogenases and/or c_3-type cytochrome with broad metal specificity coupled with the possibility to use a large spectrum of compounds as electron donors. However, theoretically, most organisms with strong hydrogenase(s) activity could be potential candidates to be used in Pd biorecovery systems. The attempt to move away from SRB reflects the fact that, to be used at industrial scale, the ideal organism should be non-pathogenic, easily culturable and fast-growing, in addition to having strong hydrogenase activity. Furthermore, H_2S, a by-product of dissimilatory sulphate reduction, is a powerful human and catalyst poison. Use of SRB as supports for making Pd catalysts hence requires extensive washing steps to remove H_2S. However, on the other hand, in the case of gold, "partial" poisoning of the catalyst by S- groups can sometimes be beneficial to catalyst activity but this is difficult to control by chemical methods and hence is not used industrially to date.

Pd(0) biodeposition has been observed in e.g. the Fe(III)-reducer *S. oneidensis* and the photosynthetic organism *Rhodobacter sphaeroides* by mechanisms which are still not elucidated (Table 1, DeWindt *et al.*, 2005; Redwood *et al.*, 2005; 2008). The facultatively anaerobic Gram-negative bacterium *E. coli* has been shown to possess the ability to reduce Pd(II) to Pd(0) in a similar manner to *D. desulfuricans* (Mikheenko, 2004) and waste *E. coli* biomass from fermentation, when palladised, was shown to be effective in the hydrogenation of soybean oil (Zhu et al., 2011). As anaerobic conditions are required for expression of hydrogenases, the use of a facultative anaerobe in a Pd recovery system would be divided into two-steps: 1) aerobic pre-growth to high biomass density followed by 2) anaerobic induction of hydrogenases prior to Pd(II) bioreduction. This possibility does not exist with SRB which are strict anaerobes. The above discussion relates Pd(0) deposition to the function of anaerobic enzymes but this is by no means exclusive. Hence, *Serratia* sp. pre-grown in an airlift fermenter produced catalytically active Pd(0) (Deplanche, 2008) while studies using *E. coli* grown aerobically attributed reduction of Pd(II) to formate dehydrogenases FDH-O and FDH-N (Foulkes, J., personal communication). Extensive studies of Pd(0) deposition on cells of the aerobic *Bacillus sphaericus* assumed metal deposition onto the cell surface S-layer (Fahmy *et al.*, 2006) and the resulting material had a similar catalytic activity in hydrogenation of itaconic acid as *Desulfovibrio* bioPd(0) (Creamer *et al.*, 2007). However, further study showed deposition of Pd(0) below the S-layer and within the inner cell surface layers (Skibar *et al.*, 2005) and an additional mechanism was implicated. Other Gram positive organisms have been used for the production of Pd(0) (Deplanche, 2008; Gauthier *et al.*, 2009) but the mechanisms of metal deposition are less well elucidated as compared to Gram negative Enterobacteria and Pseudomonads.

The bioreductive route has been successfully exploited to recover metals from various waste sources containing dilute amounts of PGMs. The metallic nanoparticles formed on the biomass are catalytically active and several studies have shown the potential of a one-step conversion of PGM wastes into bioinorganic catalysts (Table 2).

In this first case study, the ability of the common enterobacterium *Escherichia coli* to precipitate metallic species is used to demonstrate the recovery of PGMs from model solutions containing palladium and platinum mixtures. Biorecovery is then applied to a leachate obtained from a spent car catalyst containing various metallic species. Characterisation of the biorecovered material shows that discrete metallic nanoparticles (NPs) are formed on the cell surface. The catalytic activity of biomass-supported metallic NPs is described, demonstrating a one-step conversion of car catalyst waste into a new product.

Waste source	Metal(s) biorecovered	Microorganism	Catalytic test	Comments	References
Jewellery	Au	*Desulfovibrio desulfuricans, Escherichia coli*	Selective oxidation of glyerol		Deplanche *et al.*, (2007)
Electronic scrap leachate	Au, Pd, Pt	*Desulfovibrio desulfuricans*	NT	Pre-palladised cells used to recover other metals	Creamer *et al.*, (2006)
PGM reprocessing waste	Pd, Pt, Rh	*Desulfovibrio desulfuricans, Escherichia coli, Cupriavidus metallidurans*	H_2 evolution from hypophosphite, Cr(VI) reduction, Fuel cell electrocatalyst		Yong *et al.*, (2002a, 2010), Mabbett *et al.*, (2006)
PGM leachate from secondary sources	Au, Pd, Pt, Rh	*Escherichia coli*	Cr(VI) reduction	Leachate produced using microwave technology Pre-palladised cells used to recover other metals	Murray *et al.*, (2007)
Industrial effluent	Pd	*Cupriavidus necator*	Suzuki–Miyaura coupling		Gauthier *et al.*, (2010)

Table 2. Applications of precious metals bioinorganic catalysts fabricated from wastes

2.1 Materials and methods

2.1.1 Organism and culture conditions

Escherichia coli MC4100 cells were cultured in 12 litres of nutrient broth under anaerobic conditions (Mikheenko, 2004). Cells were harvested by centrifugation, washed three times in MOPS-NaOH buffer pH 7.0 and resuspended in a known volume of buffer. The cell density was checked at an optical density (OD) of 600nm; the OD_{600} was converted to bacterial dry weight by a previously determined calibration. With a dry weight of cells between 20-30 mg/ml the cell cultures were then split into six aliquots in preparation for pre-metallisation

2.1.2 Preparation of leachate from crushed automotive catalyst

Automotive catalysts are generally available in the form of monoliths. Monolithic catalysts employ a high surface area washcoat on the top of a honeycomb structure. The honeycomb is generally composed of magnesium cordierite $2MgO.2Al_2O_3$. The washcoat consists of γ-alumina as a substrate for the valuable metals (PGMs) and a variety of additives to improve the catalytic action and to stabilise alumina and active metals at exhaust operating conditions (Angelidis and Sklavounos, 1995). The channels of the monolith are small, on the order of 1 mm, which allows for a large number of channels, increasing the surface area to volume ratio and conversion rates (Depcik and Assanis, 2005).

The outer steel casing was removed and the used autocatalyst was processed by jaw crushing, ground using a Rolls crusher and then passed through a 1mm screen. Any oversize material was reground in the Rolls crusher so that all test material was of diameter $d \leq 1mm$. The automotive catalyst used for leachate production had 600 channels per square inch, thus each channel was 1.04 mm wide. Any material greater than 1mm was reground in order to avoid over-crushing but to facilitate maximum acid - washcoat interaction. *Aqua regia* (60 ml; 3 parts 37% HCl to 1 part 70% HNO_3) was added to 6 g of milled catalyst and allowed to stand in an open vessel (30 min). The vessel was then sealed and placed in a microwave (CEM Microwave Accelerated Reaction System 5) set to ramp (109°C in one min using a power of 600 W), maintain that temperature (15 min), then undergo a cooling cycle (5 min). The contents of the vessel were transferred with washings (half the volume of distilled water to *aqua regia*), centrifuged (4000 rpm; 10 min) and the supernatant was retained for biomass metallisation tests. Commercial analysis of the leachate by Engelhard Corporation gave 24 ppm Pd, 4 ppm Rh and <1 ppm Pt. Stated ICP limit of detection is 0.1 ppm for PGMs. Subsequent analysis of the leach residue solids (by copper collection and XRF of copper button) confirmed >95% Pd extraction during leaching but only 50% Rh extraction.

Analysis of the catalyst used in this study against other typical spent automotive catalysts showed that PGM levels were unusually low, with Pd being approximately 10% of the value of another catalyst processed under the same conditions for alternative testing. Thus the decision was taken to spike the leachate with Pd(II) salt (Na_2PdCl_4, Sigma Aldrich, 98% pure) to a final concentration of 400 ppm in order to increase the Pd(II) level to that more representative of a "typical" catalyst, while still providing the solubilised washcoat background matrix of a "real" leachate. For biorecovery experiments, the leachate was diluted 1:9 with dH2O and the pH was brought to pH 2.2 using 6 M NaOH.

2.1.3 Assay of metals

Removal of PGM from test solution was monitored by a spectrophotometric method using $SnCl_2$ (Dasages, 1978). The reagent solution contained 5.98 g of $SnCl_2$/100 ml of concentrated HCl. For PGM assay sample (200 μl) was added to $SnCl_2$ solution (800 μl) and A_{463} for Pd(II) and A_{401} for Pt(IV) was determined after one hour. It was not possible to estimate accurately the concentrations of metals in mixed metal solutions due to cross-interferences, but the assay was a convenient indicator of the presence of residual non-reduced metal in solution.

2.1.4 Bioreduction experiments

Solutions of 2 mM Pd and Pt were prepared in 1 mM HNO_3 using Na_2PdCl_4 and K_2PtCl_4 salts respectively. Cells were metallised with Pd or Pt (see below) as follows. The required volume of metal solution was calculated then added to aliquots of cells to achieve the desired varied metal loadings, H_2 was bubbled through the suspension for 30 minutes and suspensions were incubated at 30°C to allow reduction of metal onto the cells. Metal reduction was confirmed as loss from solution using the $SnCl_2$ assay of sample supernatants as described previously (Creamer *et al.*, 2008). Following full reduction of metals cells were harvested by centrifugation, washed once using distilled water and then resuspended in 30 ml of distilled water.

For PGM pre-loading as above, batches of cells were split into 6 aliquots and exposed to 2 mM solutions of Pd and Pt in the correct volume as above to give metal loadings of 5%, 2%

and 1% on the cells. With the cells pre-loaded to the desired levels samples (16 mg of pre-metallised cells; weight before metal deposition) were then exposed to a mixed solution of 0.34 mM Pt and 0.42 mM Pd (model leachate), which gave final metal loadings after reduction of 15 wt%, 16 wt% and 20 wt%. Hydrogen was used as the reducing agent, here via catalysis due to the metals already present on the cells rather than hydrogenase activity. Samples were taken at 0, 0.5, 1, 2 and 5 minutes, with $SnCl_2$ assay of sample supernatants. The results were expressed as percentage of target metal reduction against time. Following full reduction the cells were harvested by centrifugation, washed once in H_2O and once in acetone, dried and ground in an agate mortar. The resultant powder was passed through a 100 micron sieve to obtain a fine powder catalyst.

For experiments on PGM recovery from the spent car catalyst leachate, 1 ml aliquot of the best candidate catalyst (5% BioPd as identified via preliminary catalytic testing using Cr(VI) was suspended in 77.5 ml of leachate as to give a final total metal loading of 20% w/w on cells.

2.1.5 Transmission electron microscopy (TEM)

Pellets of Pd-loaded bacteria were rinsed twice with distilled water, fixed in 2.5% (wt/vol) glutaraldehyde, centrifuged, resuspended in 1.5 ml of 0.1 M Na-cacodylate buffer (pH 7) and stained in 1% osmium tetroxide in 0.1 M phosphate buffer, pH 7 (60 min) for transmission electron microscopy (TEM). Cells were dehydrated using an ethanol series (70, 90, 100, 100, 100% dried ethanol, 15 min each) and washed twice in propylene oxide (15 min, 9500 g). Cells were embedded in epoxy resin and the mixture was left to polymerize (24 h; 60 °C). Sections (100-150 nm thick) were cut from the resin block, placed onto a copper grid and viewed with a JEOL 1200CX2 TEM; accelerating voltage 80 kV. The identity of electron opaque deposits was previously confirmed as Pd using energy dispersive X-ray analysis (Lloyd *et al.*, 1998).

2.1.6 Cr(VI) reduction tests

The catalytic activity of dried, ground catalyst made from each solution (i.e. pre-metallised cells with subsequent loading of additional metal from either model leachate or real leachate) was estimated by the reduction of Cr(VI) to Cr(III). The catalyst (10 mg) prepared as described above was accurately weighed and added to a 12 ml serum bottle. 5 ml 0.5 mM $N_aC_rO_4.4H_2O$ in 20 mM MOPS-NaOH buffer pH 7.0 was then added and reactors were sealed with butyl rubber stoppers. Recators were degassed under vacuum, sparged with nitrogen for anaerobiosis and placed onto a rotary shaker (10 min) to ensure good mixing and distribution of catalyst. Sodium formate (1 ml; to 25 mM) was added, reactors were left bubbling under nitrogen to maintain anaerobic conditions and placed on a shaker (180 rpm). Samples were taken at 30 minute intervals, centrifuged (13000 rpm; 4 min) and residual Cr(VI) in solution was analysed using a diphenylcarbazide (DPC) assay method as described previously (Humphries *et al.*, 2006).

2.2 Results and discussion

2.2.1 Biorecovery of Pd and Pt from model solutions

Leachates produced from spent car catalysts are too aggressive to apply the bioreductive route directly using live cells. To overcome this limitation, we designed a 2-step process in which resting cells of *E. coli* are first pre-metallised under biocompatible conditions with a low amount of Pd or Pt. This bioinorganic catalyst is then used to reduce and precipitate chemically PGMs from spent leachates.

A first series of experiments aimed to identify the best system to recover PGMs from model solutions (Pd, Pt). *E. coli* cells were pre-metallised at 1, 2 and 5% using Pd or Pt. Aliquots of each suspension were subsequently exposed to a solution containing both Pd(II) and Pt(IV) at concentrations of 0.42 mM and 0.34 mM respectively to reflect the amounts found in typical car catalyst leachates. The removal of Pd(II) and Pt(IV) from the mixture was rapid; no metallic species were detected by assaying the supernatants of reactors with $SnCl_2$ after 1 min of exposure to H_2. There was no significant difference of reduction rates between cells pre-metallised with Pd or Pt, i.e. both metals autocatalytically reduced free metal ions in solution. Similarly, no significant increase in Pd(II) or Pt(IV) reduction rate was observed when the initial loading on cells was increased from 1 to 5%. This suggests that *E. coli* cells loaded with metal "seeds" can precipitate large amounts of metals. The final PGM loadings on cells following complete reduction of Pd and Pt from the mixture was estimated at 15 wt%, 16 wt% and 20 wt% for initial loadings of 1 wt%, 2 wt% and 5 wt% respectively. TEM micrographs of pre-metallised *E. coli* cells before and after contact with the PGM mixture is shown in Fig. 3. The precipitated metals can be seen as discrete electron opaque particles before treatment (intial loading of 5wt% on cells, fig 3A) and as more numerous, larger aggregates following PGM precipitation from the Pd/Pt mixture (fig. 3B, arrowed)

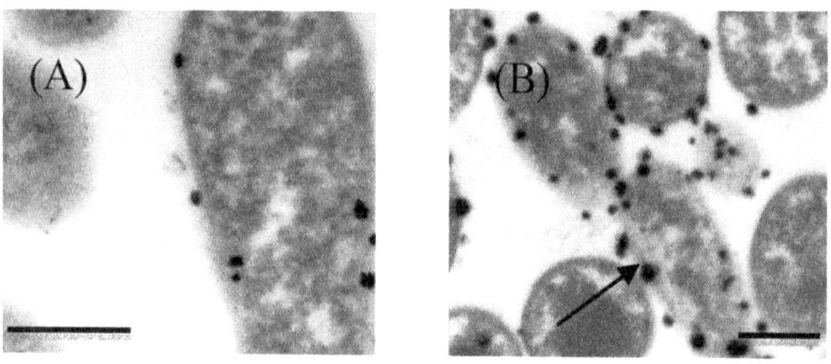

Fig. 3. TEM micrographs of pre-palladised *E. coli* cells before and after the recovery of PGMs from a model solution. The initial Pd loading was set to 5 wt% (A) and increased to 20 wt% (total metal loading) following exposure to the Pt/Pd mixture (B). Scale bars are 500 nm.

From these results the use of 1% Pd pre-loaded cells would be optimum for future use as they offer a near identical reduction rate to that of the cells with higher pre-loadings, with minimal 'sacrificial' metal and with Pd being the metal of choice as it is generally a lower cost than Pt. However, a previous study showed that catalysts made from cells pre-metallised with 2% or 5% Pd were significantly more active catalytically in the reduction of Cr(VI) to Cr(III) (Taylor *et al.*, 2011). Hence, cells pre-loaded with 5% Pd were retained for use in subsequent tests on PGM recovery from a real car catalyst leachate and catalytic testing of the resulting material.

2.2.2 Biorecovery of Pd and Pt from spent car catalyst

The majority of previous PGM biorecovery research has focused on either using model solutions or "real" leachates from mixed metal wastes e.g. spent furnace lining that contained many different metals in addition to the desired PGMs (Murray *et al.*, 2007).

Model solutions are considered too simple (and therefore non-representative of real world wastes) as they contain only the metals of interest. Conversely complex "real" leachates are difficult to study due to interference effects in accurate analysis of the metals and hence the difficulty and expense of the latter. The leaching of autocatalysts provides a good intermediate step between model solutions and complex waste leachates as the cordierite honeycomb is insoluble and hence the leachate is composed of the alumina washcoat and the solubilised PGMs i.e. it is relatively simple with respect to the number of components.

The microwave leaching process used in this study yielded 30 ppm PGM in 67% *aqua regia*. The PGM concentration was unusually low for this type of catalyst and, consequently, the leachate was spiked to 400 ppm with Pd(II) (see materials and methods). The leachate was then diluted 10x in H_2O to lower the concentration of acids and brought to pH 2.2 with 6 M NaOH. Pre-palladised cells of *E. coli* (1 ml, 5 wt% initial Pd loading) were added to 77 ml of leachate and H_2 was bubbled through this mixture for two hours. Samples were taken at 12 h intervals and the residual concentration of PGM was estimated using the $SnCl_2$ assay. Figure 4 shows the time course of PGM reduction from the spent automotive catalyst leachate.

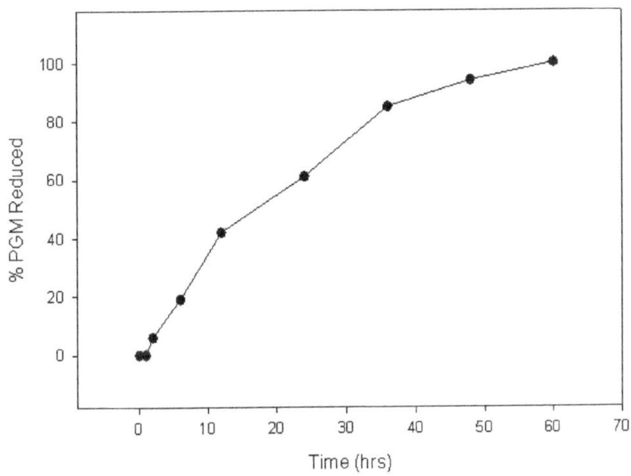

Fig. 4. Time course of PGM reduction from spent automotive catalyst leachate using *E. coli* cells pre-palladised to 5 wt%.

In contrast to model solutions (where total metal removal was obtained within 1 min), PGM reduction from the catalyst leachate by the pre-palladised *E. coli* cells was slow, and proceeded in three distinct phases (Fig. 4). An initially rapid rate of metal removal (0-10h) was followed by a ~ halving of the rate between 10-35 h; selectivity of metal removal was not tested. Removal of the final ~20% of the metals was very slow over the final 20h. Full disappearance of Pd(II) in solution was achieved after ~60 hours of contact

In order to implicate the compound responsible for the inhibition of PGM reduction a simple test was carried out. Model leachates (Pd(II) and Pt(IV)) were prepared as before using fresh *aqua regia* and aliquots were spiked with Pd(II) (400 ppm final concentration), neutralised and diluted as before. The pH was adjusted to 2.0. Aliquots of model leachates were spiked with silica (SiO_2 and Al_2O_3 to 173 ppm final concentration) and a mixture of both. The bioinorganic catalyst was added in each reactor and the rates of PGM reduction were followed as previously. Addition of either Al and/or Si promoted a significant decrease of the rate of PGM reduction; Pd(II)/Pt(IV) disappearance from the model solution was observed after 6 and 14 hours of contact with the bioinorganic catalyst respectively. Full PGM removal was not observed in the solution supplemented with both Si and Al even after 48 h exposure, whereas metal was removed from the control (leachate + H_2O) within 5 mins as per the model solutions. These results suggest that the presence of Al and Si inhibits PGM recovery and they are responsible for a substantial increase in reduction time observed with the spent car catalyst leachate.

The catalytic activity of the material obtained following PGM recovery from spent car catalyst leachate was estimated in the reduction of Cr(VI) to Cr(III) in the presence of formate. This was tested alongside the catalyst made from PGM recovered from model solutions (see above). Both catalysts were manufactured using *E. coli* cells pre-palladised to 5 wt% and had a similar final metal loading (after PGM recovery) estimated to be 20 wt% PGM.

Both catalysts were active in Cr(VI) reduction tests (Fig. 5) and showed similar initial reaction rates. Near-complete Cr(VI) reduction was obtained with the catalyst made from model leachate after 120 min whereas the catalyst obtained from real leachate showed a slower rate after 30 min, probably attributable to the presence of non-PGM contaminants (possibly Si and Al) which could partially poison catalytic PGM nanoparticles. Nevertheless, more than 90% of the Cr(VI) was reduced by 180 min by the biorecovered material.

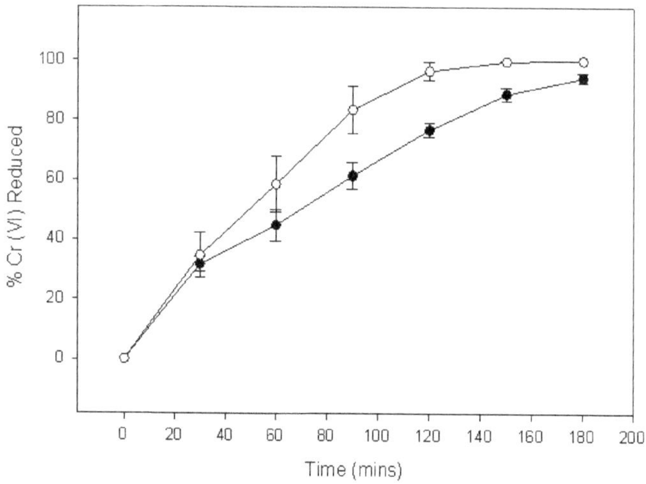

Fig. 5. Time course of the catalytic reduction of Cr(VI) to Cr(III) over two bioinorganic catalysts fabricated from (○) a model leachate and (●) a spent car catalyst leachate. Results are expressed as percentage Cr(VI) reduced over time. Data are from two separate experiments. When no error bars are shown, these were within the dimensions of the symbols.

These experiments demonstrate the potential of bioreductive processes towards PGM recovery and a one-step conversion of waste material into novel bioinorganic catalysts. The range of reactions amenable to bioPGM catalysis is constantly being expanded (Table 1) and work is on-going to optimise the formulation of bioinorganic catalysts made from wastes for specific applications. It should be noted that in some cases (Macaskie *et al.*, 2011) the presence of metallic impurities in the biorecovered materials can increase the catalytic activity (c.f. Fig. 5) and a full survey of bionanocatalysts from wastes for particular applications needs to be approached systematically on a case by case basis, applying metal thrifting by decreasing as far as possible the initial metal expenditure in the seeding step.

3. Case study 2: Biorecovery of Rare Earth Elements (REE)

In contrast to PGMs, the REE are not scarce due to low natural abundance but to the complexity of their extraction and refining. In this, China is world-leading and this near monopoly means that REE abundance is controlled socio-economically and REEs, due to their numerous high-tech applications, could become a strategic geopolitical resource.

Traditional extraction/recycling processes of REE require several steps of pre-treatments with strong acids and alkali followed by extraction using organic solvents which, if not recovered, can create extensive environmental damage often leading to the contamination of natural water streams.

Although pre-2010 there was no major study of focused microbial REE recovery the problem of legacy nuclear wastes, the decline of nuclear power in the 1990s and the current decommissioning of first generation nuclear reactors prompted an early focus on the use of microorganisms to remove radionuclides from waste streams. Due to the difficulties of handling the latter, extensive use was made of surrogate elements whose behaviour mimics the radionuclide of interest. Hence, La^{3+} was chosen to provide a convenient surrogate for Am^{3+} and, indeed, the trivalent states of Np and Pu where these would occur (under reducing conditions). An early study (Plummer & Macaskie 1990) established that by using the bacterial phosphatase mechanism (see earlier) La^{3+} was accumulated to 17% of the bacterial dry weight after only 4 h at pH 7, being twice as effective as uranium recovery. This was attributed to the reduced ability of La^{3+} to bind to citrate as compared to UO_2^{2+}; many metal ions form insoluble colloidal hydroxides in solution and incorporation of citrate reduces this. In the 'real world' a pure solution is improbable and most would contain various species capable of forming complexes with metals. Such complex formation holds the metal in a soluble form but, on the other hand, the lower free metal ion concentration (defined by the binding constant) also retards the rate of metal phosphate formation on the bacteria (see Plummer & Macaskie 1990 for discussion). The elemental ratio of La:P in the precipitate was determined as 1:1 by proton induced X-ray emission and the material was suggested to comprise $LaPO_4$; the same study also showed removal of Eu^{3+} (Macaskie *et al.*, 1994). Yong *et al.*, (1998) obtained X-ray powder diffraction data of bio-$LaPO_4$ to show that this contained crystalline components but no further analysis was attempted. The peaks were not well defined, suggesting that the material was possibly nanocrystallline in nature. Working towards development of a continuous process for Pu removal, Yong & Macaskie (1998) used polyacrylamide gel-immobilised cells in a flow through column, where the imposition of a time constraint in each voxel (volumetric element) by the application of increasing flow rate (F; t = 1/F; F = flow rate in ml/min) showed that at a flow rate of 1 ml/min (total column volume ~ 30ml) and with 0.3 mM La^{3+} or UO_2^{2+} (pH 7 with 5 mM citrate) the removal of La^{3+} and UO_2^{2+} was 80% and 60% respectively, which is less of a difference than that reported by Plummer and Macaskie (1990; above) using cell suspensions. However increasing the flow rate by 50% (i.e. halving the flow residence time per voxel: Yong *et al.*, 1998) gave corresponding removals of 70% and 40%, suggesting that there is a time element to the metal deposition process (probably the time required for metal phosphate deposition from the liquid phase); Plummer and Macaskie (1990) did not give full metal removal profiles over time.

Additional caution needs to be exercised in applying directly the results obtained using one REE to another. For example Bonthrone & Macaskie (in Diels *et al.*, 1996), using cell suspensions, found the UO_2^{2+} removal after 2 h to be 7 times greater than that of Y^{3+} at both high (~ 900 units) and low (34 units) phosphatase activities, showing that phosphate availability was not rate-limiting. This study also reported the continuous removal of Y^{3+} by cells immobilized in polyacrylamide gel but details were not given.

La^{3+} is a very useful model REE element in another key aspect. Unlike, for example, Gd^{3+}, its nucleus is 'NMR-silent'. Use was made of this to examine the flow-through column system in three and four dimensions to follow the progression of flows and blockages within the column at any spatial position using a flow-through bioreactor containing biofilm of the metal-accumulating *Serratia* sp. (reassigned from *Citrobacter*) using magnetic resonance imaging (MRI) (Fig. 2: Paterson-Beedle *et al.*, 2001; Nott *et al.*, 2001; 2005 a, b). The methodology was described by Paterson-Beedle *et al.* (2001). Nott *et al.* (2001) chose polyurethane reticulated foam as the best support (as compared to glass or ceramic) as this gave the best signal to noise ratio for MRI. An initial study showed the suitability of MRI for obtaining spatial information (Nott *et al.*, 2001) and subsequent studies using the columns in a flow through system gave numerical information which was described in terms of a chemical engineering model (Nott *et al.*, 2005 a,b). For design of a metal removing and recovery process it is vital to a have a robust mathematical description as scale-up predictions can be made from small pilot studies. In particular, for REE recovery from wastes (see later) a low pH may be unavoidable and, by using the tools of MRI, it would become possible to predict the outcome of low pH or high salt (and competing ions) with a greater degree of insight than by measuring column inputs and outputs alone.

Moving towards the goal of metal recovery at low pH Tolley *et al.*, (1995) investigated La^{3+} removal at pH 5, prompted by an earlier study (Macaskie *et al.*, 1990) that suggested that a reduction in metal removal efficiency at pH 5 was attributed to increased metal solubility at low pH. Like the MRI work (above) this study placed metal removal into a mathematical definition, showing that decreased metal removal at acidic pH could not only be compensated by an increased phosphate supply, but the extent to which this was achievable; essential tools for the design of an industrial REE recovery process. Tolley *et al.*, (1995) found that La^{3+} removal was abolished at pH 4 (at which pH the activity of the mediating phosphatase was reduced by 50%). However another study looking at uranium removal from acidic mine wastewater (Macaskie *et al.*, 1996; 1997) showed that, following a period for metal phosphate nucleation, the initially low metal removal gave way to a sustained removal of ~70% of the input UO_2^{2+} (35 ppm; ~ 0.15 mM) (Macaskie *et al.*, 1997) at a pH as low as 3.5 (Macaskie *et al.*, 1996). Importantly, the process could be accurately defined mathematically (Macaskie *et al.*, 1997) but an extensive modeling study in real U-wastewater was not done. In the case of La^{3+} removal at pH 5 the enzymatically-mediated release of phosphate from the organic donor molecule was identical to that at pH 7 but La^{3+} removal was reduced (Tolley *et al.*, 1995). Using a flow through column system and polyacryamide gel-immobilised cells the metal removal (input concentration 1 mM) was 83% and 55% at pH 7 and 5, respectively in 5 mM citrate buffer at the same flow rate

(308 ml/h). Reasoning that the solubility of metal phosphate would be higher as the pH was reduced (i.e. less tendency to form mineral deposits) the activity of the phosphatase was increased by the use of a phosphatase overproducing mutant (approx. twice the activity of the parent strain) and by the use of carbon-limited continuous culture, which increased the activity to ~ 2000 units. The latter compensated to some extent; the respective $FA_{1/2}$ (that flow rate at which 50% metal removal was achieved) values at pH 5 and 7 were 120 ml/h and 600 ml/h respectively. This meant that, in this example, increasing the phosphatase production by four-fold resulted in an increase in bioreactor efficiency of two-fold at the lower pH. This work defined the lowest useful operating pH of the enzymatically-mediated metal recovery system as ~ pH 4 and also concluded that the metal solution chemistry and desolubilisation behavour are the limiting factors at low pH, which is of major importance in the application to recovery of REE from acidic leachates.

However it is also worth noting that the choice of metal-donating substrate is also contributory; the effect of high flow rate on the phosphate release is attributable in part to the affinity of the enzyme for its substrate (K_m; that substrate concentration giving half-maximal reaction rate). The K_m in the column system is, in part, responsible for the 'resistance' of the column to high flow rates and it was found (Tolley *et al.*, 1995) that the effect of the pH on the apparent K_m is, to some extent, substrate-dependent. Since the model substrate glycerol 2-phosphate is unattractive at industrial-scale a cheaper substrate like phytic acid (inositol phosphate) would be more appropriate. The latter was shown to support metal removal (Paterson-Beedle *et al.*, 2010) and it contains 3 moles phosphate per mole, with the phosphate groups removed sequentially via phytase activity (see Paterson-Beedle *et al.*, 2010). Hence a phytic acid-based column system may have three K_m values and be more difficult to describe mathematically, but this represents a potentially attractive route for potential metal recovery from wastes.

A recent study (Jiang *et al.*, 2010) has shown another route to REE biomineralisation. This used the yeast *Saccharomyces cerevisiae* and Ce(III). (Ce(III)/(IV) is a surrogate for Pu(III)/IV) and Np(III)/(IV), and a comprehensive suite of analytical methods, including synchrotron-based X-ray absorption fine structure (XAFS), showed conclusively that Ce(III) was deposited on the yeast cell surface as needle-shaped Ce(III) phosphate nanocrystals with a monazite structure. Importantly, no exogenous phosphate was added and the mobilisation of intracellular phosphate reserves to supply inorganic phosphate into the growing crystals was concluded (Jiang *et al.*, 2010). Since yeast is well known to store phosphate in the form of polyphosphate (which is known to support metal desolubilisation when mobilised: see earlier) it seems likely that the yeast phosphate was derived from this source, with metal phosphate deposition via polyP mobilisation and phosphate efflux processes.

The yeast system of Jiang *et al.*, (2010), unlike *Acinetobacter* spp, (Boswell *et al.*, 2001) needs no aerobic/anaerobic cycling and is thus intrinsically a more useful process for industrial REE recovery but the limit of metal deposition needs to be established. Importantly, and as also found for the bacterial system with UO_2^{2+} at pH 4 (Macaskie *et al.*, 2007), removal of Ce(III) (0.18 mM) by the yeast was initially negligible (up to 96 h at pH 3) and then proceeded abruptly to 60% of the initial concentration after 120 h. Phosphate was produced by the cells at pH 3, again indicating that metal phosphate solubility and not biochemical activity, is the limiting factor in biomineral formation at low pH. Since, following a delay, metal removal can proceed effectively at pH 3 (Macaskie *et al.*, 1997; Jiang *et al.*, 2010) is can be suggested that nucleation, and not biomineral formation, is the limiting step. This was also suggested in a study of bio-hydroxyapatite (HA) formation by *Serratia* sp. where a pH of above 8 was needed for effective HA-biomineral formation. However, following a nucleation period Ca^{2+} was removed from a phosphate-containing wastewater into bio-HA at neutral pH (Yong *et al.*, 2004).

Ce(III) removal at low pH by *S. cerevisiae* (Jiang *et al.*, 2011) was similar to La(III) removal by *Serratia* sp. (Tolley *et al.*, 1995), showing good REE removal at pH 5 and less so at pH 4. Production of phosphate by the yeast cells was very similar at pH 3 and at pH 5, indicating that the anomalous behaviour was attributable to metal speciation, as was concluded in the case of La^{3+} in the bacterial system (above). Notably, phosphate release was promoted from the yeast cells by the addition of H^+ ions (Jiang *et al.*, 2010); this may reflect possible different physiological functions of polyP: as a pH control agent in yeast (phosphate is a strong buffer with three pK_a values) and as a reserve of cellular energy in the bacterial system, providing the means to metabolise anaerobically where the cells are not able to grow (see Boswell *et al.*, 2001 and references therein).

The development of methodologies using Ce(III) and La(III) as surrogates for the tri/tetravalent actinide elements has provided a good platform from which to develop effective methods targeted specifically towards recovery of REE. A study by Jiang *et al.*, (2010) has shown formation of cerium (Ce(III)) nanoparticles on the surface of the *Saccharomyces* cells (Fig. 6).

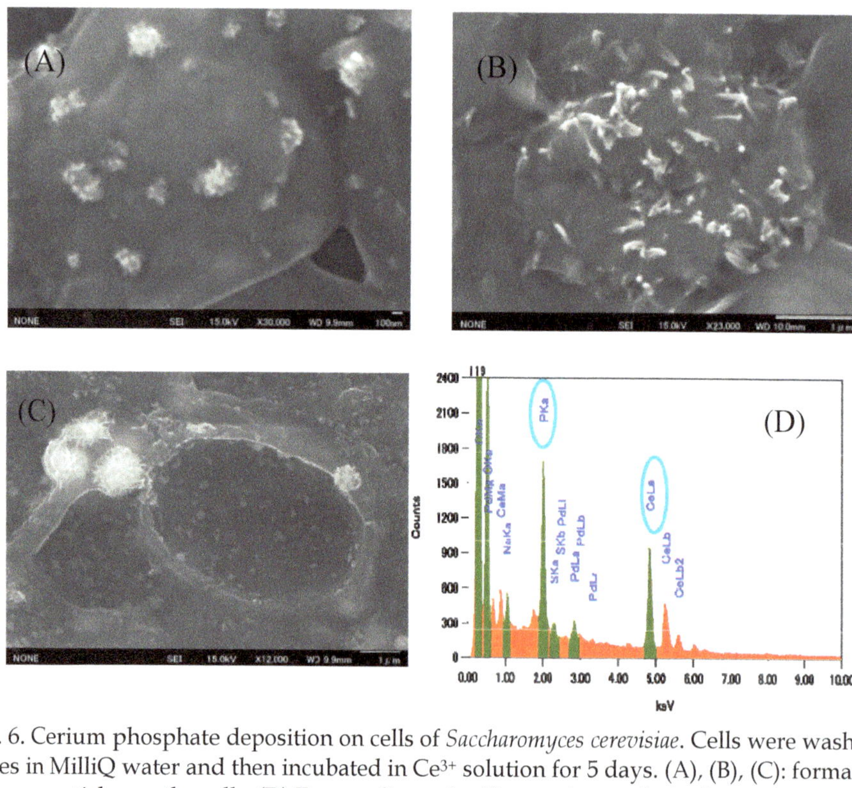

Fig. 6. Cerium phosphate deposition on cells of *Saccharomyces cerevisiae*. Cells were washed 3 times in MilliQ water and then incubated in Ce^{3+} solution for 5 days. (A), (B), (C): formation of nanoparticles on the cells. (D) Energy dispersive X-ray microanalysis shows the cellular deposits to comprise Ce and phosphate.

An SEM and TEM study of $CePO_4$ crystal development is shown in Figs 6 and 7. The deposits formed after 4-5 days, identified as containing cerium and phosphorus by energy dispersive X-ray microanalysis (Fig. 6D), did not give an interpretable X-ray diffraction pattern and were concluded to be nanocrystalline, as suggested by the high resolution TEM images (Fig. 7A,B). An unpublished study (M. Jiang *et al.,*) of Yb-phosphate nanocrystallisation on *S. cerevisiae* included X-ray extended fine structure (EXAFS) analysis which showed subtle differences in 'early' deposits (30 mins) at the three pH values, with the Yb L_{III}-edge appearing similar for samples at pH 4 and 5, but not at pH 3, while with more 'mature' samples (120h) the pH 5 material showed a similarity in its Yb-P site to a $YbPO_4$ reference which was less pronounced in samples at the lower pH values. A 48 h sample at pH 5 appeared similar to a pH 4 sample at 120 h, suggesting an evolution of the crystals which was time- and pH-dependent.

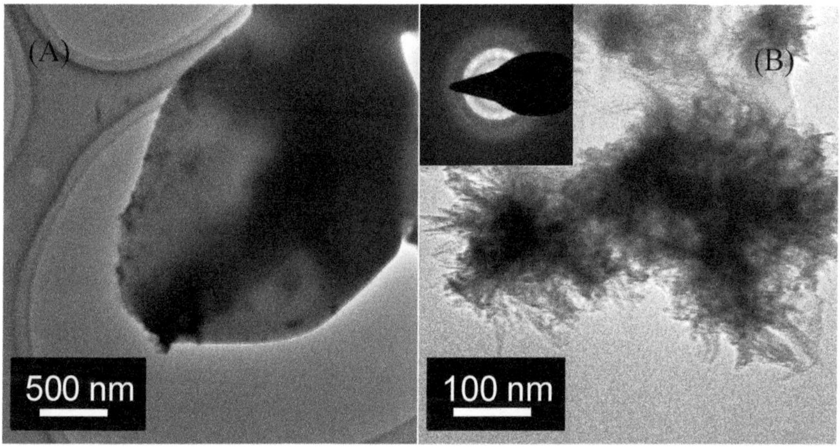

Fig. 7. (A) Development of $CePO_4$ crystals (monazite) on the surface of *Saccharomycese cerevisiae* cells after 48 hours at pH 5; enlarged image of a single particle reveals that this particle consists of needle-shaped Ce-phosphate nanocrystallites (B) (Jiang *et al.,* 2010).

In this second case study, we report pilot experiments on the continuous recovery of two model REE (europium and neodymium) in flow-through columns containing immobilised cells of a *Serratia* sp. The recovered metals are available for further reprocessing in yields exceeding by several-fold the mass those of the bacteria themselves.

3.1 Materials and methods

3.1.1 Microorganism, support and biofilm production

Serratia sp. (NCIMB 40259) was used under license from Isis Innovation, Oxford, UK. Cells were grown as biofilm on polyurethane reticulated foam (Filtren TM30, discs: diameter 2 cm and depth = 0.5 cm, 1.57 cm^3), supplied by Recticel, Belgium. The biofilm was grown in an air-lift fermenter under carbon limiting continuous culture with the phosphatase activity of the cells from the fermenter outflow determined as described previously (Paterson-Beedle & Macaskie 2004). Even colonisation of the foam within the matrix and metal deposition thoughout was confirmed using magnetic resonance imaging (Macaskie *et al.,* 2005).

3.1.2 Preparation of reactor packed-bed systems for neodymium and europium bioaccumulation

Serratia sp. biofilm immobilised onto polyurethane foam discs (Fig. 8) was packed in cylindrical glass reactors (height, 5 cm, i.d. 2 cm). Reactors were exposed to the solutions below (all experiments were done in duplicate):

Solution 1: $N_3NdO_9.6H_2O$ (1 mM), glycerol-2-phosphate (G2P) (5 mM) and sodium citrate buffer (2 mM), pH (5.5). Solution 2: as solution 1 but substituting $EuN_3O_9.xH_2O$ (1 mM) for the neodymium salt. Reactors were exposed to a continuous upward flow (ca. 5 ml/h, 54.5 h) via an external peristaltic pump (Watson-Marlow, 323).

The outflow solutions of reactors were analysed for either Nd or Eu as appropriate (see below). Controls were set up as above using biofilm autoclaved for 15 min (121°C) prior to metal treatment. The controls determined the amount of metal removal attributable to biosorption alone.

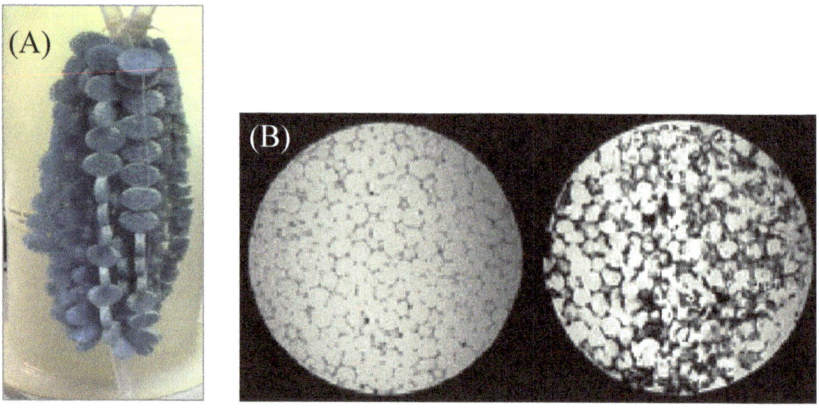

Fig. 8. (A) *Serratia* sp. biofilm grown on polyurethane reticulated foam discs. (B) Magnetic resonance imaging of discs with or without *Serratia* biofilm. MRI image of discs (diameter 20 mm, left) containing polyurethane foam only and *Serratia* sp biofilm coated polyurethane foam (right). The biofilm in a T_2-weighted spin-echo image is the darker regions, due to a reduced T_2 of water (P. Yong, M.L. Johns and L.E. Macaskie, unpublished).

3.1.3 Spectrophotometric analyses of metals

Spectrophotometric determination of Nd and Eu from column outflows was as described by Onishi (1986) using the Arsenazo(III) method with some modification, as follows. For analysis, Nd or Eu sample or standard (standard concentrations 30 μl) was added to a 1.5 ml cuvette followed by ammonium acetate buffer (1.97ml, 1M, pH 3.3). Metal was visualised after mixing by the addition of 0.1 ml of 0.15% (w/v) arsenazo (III). The absorbance was measured at A_{660} against a reagent blank.

3.2 Results and discussion

Reactors containing immobilised biofilms of *Serratia* sp. were challenged with solutions containing Eu or Nd (1 mM) for 54.5 hours at a flow rate of 5 mL/h. The outflow from each reactor was assayed for residual metals and the percentage of metal removed is shown in (Fig. 9).

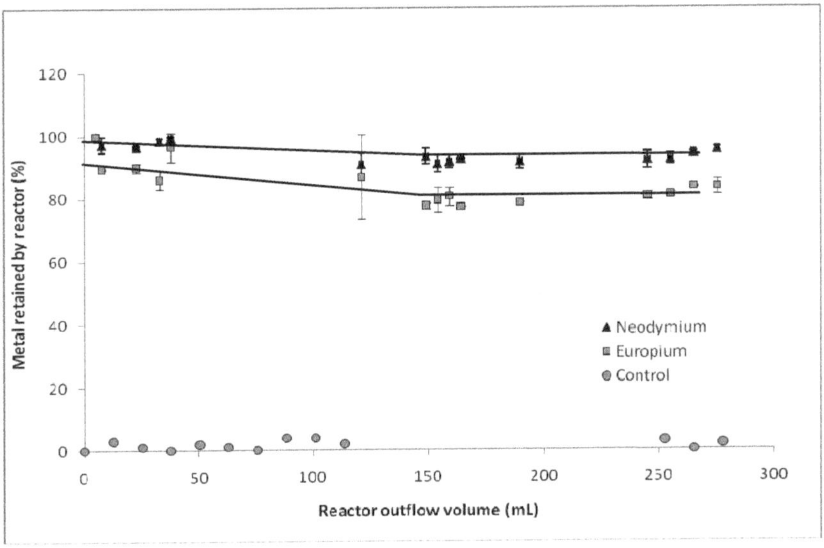

Fig. 9. Biorecovery of neodymium (▲) and europium (□) using cells of *Serratia* sp. in a flow-through reactor with corresponding killed-biofilm controls (O). Results were taken from two reactors, each containing 8 discs of foam with immobilised *Serratia* sp. biofilm. Reactors were challenged with a solution containing neodymium (III) or europium (III) (1 mM) as described at pH 5.5 for 54.5 hours at a flow rate of 5 ml/h. Where no error bars are shown these are within the dimensions of the symbols.

The removal of both metals fell slightly initially (c.f. earlier discussion about the need for a nucleation period at acidic pH) and then stabilised (from 100-150h) at steady-state removal of ~ 85% (Eu^{3+}) and more than 90% (Nd^{3+}), with no loss of activity at completion of the tests. Control reactors containing killed biofilm removed little metal, showing that live *Serratia* cells were required for metal deposition. Solid-state analysis of the metallised biofilms and comparison with, and co-development of, the bacterial and yeast systems, is ongoing in the UK and Japanese laboratories.

In particular, the promising results obtained in the preliminary studies we describe here justify a more prolonged study to determine the total biomass and column capacities since Fig. 10 shows that by the 300 ml stage (54.5 hours exposure), only the bottom layers of the column show the accumulation of metal deposits, indicating substantial unused capacity.

An early study using U showed the total capacity of the bacterial column to be 9 g uranium per g bacterial dry weight (Macaskie, 1990). The limitation of this process is usually an unacceptable back-pressure due to column blockage at high local metal loadings (i.e. at the base of the column, Fig. 10). This probably can be overcome if several column are linked in a series with the first column removed periodically and replaced with a fresh column distally. More critical to industrial application is the recovery of REE from real industrial wastes and the downstream processing required to convert the high yield product into new material. The presence of Fe in wastes may be problematic since the *Serratia* system described here removes all metals with insoluble phosphates. However previous studies treating uranium mine water containing a large excess of iron showed that Fe removal could be achieved by a simple increase in pH to precipitate-out $Fe(OH)_3$. The bioreactor was then run at pH 3.5-4.0

and uranium phosphate product was recovered (Macaskie *et al.*, 1997). A similar strategy could be envisaged for REE recovery. A study has not been made into the nature of the REE precipitate but on the basis of the La^{3+} study reported previously (see earlier) and the detailed solid state analysis of the corresponding $YbPO_4$ on yeast cells (above) it seems likely that the deposit comprises nanocrystalline REE phosphate. It is not yet known what, if any, particular advantages could be conferred by recovery of REE nanoparticles (as compared to large crystals); these are under current consideration in current work. (L.E. Macaskie and T. Ohnuki, current studies).

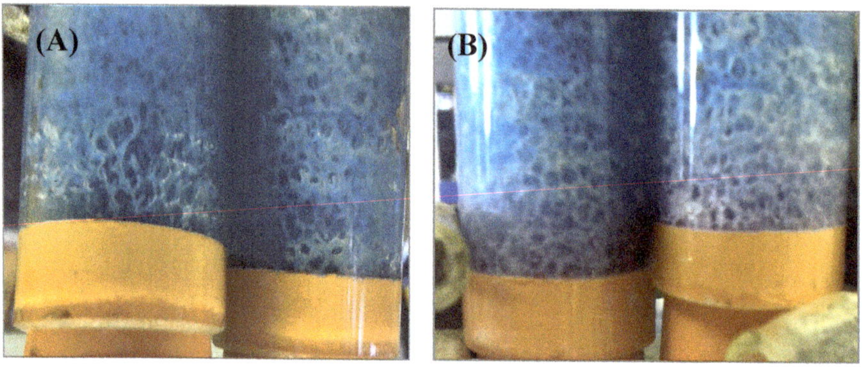

Fig. 10. Reactors containing *Serratia* sp. biofilm on polyurethane foam (A) before and (B) after treatment with $N_3NdO_9.6H_2O$ (54.5 hours); a purple colour can be seen at the bottom of the reactors, in B.

4. Conclusion

In this chapter, we describe two distinct biological routes for the recovery of PGM and REEs. First, the ability of bacterial cells to precipitate metallic ions to the elemental state *via* specific enzymes (bioreduction) is shown to be a potential alternative route for PGM recovery from dilute waste streams. The use of pre-metallised cells as catalysts for metal precipitation allows the process to be applied on aggressive leachates under conditions which are non-permissive for traditional biological treatments. The recovered material (PGM nanoparticles supported on biomass) can be utilised as a catalyst in reactions of increasing complexity.
In the second case study using a yeast and a bacterial system, the recovery of four model REE (Ce, Nb, Eu and Nd) REE was successfully demonstrated. All were precipitated as metal phosphates on the surface of the cells and immobilised cells of *Serratia* sp. recovered Nd and Eu continuously in flow through columns.

5. Acknowledgements

This work was supported extensively by grants from the Biotechnology and Biological Sciences (BBSRC) and Engineering and Physical Sciences (EPSRC) Research Councils and a Royal Society Fellowship to L.E.M. We thank Dr T. Ohnuki for permission to include unpublished work and also acknowledge, with thanks, the financial support of JAEA ('REIMEI' award to LEM) to enable future progression of the metal recovery work and its application to nuclear waste decontamination in Japan in collaboration with T. Ohnuki.

6. References

Angelidis, T.N. & Sklavounos, S.A. (1995) A SEM-EDS study of new and used automotive catalysts. *Applied Catalysis A: General*, vol. 133 pp 121-132.

Anon. (2004). Earthworks report: dirty metals; Mining, communities and the environement. http://www.scribd.com/doc/33843525/Dirty-Metals-Report.

Anon. (2006). UK Department for Transport, Platinum and hydrogen for fuel cell vehicles.

Anon. (2007). Earth's natural wealth: an audit. *New Scientist*, published 23rd May 2007.

Anon. (2008). Resource efficiency KTN, Material security - Ensuring resource availability for the UK economy.

Baxter-Plant, V.; Mikheenko, I.P. & Macaskie, L.E. (2003). Sulphate-reducing bacteria, palladium and the reductive dehalogenation of chlorinated aromatic compounds. *Biodegradation*, vol 14, No 2 pp 83-90.

Baxter-Plant, V.S.; Mikheenko, I.P.; Robson, M.; Harrad, S.J. & Macaskie, L.E. (2004). Dehalogenation of chlorinated aromatic compounds using a hybrid bioinorganic catalyst on cells of *Desulfovibrio desulfuricans*. *Biotechnology Letters*, vol 26, No 24 pp1885-90.

Bennett, J.A.; Creamer, N.J.; Deplanche, K.; Macaskie, L.E.; Shannon, I.J. & Wood, J. (2010). Palladium supported on bacterial biomass as a novel heterogeneous catalyst: A comparison of Pd/Al_2O_3 and bio-Pd in the hydrogenation of 2-pentyne. *Chemical Engineering Science*, vol 22, No 1 pp 282-291.

Bernardis, FL.; Grant RA. & Sherrington, DC. (2005). A review of methods of separation of the platinum-group metals through their chloro-complexes. *Reactive and Functional Polymers*, vol. 65 pp. 205–217.

Beveridge, T.J. (1989). The role of cellular design in bacterial metal accumulation and mineralization. *Annual Review of Microbiology*, vol 43 pp 147-171.

Bonthrone, K. M., Basnakova, G., Lin, F & Macaskie, L. E. (1996). Bioaccumulation of nickel by intercalation into polycrystalline hydrogen uranyl phosphate deposited via an enzymatic mechanism. *Nature Biotechnology*, vol 14 pp 635-638.

Bonthrone, K.M., Quarmby, J., Hewitt, C., Allan, V.J.M., Paterson-Beedle, M., Kennedy, J.F & Macaskie, L.E. (2000). The effect of the growth medium on the composition and metal binding behaviour of the extracellular polymeric material of a metal-accumulating *Citrobacter* sp. *Environmental Technology*, vol 21 pp 123-134.

Boswell, C.D.; Dick, R.E.; Eccles, H. & Macaskie, L.E. (2001). Phosphate uptake and release by *Acinetobacter johnsonii* in continuous culture and coupling of phosphate release to heavy metal accumulation. *Journal of Industrial Microbiology*, vol 26 pp 333-340.

Bunge, M.; Sobjerg, L.S.; Rotaru, A.E.; Gauthier, D., Lindhardt, A.T.; Hause, G.; Finster, K.; Kingshott, P.; Skrydstrup, T. & Meyer, R.L. (2010). Formation of palladium(0) nanoparticles at microbial surfaces. *Biotechnology and Bioengineering*, vol 107, No 2 pp 206-215.

Cawthorn, RG. (1999) The platinum and palladium resources of the Bushveld complex. *South African Journal of Science*, vol. 95 pp 481-489.

Chen, J.M. & Hao, O.J. (1998). Microbial chromium (VI) reduction. *Critical Reviews in Environmental Science and Technology*, vol 28, No 3 pp 219-251.

Chidambaram, D.; Hennebel, T.; Taghavi, S.; Mast, J.; Boon, N.; Verstraete, W.; van der Lelie, D & Fitts, J.P. (2010). Concomitant microbial generation of palladium nanoparticles and hydrogen to immobilize chromate. *Environmental Science and Technology*, vol 44 pp 7635-7640.

Churchill, S.A.; Walters, J.V. & Churchill, P.F. (1995). Sorption of heavy metals by prepared bacterial cell surfaces. *Journal of Environmental Engineering*, vol 121 pp 706-711.

Creamer, N.J.; Baxter-Plant, V.S.; Henderson, J.; Potter, M. & Macaskie, L.E. (2006). Palladium and gold removal and recovery from precious metal solutions and electronic scrap leachates by *Desulfovibrio desulfuricans*. *Biotechnology Letters*, vol 28 pp 1475-1484.

Creamer, N.J.; Mikheenko, I.P.; Yong, P.; Deplanche, K.; Sanyahumbi, D.; Wood, J.; Pollmann, K.; Merroun, M.; Selenska-Pobell, S. & Macaskie, L.E. (2007). Novel supported Pd hydrogenation bionanocatalyst for hybrid homogeneous/heterogeneous catalysis. *Catalysis Today*, vol 128, No 1-2 pp 80-87.

Creamer, N.J; Deplanche, K.; Snape, T.J.; Mikheenko, I.P.; Yong, P.; Sanyahumbi, D.; Wood, J.; Pollmann, K.; Selenska-Pobell, S. & Macaskie, L.E. (2008). A biogenic catalyst for hydrogenation, reduction and selective dehalogenation in non-aqueous solvents. *Hydrometallurgy*, vol 94 pp 138-143.

Dimitriadis, S.; Nomikou, N & McHale, A. P. (2007). Pt-based electro-catalytic materials derived from biosorption processes and their exploitation in fuel cell technology. *Biotechnology Letters*, vol 29 pp 545-551.

Demopoulos, G. (1989). Refining of platinum group metals. *CIM Bulletin* (March) pp 165-171.

Depcik, C. & Assanis, D. (2005) One-dimensional automotive catalyst modelling. *Progress in Energy and Combustion Science*, vol. 31 pp 308-369.

Deplanche, K.; Attard, G.A. & Macaskie L.E. (2007). Biorecovery of gold from jewellery wastes by *Escherichia coli* and biomanufacture of active Au-nanomaterial. *Advanced Materials Research*, vol 20-21 pp 647-650.

Deplanche, K. (2008). New nanocatalysts made by bacteria from metal solutions and recycling of metal wastes. *PhD thesis*, The University of Birmingham, UK.

Deplanche, K. & Macaskie, L.E. (2008). Biorecovery of gold by *Escherichia coli* and *Desulfovibrio desulfuricans*. *Biotechnology and Bioengineering*, vol 99, No 5 pp 1055-1064.

Deplanche, K.; Snape, T.J.; Hazrati, S.; Harrad, S. & Macaskie L.E. (2009). Versatility of a new bioinorganic catalyst: palladized cells of *Desulfovibrio desulfuricans* and application to dehalogenation of flame retardant materials. *Environmental Technology*, vol 30, No 7, pp 681-692.

Deplanche, K.; Caldelari, I.; Mikheenko, I. P.; Sargent, F. & Macaskie, L. E. (2010a). Involvement of hydrogenases in the formation of highly catalytic Pd(0) nanoparticles by bioreduction of Pd(II) using *Escherichia coli* mutant strains. *Microbiology*, vol 156 pp 2630-2640.

Deplanche, K.; Mikheenko, I.P.; Bennett, J.A.; Merroun, M.; Mounzer, H.; Wood, J. & Macaskie, L.E. (2011). Selective oxidation of benzyl-alcohol over biomass-supported Au/Pd bioinorganic catalysts. *Topics in Catalysis*, in press.

Deplanche, K. & Macaskie, L.E. (2010). Improved catalyst. Patent application number PCT/GB2011/000018.

DeVargas, I.; Macaskie, L.E. & Guibal, E. (2004). Biosorption of palladium and platinum by sulfate reducing bacteria. *Journal of Chemical Technology and Biotechnology*, vol 79 pp 49-56.

DeVargas, I.; Sanyahumbi, D.; Ashworth, M.A.; Hardy, C.M. & Macaskie, L.E. (2005). Use of X-ray photoelectron spectroscopy to elucidate the mechanisms of palladium and platinum biosorption by *Desulfovibrio desulfuricans* biomass. In: S.T.L. Harrison, D.E. Rawlings and J. Petersen (Editors), 16th International *Biohydrometallurgy Symposium*. 16th International Biohydrometallurgy Symposium. Cape Town; South Africa, pp: 605-616.

DeWindt, W.; Aelterman, P. & Verstraete, W. (2005). Bioreductive deposition of palladium(0) nanoparticles on *Shewanella oneidensis* with catalytic activity towards reductive dechlorination of polychlorinated biphenyls. *Environmental Microbiology*, vol 7, No 3 pp 314-325.

DeWindt, W.; Boon, N.; Bulcke, J.; Rubberecht, L.; Prata, F.; Mast, J.; Hennebel, T. & Verstraete, W. (2006). Biological control of the size and reactivity of catalytic Pd(0) produced by *Shewanella oneidensis*. *Antonie van Leeuwenhoek*, Vol 90, No 4, November pp. 377-389.

Dick, R.E.; Boswell, C.D. & Macaskie, L.E. (1995). Uranyl phosphate accumulation by *Acinetobacter* spp. Proceedings of International Conference of Biohydrometallurgical Processing, Vina del Mar, Chile, University of Chile, Santiago.

Diels, L. ; Macaskie, L.E. ; Tzesos, M. ; Pümpel, T. & Glombitza, F. (1996). Modelling of genetic, biochemical, cellular and microenvironmental parameters determining bacterial sorption and mineralization processes for recuperation of heavy or precious metals Final report EU contract BE-5350, The European Commission, Brussels.

Eccles, H.; Garnham, G.W.; Lowe, C.R. & Bruce, N.C. (1997). Biosensors for detecting metal ions capable of being reduced by reductase enzymes. *Biotechnology Advances*, vol 15, No 1 pp 189-190.

Ek, KH; Morrison, GM. & Rauch, S. (2004). Environmental routes for platinum group elements to biological material - a review. *Science of the Total Environment*, vol. 334-335 pp 21-38.

Fahmy K, Merroun M, Pollmann K, Raff J, Savchuk O, Hennig C, Selenska-Pobell S., Secondary structure and Pd(II) coordination in S-layer proteins from *Bacillus sphaericus* studied by infrared and X-ray absorption spectroscopy. 2006. *Biophys J.* 2006 Aug 1;91(3):996-1007

Forster, C & Wase, J. (2003). Biosorbents for Metal Ions. Taylor & Francis Ltd. 1-229.

Fu, J.K.; Liu, Y.Y.; Gu, P.Y.; Tang, D.L.; Lin, Z.Y.; Yao, B.X; & Weng, S.Z. (2000). Spectroscopic characterization on the biosorption and bioreduction of Ag(I) by *Lactobacillus* sp. A09. *Acta Physica Chimica Sinica*, vol 16 pp 779-782.

Gadd, G.M. (2009). Biosorption: critical review of scientific rationale, environmental importance and significance for pollution treatment. *Journal of Chemical Technology and Biotechnology* Vol. 84, No 1 pp 13–28.

Gavrilescu, M. (2004). Removal of heavy metals from the environment by biosorption. *Engineering and Life Sciences*, vol 4, No 3 pp 219-232.

Gauthier, D.; Søbjerg, L.S.; Jensen, K.M.; Lindhardt, A.T.; Bunge, M.; Finster, K.; Meyer, R.L. & Skrydstrup. T. (2010). Environmentally benign recovery and reactivation of palladium from industrial waste by using gram-negative bacteria. *ChemSusChem*, vol 3, No 9 pp 1036–1039.

Guibal, E.; Vincent, T. & Roussy, J. (2009). Interaction of chitosan with metal ions: from environmental applications to the elaboration of new materials. *Advanced Materials Research*, vol 71-73, pp 519-526.

Gordon, RB. ; Bertram, M. & Graedel, TE. (2006) Metal stocks and sustainability. *Proceedings of the National Academy of Sciences of the United States of America*, vol. 103 pp. 1209-1214.

Handley-Sidhu, S.; Renshaw, J.C.; Moriyama, S.; Stolpe, B.; Mennan, C.; Bagheriasl, S.; Yong, P.; Stamboulis, A.; Paterson-Beedle, M.; Sasaki, K.; Pattrick, R.A.D.; Lead, J.R. & Macaskie, L.E. (2011). Uptake of Sr^{2+} and Co^{2+} into biogenic hydroxyapatite: Implications for biomineral ion exchange synthesis. *Environmental Science and Technology*, in press.

Harrad, S.; Robson, M.; Hazrati, S.; Baxter-Plant, V.S.; Deplanche, K.; Redwood, M.D. & L. E. Macaskie (2007). Dehalogenation of polychlorinated biphenyls and polybrominated diphenyl ethers using a hybrid bioinorganic catalyst. *Journal of Environmental Monitoring*, vol 9 pp 314-318.

Hennebel, T.; Benner, J.; Clauwaert, P.; Vanhaecke, L.; Aelterman, P.; Callebaut, R.; Boon, N & and Verstraete, W. (2011). Dehalogenation of environmental pollutants in microbial electrolysis cells with biogenic palladium nanoparticles. *Biotechnology Letters*, vol 33, No pp 89-95.

Humphries, A.C.; Nott, K.P.; Hall, L.D. & Macaskie, L.E. (2004). Continuous removal of Cr(VI) from aqueous solutions catalyzed by palladized biomass of *Desulfovibrio vulgaris*. *Biotechnology Letters*, vol 26 pp 1529-1532.

Humphries, A.C.; Mikheenko, I.P. & Macaskie, L.E. (2006), Chromate reduction by immobilized palladized sulfate-reducing bacteria. *Biotechnology and Bioengineering*, vol 94, No 1 pp 81-90.

Jiang, M.; Ohnuki, T.; Kozai, N.; Tanaka, K.; Suzuki, Y.; Sakamoto, F.; Kamiishi, E. & Utsunomiya, S. (2010). Biological nano-mineralization of Ce phosphate by *Saccharomyces cerevisiae*. *Chemical Geology*, vol 277 pp 61-69.

Kashefi, K.; Tor, J.M.; Nevin, K.P. & Lovley, D.R. (2001). Reductive precipitation of gold by dissimilatory Fe(III)-reducing Bacteria and Archea. *Applied and Environmental Microbiology*, vol 67 pp 3275-3279.

Konishi, Y.; Tsukiyama, T.; Ohno, K.; Saitoh, N.; Nomura, T. & Nagamine, S. (2006). Intracellular recovery of gold by microbial reduction of $AuCl_4^-$ ions using the anaerobic bacterium *Shewanella algae*. *Hydrometallurgy*, vol 81 pp 24-29.

Larsson, P.; Olka, L. & Tranvik, L. (1988). Microbial degradation of xenobiotic, aromatic pollutants in humic water. *Applied and Environmental Microbiology*, vol 54, No 7 pp 1864- 1867.

Lloyd, J.R.; Harding, C.L. & Macaskie, L.E. (1997). Tc(VII) reduction and accumulation by immobilized cells of *Escherichia coli*. *Biotechnology and Bioengineering*, vol 55 pp 505-510.

Lloyd, J.R.; Yong, P. & Macaskie, L.E. (1998). Enzymatic recovery of elemental palladium by using sulfate-reducing bacteria. *Applied and Environmental Microbiology*, vol 64 pp 4607-4609.

Lloyd, J.R.; Ridley, J.; Khizniak, T.; Lyalikova, N.N. & Macaskie, L.E. (1999a). Reduction of technetium by *Desulfovibrio desulfuricans*: Biocatalyst characterization and use in a flowthrough bioreactor. *Applied and Environmental Microbiology*, vol 65, No 6 pp 2691-2696.

Lloyd, J.R.; Thomas, G.H.; Finlay, J.A.; Cole, J.A. & Macaskie, L.E. (1999b) Microbial reduction of technetium by *Escherichia coli* and *Desulfovibrio desulfuricans*: enhancement via the use of high activity strains and effect of process parameters. *Biotechnology and Bioengineering*, vol 66 pp 122-130.

Lloyd, J.R.; Yong, P. & Macaskie, L.E. (2000) Biological reduction and removal of pentavalent Np by the concerted action of two microorganisms. *Environmental Science and Technology*, vol 34 pp 1297-1301.

Lloyd, J.R. (2003). Microbial reduction of metals and radionuclides. *FEMS Microbiology Reviews*, vol 27 pp 411-425.

Lovley, D.R. (1991). Dissimilatory Fe(III) and Mn(IV) reduction. *Microbiology and Molecular Biology Reviews*, vol 55, No 2 pp 259-287.

Lovley, D.R.; & Phillips, E.J.P. (1992). Reduction of uranium by *Desulfovibrio desulfuricans*. *Applied and Environmental Microbiology*, vol 58, No 3 pp 850-856.

Lovley, D.R. (1994). Biormediation of organic and metal contaminants with dissimilatory metal reduction. *Journal of Industrial Microbiology*, vol 14 pp 85-93.

Mabbett, A.N. & Macaskie, L.E. (2002) A new bioinorganic process for the remediation of Cr(VI). *Journal of Chemical Technology and Biotechnology*, vol 77, No 10 pp 1169-1175.

Mabbett, A.N.; Lloyd, J.R. & Macaskie, L.E. (2002). Effect of complexing agents on reduction of Cr(VI) by *Desulfovibrio vulgaris* ATCC 29579. *Biotechnology and Bioengineering*, vol 79, No 4 pp 389-397.

Mabbett, A.N.; Yong, P.; Farr, P.G. & Macaskie, L.E. (2004). Reduction of Cr(VI) by "palladized" biomass of *Desulfovibrio desulfuricans* ATCC 29577. *Biotechnology Bioengineering*, vol 5, No 1 pp 104–109.

Mabbett, A.N.; Sanyahumbi, D.; Yong, P. & Macaskie, L.E. (2006) Biorecovered precious metals from industrial wastes. Single step conversion of a mixed metal liquid waste to a bioinorganic catalyst with environmental applications. *Environmental Science and Technology*, vol 40 pp 1015-1021.

Macaskie, L. E. & Dean, A. C. R. (1982). Cadmium accumulation by micro-organisms. *Environmental Technology Letters*. 3: 49-56.

Macaskie, L.E. & Dean, A.C. R. (1984). Cadmium accumulation by immobilised cells of *Citrobacter* sp., *Environmental Technology Letters*. 5: 177–186.

Macaskie, L.E. & Dean, A.C. R. (1985). Strontium accumulation by immobilised cells of a Citrobacter sp., *Biotechnology Letters*. 7: 457–462.

Macaskie, L.E. (1990). An immobilized cell bioprocess for the removal of heavy metals from aqueous flows. *Journal of Chemical Technology and Biotechnology*, vol 49 pp 357-379.

Macaskie, L. E., Empson, R. M., Cheetham, A. K., Grey, C. P. & Skarnulis, A. J. (1992). Uranium bioaccumulation by a *Citrobacter* sp. as a result of enzymically-mediated growth of polycrystalline HUO_2,PO_4. *Science*, vol 257 pp 782-784.

Macaskie, L. E., Jeong, B. C. & Tolley, M. R. (1994). Enzymically accelerated biomineralization of heavy metals: application to the removal of americium and plutonium from aqueous flows. *FEMS Microbiology Reviews*, vol, 14 pp 351-368.

Macaskie, L.E.; Lloyd, J.R.; Thomas, R.A.P. & Tolley, M.R. (1996). The use of microorganisms for the remediation of solutions contaminated with actinide elements, other radionuclides and organic contaminants generated by nuclear fuel cycle activities. *Nuclear Energy*, vol 35 pp 257-271.

Macaskie, L.E.; Yong, P.; Doyle, T.C.; Roig, M.G., Diaz, M. & Manzano, T. (1997). Bioremediation of uranium-bearing wastewater: biochemical and chemical factors influencing bioprocess application. *Biotechnology and Bioengineering*, vol 53 pp 100-109.

Macaskie, L.E., Bonthrone, K.M., Yong, P & Goddard, D.T. (2000). Enzymically mediated bioprecipitation of uranium by a *Citrobacter* sp. : a concerted role for exocellular lipopolysaccharide and associated phosphatase in biomineral formation. *Microbiology*, vol 146 pp 1855-67.

Macaskie, L.E.; Yong, P.; Paterson-Beedle, M.; Thackray, A.C.; Marquis, P.M.; Sammons, R.L.; Nott, K.P. & Hall, L.D. (2005a). A novel non line of-sight method for coating hydroxyapatite onto the surfaces of support materials by biomineralization. *Journal of Biotechnology*, vol 118 pp 187-200.

Macaskie, L.E.; Baxter-Plant, V.S.; Creamer, N.J.; Humphries, A.C.; Mikheenko, I. P.; Mikheenko, P.M.; Penfold, D.W. & Yong, P. (2005b) Applications of bacterial hydrogenases in waste decontamination, manufacture of novel bionanocatalysts and in sustainable energy. *Biochemical Society Transactions*, vol 33, No 1 pp 76-79.

Mertens, B.; Blothe, C.; Windey, K.; De Windt, W. & Verstraete, W. (2007). Biocatalytic dechlorination of lindane by nano-scale particles of Pd(0) deposited on *Shewanella oneidensis*. *Chemosphere*, vol 66, No 1 pp 99-105.

Mikheenko, I.P. (2004). Nanoscale palladium recovery, *PhD thesis*, The University of Birmingham, UK.

Mikheenko, I.P.; Baxter-Plant, V.S.; Rousset, M.; Dementin, S.; Adryanczyk-Perrier, G. & Macaskie, L.E. (2003). Reduction of Pd(II) with *Desulfovibrio fructosovorans*, its [Fe]-only hydrogenase negative mutant and the activity of the obtained hybrid bioinorganic catalysts. In: S.T.L. Harrison, D.E. Rawlings and J. Petersen (Editors), 16th International Biohydrometallurgy Symposium. 16th International biohydrometallurgy symposium, Cape Town; South Africa, pp 383-387.

Macaskie, L.E., I.P. Mikheenko, P. Yong, K. Deplanche, A.J. Murray, M. Paterson-Beedle, V.S. Coker, C.I. Pearce, R. Cutting, R.A.D. Pattrick, D. Vaughan, G. van der Laan, J.R. Lloyd (2011) Today's wastes, tomorrow's materials for environmental protection in *Comprehensive Biotechnology* (2nd Edition) Eds M. Moo-Young, M. Butler, C. Webb, A. Moreira, B. Grodzinski, Z F Cui & S. Agathos, Elsevier, Amsterdam, pp 719-725

Mikheenko, I.P.; Rousset, M.; Dementin, S. & Macaskie, L.E. (2008). Bioaccumulation of palladium by *Desulfovibrio fructosovorans* and hydrogenase deficient mutants. *Applied and Environmental Microbiology*, vol 74, No 19 pp 6144–6146.

Murray, A.J.; Mikheenko, I.P.; Goralska, E.; Rowson, N.A. & Macaskie L.E. (2007). Biorecovery of platinum group metals from secondary sources. *Advanced Materials Research*, vol. 20-21 pp. 651-654.

Niu, H. & Volesky, B. (1999). Characteristics of gold biosorption from cyanide solution. *Journal of Chemical Technology and Biotechnology*, vol 74 pp 778-784.

Nott , K. P.; Paterson-Beedle, M.; Macaskie, L.E. & Hall, L.D. (2001). Visualisation of metal deposition in biofilm reactors by three dimensional magnetic resonance imaging. *Biotechnology Letters*, vol 23 pp 1749-1757.

Nott, K.P.; Hall, L.D.; Macaskie, L.E. & Paterson-Beedle, M. (2005a). Measurement of flow field in biofilm reactors by 3-D magnetic resonance imaging. *American Institute of Chemical Engineers Journal*, vol 51 pp 3072-3079.

Nott, K.P.; Heese, F.P.; Paterson-Beedle, M., Macaskie, L.E. & Hall, L.D. (2005b). Visualization of the function of a biofilm reactor by magnetic resonance imaging. *Canadian Journal of Chemical Engineering*, vol 83 pp68-79.

Onishi, H. (1986). Photometric determination of traces of metals, Part IIA: Individual metals, aluminium to lithium, 4th edn. John Wiley & Sons, New York.

Palmieri, C.M.; Garcia, O. & Melnikov, P. (2000). Neodymium biosorption from acidic solutions in batch systems. *Process Biochemistry*, vol 36 pp 441-444.

Paterson-Beedle, M.; Nott, K.P.; Macaskie, L.E. & Hall, L.D. (2001). Study of biofilm within a packed bed reactor by three dimensional magnetic resonance imaging. *Methods in Enzymology*, vol 337 pp 285-305.

Paterson-Beedle, M. & Macaskie, L.E. (2004). Use of PhoN phosphatase to remediate heavy metals. In: Barredo JL (ed) Methods in Biotechnology: Microbial products and Biotransformations, Humana Press, Totowa, vol 18 pp 413-436.

Paterson-Beedle, M & Macaskie, L.E. (2004). Removal of Co, Sr and Ce from aqueous solutions using native biofilm of *Serratia* sp. and biofilm pre-coated with hydrogen uranyl phosphate. In Biohydrometallurgy: A Sustainable Technology in Evolution, eds., M. Tsezos, A. Hatzikioseyian and E. Remoundaki, National Technical University of Athens, Athens, pp. 1155-1162.

Paterson-Beedle, M; Macaskie, L.E.; Lee, C.H.; Hriljac, J.A.; Jee, K.Y. & Kim, W.H. (2006). Utilisation of a hydrogen uranyl phosphate-based ion exchanger supported on biofilm for the removal of cobalt, strontium and caesium from aqueous solutions. *Hydrometallurgy*, vol 83 pp 141-145.

Paterson-Beedle, M.; Readman, J.E.; Hriljac, J.A. & Macaskie, L.E. (2010). Biorecovery of uranium from aqueous solution at the expense of phytic acid. *Hydrometallurgy*, vol 104 pp 524-528.

Pincock, Allen and Holt, (2008) The Processing of Platinum Group Metals (PGM) - Part 1. Pincock Perspectives. March 2008.

Plummer, E.J &, Macaskie, L.E. (1990). Actinide and lanthanum toxicity towards a *Citrobacter* sp.: uptake of lanthanum and a strategy for the biological treatment of wastes containing plutonium. *Bulletin of Environmental Contamination and Toxicology*, vol 44 pp 173-180.

Rawlings, D.E. & Johnson, D.B. (2007). The microbiology of biomining: development and optimization of mineral-oxidizing microbial consortia. *Microbiology*, vol 153, No 2 pp 315-324.

Redwood, M.D.; Deplanche, K.; Yong, P.; Baxter-Plant, V.S. & Macaskie, L.E. (2005). Biomass-supported palladium catalysts on *Desulfovibrio* and *Rhodobacter*. In: S.T.L. Harrison, D.E. Rawlings and J. Petersen (Editors), 16th International biohydrometallurgy symposium. 16th International Biohydrometallurgy Symposium, Cape Town; South Africa, pp 335-342.

Redwood, M.D; Deplanche, K.; Baxter-Plant, V.S. & Macaskie, LE (2008). Biomass-supported palladium catalysts on *Desulfovibrio desulfuricans* and *Rhodobacter sphaeroides*. *Biotechnology and Bioengineering*, vol 99, No 5 pp 1045-54.

Renshaw, J.C.; Butchins, L.J.C.; Livens,[†] F.R.; May, I.; Charnock, J.M. & Lloyd, J.R. (2005). Bioreduction of uranium: environmental implications of a pentavalent intermediate. *Environmental Science and Technology*, vol. 39, No 15, pp 5657–5660.

Roux, M.; Sarret, G.; Pignot-Paintrand, I.; Fontecave, M. & Coves, J. (2001). Mobilization of selenite by *Ralstonia metallidurans* CH34. *Applied and Environmental Microbiology*, vol 67, No 2 pp 769-773.

Sanyahumbi, D., de Vargas, I.; Mabbett, A.N.; Yong, P. and Macaskie, L.E. (2005). Selective biosorption of palladium and platinum from bi- and multi-metallic solutions. In: S.T.L. Harrison, D.E. Rawlings and J. Petersen (Editors), 16th International Biohydrometallurgy Symposium. Cape Town; South Africa, pp: 571-579.

Schmuhl, R.; Krieg, H.M. & Keizer, K. (2001). Adsorption of Cu(II) and Cr(VI) ions by chitosan : Kinetics and equilibrium studies. *Water South Africa*, vol 27, No 1 pp 1-7.

Skibar, W.S; Pompe, W.; Fratzl, P.; Selenska-Pobell, S.; Rousset, M. & Macaskie, L.E. (2005). Novel precious metal-based bionanocatalysts from scrap Final report Contract No G5RD-CT-2002-00750. The European Commission, Brussels.

Søbjerg, S. L.; Gauthier, D.; Lindhardt, A.T.; Bunge, M.; Finster, K.; Meyer, R.L. & Skrydstrup, T. (2009). Bio-supported palladium nanoparticles as a catalyst for Suzuki–Miyaura and Mizoroki–Heck reactions. *Green Chemistry*, vol 11, 2041-2046.

Søbjerg, S. L.; Lindhardt, A.T.; Skrydstrup, T.; Finster, K. & Meyer, R.L. (2011). Size control and catalytic activity of bio-supported palladium nanoparticles. *Colloids and Surfaces B: Biointerfaces*, Vol 85, No 2, pp 373-378.

Streat, M. & Naden, D. (1983). Hydrometallurgy: research development and plant practice. Trans Metallic Society of AIME, New York, U.S.A.

Taylor, S.; Murray, A.J.; Blackburn, S. & Macaskie, L.E. (2011). The recovery of platinum group metals from model solutions and fabrication into new catalysts. *Transactions of the Nonferrous Metals Society of China,*, in press.

Tolley, M.R., Macaskie, L.E., Moody, J.C. & Strandling, G.N. (1991). Actinide and lanthanum accumulation by immobilised cells of *Citrobacter* sp., *Proceedings of the 201st Meeting American Chemical Society*, vol 31 pp 213–216.

Tolley, M.R.; Strachan, L.F. & Macaskie, L.E. (1995). Lanthanum accumulation from acidic solutions using a *Citrobacter* sp. immobilized in a flow-through bioreactor. *Journal of Industrial Microbiology*, vol 14 pp 271-280.

Tolley, M.R, PhD. Thesis University of Oxford, UK

Vieira, R.H.S.F. & Volesky, B. (2000). Biosorption: a solution to pollution? *International Microbiology*, vol 3 pp 17-24.

Volesky, B.; May, H. & Zolan, Z.R. (1994). Cadmium biosorption by *Saccharomyces cerevisiae. Biotechnology and Bioengineering*, vol 41, No 8 pp 826-829.

Volesky, B. (1999). Biohydrometallurgy and the environment toward the mining of the 21st century. *Part B. Molecular biology, biosorption, bioremediatio*n. Elsevier Science, Amsterdam, Netherlands

Volesky, B. (2007). Biosorption and me. *Water Research*, vol 41 pp 4017-4029.

Wood, J.; Bodenes, L.; Bennett, J.; Deplanche, K. & L. E. Macaskie. (2010). Hydrogenation of 2-Butyne-1,4-diol using novel bio-palladium catalysts. *Industrial Engineering and Chemistry Research*, vol 49, No 3, pp 980–988.

Whiteley, JD. & Murray, F. (2003). Anthropogenic platinum group element (Pt, Pd and Rh) concentrations in road dusts and roadside soils from Perth, Western Australia. *Science of the Total Environment*, vol. 317 pp 121-135.

Xiao, Z. & Laplante, AR. (2004) Characterizing and recovering the platinum group minerals - a review. *Minerals Engineering*, vol. 17 pp 961-979.

Yee, N.; Ma, J.; Dalia, A.; Boonfueng, T. & Kobayashi, D.Y. (2007). Se(VI) reduction and the precipitation of Se(0) by the facultative bacterium *Enterobacter cloacae* SLD1a-1 is regulated by FNR. *Applied and Environmental Microbiology*, vol 73, No 6 pp 1914-1920.

Yong, P. & Macaskie, L. E. (1998). Bioaccumulation of lanthanum, uranium and thorium and use of a model system to develop a method for the biologically-mediated removal of plutonium from solution. *Journal of Chemical Technology & Biotechnology*, vol 71 pp 15-26.

Yong, P.; Rowson, N.A.; Farr, J.P.; Harris, I.R. & Macaskie, L.E. (2002a). Bioaccumulation of palladium by *Desulfovibrio desulfuricans. Journal of Chemical Technology and Biotechnology*, vol 77 pp 1-9.

Yong, P.; Rowson, N.A.; Farr, J.P.; Harris, I.R. & Macaskie, L.E. (2002b). Bioreduction and biocrystallization of palladium by *Desulfovibrio desulfuricans* NCIMB 8307. *Biotechnology and Bioengineering*, 80: 369-379.

Yong, P.; Rowson, N.A.; Farr, J.P.; Harris, I.R. & Macaskie, L.E. (2003). A novel electrobiotechnology for the recovery of precious metals from spent automotive catalysts. *Environmental Technology*, vol 24 pp 289-297.

Yong, P.; Macaskie, L.E.; Sammons, R.L. & Marquis, P.M. (2004). Synthesis of nanophase hydroxypatite by a *Serratia* sp. from wastewater containing inorganic phosphate. *Biotechnology Letters*, vol 26, pp 1723-1730.

Yong, P.; Paterson-Beedle, M.; Mikheenko, I. P. & Macaskie, L. E. (2007). From bio-mineralisation to fuel cells: biomanufacture of Pt and Pd nanocrystals for fuel cell electrode catalyst. *Biotechnology Letters*, vol 29 pp 539-544.

Yong, P; Mikheenko, I.P.; Deplanche, K; Redwood, M.D. & Macaskie, L.E. (2010) Biorefining of precious metals from wastes: an answer to manufacturing of cheap nanocatalysts for fuel cells and power generation via an integrated biorefinery? *Biotechnology Letters*, vol 32 pp 1821-1828

Zereini, CLS. & Zereini, F. (2009). Airborne particulate matter, platinum group elements and human health: A review of recent evidence. *Science of the Total Environment*. vol. 407 pp 2493-2500.

13

Molecular Design and Supramolecular Assemblies of Novel Amphiphiles with Special Molecular Structures in Organized Molecular Films

Tifeng Jiao, Sufeng Wang and Jingxin Zhou
Hebei Key Laboratory of Applied Chemistry
College of Environmental and Chemical Engineering
Yanshan University, Qinhuangdao,
China

1. Introduction

It is well known that supramolecular chemistry has played a major role in progressing research in nanoscience and nanotechnology, leading to novel classes of materials which are capable of light or electrically stimulated chemistry and long-range organized assembly (Lehn, 1995). Interfacial supramolecular assemblies use well-characterized amphiphiles as building blocks to create assemblies on surfaces that are purposefully structured on the molecular level, while at the same time extending over supramolecular distances (Kuhn et al., 2000). Creating organized structures is an important goal in interfacial supramolecular chemistry. The Langmuir-Blodgett technique is important for the production of macroscopic materials that are organized on the molecular length scale. This approach allows amphillic molecules to be oriented at the air-water interface and then transferred sequentially onto a solid support.

The flexibility of molecular design is one of the most significant advantages of supramolecular assembly, allowing organic, inorganic and biological components to be used as building blocks. Organized molecular films deposited on solid surfaces are of great conceptual interest because their small thickness makes them 'quasi-ideal' two-dimensional systems. They constitute a novel 'bottom-up' approach to creating nanoscale structures. This approach contrasts with 'top-down' approaches that entail making existing devices so small that they eventually finish up as nanosized objects, with dimensions of no more than a few hundred nanometers.

On the other hand, amphiphiles are organic molecules that have the ability to adsorb at interfaces, thereby altering significantly the physical properties of those interfaces. Numerous variations are possible within the structure of both the head and tail group of amphiphiles (Holmberg, 1998). The head group can be charged or neutral, small and compact in size, or a polymeric chain. The tail group is usually a single or double, straight or branched hydrocarbon chain, but may also be a fluorocarbon, or a siloxane, or contain aromatic groups. Since the hydrophilic part normally achieves its solubility either by ionic interactions or by hydrogen bonding, the simplest classification is based on amphiphile head group type, with further subgroups according to the nature of the lyophobic moiety. Four basic classes therefore emerge as: the anionics and cationics, which dissociate in water into two oppositely charged species (the surfactant ion and its counterion); the non-ionics, which include a highly polar (non charged) moiety, such as polyoxyethylene or polyol groups; the zwitterionics (or amphoterics), which combine both a positive and a negative group. With the continuous search for improving amphiphile properties, new structures have recently emerged that exhibit interesting synergistic interactions or enhanced surface and aggregation properties. These novel amphiphiles have attracted much interest, and include the catanionics, bolaforms (Fuhrhop et al., 1986; Fuhrhop et al., 2006), gemini surfactants (Menger et al., 2000; Zana, 2002), polymeric and polymerisable surfactants. Characteristics and typical examples are shown in Table 1.

Classes	Structural characteristics	Example
Catanionic	Equimolar mixture of cationic and anionic surfactants (no inorganic counterion)	*n*-dodecyltrimethylammonium *n*-dodecyl sulfate (DTADS) $C_{12}H_{25}(CH_3)_3N^+\ ^-O_4S\,C_{12}H_{25}$
Bolaform	Two charged headgroups connected by a long linear polymethylene chain	Hexadecanediyl-1,16-bis(trimethyl ammonium bromide) $Br^-(CH_3)_3N^+-(CH_2)_{16}-N^+(CH_3)_3Br^-$
Gemini (or dimeric)	Two identical surfactants connected by a spacer close to or at the level of the headgroup	Propane-1,3-bis(dodecyldimethyl ammonium bromide) C_3H_6 -1,3-bis$[(CH_3)_2N^+C_{12}H_{25}Br^-]$
Polymeric	Polymer with surface active properties	Copolymer of isobutylene and succinic anhydride 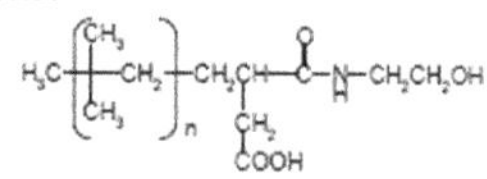
Polymerisable	Surfactant that can undergo homo-polymerisation or copolymerisation with other components of the system	11-(acryloyloxy)undecyltrimethyl ammonium bromide 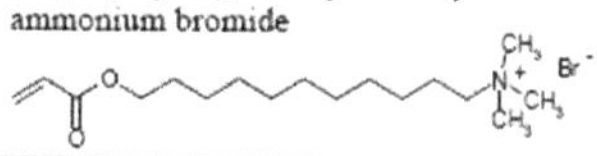

Table 1. Structural features and examples of some amphiphiles classes.

In our research, some special amphiphiles, including bolaamphiphiles, gemini and amphiphiles with functional substituted groups, have been designed and synthesized, and their organized supramolecular assemblies at the air/water interface have also been investigated, which may broaden the traditional research areas about LB film and give some perspectives and clues for the relative research in the future. In addition, some of the analytical methods, theoretical treatments and synthetic tools, which are being applied in the area of interfacial supramolecular chemistry and are driving its rapid development, will be highlighted. In this chapter, we summarize our main research contributions in recent years in three sections: (1) design and investigation of some series of bolaform amphiphiles with distinct structures; (2) supramolecular assemblies of some series of gemini amphiphiles with distinct structures; (3) design and interfacial characterization of some series of amphiphiles with functional substituted groups.

2. Design and investigation of some series of bolaform amphiphiles with distinct structures

Firstly, the spreading behavior and nano-architectures of a bolaamphiphilic diacid, 1,13-tridecandicarboxylic acid (TDA) on water surface and the subphase containing Eu(III) ion were investigated (Jiao et al., 2005a). It was found that although TDA itself could not spread on water surface, it could form an ultrathin film on the aqueous subphase containing Eu(III) ion. In addition, interesting nano-architectures were observed for the transferred film on mica surface. It was found that the formation and change of the nano-architectures were depended on the surface pressure and concentration of Eu(III) ion in the subphase. A rectangular slide morphology was observed for the film spread on an aqueous subphase containing lower concentration of Eu(III), while a walnut-like nano-architectures were observed for the film spread on a higher concentration of Eu(III) ion. Flower structure was observed at a higher surface pressure. The nano-architecture can be further regulated through mixing TDA with octadecylamine (OA) in which linear fiber nanostructure was obtained. It was revealed that while the nano-architectures were formed mainly through a three dimensional aggregation in the case of TDA/Eu(III) films, a two-dimensional aggregation occurred when TDA was mixed with OA. A series of methods such as atomic force microscope, scanning electron microscopy, Fourier transform infrared spectroscopy and X-ray diffraction were used to characterize the supramolecular structures. A possible mechanism for the formation of these nanoarchitectures was discussed, as shown in Fig. 1. This provides an important clue to construct ordered 3D nano-architectures by utilizing the Langmuir-Blodgett technique at the air/water interface. In addition, the nano-architecture can be regulated between 2D and 3D dimension by adding other appropriate amphiphiles into the bolaamphiphiles.

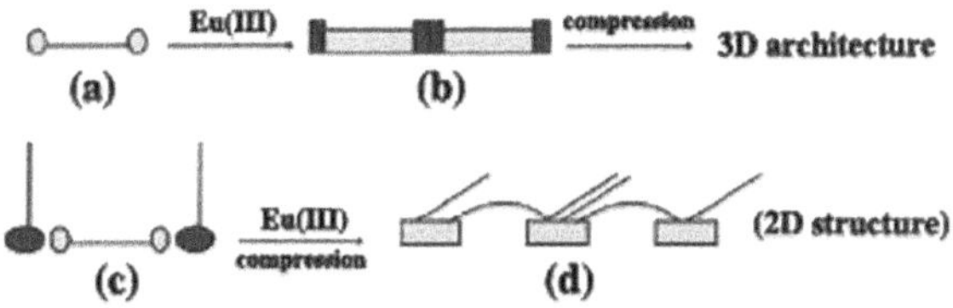

Fig. 1. Schematic illustrations on the formation of the nanoarchitectures in TDA/Eu(III) and TDA/OA films. (a) Bolaamphiphile (TDA); (b) TDA coordinated with Eu(III) to form floating "blog" on water surface, upon compression, these "blogs" will experience a three-dimensional aggregation to form a 3D nano-architecture; (c) mixed film of OA and bolaamphiphile; (d) spreading on the subphase of $EuCl_3$ caused the formation of fiber structure, the aggregation was limited in a two dimension (Reproduced from Jiao et al., 2005a, Thin Solid Films, 479, 269-276).

In addition, a bolaform Schiff base, *N,N'*-bis(salicylidene)-1,10-decanediamine (BSC10), has been synthesized and its interfacial hydrogen bond formation or molecular recognition with barbituric acid was investigated in comparison with that of a single chain Schiff base, 2-hydroxybenzaldehyde-octadecylamine (HBOA) (Jiao et al., 2005; 2011d). It has been found that while HBOA formed a monolayer at the air/water interface, the bolaform Schiff base formed a multilayer film with ordered layer structure on water surface. When the Schiff bases were spread on the subphase containing barbituric acid, both of the Schiff bases could form hydrogen bonds with barbituric acid in situ in the spreading films. For example, one-layer HBOA LB film deposited from the water surface showed the dotted domains at a surface pressure of 5 mN/m and showed some large ribbon structures at a surface pressure over the plateau region, as shown in Fig. 2. On the other hand, when HBOA was spread on the BA subphase, plank-like morphology with a clear boundary was observed, indicating a complex monolayer formation between HBOA and BA at the air/water interface. As a result, an increase of the molecular areas in the isotherms was observed. The in situ H-bonded films could be transferred onto solid substrates, and the transferred multilayer films were characterized by various methods such as UV-vis and FT-IR spectrosopies. Spectral changes were observed for the films deposited from the barbituric acid subphase, which supported the hydrogen bond formation between the Schiff bases and barbituric acid. By measuring the MS-TOF of the deposited films dissolved in $CHCl_3$ solution, it was concluded that a 2:1 complex of HBOA with barbituric acid and a 1:2 complex of BSC10 with barbituric acid were formed. Such a complex mode can be regarded as related to the molecular structures of the Schiff bases, which is illustrated in Fig. 3. For the BSC10/BA films, BSC10 molecule, having two Schiff base headgroups, can act as a tweezers and two BA molecules are embedded in the BSC10 molecule. In this case, the alkyl spacer curved at the air/water interface and the two BA molecules formed intermolecular hydrogen bonding and were encapsulated under the alkyl chain. On the other hand, when the multilayer films of both Schiff bases were immersed in an aqueous solution of barbituric acid, a similar molecular recognition through the hydrogen bond occurred. A clear conformational change of the alkyl spacer in the bolaform Schiff base was observed during the complex formation with the barbituric acid. Now we provided a new hydrogen bonding system at the air/water interface. In addition, a continuous conformational change of the alkyl spacer was clearly shown experimentally in the LB film.

Fig. 2. AFM images of one-layer HBOA LB films from pure water at surface pressures of 5 (a) and 25 mN/m (b), respectively, and from 1.0 mM aqueous BA subphase at surface pressures of 5 (c) and 25 mN/m (d), respectively
(Reproduced from Jiao et al., 2005, The Journal of Physical Chemistry B, 109, 2532-2539).

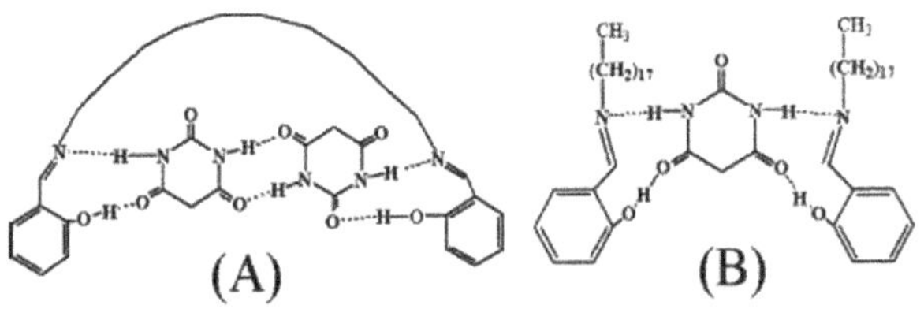

Fig. 3. Schematic Illustration of Structures of BSC10/BA (A) and HBOA/BA (B) Complexes (Reproduced from Jiao et al., 2005, The Journal of Physical Chemistry B, 109, 2532-2539).

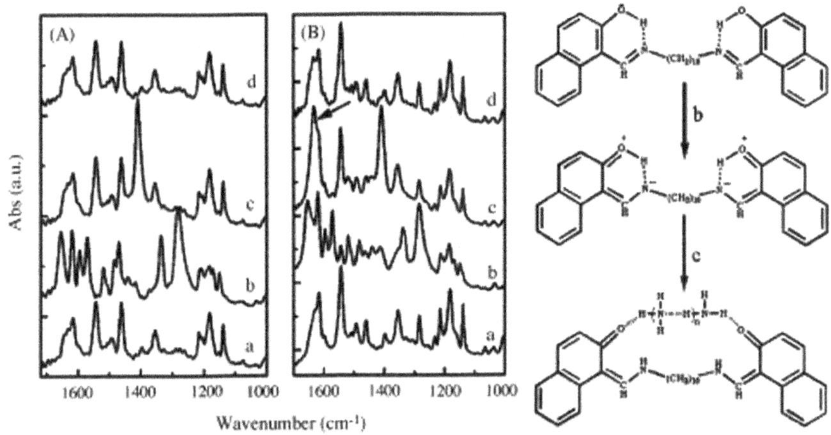

Fig. 4. FT-IR spectra of HNOA (A) and BNC10 (B) and acidichromism mechanism of BNC10: (a) LB film before gas exposure; (b) after exposure in HCl gas; (c) after exposure "b" in NH_3 gas; (d) after washing by water (Reproduced from Yin et al., 2007, Chinese Chemical Letters, 18, 30-32).

Nextly, a bolaform (BNC10) and single-headed (HNOA) amphiphilic Schiff bases containing naphthyl group were designed and their Langmuir-Blodgett films were investigated. It was found that both the LB films show acidichromism, i.e. a reversible color change upon alternatively exposing the films to HCl and NH_3 gases, respectively (Yin et al., 2007). It was further found that the bolaform Schiff bases film could trap NH_3 gas during the acidichromic process. Such difference between the single-headed and bolaform Schiff base can be explained as follows. The as-deposited films exist predominantly in the enol form, as verified by the FT-IR spectra. When the films were exposed to HCl gas, the intramolecular H-bond was destroyed and the keto form will be predominant. Upon exposing to NH_3 gas, the keto form returned to enol form for the HNOA, and thus the color changed reversibly. However, for the BNC10, when the film was exposed to NH_3 gas, NH_3 gas molecules were suggested to be trapped into the BNC10 molecules through intramolecular H-bonds, as shown in the mechanism scheme in Fig. 4. The trapped NH_3 was stabilized by the two keto form Schiff base headgroup, which is similar to the "chelating of metal ions by ligand". In conclusion, we have shown the acidichromism in the LB films of Schiff bases containing naphthyl group. The bolaform Schiff base can additionally trap NH_3 gas during the acidichromic process.

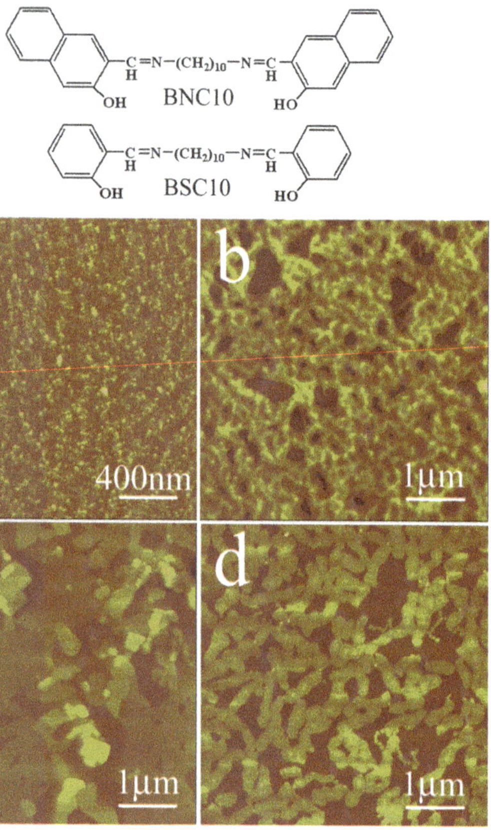

Fig. 5. Molecular structures and AFM images of one-layer BSC10 and BNC10 LB films from pure water (a, c) and aqueous 1.0 mM $Cu(Ac)_2$ (b, d) at 20 mN/m, respectively (Reproduced from Jiao, et al., 2011f, Journal of Dispersion Science and Technology, 32, in press).

In addition, two bolaform Schiff bases derivatives with different substituted headgroups have been designed and their interfacial phase behaviors and coordination with Cu(II) ions were investigated (Jiao et al., 2011f). It has been found that while one molecule with benzene headgroup formed dotted aggregations at the air/water interface, another with naphthyl moiety as headgroup formed crystalline multilayer films on water surface. When on the subphase containing Cu(II) ions, both of the Schiff bases can coordinate with Cu(II) in situ in the spreading films with the obvious conformational change of alkyl chains. For the BSC10 film transferred from aqueous Cu(II) ions, a rigid superstructure with many hole was observed, as shown in Fig. 5. This indicated that BSC10 formed a multilayer film upon reaction with Cu(II) ions. On the other hand, as for the BNC10 LB films deposited from the water surface or Cu(II) subphase, different morphologies were observed. The distinct morphologies suggested the different aggregation modes in both cases, which agreed well with the results of other sections. In comparison, the ex situ coordination process at the liquid/solid interface have also been investigated by continuous spectral measurement. Depending on the different headgroups, these amphiphiles showed different aggregation behaviors in the Langmuir-Blodgett films. Particularly, during the coordination process of ligand with Cu(II) ions in organized molecular films, great conformational change of the alkyl chains was observed. At the same time, a rational explanation about the headgroup effect on regulating the aggregation behaviors was discussed. The present work may give some insight to design and character the relationship between the molecular structures and supramolecular aggregation of bolaamphiphiles in organized molecular films.

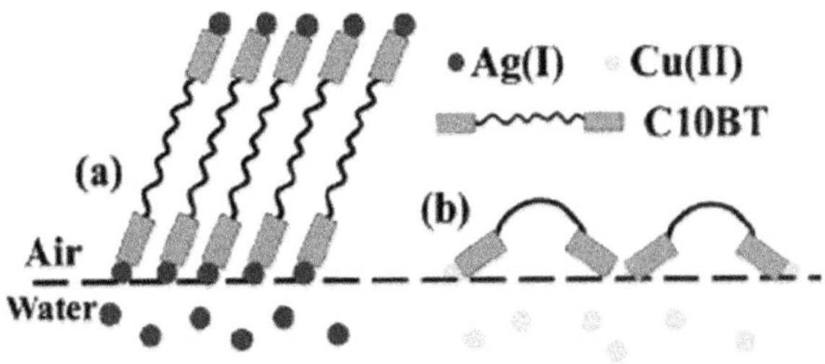

Fig. 6. Cartoon illustration of the packing modes for C10BT complex films on different subphases (Reproduced from Jiao et al., 2006b, Thin Solid Films, 503, 230-235).

Furthermore, the interfacial assemblies at the air/water interface from a newly designed tyrosine-based bolaamphiphile, 1,10-bis(O-l-tyrosine)-decane (C10BT), were investigated (Jiao et al., 2006b; 2011e). This bolaform compound could be spread on water surface and form organized ultrathin film. It was interesting to find that metal ions such as Ag^+ and Cu^{2+} in the subphase can greatly modulate the molecular packing of C10BT and the morphology of the subsequently deposited Langmuir-Blodgett (LB) films. Atomic force microscopic measurements revealed that C10BT LB film from the subphase containing Ag^+ ion showed well-ordered layered nanofibers, while Cu^{2+} ion coordinated C10BT film demonstrated dense cross-linked network. It was suggested that both the strong chelating property to the carboxylate and the different packing mode of hydrocarbon chain resulted in the distinct nanostructures. Fourier transform infrared spectra reveal the difference between the Ag-C10BT complex film and that of Cu^{2+} ion, and the mechanism of the packing mode of hydrocarbon chain was discussed. Furthermore, the X-ray diffraction and X-ray photoelectron spectra also verified the orderly layer structure and the relative molar ratios compared with different metal ions. Considering the coordinating orientation of the two metal ions and the conformation of the alkyl spacer, it is rational that in the case of Cu(II), the molecules were linked through coordination with a bended spacer and formed a polymeric compound and aggregate to construct the network morphology, while in the case of Ag(I), the molecules was aligned orderly via the cooperative interaction such as coordination with Ag(I) ions, hydrophobic force between the alkyl spacers and/or the π-π stacking interaction between the benzene moieties, as shown in Fig. 6. While many efforts have been devoted to manipulation of the nanostructures and functions of sophisticated bolaform amphiphiles, we provided a simple method of modulating the organization and morphology of C10BT films through metal ions.

3. Supramolecular assemblies of some series of gemini amphiphiles with distinct structures

Firstly, aggregations of cationic amphiphiles containing one, two and four positive charges with an anionic TPPS at the air/water interface were investigated (Zhang et al., 2006). The positive charge centers in these amphiphiles were separated by a rigid spacer. It was found that TPPS in the subphase could aggregate onto the spreading films as J-aggregate. Although the amphiphiles and the TPPS were achiral, CD signals were detected for the complex films. A possible helical sense of the TPPS stacked in the complex film was suggested to be responsible for the supramolecular chirality of the multilayer films. The charge distributions and spacers have great influence on the CD intensity as well as their split and the surface morphologies. AFM measurements revealed nanofiber structures in the surface morphologies for the complex films of TPPS with amphiphiles having one or two positive charges, while nanorod morphologies were observed for that with four positive charges. It was suggested that the charge centers in the amphiphiles limited the aggregation of TPPS and the interactions between the adjacent J-aggregates. Such effect and the differences from simple amphiphiles reported can be explained as illustrated in Fig. 7. The results provided a clue in regulating the surface morphologies and chirality of TPPS assemblies.

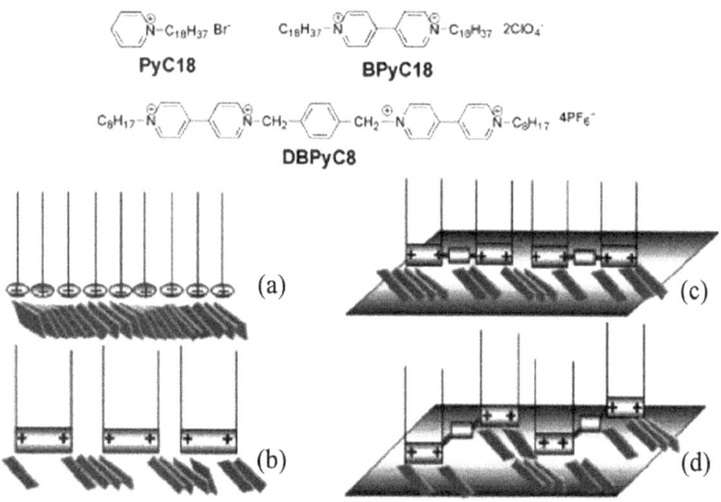

Fig. 7. Molecular structures and illustration of the possible structures of the complex film of TPPS with amphiphiles (a) PyC18, (b) BPyC18, (c and d) DBPyC8 with different conformations (Reproduced from Zhang et al., 2006, Colloids and Surfaces A: Physicochemical and Engineering Aspects, 284-285, 130-134).

In addition, the designed gemini-type amphiphile containing Schiff base groups can form stable monolayers at the air/water interface. They can coordinate with Cu(II) ions in situ at the air/water interface and form an intramolecular complex between the ligand and the metal ion (Jiao et al., 2006). For BisSBC18C2, an additional intermolecular dimerlike complex could be formed. The series of amphiphiles shows a strong spacer effect on both the interfacial properties and the assemblies. Maximum limiting molecular areas are observed for the monolayers of the gemini amphiphiles with spacer lengths of C6 and C8 on water and aqueous $Cu(Ac)_2$, respectively, which is due to the formation of the upward-curved conformation of the hydrophobic spacer. Such an effect of the spacer leads to different morphologies such as nanonail and wormlike nanoarchitectures, depending on the spacer length and the subphases, as shown in Fig. 8. Because of the covalent connection of the Schiff base groups by the alkyl spacer, the π-π^* absorption of the localized aromatic rings split into two bands that could be assigned to H and J aggregation, respectively. With the spacer length increased, the J aggregation of the aromatic ring seems to be preferred. An interdigitated bilayer structure is formed in the multilayer films both for the film from water and from the aqueous $Cu(Ac)_2$ subphase, as shown in Fig. 9. The present work provided new insight into the design and interfacial assembly of the gemini amphiphiles.

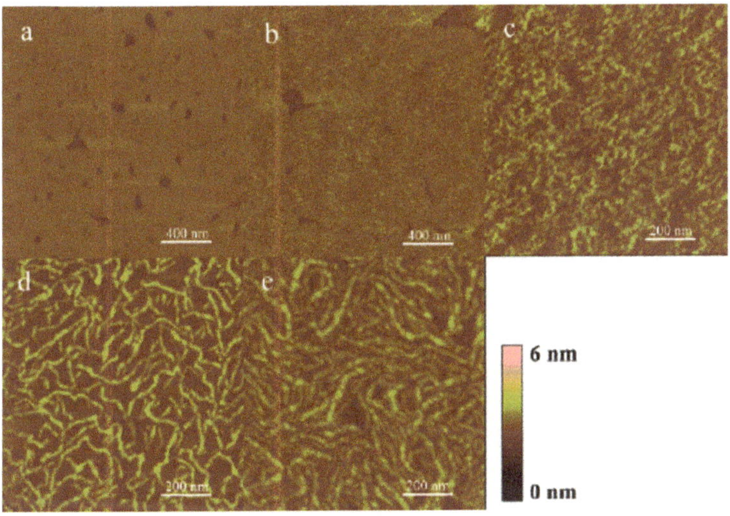

Fig. 8. AFM images of one-layer BisSBC18Cn LB films deposited from the subphase of a 1.0 mM $Cu(Ac)_2$ solution at 20 mN/m: (a) BisSBC18C2, (b) BisSBC18C4, (c) BisSBC18C6, (d) BisSBC18C8, and (e) BisSBC18C10
(Reproduced from Jiao et al., 2006, Langmuir, 22, 5005-5012).

Furthermore, a series of new gemini amphiphiles with hydrophilic spacers, GNn (n=1-3), have been designed and they can form stable monolayers on water subphase with various pH values (Jiao et al., 2007). On strong alkaline and acidic subphase, the headgroup and the spacer of the gemini amphiphiles underwent dissociation and protonation, respectively, resulting in the enlargement of the molecular areas. Flat and uniform domains were obtained for the monolayers from nearly neutral subphase; flower-like or dendritic domains were observed for the films transferred from strong acidic subphase, as shown in Fig. 10. On the other hand, when an anionic tetrakis(4-sulfonatonphenyl)porphine (TPPS) was added into an acidic subphase, an in situ complex formation between the gemini amphiphiles and TPPS occurred. The complex monolayers were transferred onto solid substrate and TPPS existed predominantly as J-aggregate in the complex films. Due to the multisited positive charges in the spacer on acidic subphase, the complex films of gemini amphiphiles with TPPS appeared as short fiber or nanorod structures and formed two-dimensional (2D) conglomerate chiral domains, as shown in Fig. 11. Compared with the features of those conventional or gemini-type amphiphiles with hydrophobic spacer, the introduction of the hydrophilic spacer in the present work provides a pH regulated interfacial properties of the film.

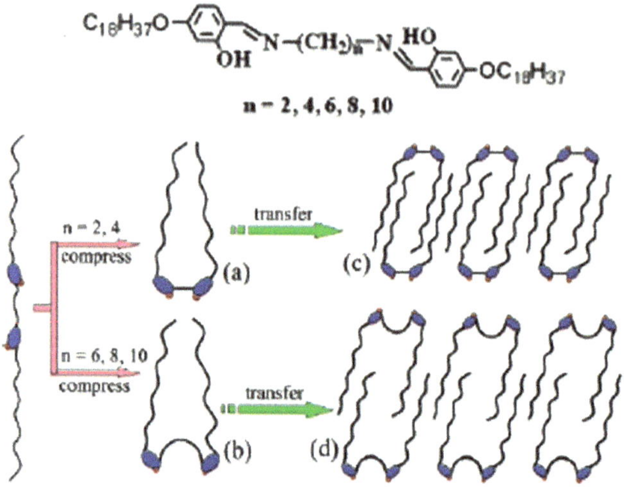

Fig. 9. Molecular structures and schematic illustration of the packing modes of BisSBC18Cn in organized films: when compressed at the air/water interface, the spacers are parallel to the water surface for a short spacer (a) and have an upwardcurved conformation with a longer spacer (b). During the transfer process, the overturning of the film leads to the formation of interdigitated films
(Reproduced from Jiao et al., 2006, Langmuir, 22, 5005-5012).

In addition, hybrid multilayer films based on a series of gemini amphiphiles BisBC18Nn (n=1,2,3) with different hydrophilic spacer lengths with phosphomolybdic acid (PMo_{12}) were constructed through the LB technique (Jiang et al., 2008). The hybrid monolayers could be deposited as multilayer films and all the hybrid films exhibited a well-defined layer structure. When exposed to UV light, the BisBC18Nn/PMo_{12} hybrid films changed from pale yellow to cyan, as shown in Fig. 12. The reverse reaction occurred in air or O_2 with the help of heat. Repeatable coloration-decoloration proved the excellent photochromic reversibility of the hybrid films. Such characteristics of absorption bands may be attributed not only to the coexistence of PMo_{12} and BisBC18Nn molecules, but also to the ordered arrangement of the molecules. Although the spacer length of the gemini amphiphiles has an obvious effect on the monolayer property, it affects the photochromic properties only slightly. The spacer length causes slight differences in the maximum absorption position of the coloured species, which may be due to subtle change of the environment around the POM. By analyzing changes in the position and relative strength of Mo-O_d and P-O_a stretching band, it was revealed that BisBC18N1/PMo_{12} hybrid films with the shortest spacer showed the best photochromic ability. The present results prove that an appropriate selection of the polyoxometalate and amphiphiles could lead to the preparation of hybrid LB films that exhibit good photochromic properties.

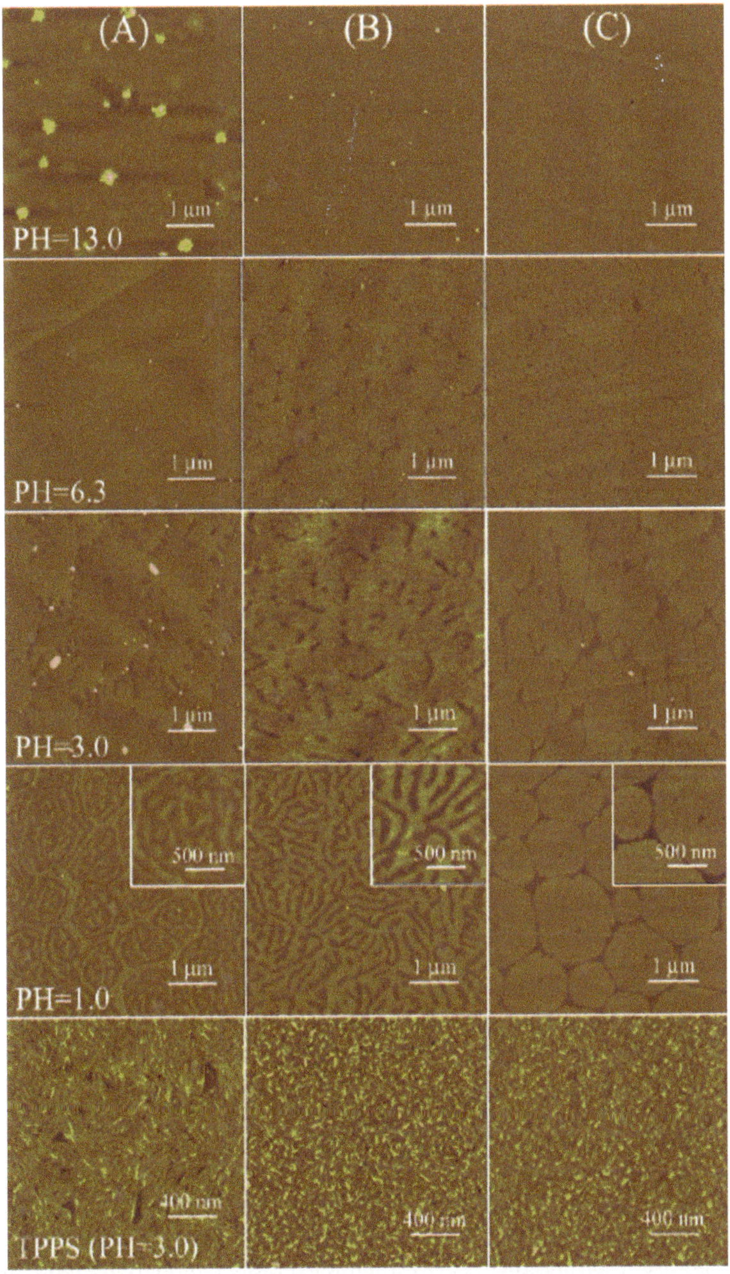

Fig. 10. AFM pictures of one-layer LB films of GN1 (A), GN2 (B), and GN3 (C) transferred from aqueous subphases with various pH values and aqueous TPPS (0.01 mM) at 20 mN/m (Reproduced from Jiao et al., 2007, The Journal of Physical Chemistry B, 111, 3090-3097).

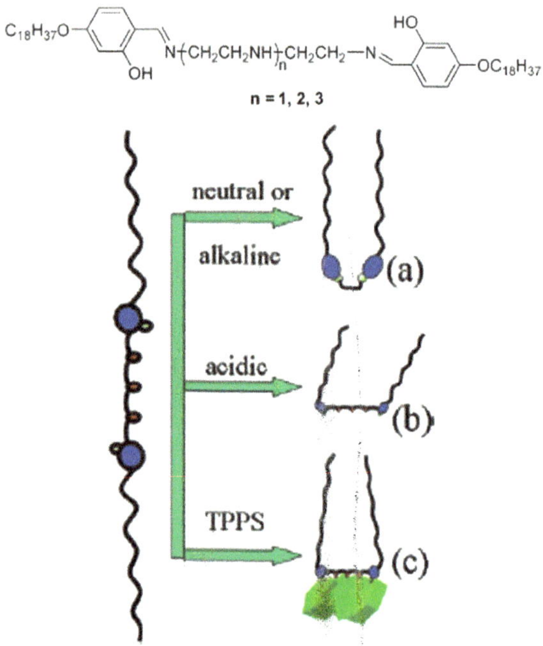

Fig. 11. Molecular structures of gemini amphiphiles and illustration on the packing of the gemini amphiphiles with their spacers on water surface. Part a, on neutral or alkaline subphase: the spacers were immersed into the subphase. Part b, on acidic subphase: the spacers were protonated and stayed on water surface. Part c, complex film with TPPS: the spacers were protonated and interacted with TPPS through electrostatic interaction (Reproduced from Jiao et al., 2007, The Journal of Physical Chemistry B, 111, 3090-3097).

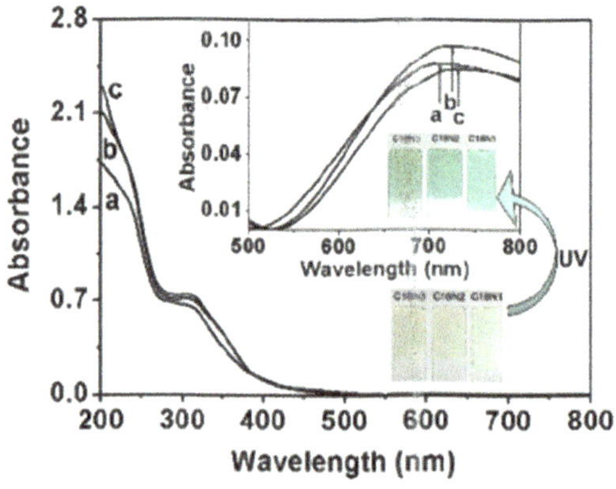

Fig. 12. UV-Vis absorption spectra of 70-layer multilayer films transferred at a constant pressure of 20 mN/m: (a) BisBC18N1/PMo_{12}, (b) BisBC18N2/PMo_{12}, (c) BisBC18N3/PMo_{12}. The inset shows UV-Vis spectra of the corresponding films after irradiation (Reproduced from Jiang et al., 2008, New Journal of Chemistry, 32, 959-965).

Nextly, this series of gemini amphiphiles in the organized molecular films were used to investigate the generation and assembly of gold nanostructures (Zhong et al., 2008). The chloroauric acid, dissolved in the aqueous subphase, was incorporated into the monolayers of the gemini amphiphiles containing ethyleneamine spacers through an interfacial assembly. The in situ formed complex monolayers were transferred onto solid substrates, and gold nanoparticles were generated in the film by a chemical or photochemical reduction, which is illustrated in Fig. 13. Discrete gold nanoparticles with an absorption maximum at 550 nm were generated in the films by photoirradiation, while different gold nanostructures were obtained by chemical reduction. Depending on the chemical reductant, various shape and assembly of gold nanostructures were obtained. When reduced by hydroquinone, a tree-branched assembly of the nanoparticles was obtained and the film showed a broad band centered at around 900 nm. When $NaBH_4$ was applied, crooked nanowires or assembly of nanoparticles were obtained, depending on concentration, and the film showed absorption at 569 or 600 nm. Furthermore, by combining the photochemical and chemical reduction methods, i.e., the chloroaurate ion-incorporated film was initially irradiated with UV light and then subjected to chemical reduction, the optical absorption of the formed gold nanostructures can be regulated. This study extends the research on the synthesis of metallic particles in films from conventional amphiphiles to other special amphiphiles and provides a very convenient method to modulate the optical properties in a large wavelength range, which will be important in designing optical devices based on different gold nanostructures.

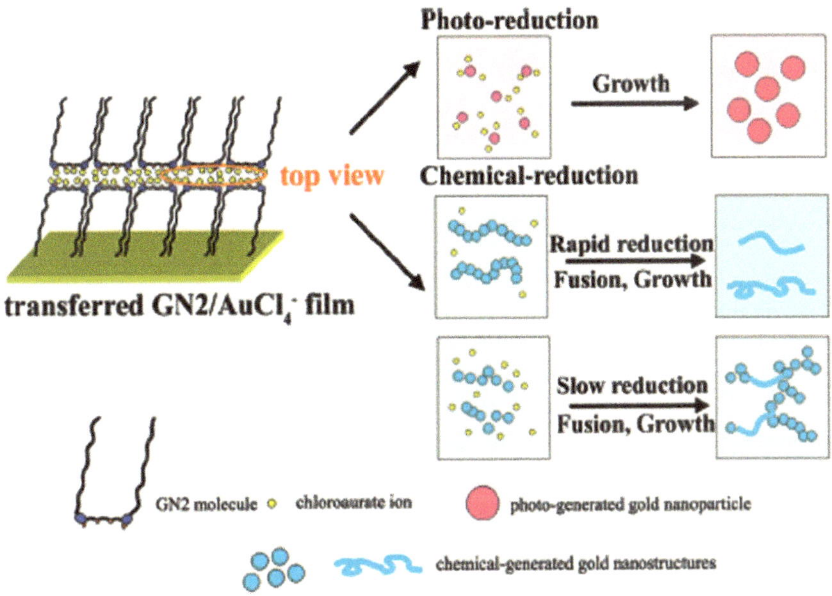

Fig. 13. Schematic illustration on the generation of gold nanoparticles by photochemical and chemical reduction in $GN2/AuCl_4^-$ complex film (Reproduced from Zhong et al., 2008, Langmuir, 24, 11677-11683).

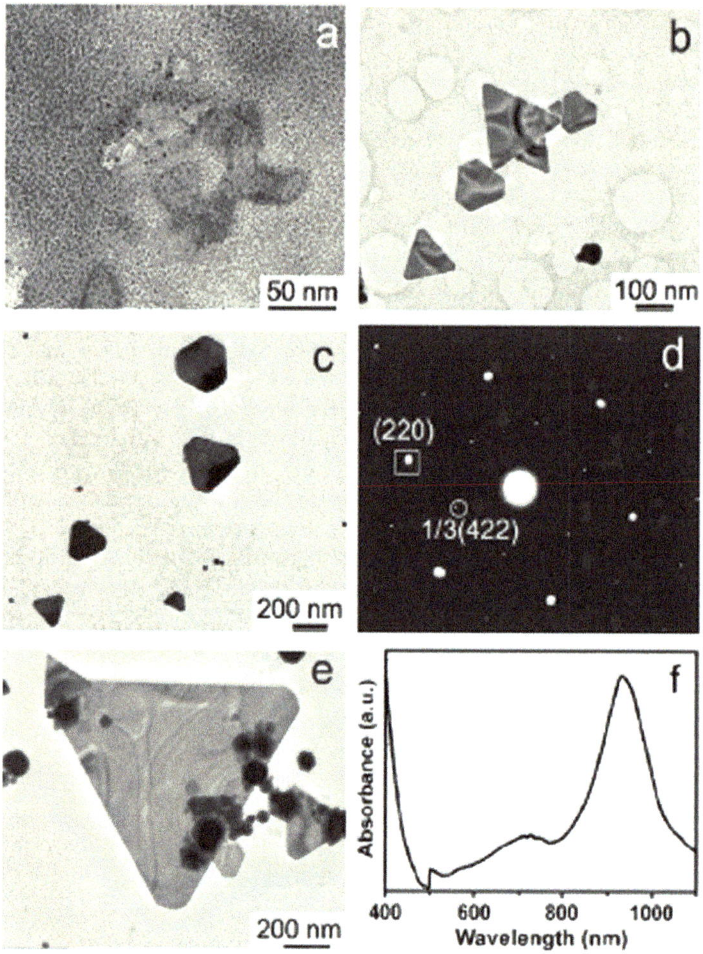

Fig. 14. TEM images of gold nanoparticles with GN2 to chloroaurate ions ratio of 1:2 after 3 hours (a), 7 hours (b), 15 hours (c) and 75 hours (e). SAED analysis of a single gold nanoprism after 7 hours on a single gold nanoprism (d). UV-vis spectrum of reduced gold nanoprisms by GN2 after 15 hours of reaction (f) (Reproduced from Zhong et al., 2009, Journal of Nanoscience and Nanotechnology, 9, 2726-2730).

On the other hand, this series of gemini amphiphiles with different lengths of hydrophilic ethyleneamine spacers can also be used to synthesize gold nanostructures at a liquid-liquid interface (Zhong et al., 2009). By stirring the aqueous solution containing $AuCl_4^-$ ions with the chloroform solution of gemini amphiphiles, $AuCl_4^-$ ions were transferred into the oil phase and reduced to gold nanostructures. UV-vis and Fourier Transform Infrared spectral measurements indicated that the gemini amphiphiles could serve as both capping and reducing agents. Crystalline gold nanoprisms were predominantly obtained if the gemini amphiphiles with appropriate spacer length and concentration were applied, as shown in Fig. 14. Thus the formation of the nanoprism can be possibly suggested as follows. First, gemini amphiphile reduced $AuCl_4^-$ ions to zerovalent Au or Au clusters. Next, the gemini amphiphiles selectively adsorbed on certain plane and nanoprism could be finally obtained. Thus appropriate spacer and concentration are both important for fabrication of anisotropic nanostructures. It is predicted that gold and other novel metal nanostructures may be produced by gemini amphiphiles, whose properties can be well-controlled by designing different headgroups, spacers or alkyl chains.

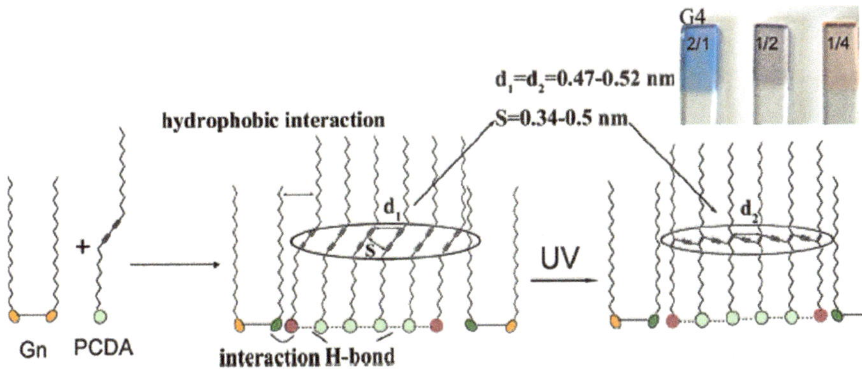

Fig. 15. Illustration of molecular packing in mixed PCDA/Gn film. The inset shows the color of the photopolymerized PCDA/G4 films with different ratios (Reproduced from Zhong et al., 2009a, The Journal of Physical Chemistry B, 113, 8867-8871).

In addition, we have investigated the photopolymerization of an amphiphilic diacetylene, 10,12-pentacosadiynoic acid (PCDA), in organized molecular films in the presence of this series of gemini amphiphiles with different spacer lengths (Zhong et al., 2009a). It has been found that, when gemini amphiphiles were mixed with the diacetylene, the film-forming properties were greatly improved and the photopolymerization could be regulated by the gemini amphiphiles. Miscibility and Fourier transform infrared spectroscopy (FT-IR) investigations revealed that the polymerization of PCDA in a mixed film was regulated by the mixing ratio and spacer length of the gemini amphiphiles. Although a slight amount of gemini amphiphile did not make the PCDA polymerize into blue films, the increment of the gemini amphiphile with the short spacer length in the mixed film caused the formation of a red film, and the intensity of red phase to blue phase can be modulated by changing the mixing ratios. When gemini amphiphiles with longer spacer lengths were mixed, blue films were predominantly obtained in all mixing ratios. A mechanism including the interaction between the headgroup of the gemini amphiphiles and the diacetylene and the regulation of the spacer was proposed, as shown in Fig. 15. It was suggested that the ionic interaction between the headgroups of geminis and diacetylene played an important role in causing such differences. The strong ionic interaction deviated the packing of PCDA molecules from the one with their carboxylic acid groups H-bonded and tended to form red film when the spacer length was short, while a longer spacer in the gemini can weaken the ionic interaction and keep the film packing in favor of the blue film. The present work provides an efficient way of regulating the photopolymerization of diacetylene through simple mixing with gemini amphiphiles.

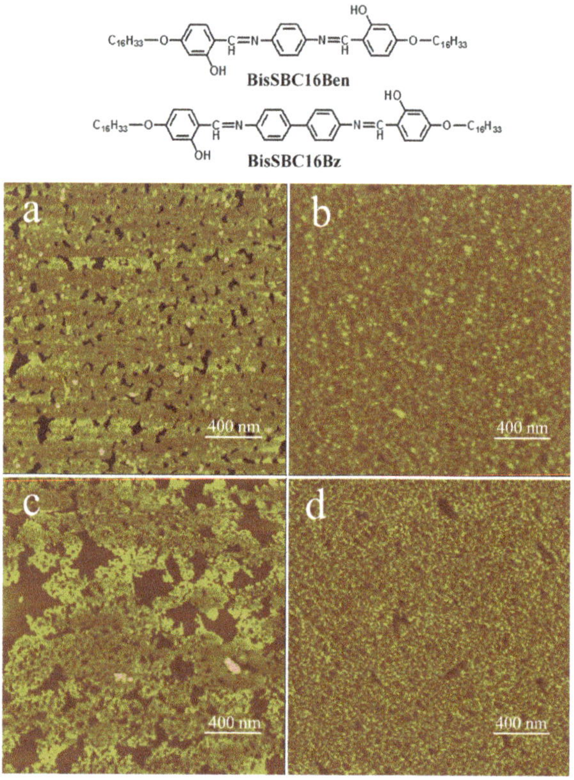

Fig. 16. Chemical structures and AFM images of one-layer LB films of BisSBC16Ben and BisSBC16Bz on pure water surface (a and c, respectively) and the aqueous 1 mM $Cu(Ac)_2$ subphase (b and d, respectively) at surface pressures of 20 mN/m (Reproduced from Jiao et al., 2011g, unpublished results).

Recently, two gemini-type Schiff base amphiphiles containing aromatic spacers have been synthesized and characterized in LB films (Jiao et al., 2011g). These compounds could form stable multilayer films and show distinct phase behaviors on water surface and ion subphase. When the compounds were spread on Cu(II) ions subphase, an in situ coordination can occur for both ligands, while there have different morphologies and molecular rearrangement, as shown in Fig. 16. For both process, the aromatic spacers in gemini amphiphiles have predominant effect in regulating the aggregation mode and spectral changes in organized molecular films. The present work may give some insight to design and character the relationship between the molecular structures and supramolecular aggregation of gemini amphiphiles in organized molecular films.

4. Design and interfacial characterization of some series of amphiphiles with functional substituted groups

Firstly, diblock copolymer of $PMMA_{291}$-b-PEO_{114} and triblock copolymer of $PMMA_{120}$-b-PEO_{227}-b-$PMMA_{120}$ were synthesized and their interfacial properties at the air-water interface were investigated, where both blocks of polyethylene oxide (PEO) and poly(methyl methacrylate) (PMMA) are surface active but the former is soluble in water while the latter is not (Cheng et al., 2005). Both the block copolymers could form monolayers with two obvious transition regions. The first transition point is around 10 mN/m, which could be assigned to the pancake-brush change of the PEO chains. The other is around 18 mN/m, which could be ascribed to the condensed packing of PMMA. The surface morphological changes during the compression of the Langmuir monolayers are investigated by measuring the AFM and SEM pictures for the films deposited at difference surface pressure or molecular areas. At a lower surface pressure, morphology typical of PEO-containing lipopolymers is observed. Upon compression, sphere-dominant morphologies were observed, as shown in Fig. 17. While the diblock copolymer is easy to folding, the triblock copolymer is in favor of formation of circular domains through vesiculation.

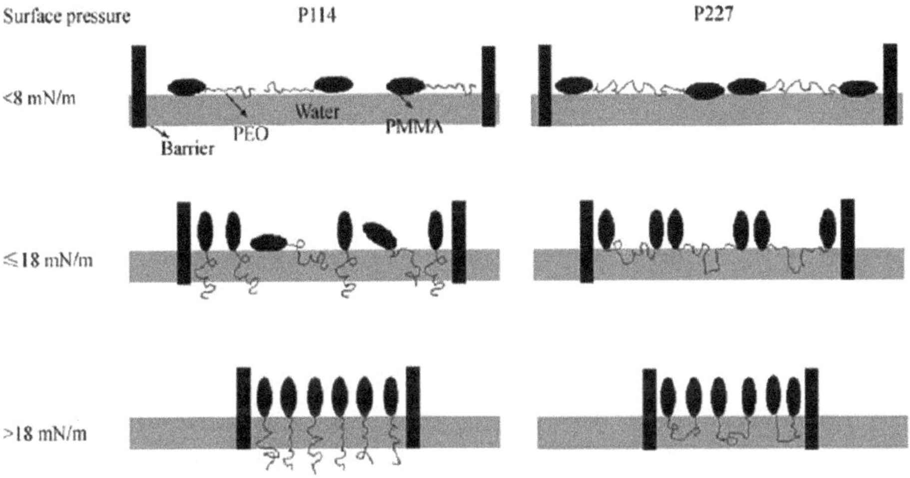

Fig. 17. Schematic drawing of possible PEO chain conformation from diblock and triblock copolymer at air/water interface (Reproduced from Cheng et al., 2005, Science in China Series B-Chemistry, 48, 567-573).

In addition, we have investigated the interfacial behavior of a novel ABC of PEO_{45}-*b*-$PDMA_{69}$ at various compression stages by the LB balance and morphology study (Cheng et al., 2006). Accompanied by a pancake-to-brush transition of PEO conformation, the wormlike surface micelles are compression induced. The micelles as the building blocks can be arrayed parallel to generate a long-range ordered structure, further bended and twisted upon compression, as shown in Fig. 18. Unusual 3D large aggregates of "springs" and "disks" appear when the LB film completely collapses. Such a rich phenomenon of hierarchical nanostructure evolution should be attributed to the coil-semirod architecture of the diblock and the unique assembly ability of the Percec-type dendronized PDMA. This provides a clue in designing block copolymers to control the nanostructures at the air/water interface.

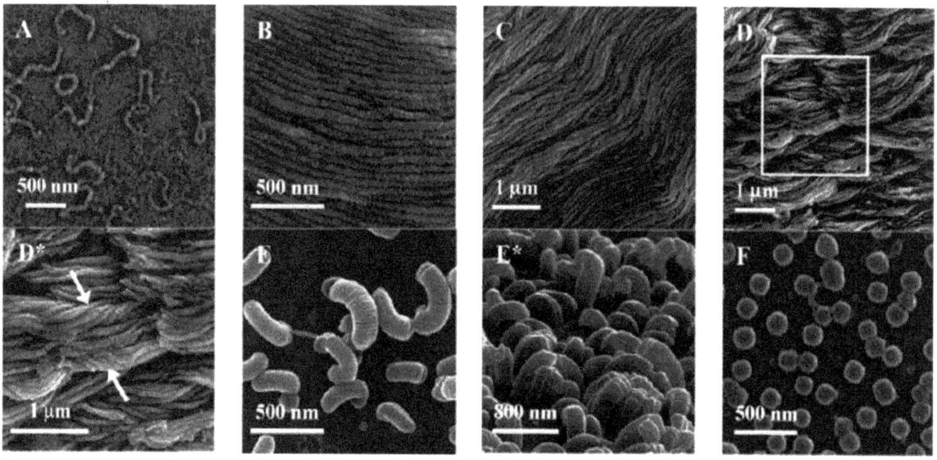

Fig. 18. A set of SEM images of the LB films at various molecular areas. The images A-F correspond to the points A-F. D* is the enlarged image of the area indicated by the square in D, where the twisted micelles are pointed by the arrows. E* shows the "springs" at point E with a viewing angle (Reproduced from Cheng et al., 2006, Macromolecules, 39, 6327-6330).

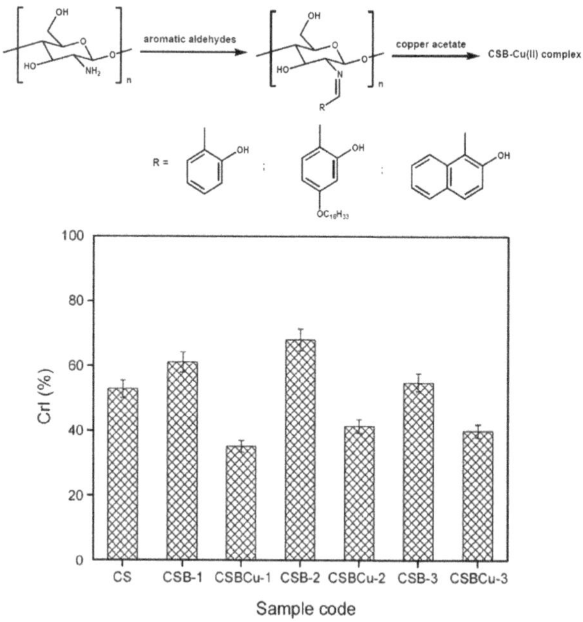

Fig. 19. The synthetic route and variation of crystallinity of chitosan-based Schiff bases and their copper (II) complexes (Reproduced from Jiao et al., 2011, Iranian Polymer Journal, 20(2), 123-136).

It is well known that Chitosan is the product of *N*-deacetylation of chitin. For its innocuous, renewable, biocompatible property chitosan is applied in many fields e.g., pharmaceutical, food, catalysis, material. In this work, three chitosan-based Schiff-based (CSB) compounds with aromatic substituent groups were synthesized from the reaction of chitosan with different aromatic aldehydes i.e., salicylaldehyde, 4-hexadecyloxy-2-hydroxybenzaldehyde and 2-hydroxy-1-naphthaldehyde (Jiao et al., 2011). The chitosan-based Schiff base copper (II) complexes (CSBCu) were subsequently obtained through the reaction of relative Schiff bases with copper acetate. Generally, elemental analysis data may confirm the formation of chitosan-Schiff base as well as the coordination reaction of CSB with copper ions. FTIR analysis indicated that Schiff base and coordination reaction take place in Schiff base skeleton. Moreover, with the difference in substituent groups and spacer, FTIR spectra showed the clear variety. Thermal analysis showed that the thermal stability of CSB increased slightly, while that of the copper complexes was reduced significantly in comparison with chitosan. The XRD results demonstrated the appearance of a new crystallization peak of CSB in the vicinity of five degree and the lower crystallinity of CSBCu, as shown in Fig. 19. The change of crystallinity is mainly dependent on the formation of Schiff base and to some extent on other factors, such as spacial hindrance, hydrophobic force and π-π stacking. In the present case, these chitosan-based Schiff base compounds have different aromatic substituent groups and/or long hydrophobic alkyl chain. Thus, it seems that the crystallinity and designed structures can be regulated by the modification of substituent groups in chitosan molecule. The present results show that the specific properties of chitosan-based Schiff base derivatives can be altered by modifying the molecular structures of objective compounds with proper substituent groups.

In another case, three amphiphilic Schiff bases containing naphthylidene group (abbreviated as HNOA-1, HNOA-2, and HNOA-3) were designed and their interfacial assemblies were investigated (Jiao et al., 2006a; 2011a). These amphiphiles have different substitution groups in the Schiff base moiety. It was observed that HNOA-1, being in lack of second aromatic ring, formed multilayer films at the air/water interface, while the other two compounds, with another aromatic ring and some hydrophilic groups, can be spread as monolayers on water surface. All of Schiff bases could coordinate with $Cu(Ac)_2$ in situ in the spreading films. Both the spreading films from water and aqueous $Cu(Ac)_2$ subphases were transferred onto solid substrates and their surface morphologies as well as molecular packing modes were investigated by a series of methods such as atomic force microscopy, Fourier transform infrared spectra and X-ray diffraction. Depending on the different substitutions, these amphiphiles showed different orientations in the Langmuir-Blodgett films, as shown in Fig. 20. Particularly, during the process of complex formation at the air/water interface, great conformational change of the alkyl chain was observed for HNOA-2 in comparison with that of other compounds. In addition, nanofiber structures were observed for the Cu(II)-complexed HNOA-1 and HNOA-2 films. It was concluded that both the molecular orientation and the resulted nanostructures in the films could be controlled by modifying the molecular structure through the substitution groups.

The phenomena of self-sorting, defined as the ability to high efficiently distinguish between self and non-self within complex mixtures, has drawn more attention in recent years because of its commonality in biological system and potential applications in fields such as chemical sensors, artificial regulatory systems, and synthetic self-organizing systems, as well with biomimetic devices. To date, research about self-sorting has mainly focused on the systems with different functional groups in bulk solution, especially from H-bonded aggregates, which is easily identified by nuclear magnetic resonance (NMR) measurement. In our present system, a series of Cu(II)-coordinated Schiff base complexes with alkyl or aromatic spacers have been synthesized and characterized by utilizing monolayer and LB techniques (Jiao et al., 2010a). It is found that the substitutional spacer in complexes play an important role in the morphological regulation and spectral change, which will provide distinct clue for designing similar organic/inorganic hybrid compounds at air/water interface. In addition, the self-sorting appears when the binary or ternary complexes mixtures were spread onto air/water interface. The situation of complex mixture with alkyl/aryl spacers shows more distinct interfacial morphologies than the case with only aromatic spacers, as shown in Fig. 21. The present work demonstrates the high versatility in designing similar hybrid organic-inorganic systems and investigating supramolecular assembly.

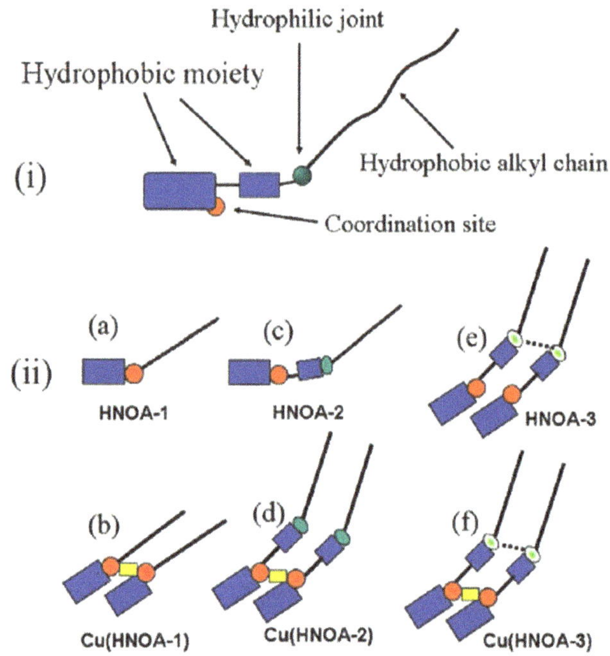

Fig. 20. Schematic illustrations on the designed common molecular structure and aggregation modes of the amphiphilic Schiff bases at air/water interface. (i) The illustrated molecular structure with the depicted function for each moiety; (ii) the detailed aggregation modes for different compounds: (a) for HNOA-1 with simple long alkyl chain, there was no distinct conformational change between water and $Cu(Ac)_2$ surface (b); as for HNOA-2 (c), due to the hindrance of aromatic moiety and coordination matching, the long alkyl chain became more vertical to the surface distinctly (d); because of the intermolecular hydrogen banding, the molecular skeleton of HNOA-3 was almost not changed both on water surface (e) and on the subphase of $Cu(Ac)_2$ solution (f) (Reproduced from Jiao et al., 2006a, Journal of Colloid and Interface Science, 299, 815-822).

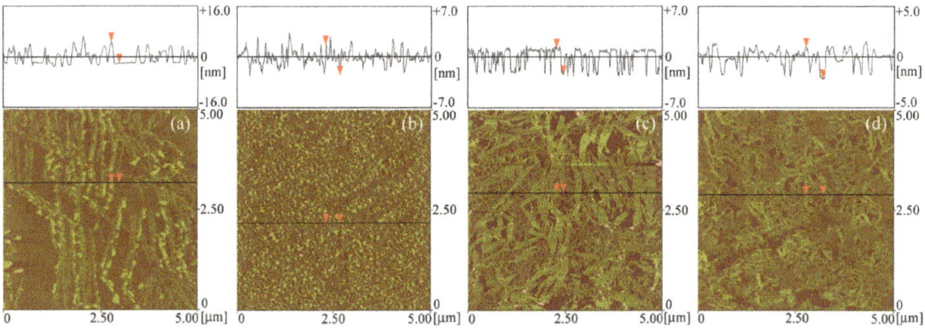

Fig. 21. AFM images of one-layer mixed binary/ternary complexes mixture LB films with alkyl/aryl spacers on pure water interface at 20 mN/m: a) Cu(c10-SB):Cu(o-SB) (1:1); b) Cu(c10-SB):Cu(m-SB) (1:1); c) Cu(c10-SB):Cu(p-SB) (1:1); d) Cu(c10-SB):Cu(o-SB):Cu(m-SB):Cu(p-SB) (1:1:1:1) (Reproduced from Jiao et al., 2010a, Journal of Dispersion Science and Technology, 31, 1120-1127).

It is well known that phthalocyanines (Pcs) are an important class of pigments that have fascinated scientists for many years because of their applications in various disciplines. A wide range of methods recently have been reported for modifying the supramolecular architecture of aggregated dyes such as the symmetrical Pcs in both solutions and thin films formation. Langmuir-Blodgett (LB) thin film deposition or solution casting strategies are made possible by side chain modifications. At present case, the Langmuir-Blodgett films of the azo group-modified phthalocyanines bearing longer alkyl groups on the periphery were successfully prepared (Li et al., 2007). The acidichromism of the LB thin films of both compounds has been studied, and the two compounds show a reversible acidichromic property, as shown in Fig. 22. It should be pointed out that the easy formation of their LB films makes the phthalocyanines to be conveniently fabricated into devices, and the new Pcs derivatives may find interesting applications as chemical sensors.

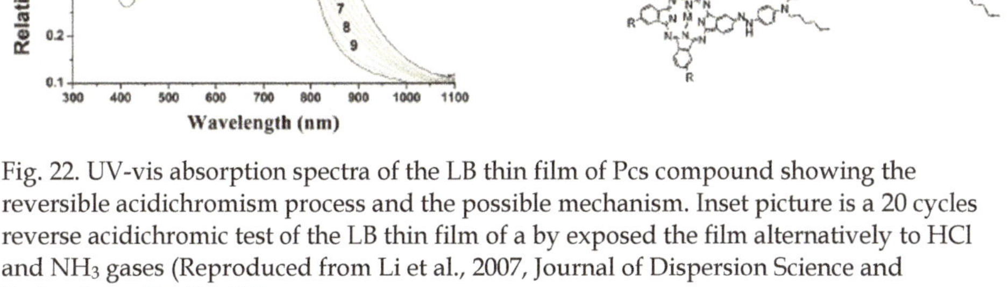

Fig. 22. UV-vis absorption spectra of the LB thin film of Pcs compound showing the reversible acidichromism process and the possible mechanism. Inset picture is a 20 cycles reverse acidichromic test of the LB thin film of a by exposed the film alternatively to HCl and NH_3 gases (Reproduced from Li et al., 2007, Journal of Dispersion Science and Technology, 28, 603-605).

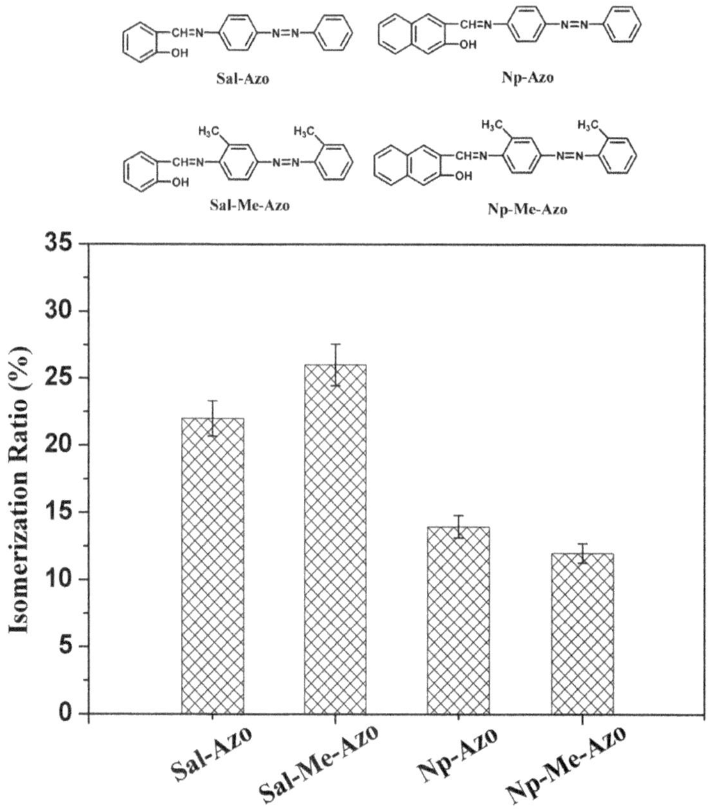

Fig. 23. Chemical structures and isomerization ratio of azobenzene derivatives in ethanol solution upon UV light irradiation (Reproduced from Jiao et al., 2011h, Applied Mechanics and Materials, in press).

Some functional azobenzene derivatives with aromatic substituted groups have been synthesized and their photoisomerization have also been investigated (Jiao et al., 2011). It has been found that depending on different substituted groups, such as phenyl or naphthyl segments, the formed azobenzene derivatives showed different properties, indicating distinct regulation of molecular skeletons. Spectral data confirmed commonly the characteristic absorption of substituted groups and aromatic segments in molecular structures. In addition, the photoisomerization of all compounds in solution can show *trans*-to-*cis* photoisomerization by UV light irradiation, and demonstrate distinct isomerization ratio depending on effect of different substituted headgroups, as shown in Fig. 23. The difference is mainly attributed to the aromatic substituted headgroups and methyl group in molecular structure. The present results have showed that the special properties of azobenzene derivatives could be effectively turned by modifying molecular structures of objective compounds with proper substituted groups, which show potential application in sensor and functional material field.

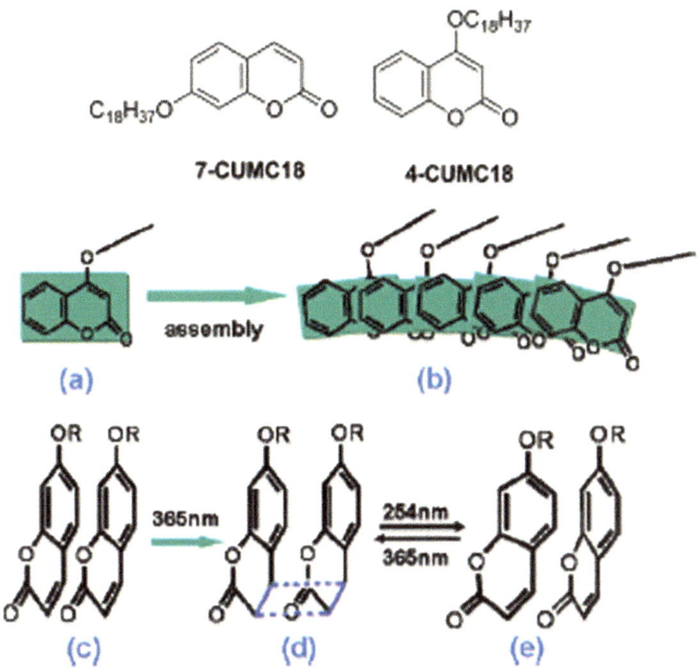

Fig. 24. Molecular structures and illustration of the possible packing and photoreaction in the LS films. Only one layer is presented in the scheme. The top part illustrates the possible orientation of 4-CUMC18 on the water surface (a) and in films (b). The bottom part shows the packing of as-deposited 7-CUMC18 film (c) and the reversible reaction between the dimer (d) and monomer (e) (Reproduced from Guo et al., 2007, Langmuir, 23, 1824-1829).

Coumarin is one of the important parent compounds found in many plants and could be used as the important component of various functional materials. In this work, the Langmuir and Langmuir-Schaefer (LS) films of two coumarin derivatives, 4-octadecyloxylcoumarin (4-CUMC18) and 7-octadecyloxylcoumarin (7-CUMC18) were newly synthesized, and their interfacial assemblies were investigated (Guo et al., 2007). Owing to the different substituent position of the long octadecyloxy chain in the coumarin parent, the two compounds showed completely different behaviors in the interfacial assemblies. When they were spread at the air/water interface, 7-CUMC18 formed a monolayer while 4-CUMC18 formed a multilayer film on the water surface. The spreading films on the water surface were transferred onto solid substrates by a Langmuir-Schaefer method, and the transferred films were characterized by UV-vis, Fourier transfer infrared, X-ray diffraction, circular dichroism, and atomic force microscopy spectra. Different packing of the molecules in the multilayer films was observed, as shown in Fig. 24. Furthermore, distinct properties of the multilayer films were observed. It is revealed that a reversible [2+2] photodimerization and photocleavage could be induced in the LS film of 7-CUMC18 under photoirradiation with UV light of 365 and 254 nm, respectively. Although 4-CUMC18 was achiral, it showed supramolecular chirality in the transferred films due to the formation of macroscopic 2D conglomerates or chiral racemates. These results provided a way of regulating the properties of the molecular assemblies by precisely designing the molecules with subtle substituent position.

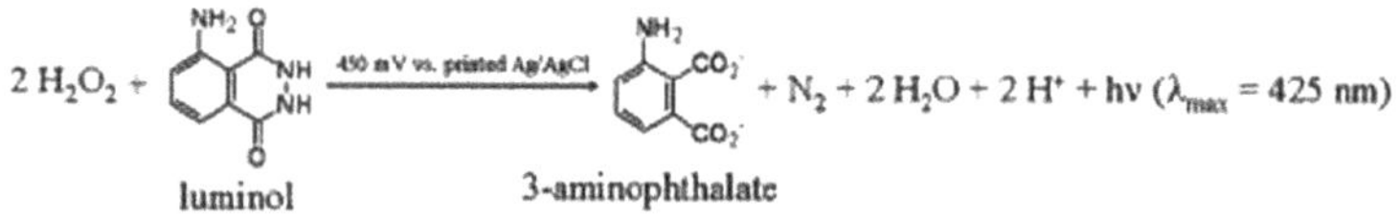

Fig. 25. Luminol electrochemiluminescence (ECL) reaction in the presence of H_2O_2 (Reproduced from Jiao et al., 2008a, Colloids and Surfaces A: Physicochemical and Engineering Aspects, 321, 143-146).

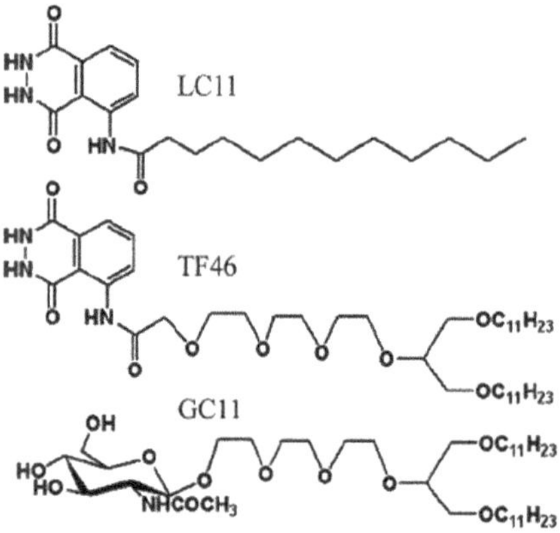

Fig. 26. Chemical structures of luminol derivatives (LC11, TF46) and glycolipid (GC11) (Reproduced from Jiao et al., 2008, Colloids and Surfaces A: Physicochemical and Engineering Aspects, 321, 137-142).

It is well-known that luminol ECL in the presence of hydrogen peroxide can be produced through the action of a chemical catalyst like ferricyanide or a biocatalyst such as peroxidase, as shown in Fig. 25. The application of a potential to oxidize luminol can successfully replaced a catalyst to provoke luminol electrogenerated chemiluminescence (ECL) with inherent high sensitivities and wide linear working ranges. The possibility to insert amphiphilic luminol derivatives in LB films as supports for ECL detection could give the opportunity to develop reagentless biomimetic nanosensors.

Firstly, two new synthetic luminol derivatives with different substituted chains, LC11 and TF46, as shown in Fig. 26, were mixed with glycolipid GC11 to form monolayers at the air-water interface (Jiao et al., 2008). The pure and mixed interfacial Langmuir films were studied by measuring the surface pressure-molecular area isotherms and their morphologies were characterized by Brewster angle microscopy (BAM). The surface pressure-area isotherms showed differences between both luminol compounds. LC11 (single strand acyl chain) displayed a condensed isotherm with a kink point and a high collapse pressure, whereas TF46 (double strand hydrocarbon chains connected to a triethylene glycol (TEG) moiety) displayed a typical liquid-expanded (LE) curve with a low collapse pressure. Dot-like domains were observed by BAM for TF46 Langmuir films, contrariwise to some strip-like aggregated domains in the case of LC11, as shown in Fig. 27. These different morphologies may be attributed to distinct aggregation modes induced by differences in molecular structure. These results clearly demonstrate that changes of molecular structure induce different aggregation modes that markedly influence the interfacial phase behavior of their Langmuir films.

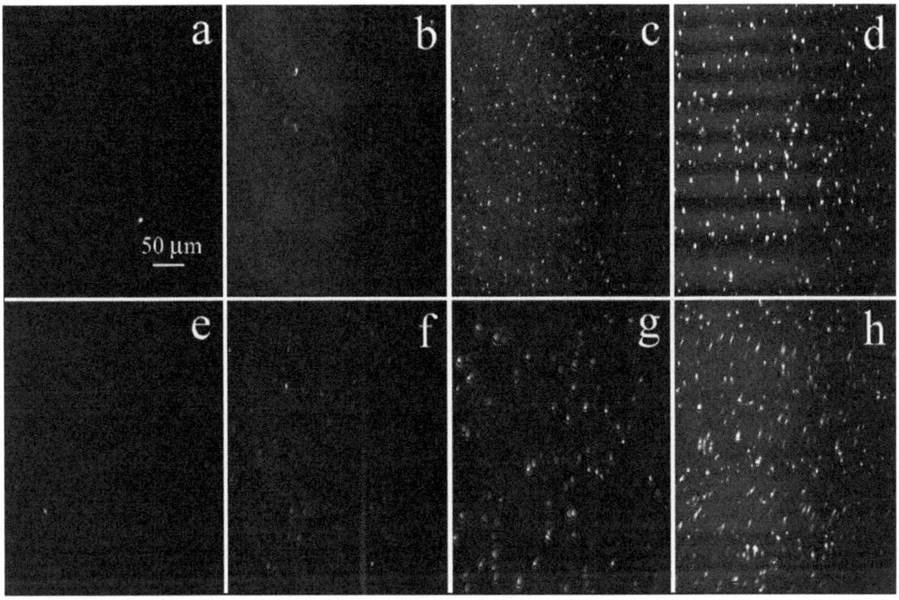

Fig. 27. BAM images of TF46 and mixed TF46/GC11 monolayers at different surface pressures. For TF46 monolayer: (a) 2 mN/m; (b) 12 mN/m; (c) 24 mN/m; (d) 28 mN/m. For mixed TF46/GC11 monolayer: (e) 25 mN/m; (f) 30 mN/m; (g) 32 mN/m; (h) 35 mN/m. The scale bar is 50μm (Reproduced from Jiao et al., 2008, Colloids and Surfaces A: Physicochemical and Engineering Aspects, 321, 137-142).

In addition, both amphiphilic luminol derivatives have been evaluated for electrochemiluminescence (ECL) measurements with a view to develop new lipid membranes as sensing element of nanobiosensors (Jiao et al., 2008a). The design and synthesis of newbiospecific luminol derivatives with various substitution groups have been then undertaken. In a first approach, the possibility to use polarized screen-printed electrodes to trigger ECL in a buffer solution containing luminol derivatives and hydrogen peroxide has been studied, as shown in Fig. 28. The ECL response to H_2O_2 in the presence of these luminescent derivatives has been investigated taking into account crucial factors such as the applied potential value, concentration of luminol derivatives and the substitution group in molecular structures. The lower hydrogen peroxide concentration detected varied from 1×10^{-7} M to 5×10^{-5} M, depending on the substitution group in the luminol derivative structures. Finally, the effect of acyl group in the molecular structure displayed better ECL performances with TF46 (dialkylchain and TEG spacer) than with LC11 (single alkyl chain).

The present results allow to consider the insertion of the luminol derivatives in a lipid-sensing layer, and to develop newoptoelectronic devices based on ECL detection.

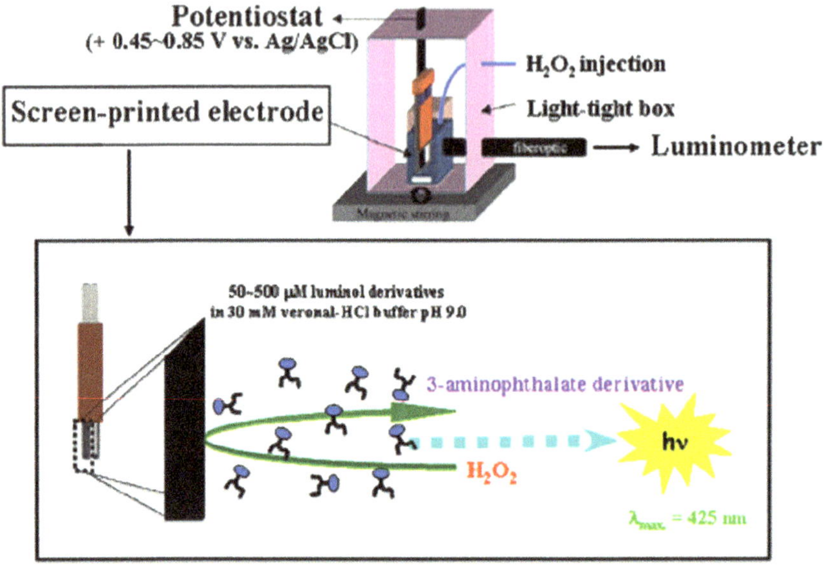

Fig. 28. Schematic representation of the measurement device for ECL detection using a screen-printed electrode. The potential poised at a fixed value between 0.45V and 0.85V vs. printed Ag/ AgCl reference electrode allows triggering ECL in the presence of H_2O_2 (Reproduced from Jiao et al., 2008a, Colloids and Surfaces A: Physicochemical and Engineering Aspects, 321, 143-146).

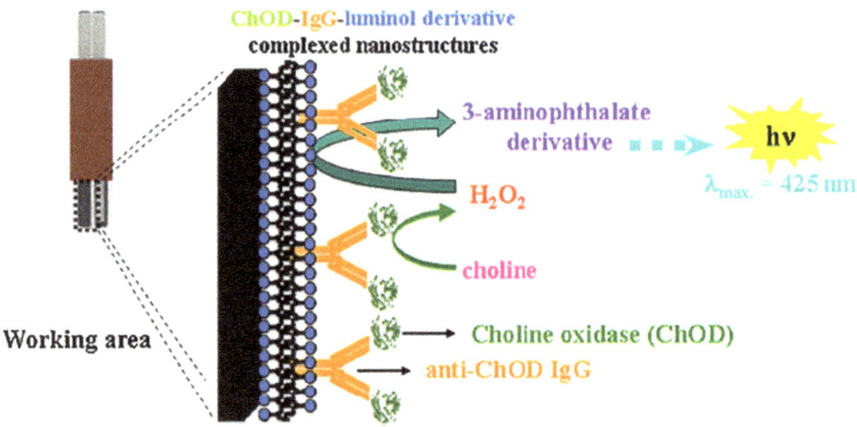

Fig. 29. Reagentless detection principle of choline oxidase activity by ECL reaction triggered directly in the sensing layer (Reproduced from Jiao et al., 2010, Colloids and Surfaces A: Physicochemical and Engineering Aspects, 354, 284-290).

Nextly, this work explores the potentiality of two neosynthesized amphiphilic luminol derivatives to form a lipid bilayer serving as a matrix used for both antibody insertion and ECL detection in order to develop a newsensing layer allowing a reagentless detection (Jiao et al., 2010). As a model, choline oxidase activity has been detected, as shown in Fig. 29. After enzyme immobilization at the surface of the luminol derivative LB bilayer by the way of specific recognition of a non-inhibitory antibody, *in situ* catalytic generation of hydrogen peroxide is able to trigger ECL reaction in the sensing layer interfaced with an optoelectronic device leading to a reagentless detection of choline oxidase activity. Since the sensing layer is based on self-molecular organization of luminol derivatives, it is designed on the nanoscale, and hence, the detection area can be further miniaturized. In this sense, luminol derivative sensing layers exhibit great potentialities for further development toward sensor miniaturization. The main accomplishment of this work was the demonstration of a reagentless detection of oxidase activity based on electrochemiluminescent reaction. This can be regarded as a new achievement in the exploration of signal transduction of complex membrane structures. To our knowledge, only few enzyme biocatalysis studies have been achieved using LB films. The results reported in this paper are thus promising not only to perform new developments in the field of biomimetic miniaturized sensors but also to investigate biological processes involving oxidase activities.

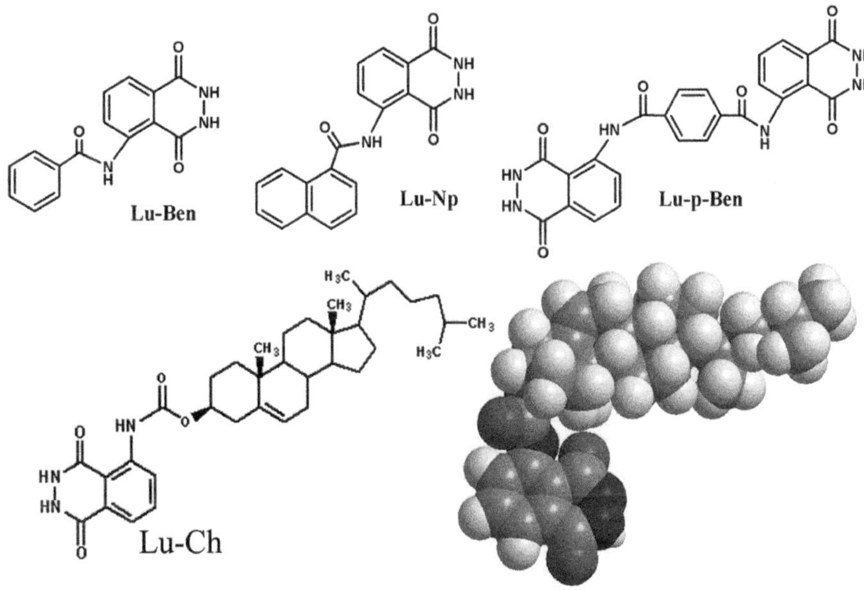

Fig. 30. Chemical structures and abbreviations of new luminol-based derivatives (Reproduced from Jiao et al., 2011b, Advanced Materials Research, 197-198, 606-609; 2011i, *Materials Science Forum*, 694, 565-569.).

Some functional luminol derivatives with cholesteryl or aromatic substituted groups have been designed and synthesized from the reaction of the corresponding aromatic acyl chloride precursors with luminol (Jiao et al., 2011b; 2011i). It has been found that depending on the size of aromatic groups, the formed luminol derivatives showed different properties, indicating distinct regulation of molecular skeletons. UV and IR data confirmed commonly the formation of imide group as well as aromatic segment in molecular structures. Thermal analysis showed that the thermal stability of luminol derivatives with p-phthaloyl segment was the highest in those derivatives. The difference of thermal stability is mainly attributed to the formation of imide group and aromatic substituent groups in molecular structure. The present results have demonstrated that the special properties of luminol derivatives can be turned by modifying molecular structures of objective compounds with proper substituted groups, which show potential application in functional material field and ECL sensor.

5. Present situation and future prospects

We are working on molecular design and supramolecular assemblies of novel amphiphiles with special molecular structures in organized molecular films. In this chapter, various kinds of amphiphiles, including bolaamphiphiles, gemini and amphiphiles with functional substituted groups, have been designed and synthesized, and their organized supramolecular assemblies at the air/water interface have also been investigated. First, some series of bolaform amphiphiles with distinct structures have been designed, and their interfacial assembling behaviors were investigated. Various factors, such as coordination mode, spacer length, and headgroup size, can distinctly influence the conformation of spacer, showing stretched or U-shaped arrangement. Second, some special gemini amphiphiles without charge have been also designed. It was found that the length, rigidity/flexibility and hydrophobicity/hydrophilicity of spacer, as well as the components in subphase (metal ions, dyes, and nanoparticles), can regulate the assembly mode and spectral characters, and induce the distinct interfacial morphologies. Moreover, we have done some groping research work about the interfacial behaviors of amphiphiles with functional substituted groups. The above work may give the potential perspective for the fabrication of nano-materials and biosensors. The present work may broaden the traditional research areas about LB film and give some perspectives and clues for the relative research in the future.

In closing, supramolecular assembly at interface is a key physical chemistry subject due to its close relationship to many fundamental and application scientific questions like catalysis, chirality, electron and energy transfer, single molecule science, and organic electronics. With the development of nanoscience and nanotechnology, supramolecular assembly is increasingly being recognized as a representative approach in "bottom-up" technique for the fabrication of organized nanostructure. The detailed understanding of the driving force for the assembly process makes it possible to tailor the assembly process. The sophisticated nanostructures with multiple components have deepened the insight of supramolecular assembly and enriched the content of supramolecular chemistry. The results mentioned here only provide a cursory browse of some progress on supramolecular assemblies of novel amphiphiles in organized molecular films. The input from other fields, such as theoretical modeling and surface spectroscopic techniques, is highly demanded to help us get quantitative information of the supramolecular assembly. It can be predicted that as the growth of understandings of the rules in the nanoscale, our dream to manipulate molecules could be realized in future.

6. Acknowledgments

The authors would like to extend their thanks to the National Natural Science Foundation of China (No. 20903078); the Natural Science Foundation of Hebei Province (No. B2009000347); the Support Program for Hundred Excellent Innovation Talents from Universities and Colleges of Hebei Province (No. CPRC020); the Open Foundation of State Key Laboratory of Solid Lubrication (Lanzhou Institute of Chemical Physics, CAS) (No. 1002); the Scientific Research Plan of Education Department of Hebei Province (No. Z2009146); and the Doctoral Foundation of Yanshan University (No. B360) for providing funds for this research. This work was also partially supported by Hebei Key Laboratory of Applied Chemistry. Specially, author, Tifeng Jiao would like to express his heartful appreciation to his advisor, prof. Minghua Liu for his guidance, encouragement, and critical comments during his Ph.D research work. Moreover, sincere thanks are also extended to Prof. Agnes P. Girard-Egrot, Prof. Paul Boullanger, Prof. Loïc J. Blum for their invaluable discussion and support in this research work during CNRS (Centre National de la Recherche Scientifique) post-doctoral fellowship.

7. References

Cheng, C.; Jiao, T.; Tang, R.; Liu, M. & Xi, F. (2005). Interfacial behaviors of PMMA-PEO block copolymers at the air/water interface. *Science in China Series B-Chemistry*, 48, 567-573.

Cheng, C.; Jiao, T.; Tang, R.; Chen, E.; Liu, M. & Xi, F. (2006). Compression-induced hierarchical nanostructures of a poly(ethylene oxide)-block-dendronized polymethacrylate copolymer at the air/water interface. *Macromolecules*, 39, 6327-6330.

Fuhrhop, J.H. & Fritsch, D. (1986). Bolaamphiphiles form ultrathin, porous and unsymmetric monolayer lipid membranes. *Accounts of Chemical Research*, 19, 130-137.

Fuhrhop, J.H. & Wang, T. (2006). Bolaamphiphiles. *Chemical Reviews*, 104, 2901-2938.

Guo, Z.; Jiao, T. & Liu, M. (2007). Effect of substituent position in coumarin derivatives on the interfacial assembly: reversible photodimerization and supramolecular chirality. *Langmuir*, 23, 1824-1829.

Holmberg, K. (1998). *Novel Surfactants*, Marcel Dekker, New York, US.

Jiang, M.; Jiao, T. & Liu, M. (2008). Photochromic Langmuir-Blodgett films based on polyoxomolybdate and gemini amphiphiles. *New Journal of Chemistry*, 32, 959-965.

Jiao, T. & Liu, M. (2005). Supramolecular assemblies and molecular recognition of amphiphilic Schiff bases with barbituric acid in organized molecular films. *The Journal of Physical Chemistry B*, 109, 2532-2539.

Jiao, T. & Liu, M. (2005a). Supramolecular nano-architectures and two-dimensional/three-dimensional aggregation of a bolaamphiphilic diacid at the air/water interface. *Thin Solid Films*, 479, 269-276.

Jiao, T. & Liu, M. (2006). Supramolecular assemblies of a new series of gemini-type Schiff base amphiphiles at the air/water interface: interfacial nanoarchitectures, in situ coordination and spacer effect. *Langmuir*, 22, 5005-5012.

Jiao, T. & Liu, M. (2006a). Substitution controlled molecular orientation and nanostructure in the Langmuir-Blodgett films of a series of amphiphilic naphthylidene-containing Schiff base derivatives. *Journal of Colloid and Interface Science*, 299, 815-822.

Jiao, T.; Cheng, C.; Xi, F. & Liu, M. (2006b). Metal ion modulated ultrathin films and nanostructures of tyrosine-based bolaamphiphile at the air/water interface. *Thin Solid Films*, 503, 230-235.

Jiao, T.; Zhang, G. & Liu, M. (2007). Design and interfacial assembly of a new series of gemini amphiphiles with hydrophilic poly(ethyleneamine) spacers. *The Journal of Physical Chemistry B*, 111, 3090-3097.

Jiao, T.; Leca-Bouvier, B.D.; Boullanger, P.; Blum, L.J. & Girard-Egrot, A.P. (2008). Phase behavior and optical investigation of two synthetic luminol derivatives and glycolipid mixed monolayers at the air-water interface. *Colloids and Surfaces A: Physicochemical and Engineering Aspects*, 321, 137-142.

Jiao, T.; Leca-Bouvier, B.D.; Boullanger, P.; Blum, L.J. & Girard-Egrot, A.P. (2008a). Electrochemiluminescent detection of hydrogen peroxide using amphiphilic luminol derivatives in solution. *Colloids and Surfaces A: Physicochemical and Engineering Aspects*, 321, 143-146.

Jiao, T.; Leca-Bouvier, B.D.; Boullanger, P.; Blum, L.J. & Girard-Egrot, A.P. (2010). A chemiluminescent Langmuir–Blodgett membrane as the sensing layer for the reagentless monitoring of an immobilized enzyme activity. *Colloids and Surfaces A: Physicochemical and Engineering Aspects*, 354, 284-290.

Jiao, T.; Zhou, J.; Liu, M.; Wang, S.; Zhou, J.X. & Zhang F. (2010a). Supramolecular assemblies and self-sorting of a series of Cu(II)-coordinated Schiff bases complexes at the air/water interface. *Journal of Dispersion Science and Technology*, 31, 1120-1127.

Jiao, T.; Zhou, J.; Zhou, J.X.; Gao, L.; Xing, Y. & Li, X. (2011). Synthesis and characterization of chitosan-based Schiff base compounds with aromatic substituent groups. *Iranian Polymer Journal*, 20(2), 123-136.

Jiao, T.; Zhou, J.; Zhou, J.X.; Wang, Q. & Luo, X. (2011a). Research on some novel trigonal Schiff base compounds with aromatic core and different substituted groups. *Advanced Materials Research*, 197-198, 598-601.

Jiao, T.; Xing, Y.; Zhou, J.X. & Wang W. (2011b). Synthesis and characterization of some functional luminol derivatives with aromatic substituted groups. *Advanced Materials Research*, 197-198, 606-609.

Jiao, T.; Li, X.; Zhou, J.X.; Liang, J. & Ren, J. (2011c). Synthesis and characterization of some functional Schiff base derivatives with azobenzene substituted groups. *Advanced Materials Research*, 197-198, 623-626.

Jiao, T.; Zhou, J.; Huang, L. & Zhou, J.X. (2011d). Spectral investigation of bolaform Schiff bases with barbituric acid. *Advanced Materials Research*, 160-162, 590-593.

Jiao, T.; Xing, Y. & Zhou, J.X. (2011e). Interfacial coordination and nanofiber fabrication of a tri-pyridine-based derivative with platinum (II) ions. *Advanced Materials Research*, 160-162, 1044-1049.

Jiao, T.; Zhou, J.X.; Zhang L. & Liu, M. (2011f). Supramolecular assembly and headgroup effect in interfacial organized films: a study of some bolaamphiphiles. *Journal of Dispersion Science and Technology*, 32, in press.

Jiao, T.; Zhou, J. & Zhou, J.X. (2011g). Interfacial assembly of two gemini-type Schiff base amphiphiles with aromatic spacers in organized molecular films. unpublished results.

Jiao, T.; Li, X.H.; Li, Q.R. & Zhou, J.X. (2011h). Optical property and photoisomerization of some functional azobenzene derivatives with aromatic substituted groups. *Applied Mechanics and Materials*, in press.

Jiao, T.; Xing, Y.Y. & Zhou, J.X. (2011i). Synthesis and characterization of functional cholesteryl substituted luminol derivative. *Materials Science Forum*, 694, 565-569.

Kuhn, H. & Forsterling, H.D. (2000). *Principles of Physical Chemistry: Understanding Molecules, Molecular Assemblies and Supramolecular Machines*, Wiley, Chichester, UK.

Lehn, J.M. (1995). *Supramolecular Chemistry*, VCH, Weinheim, Germany.

Li, Y.; Jiao, T.; Liu, M. & Yang, L. (2007). Fabrication and efficient acidichromism of Langmuir-Blodgett films of azo group-modified phthalocyanines. *Journal of Dispersion Science and Technology*, 28, 603-605.

Menger, F.M.; Keiper, J.S. & Azov, V. (2000). Gemini surfactants with acetylenic spacers. *Langmuir*, 16, 2062-2067.

Yin, M.; Jiao, T. & Liu, M. (2007). Acidichromism in the LB film of bolaform Schiff base. *Chinese Chemical Letters*, 18, 30-32.

Zana, R. (2002). Dimeric and oligomeric surfactants. Behaviors at the interface and in aqueous solution: a review. *Advances in Colloid and Interface Science*, 97, 205-253.

Zhang, L.; Jiao, T.; Shao, X.; Li, Z. & Liu, M. (2006). Aggregation of TPPS on spreading films of achiral cationic amphiphiles: effect of the charge and rigid spacer on the morphologies and supramolecular chirality. *Colloids and Surfaces A: Physicochemical and Engineering Aspects*, 284-285, 130-134.

Zhong, L.; Jiao, T. & Liu, M. (2008). Synthesis and assembly of gold nanoparticles in organized molecular films of gemini amphiphiles. *Langmuir*, 24, 11677-11683.

Zhong, L.; Jiao, T. & Liu, M. (2009). A facile method to the synthesis of gold nanoprisms using a series of gemini amphiphiles. *Journal of Nanoscience and Nanotechnology*, 9, 2726-2730.

Zhong, L.; Jiao, T. & Liu, M. (2009a). Gemini amphiphiles regulated photopolymerization of diacetylene acid in organized molecular films. *The Journal of Physical Chemistry B*, 113, 8867-8871.

Permissions

The contributors of this book come from diverse backgrounds, making this book a truly international effort. This book will bring forth new frontiers with its revolutionizing research information and detailed analysis of the nascent developments around the world.

We would like to thank Prof. Mohammed Muzibur Rahman, for lending his expertise to make the book truly unique. He has played a crucial role in the development of this book. Without his invaluable contribution this book wouldn't have been possible. He has made vital efforts to compile up to date information on the varied aspects of this subject to make this book a valuable addition to the collection of many professionals and students.

This book was conceptualized with the vision of imparting up-to-date information and advanced data in this field. To ensure the same, a matchless editorial board was set up. Every individual on the board went through rigorous rounds of assessment to prove their worth. After which they invested a large part of their time researching and compiling the most relevant data for our readers. Conferences and sessions were held from time to time between the editorial board and the contributing authors to present the data in the most comprehensible form. The editorial team has worked tirelessly to provide valuable and valid information to help people across the globe.

Every chapter published in this book has been scrutinized by our experts. Their significance has been extensively debated. The topics covered herein carry significant findings which will fuel the growth of the discipline. They may even be implemented as practical applications or may be referred to as a beginning point for another development.

The editorial board has been involved in producing this book since its inception. They have spent rigorous hours researching and exploring the diverse topics which have resulted in the successful publishing of this book. They have passed on their knowledge of decades through this book. To expedite this challenging task, the publisher supported the team at every step. A small team of assistant editors was also appointed to further simplify the editing procedure and attain best results for the readers.

Our editorial team has been hand-picked from every corner of the world. Their multi-ethnicity adds dynamic inputs to the discussions which result in innovative outcomes. These outcomes are then further discussed with the researchers and contributors who give their valuable feedback and opinion regarding the same. The feedback is then collaborated with the researches and they are edited in a comprehensive manner to aid the understanding of the subject.

Apart from the editorial board, the designing team has also invested a significant amount of their time in understanding the subject and creating the most relevant covers. They scrutinized every image to scout for the most suitable representation of the subject and create an appropriate cover for the book.

The publishing team has been involved in this book since its early stages. They were actively engaged in every process, be it collecting the data, connecting with the contributors or procuring relevant information. The team has been an ardent support to the editorial, designing and production team. Their endless efforts to recruit the best for this project, has resulted in the accomplishment of this book. They are a veteran in the field of academics and their pool of knowledge is as vast as their experience in printing. Their expertise and guidance has proved useful at every step. Their uncompromising quality standards have made this book an exceptional effort. Their encouragement from time to time has been an inspiration for everyone.

The publisher and the editorial board hope that this book will prove to be a valuable piece of knowledge for researchers, students, practitioners and scholars across the globe.

List of Contributors

Sher Bahadar Khan and Abdullah M. Aisiri
The Center of Excellence for Advanced Materials Research, King Abdulaziz University, Jeddah, Kingdom of Saudi Arabia
Chemistry Department, Faculty of Science, King Abdulaziz University, Jeddah, Kingdom of Saudi Arabia

Aslam Jamal and Mohd Faisal
Centre for Advanced Materials and Nano-Engineering (CAMNE), Department of Chemistry, Faculty of Sciences and Arts, Najran University, Najran, Kingdom of Saudi Arabia

Mohammed M. Rahman
The Center of Excellence for Advanced Materials Research, King Abdulaziz University, Jeddah, Kingdom of Saudi Arabia

H. Mehranpour and Masoud Askari
Department of Material Science and Engineering, Sharif University of Technology, Tehran, Iran

M. Sasani Ghamsari
Solid State Lasers Research Group, Laser & Optics Research School, NSTRI, Tehran, Iran

Raphaël Schneider and Fadi Aldeek
Laboratoire Réactions et Génie des Procédés (LRGP), Nancy-University, CNRS, Nancy, France

Lavinia Balan
Institut de Science des Matériaux de Mulhouse (IS2M), Mulhouse, France

A. D. Rud, N. I. Kuskova, L. I. Ivaschuk, L. Z. Boguslavskii and A. E. Perekos
G. V. Kurdyumov Institute for Metal Physics of NASU, Institute of Pulse Research and Engineering of NASU, Ukraine

Krzysztof Lukaszkowicz
Silesian University of Technology, Poland

Nicoleta Petrea, Petrişor Zamora Iordache, Rodica Mihaela Lungu, Ioan Safta, Razvan Petre and Andrada Pretorian
Scientific Research Center for CBRN Defense and Ecology, Romania

Jeong Hyun Yeum, Jae Hyeung Park and Jae Young Choi
Department of Advanced Organic Materials Science & Engineering, Kyungpook National University, Daegu, Korea

Jong Won Kim
R&D Division, Korea Dyeing Technology Center, Daegu, Korea

Sung Kyou Han
R&D Division, Taihan Textile Co. Ltd., Daegu, Korea

Weontae Oh
Department of Materials and Components Engineering, Dong-eui University, Busan, Korea

Esther G. Calvo, J. Ángel Menéndez and Ana Arenillas
Instituto Nacional del Carbón (CSIC), Oviedo, Spain

Sujoy K. Das and Enrico Marsili
School of Biotechnology, Dublin City University, Dublin, Ireland

Joaquim G. G. Esteves da Silva
Centro de Investigação em Química da Universidade do Porto (CIQ-UP), Department of Chemistry and Biochemistry, Faculty of Sciences, University of Porto, Porto, Portugal

Kevin Deplanche, Angela Murray, Claire Mennan, Scott Taylor and Lynne Macaskie
The University of Birmingham, United Kingdom

Tifeng Jiao, Sufeng Wang and Jingxin Zhou
Hebei Key Laboratory of Applied Chemistry, College of Environmental and Chemical Engineering, Yanshan University, Qinhuangdao, China